Lecture Notes in Nanoscale Science and Technology

Volume 15

For further volumes:
http://www.springer.com/series/7544

Quantum Dot Solar Cells

Jiang Wu • Zhiming M. Wang
Editors

Quantum Dot Solar Cells

Springer

Editors
Jiang Wu
University of Electronic Science
and Technology of China
State Key Laboratory of Electronic
Thin Film and Integrated Devices
Chengdu, China, People's Republic

Zhiming M. Wang
University of Electronic Science
and Technology of China
State Key Laboratory of Electronic
Thin Film and Integrated Devices
Chengdu, China, People's Republic

ISSN 2195-2159 ISSN 2195-2167 (electronic)
ISBN 978-1-4614-8147-8 ISBN 978-1-4614-8148-5 (eBook)
DOI 10.1007/978-1-4614-8148-5
Springer New York Heidelberg Dordrecht London

Library of Congress Control Number: 2013949162

Printed on acid-free paper

Springer is part of Springer Science+Business Media (www.springer.com)

Preface

The world's primary energy sources are currently fossil fuels in the form of oil, coal, and natural gas. With rapidly growing global energy consumption, and for the first time in a century, the world faces serious energy, environmental, and economic crises as a result of depleted stocks of fossil fuels, pollution, climate change, etc. Development of sustainable and clean energy sources has taken on increased urgency. Solar energy is one of the most abundant energy sources on the planet, but despite its great promise, the cost and efficiency of current photovoltaic cells present great challenges to implement solar energy on a large scale. After over 30 years of slow development, the contribution of solar energy, including photovoltaic electricity, to global energy consumption is still marginal [1].

So far, one of the major challenges for widespread deployment of solar cells is the high cost of solar cells compared with fossil fuels. Further substantial improvements in current photovoltaic technologies are needed in terms of cost and efficiency. To this end, research efforts have divided mainly into two directions: exploring new cost-effective photovoltaic materials and development of new high efficiency device architectures. In the last decades, new classes of photovoltaic materials have been investigated, including amorphous silicon (a-Si), organic polymers, and CIGS [$CuIn_xGa_{1-x}Se(S)_2$], with the aim of cost-effective solar electricity generation. As a new technology becomes mature, material cost puts an ultimate limit on the effort to bring down the price-per-watt for power generation [2].

Development of new photovoltaic cells with substantially increased power conversion efficiency is critical to convert solar energy from a promising clean renewable energy to a competitive primary energy source. Nanostructured materials developed during the last 20 years are promising for fulfilling this task. Solar cells based on nanomaterials, including multiple exciton generation solar cells [3, 4], intermediate band solar cells [5, 6], and hot electron extraction solar cells [7, 8], have shown rapid improvement. While there are a number of books covering nanostructured solar cells, very few cover the recently fast-developing quantum dot solar cells. It is therefore the goal of this book to present a comprehensive overview of the current status of quantum dot solar cells and related technologies.

The main body of the book is comprised of contributions that focus on photovoltaic cells based on quantum dot materials. Various novel solar cell materials and structures are covered in different chapters. Specifically, Chaps. 1–4 offer a comprehensive perspective on hybrid colloidal quantum dot solar cells. Chapter 1 reviews recent advances in colloidal quantum dot sensitized quantum dot solar cells. Chapter 2 presents two unique hierarchically nanostructured quantum dot sensitized solar cells with improved light harvesting and charge collection. A unique hybrid solar cell structure with colloidal quantum dots deposited on traditional solar cells to boost efficiency is introduced in Chap. 3. Due to the equal importance of electricity generation and light generation, Chap. 3 also reports on improved light emission diode efficiency with colloidal quantum dots. In Chap. 4, photoinduced charge transfer in hybrid nanosystems with colloidal quantum dots is explored.

Chapters 5–7 introduce studies of quantum dot-based intermediate band solar cells. Chapters 5 and 6 are theoretical contributions which model electronic structure and optical properties and address the optimization of material selection for quantum dots by the finite element method, respectively. Chapter 7 presents a study on quaternary quantum dot solar cells and shows that two-photon absorption through the sub-bandgap indicates intermediate band formation.

Unlike quantum dot sensitized quantum dot solar cells and quantum dot intermediate band solar cells which aim for wide spectral absorption, hot electron solar cells and multiple exciton generation solar cells convert the excess carrier energies into electricity. Chapter 8 presents new models of operating principles of quantum dot hot-carrier solar cells using detailed balance of particle and energy fluxes and a search for the requisites of high conversion efficiency. Chapter 9 reviews recent progress in multiple exciton generation in quantum dots and solar cells benefiting from multiple exciton generation. Chapter 10 deals with multiple exciton generation dynamics in quantum dots using an improved transient grating technique.

Chapter 11 presents a comprehensive review of progress in low-cost graphene quantum dot solar cells. In Chap. 12, the synthesis of graphene and graphene quantum dots is introduced. Chapter 12 also addresses the application of graphene and quantum dots to photovoltaic devices.

While the above chapters focus on photovoltaic materials and devices based on quantum dots, the last two chapters cover a novel characterization technique and device architecture, respectively. Chapter 13 is devoted to a detailed analysis of an ultrafast photovoltammetry technique. This technique can be used to investigate surface charge carrier dynamics at the nanometer scale, such as in quantum dot solar cells. The authors of Chap. 14 discuss enhanced solar cell performance through light trapping and optical confinement from photonic and plasmonic structures. This issue is particularly important for quantum dot solar cells, which generally consists of optically thin active regions.

The editors thank all the contributors of this book for their remarkable chapters. We owe special thanks to Dr. David Packer, executive editor at Springer, for supporting this book. Last but not least, we would like to thank Mr. Peng Yu who provided indispensable editorial assistance and support. The editors acknowledge the financial support of the National Natural Science Foundation of China through Grant NSFC-61204060.

Chengdu, China, People's Republic — Jiang Wu
Chengdu, China, People's Republic — Zhiming M. Wang

References

1. Kerr, R.A.: Do we have the energy for the next transition? Science **329** (5993), 780–781 (2010)
2. Green, M.A.: In: Anonymous (ed.) Third Generation Photovoltaics: Advanced Solar Conversions, p. 160. Springer, New York (2005)
3. Semonin, O.E., et al.: Peak external photocurrent quantum efficiency exceeding 100% via MEG in a quantum dot solar cell. Science **334** (6062), 1530–1533 (2011)
4. Sambur, J.B., Novet, T., Parkinson, B.A.: Multiple exciton collection in a sensitized photovoltaic system. Science **330** (6000), 63–66 (2010)
5. Luque, A., Martí, A.: Increasing the efficiency of ideal solar cells by photon induced transitions at intermediate levels. Phys. Rev. Lett. **78** (26), 5014 (1997)
6. Luque, A., Marti, A.: The intermediate band solar cell: progress toward the realization of an attractive concept. Adv. Mater. **22** (2), 160–174 (2010)
7. Tisdale, W.A., et al.: Hot-electron transfer from semiconductor nanocrystals. Science **328** (5985), 1543–1547 (2010)
8. Pandey, A., Guyot-Sionnest, P.: Slow electron cooling in colloidal quantum dots. Science **322** (5903), 929–932 (2008)

Contents

Contributors

A. Al-Otaify School of Physics and Astronomy & Photon Science Institute, University of Manchester, Manchester, UK

Irene Barceló Institut Universitari d'Electroquímica i Departament de Química Física, Universitat d'Alacant, Alacant, Spain

D.J. Binks School of Physics and Astronomy & Photon Science Institute, University of Manchester, Manchester, UK

M. Cadirci School of Physics and Astronomy & Photon Science Institute, University of Manchester, Manchester, UK

Suresh Chand Physics of Energy Harvesting Division, Organic and Hybrid Solar Cell Group, CSIR-National Physical Laboratory, New Delhi, India

Kiseok Chang Department of Physics and Astronomy, Michigan State University, East Lansing, MI, USA

Yong-June Chang Department of Materials Science and Engineering, and Division of Advanced Materials Science, Pohang University of Science and Technology (POSTECH), Pohang 790-784, Republic of Korea

Mircea Cotlet Center for Functional Nanomaterials, Brookhaven National Laboratory, Upton, NY, USA

Alfred Forchel Technische Physik, Physikalisches Institut and Wilhelm Conrad Röntgen-Research Center for Complex Material Systems, Universität Würzburg, Am Hubland, Würzburg, Germany

Robert Gilmore Department of Physics, Drexel University, Philadelphia, PA, USA

Roberto Gómez Departament de Química-Física i Institut Universitari d'Electroquímica, Universitat d'Alacant, Alacant, Spain

Néstor Guijarro Departament de Química-Física i Institut Universitari d'Electroquímica, Universitat d'Alacant, Alacant, Spain

Vinay Gupta Physics of Energy Harvesting Division, Organic and Hybrid Solar Cell Group, CSIR-National Physical Laboratory, New Delhi, India

Tzong-Ru T. Han Department of Physics and Astronomy, Michigan State University, East Lansing, MI, USA

Corey R. Hine Department of Chemistry, Syracuse University, Syracuse, NY, USA

Sven Höfling Technische Physik, Physikalisches Institut and Wilhelm Conrad Röntgen-Research Center for Complex Material Systems, Universität Würzburg, Würzburg, Germany

Hyun Myung Jang Department of Materials Science and Engineering, and Division of Advanced Materials Science, Pohang University of Science and Technology (POSTECH), Pohang 790-784, Republic of Korea

Steven Jenks Department of Physics, Drexel University, Philadelphia, PA, USA

Martin Kamp Technische Physik, Physikalisches Institut and Wilhelm Conrad Röntgen-Research Center for Complex Material Systems, Universität Würzburg, Würzburg, Germany

Kenji Katayama Faculty of Science and Engineering, Department of Applied Chemistry, Chuo University, Tokyo, Japan

Kazunori Koga Graduate School of Information Science and Electrical Engineering, Kyushu University, Fukuoka, Japan

Eui-Hyun Kong Department of Materials Science and Engineering, and Division of Advanced Materials Science, Pohang University of Science and Technology (POSTECH), Pohang 790-784, Republic of Korea

Stefan Kremling Technische Physik, Physikalisches Institut and Wilhelm Conrad Röntgen-Research Center for Complex Material Systems, Universität Würzburg, Würzburg, Germany

Teresa Lana-Villarreal Departament de Química-Física i Institut Universitari d'Electroquímica, Universitat d'Alacant, Alacant, Spain

Chien-Chung Lin Institute of Photonic System, College of Photonics, National Chiao Tung University, Tainan, Taiwan

Siyuan Luo State Key Laboratory of Electronic Thin Film and Integrated Devices, University of Electronic Science and Technology of China, Chengdu, People's Republic of China

Mathew M. Maye Department of Chemistry, Syracuse University, Syracuse, NY, USA

N. McElroy School of Physics and Astronomy & Photon Science Institute, University of Manchester, Manchester, UK

Qinpeng Meng Center for Functional Nanomaterials, Brookhaven National Laboratory, Upton, NY, USA

Jeremy N. Munday University of Maryland ECE and IREAP Energy Research Facility, College Park, MD, USA

Ryan A. Murdick Department of Physics and Astronomy, Michigan State University, East Lansing, MI, USA

Joseph Murray Department of Electrical and Computer Engineering, University of Maryland, College Park, MD, USA

Yoshitaka Okada Research Center for Advanced Science and Technology (RCAST), The University of Tokyo, Tokyo, Japan

R. Page School of Physics and Astronomy & Photon Science Institute, University of Manchester, Manchester, UK

Chong-Yu Ruan Department of Physics and Astronomy, Michigan State University, East Lansing, MI, USA

Christian Schneider Technische Physik, Physikalisches Institut and Wilhelm Conrad Röntgen-Research Center for Complex Material Systems, Universität Würzburg, Würzburg, Germany

Hyunwoong Seo Graduate School of Information Science and Electrical Engineering, Kyushu University, Fukuoka, Japan

Qing Shen Faculty of Informatics and Engineering, Department of Engineering Science, The University of Electro-Communications, Tokyo, Japan

Masaharu Shiratani Graduate School of Information Science and Electrical Engineering, Kyushu University, Fukuoka, Japan

Yasuhiko Takeda Toyota Central Research and Development Laboratories, Inc., Aichi, Japan

Zhen-Sheng Tao Department of Physics and Astronomy, Michigan State University, East Lansing, MI, USA

Stanko Tomić Joule Physics Laboratory, School of Computing, Science and Engineering, University of Salford, Salford, UK

Taro Toyoda Faculty of Informatics and Engineering, Department of Engineering Science, University of Electro-Communications, Tokyo, Japan

Giichiro. Uchida Graduate School of Information Science and Electrical Engineering, Kyushu University, Fukuoka, Japan

Tanvi Upreti Physics of Energy Harvesting Division, Organic and Hybrid Solar Cell Group, CSIR-National Physical Laboratory, New Delhi, India

Yuting Wang Graduate School of Information Science and Electrical Engineering, Kyushu University, Fukuoka, Japan

Zhiming M. Wang University of Electronic Science and Technology of China, State Key Laboratory of Electronic Thin Film and Integrated Devices, Chengdu, China, People's Republic

Jiang Wu University of Electronic Science and Technology of China, State Key Laboratory of Electronic Thin Film and Integrated Devices, Chengdu, China, People's Republic

Xukai Xin School of Materials Science and Engineering, Georgia Institute of Technology, Atlanta, GA, USA

Yunlu Xu Department of Electrical and Computer Engineering, University of Maryland, College Park, MD, USA

Zhihua Xu Center for Functional Nanomaterials, Brookhaven National Laboratory, Upton, NY, USA

Peng Yu State Key Laboratory of Electronic Thin Film and Integrated Devices, University of Electronic Science and Technology of China, Chengdu, People's Republic of China

Fei Yuan Department of Physics and Astronomy, Michigan State University, East Lansing, MI, USA

Chapter 1
Recent Progress in Colloidal Quantum Dot-Sensitized Solar Cells

Irene Barceló, Néstor Guijarro, Teresa Lana-Villarreal, and Roberto Gómez

Abstract Among the third-generation solar cells based on quantum dots, the quantum dot-sensitized solar cells (QDSSCs) have received substantial attention probably because they are a straightforward extension of the very popular dye-sensitized solar cells. The nature of the quantum dots (QDs) and their mode of attachment to a wide band gap semiconductor (oxide) are important factors determining the efficiency of the corresponding devices. In this respect, the QDSSCs can be classified into two main groups: those prepared by in situ deposition of the QDs on the oxide template and those resulting from the adsorption of pre-formed, colloidal quantum dots on the oxide surface. Most of the studies on these cells appeared in the last few years have focused on the former type, while the effort devoted to colloidal QDSSCs has been comparatively minor. However, the latter allow for a better control of the device structure and for a separate optimization of different aspects. In addition, there are fundamental reasons to believe that colloidal QDSSCs may have advantages over those prepared by in situ QD deposition, particularly in what refers to recombination. In this chapter, we review and discuss the state of the art in colloidal QDSSCs with a focus on the results obtained in our laboratory.

1.1 Introduction

Since the advent of the quantum dots (QDs) in the fields of science and technology, they have been the subject of an ever increasing interest within the scientific community. This is certainly linked to their fascinating properties, many of which depend in a dramatic way on their size [1]. This fact makes them a central object

I. Barceló • N. Guijarro • T. Lana-Villarreal • R. Gómez (✉)
Institut Universitari d'Electroquímica i Departament de Química Física,
Universitat d'Alacant, Apartat 99, 03080 Alacant, Spain
e-mail: Irene.Barcelo@ua.es; Nestor.Guijarro@ua.es; Teresa.Lana@ua.es; Roberto.Gomez@ua.es

J. Wu and Z.M. Wang (eds.), *Quantum Dot Solar Cells*, Lecture Notes in Nanoscale Science and Technology 15, DOI 10.1007/978-1-4614-8148-5_1,

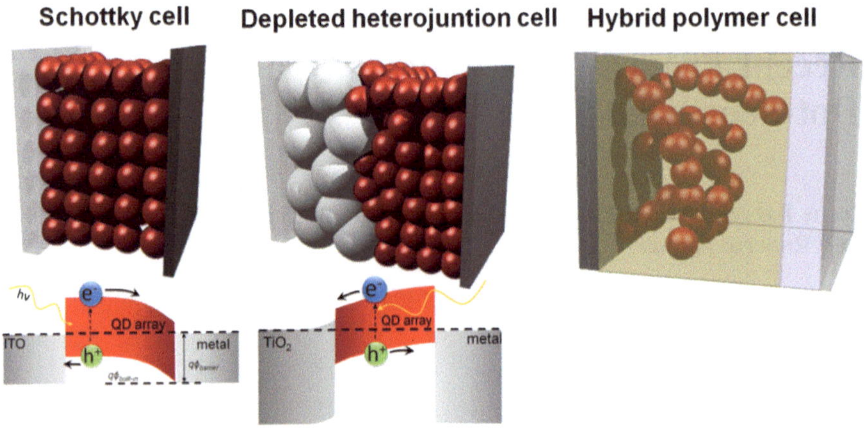

Fig. 1.1 Sketches illustrating the different types of solar cells based on colloidal QDs, apart from the QDSSC: Schottky cell, depleted heterojunction cell and hybrid bulk heterojunction cell. Energy-band diagrams for the first two devices are also included

revealing in a clear-cut way the potentialities of nanoscience and nanotechnology. Among the properties that can be tuned by changing the QD size, we could highlight the potential energy of the valence band (VB) and conduction band (CB) edges (HOMO and LUMO energies, respectively) of the QDs. The magnitude of such electronic level shift as a result of a change in size will depend on the effective masses of electrons and holes. In any case, as a result of quantization, for decreasing QD sizes, the HOMO level shifts downward in the potential energy scale, while the LUMO level shifts upward. Obviously, this leads not only to an increase in the band gap but also to an increase in the driving force for electron and hole injection into the surrounding phases, which is crucial for their use as sensitizers in solar cells.

Several configurations of solar cells based on colloidal QDs have been proposed and demonstrated (Fig. 1.1) [2–4]. The Schottky cell is a simple device in which an ultrathin film of colloidal QDs is sandwiched between two contacts for the extraction of photogenerated electrons and holes. Typically, p-type QDs are employed supported on indium tin oxide (ITO) conducting glass, which acts as a contact for the extraction of carriers. A low work function metal is evaporated on the quantum dot layer. Charge transfer between the metallic contact and the QDs generates band bending in the QD layer together with a Schottky barrier that makes the contact selective. Impressive photocurrents have been achieved, although Fermi level pinning at the metal/QD contact leads to low open circuit voltages [5]. The depleted heterojunction cell consists of a nanostructured wide band gap oxide (TiO_2 or ZnO) supported on conducting glass. On top of this layer, a multilayer of QDs is deposited and the other contact of the device is provided by deposition of a metal (Au). The nanostructured oxide film acts as an electron conducting phase and helps the selective extraction of electrons from the QDs. In this case, the metal contact extracts the majority carriers. Although the photocurrents achieved in this case are smaller than for Schottky cells, improved open circuit voltages are obtained [6]. Another configuration that has been explored for more than 15 years is that of the so-called hybrid bulk heterojunction solar cell [7]. These cells are conceptually similar to the organic solar cells, although in this case the typical electron accepting phase would be substituted by QDs that would act as both electron transporting phase and light absorber (together with the polymer). These cells have given so far maximum conversion efficiencies substantially lower than those of the above-mentioned QD solar cells. This is probably due to the tendency to aggregation of the QDs and the lack of continuous pathways for the transport of charge. In addition, the design of the interface between the QDs and the hole conducting polymer is far from optimized. Importantly, all these cells are fabricated with colloidal QDs, being critical the quality of their surfaces.

The QDSSCs follow an operational principle different from that of the previous devices, as in this case the QDs do not play the role of carrier transporting phase. They could be considered as an offshoot of the ubiquitous dye-sensitized solar cells (DSSCs) appeared in 1991 [8]. Some years later, some researchers came up with the idea of replacing dyes by QDs [9]. Initially, the so-called QDSSCs were developed according to the roadmap already established for DSSCs, i.e. using the same components (electron acceptor, electrolyte, and counter electrode) that had demonstrated remarkable efficiencies and robustness in DSSCs. Unfortunately, it was soon observed that differences between dyes and QDs would require a drastic makeover of the architecture and components to effectively exploit the potential of QDs. As an example, it is quite illustrative the fact that whereas the first DSSCs exhibited efficiencies near 8 % without particular optimization [8], the best efficiencies reported for QDSSCs are still lagging behind this value nowadays [10].

In QDSSCs, QDs behave as light absorbers. Their photoexcitation leads to the generation of carriers that are selectively transferred to the surrounding phases. On the one hand, the photogenerated electrons are most commonly transferred to a wide band gap oxide (TiO_2, SnO_2, ZnO, etc.) on which the QDs are adsorbed. The electrons are finally extracted through the conducting glass that supports the nanostructured oxide. On the other hand, the photogenerated holes are transferred to either a liquid electrolyte or a solid hole transporting material (HTM) (Fig. 1.2). In the first case, the electrolyte contains a reversible redox couple whose reduced form accepts holes from the QDs. The counter electrode can be either a metal or a semiconductor electrode with a fast kinetics for the redox couple. The solid hole transporting phase is either a molecular material (such as 2,2′,7,7′-tetrakis-(*N*,*N*-di-pmethoxyphenylamine)9,9′-spirobifluorene, usually called spiro-OMeTAD) or a hole transporting conducting polymer, such as different polythiophenes. The contact for the extraction of holes is commonly a metal such as gold, forming an ohmic contact with the hole conducting phase. For the fabrication of these cells we may employ either presynthesized, colloidal quantum dots or alternatively, grow them on the nanostructured film. In the case of colloidal QDSSCs, several ways of attaching the QDs to the oxide structure have been employed. In some cases, the capped colloidal QDs self-assemble spontaneously on the nanostructured oxide layer (direct adsorption, DA) or are brought into contact with the oxide by

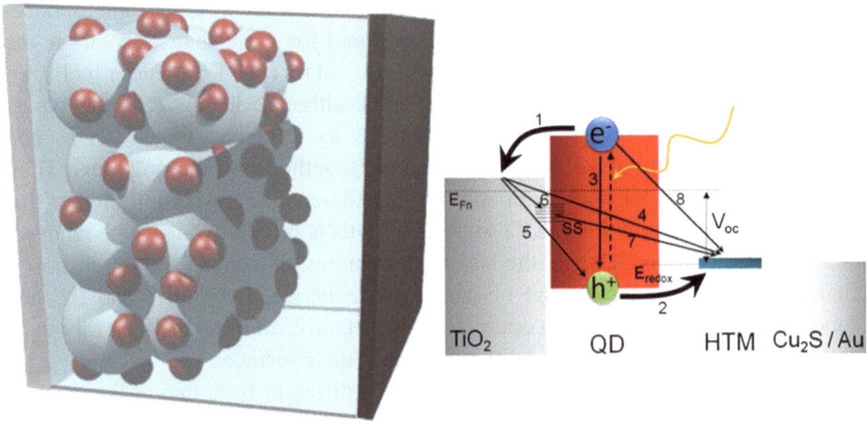

Fig. 1.2 Sketch depicting a QDSSC and the corresponding energy-band diagram illustrating the main processes: (*1*) electron injection; (*2*) hole transfer to the HTM; (*3*) QD band-to-band recombination; (*4*) recombination of TiO_2 CB electrons with the oxidized HTM; (*5*) back electron transfer; (*6*) electron trapping in interfacial states; (*7*) recombination of TiO_2 trapped electrons with the oxidized HTM; (*8*) recombination of QD CB electrons with the oxidized HTM

electrophoresis. In any case, no additional components are introduced in the system to anchor the QDs, as opposed to the so-called linker-assisted adsorption in which a linker molecule is included to act as a molecular cable between the oxide and the QD. These linkers are bifunctional molecules with terminal functional groups that have specific avidity for either oxide or QD surface sites.

Obviously, fabricating efficient QDSSCs requires the achievement of large incident-photon-to-current efficiencies (IPCEs), high open circuit voltages, and high fill factors. We address in the following these factors and identify the potential differences between colloidal and non-colloidal QDs. As for the IPCE (and thus the photocurrent), one should take into account that [11]:

$$\mathrm{IPCE}(\lambda) = \mathrm{LHE}(\lambda)\phi_{\mathrm{inj}}\eta_{\mathrm{coll}} \tag{1.1}$$

where LHE(λ) is the light-harvesting efficiency, ϕ_{inj} the electron injection efficiency, and η_{coll} the electron collection efficiency. The LHE is determined by the QD loading and band gap. The QD loading depends on the thickness of the oxide template, the oxide particle size, and the degree of coverage achieved in each case. The electron injection efficiency will depend on the nature and type of attachment of the QDs to the oxide (direct, linker-assisted, in situ grown, etc.). In addition, in the case of linker-assisted adsorption, ϕ_{inj} will strongly depend on the nature of the linker species. The existence of a multilayer of QDs will obviously affect the electron injection efficiency. Finally, electron collection efficiency will very much depend on the nature and morphology of the carrier transporting phases and the corresponding interfaces. Concretely, the efficiency will benefit from high carrier mobilities and slow recombination processes.

It is important to identify the different recombination processes that limit η_{coll} as their rate will depend, among other things, on the nature (colloidal vs. in situ grown) of the QD sensitizers. Figure 1.2 summarizes the different recombination processes that may operate in these devices. On the one hand, we have the recombination in the quantum dot, which can be either band-to-band or mediated by band gap states (surface or interfacial states or traps). It is important to recognize that the QD/QD interfacial area will be much larger in the case of chemically grown quantum dots than in the case of colloidal quantum dots, where the typical coverages are smaller. Obviously, the recombination via surface states will also be different in the case of chemically deposited QDs, which are unprotected, and in that of colloidal QDs, which are typically capped. Another pathway for recombination occurs via the electrolyte (or HTM). In this case either the electrons in the QDs or those injected into the oxide can recombine with holes in the hole transporting phase, whether solid or liquid (in the case of a liquid electrolyte, the holes would be the oxidized species of the redox couple). In both cases, the electrons can be transferred from either the conduction band or surface states that capture electrons prior to their transfer to the hole transporting phase. When the electrons photogenerated in the QDs are directly involved in the recombination process, its rate is expected to strongly depend on the structure of the QD/electrolyte interface and thus on the QD nature. Finally, recombination could occur between electrons injected in the oxide and holes remaining at the QDs (back electron transfer). The states at the interface between the oxide and the QDs could act as recombination centers. The density of these states is smaller in the case of colloidal QDs and, in addition, their nature should be different. This should also lead to different recombination characteristics in both cells [12].

The recombination processes cited above also affect the fill factor of the cell together with transport/kinetic limitations at the hole transporting phase/counter electrode. On the other hand, the maximum attainable open circuit voltage will depend on factors such as the energy level of the oxide CB edge and the HOMO level of the hole transporting material (or redox couple potential). No particular specificities resulting from the use of colloids instead of chemically deposited sensitizer nanoparticles are expected.

Among the factors that make quantum dot solar cells attractive, we find the possibility of taking advantage of hot carriers [13]. One of the mechanisms for doing so relies on the so-called multiple exciton generation (MEG), by which the excess kinetic energy that a hot electron possesses (as long as it is larger than the material band gap) is employed in the promotion of a second electron (by impact ionization) from the valence to the conduction band. By means of such a mechanism, the absorption of one photon would lead to the generation of more than one exciton, which would translate into maximum quantum yields over 100 %. MEG has been reported for rutile single crystal photoanodes sensitized with PbS quantum dots [14]. According to Nozik, the cooling process of hot carriers is slower as a result of confinement [13]. The fact that confinement is only partially present in the case of in situ deposited QDs, together with the high coverage and density of non-passivated surface states, makes less probable MEG, as these defects could act as trapping centers. Actually, MEG has been observed only in systems based on colloidal QDs, which supports this idea [14, 15]. Inefficient cooling of charge carriers could also lead to their transfer to the surrounding media, avoiding the limitations imposed by the relative disposition of the energy levels.

This chapter is organized as follows. Recent results on QD-sensitized photoanodes and their integration in liquid electrolyte QDSSCs are presented first, to turn then our attention toward the promising field of solid-state devices. Next, and very briefly, a few studies on QD-sensitized photocathodes are presented as well as their potential use in tandem cells (with QD-sensitized photoanode and photocathode). Finally, some ideas on the rationalization of the results are given in a separate section in which selected results obtained by means of time-resolved techniques are reviewed. The chapter ends with a conclusion section that summarizes the main ideas and gives some clues on the way forward in this field.

1.2 Liquid Phase Solar Cells with QD-Sensitized Photoanodes

In this section the main components of liquid-based QDSSCs will be reviewed, exposing their evolution in the last years and how they have drifted apart from those of DSSCs. Note that although recent efforts are aimed to replace liquid electrolyte by solid hole conductors (see below), the research on liquid-based QDSSCs is still predominant in the field.

1.2.1 Materials Used as Electron Transporting Phase

One of the major achievements of the DSSC concept is the separation of the roles of photocarrier generation and transport in different materials, in stark contrast with the design of thin film technologies [16]. In such a way, the deleterious recombination between photogenerated carriers is minimized. In the same way in QDSSCs, the electron acceptor phase plays the key role of rapidly extracting photogenerated electrons from the QDs, funneling them, subsequently, toward the collector. Ideally, a suitable electron acceptor for QDSSCs should fulfill the following conditions (1) appropriate conduction band edge position for enabling electron injection, (2) moderate-high surface area to load enough QDs to harvest most of the irradiated photons, (3) open structure (wide and interconnected pores) allowing for a facile penetration/adsorption of colloidal QDs and filling with hole transporting phase, (4) high electron mobility to efficiently collect photoinjected electrons, (5) poor light absorption in the visible region together with scattering effects to favor QD light-harvesting, and (6) easy fabrication. Driven by the inertia of DSSCs, first electron acceptor materials in QDSSCs were TiO_2 and ZnO, which had already demonstrated the feasibility of being sensitized by QDs for applications in photocatalysis [17, 18]. Soon afterward, advances in materials nanoengineering led to the application of new structures and exploration of new materials.

1.2.1.1 Titanium Dioxide: Effects of Nanostructuring

TiO_2 is probably the electron acceptor with the most promising properties, as indicated by both the numerous works where it is employed and the fact that it seems to be essential for preparing highly efficient devices [3]. Commonly, homemade or commercial TiO_2 pastes are doctor bladed or screen-printed over a transparent conductive substrate and thermally annealed for preparing films. However, it has been observed that the micro/nanostructure of these films dramatically affects the performance of photoanodes. In this respect, it has been demonstrated that nanostructured TiO_2 films composed of large particles lead to higher conversion efficiencies in QDSSCs [19]. This result is probably due to the fact that a more open structure of the TiO_2 layer facilitates the transport of colloidal QDs prior to adsorption, leading to higher QD loadings, and, in turn, preventing the blockage of nanochannels (mesopores), which would make inactive part of the electrode (Sect. 1.5.3 of this chapter). Nowadays, the tendency is to prepare QDSSCs combining two different kinds of TiO_2 pastes to better meet the requirements of the electron acceptor substrate. Commonly, one transparent TiO_2 layer is deposited on the substrate to guarantee a high surface area and QD loading, whereas a second one, formed by larger nanoparticles, is applied subsequently to ensure light scattering. The best efficiency reported for a colloidal QDSSCs so far (5.4 %) follows this double-layer TiO_2 scheme, using a 8 μm thick transparent layer (30 nm particle size) combined with a 4 μm thick scattering layer (200–400 nm particle size) on top [10].

In spite of the good behavior of mesoporous substrates, the electron transport through a disordered assembly of TiO_2 nanoparticles appears to be an aspect that can potentially be improved. In fact, restructuring the TiO_2 in pursuit of more ordered one-dimensional morphologies is desirable for providing more directionality to the electron pathway from the site of electron injection to the collector (Fig. 1.3). In this context, TiO_2 nanotube [20] and nanorod [21] arrays have emerged as promising alternatives.

Interestingly, Kongkanand et al. showed that colloidal CdSe QD-sensitized TiO_2 films composed of nanotubes yield higher IPCE values than those composed of TiO_2 nanoparticles (*a* ~ 10 % improvement) as shown in Fig. 1.4 [22]. The authors ascribed this enhancement of charge collection to the nanotube architecture, which would facilitate electron transport. By contrast, in nanoparticulate TiO_2 films, the "random walk" and particle-to-particle hopping occurring during electron transport, together with electron trapping at grain boundaries slow down carrier transport, thus decreasing carrier collection [23]. Unfortunately, the overall light-to-electric energy conversion yield obtained with nanotube arrays is still lower than that obtained with nanoparticulate films, presumably due to the lower surface area for the same film

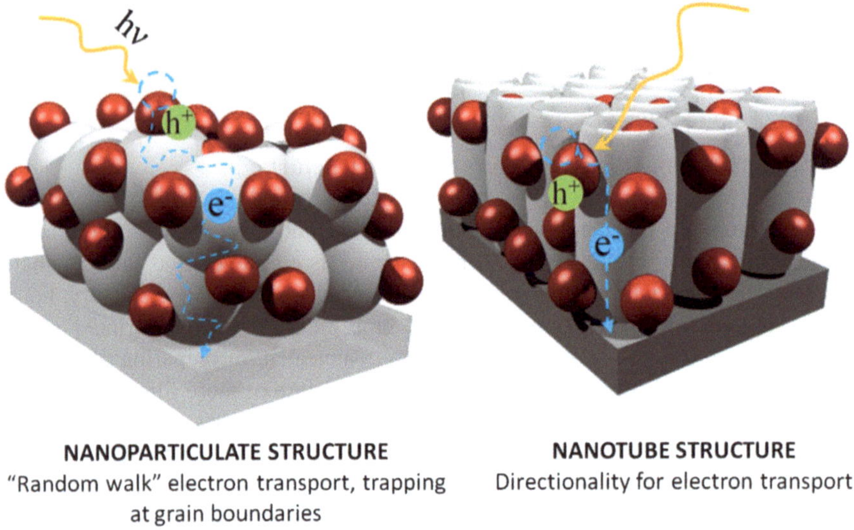

Fig. 1.3 Sketches of nanoparticulate (*left*) and nanotube (*right*) architectures, depicting the corresponding pathways for electron transport

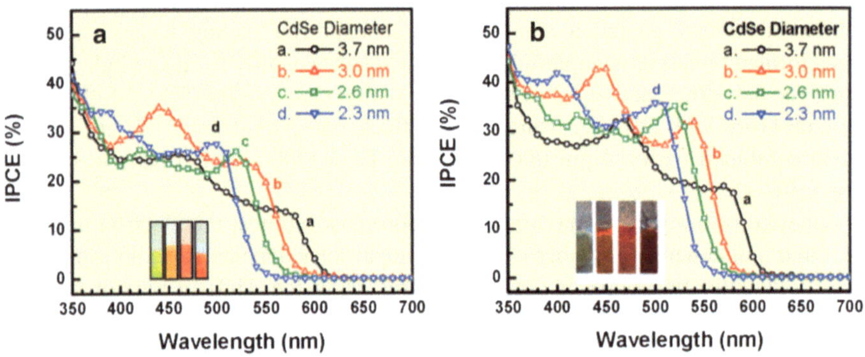

Fig. 1.4 IPCE spectra recorded for (**a**) OTE/TiO_2(Nanoparticulate)/CdSe and (**b**) Ti/TiO_2 (Nanotubes)/CdSe electrodes, as a function of the CdSe QD diameter (*a*) 3.7, (*b*) 3.0, (*c*) 2.6, and (*d*) 2.3 nm. Reprinted with permission from [22]

thickness (lower QD loading) and the fact that only one electrode side is accessible for illumination, since the nanotube films were directly grown on opaque titanium foil by anodization.

It is worth mentioning that other more open TiO_2 nanostructures such as hollow fibers or inverse opals have been tested in devices sensitized with in situ grown QDs [24, 25]. However, the conversion efficiencies are still lagging behind those obtained with nanoparticulate electrodes. As far as we know, these types of nanostructured electrodes have not been employed in the case of colloidal QDSSCs.

1.2.1.2 Other Oxides

Apart from TiO_2, other oxides such as ZnO and SnO_2 have been examined for their application as electron acceptors in QDSSCs. ZnO is a semiconductor with a wide band gap of 3.37 eV that, in comparison with TiO_2, possesses higher electron mobility and can be crystallized more easily in different morphologies, especially 1D-architectures, although its capability for adsorption is lower. Barceló et al. [26] demonstrated that the sensitization of nanoparticulate films of ZnO with colloidal CdSe QDs was feasible, showing that IPCE values are tantamount to those obtained when sensitizing nanoparticulate TiO_2 films. To improve electron mobility, ZnO nanorod/nanowire arrays have been exploited as electron acceptors in QDSSCs. In general, these structures are easily grown from a nanometer ZnO seed layer deposited on a substrate using a hydrothermal method [27, 28]. Interestingly, the conversion efficiency obtained for colloidal QDs attached to these substrates is rather similar to that attained with TiO_2 nanotubes. As an example of the versatility of the ZnO morphologies, Chen et al. reported on a bilayer formed by ZnO nanorods ended with a flower-shaped nanostructure [29]. These authors pointed out that, in such a way, the QD loading could be increased with respect to the native nanorods, leading to higher overall conversion efficiencies. Lastly, it is worth mentioning the hybrid architecture based on ZnO recently reported by Sudhagar et al. In this case, the authors designed a 3D-hierarchical structure composed of silicon nanowires decorated with ZnO nanorods to support the QDs [30]. Although QD sensitization was carried out by means of in situ deposition methods, the high performance obtained with such "pine tree" architecture appears promising for colloidal QDSSCs, due to its high surface area, and its scattering effects together with its open and ordered structure.

The reports dealing with SnO_2 as electron transporting phase are much scarcer, and in most of them in situ grown QDs are employed as sensitizers instead of colloidal ones [31]. The use of SnO_2 could be advantageous in some cases as its conduction band edge is located at an energy lower than that of TiO_2, which would facilitate electron transfer from the photoexcited QD [32]. This was clearly evidenced by Leventis et al., proving that, for samples sensitized with colloidal PbS QDs, the replacement of nanostructured TiO_2 by SnO_2 significantly enhances the yield of charge-separated species [33]. Similarly, Hossain et al. compared the performance of QDSSCs based on TiO_2 or SnO_2 nanoparticulate films [32]. Although in this work the sensitization was carried out by in situ deposition methods, the results are relevant for a possible implementation with colloidal QDs. Concretely, the photocurrents obtained with SnO_2 were higher than those with TiO_2, pointing to a better injection and collection efficiency. However, the lower CB edge location of SnO_2 limits the open circuit potential value. In any case, the fact that similar overall conversion efficiencies are attained with both materials provides compelling evidence that SnO_2 might replace TiO_2. Finally, concerning the morphology of the SnO_2 substrate, mesoporous spheres and upright-standing SnO_2 nanowalls have also been reported for the preparation of sensitized devices, although the efficiencies are far from those obtained with conventional nanoparticulate electrodes [34, 35].

1.2.1.3 Other Materials

In recent years, breakthroughs made in the preparation of carbon nanotubes and graphene have paved the way for their implementation in QDSSCs, primarily as electron acceptors to take advantage of their extremely high electron mobility through the highly delocalized π cloud of the graphitic carbon sheets. The sensitization of carbon nanotubes (CNT) with colloidal QDs dates back from 2005, when Sheeney-Haj-Khia et al. supported, on a gold electrode, nanotubes modified with CdS QDs [36]. Photocurrent action spectra mimicking the absorption of the QDs were recorded. A similar approach to implement carbon nanotubes in QDSSCs was reported by Farrow et al. In this case, the photoanode was prepared by depositing electrophoretically the CNT–CdSe composite on a conductive substrate [37]. In spite of the potentiality of this material, the poor overall conversion efficiency (<0.1 %) underlines several shortcomings of this design that must be addressed: (1) CNTs absorb in the UV–vis range; (2) the low energy of the acceptor level leads to poor open circuit voltages (as with SnO_2); and (3) the high degree of aggregation of the adsorbed QDs favors recombination rather than injection.

On the other hand, graphene offers higher surface area and electron mobility than CNTs, along with the prospect of tuning the band gap [38]. Unfortunately, given that the fabrication of graphene is prohibitively costly, much attention has been focused on the use of reduced graphene oxide (rGO). This material is composed by a few layers of graphene partially functionalized, which leads to inferior properties in comparison with "pure" graphene [39]. Several authors have managed to use rGO as electron acceptor in colloidal QDSSCs. Chen et al. [40] assembled by electrophoresis photoactive rGO–CdSe composites on a transparent conductive substrate, achieving overall solar-to-current conversion efficiencies of 0.58 %, and IPCE values near 12 % for the excitonic peak wavelength. Similarly, Lightcap et al. deposited an rGO–CdSe composite on conductive substrates with TiO_2, achieving IPCE values of 15 % at the excitonic peak wavelength [41]. Other authors have suggested the fabrication of photoanodes by layering graphene and QDs successively [38]. In this way, by using in situ grown QDs, IPCE values near 11 % have been attained.

Briefly, a few attempts have been undertaken to exploit C_{60} molecules as electron acceptors in QDSSCs [42, 43]. In general, the approach involves capping the colloidal QDs with a shell made of C_{60}, followed by deposition of the C_{60}-QD clusters on a conductive substrate by electrophoretic methods. However, IPCE values well below those obtained with other carbon nanostructures have been obtained (<5 %).

1.2.2 *QDs Employed as Sensitizers: Synthesis and Attachment Methods*

The sensitizer in QDSSCs should meet the following requirements (1) it should efficiently absorb most of the photons in the UV–vis range (and IR if possible) and (2) the energy of the photogenerated electron should lie above the CB edge position of the electron acceptor. In the last decades, advances in nanomaterials synthesis have released successful procedures for the preparation of a myriad of quantum dots of different nature. Up to now, just a few of them have been tested as sensitizers. Broadly speaking, the QDs implemented in sensitized devices can be classified as "single-core," "core/shell," and "alloy" structures, depending on their composition. Within the first group, the most widely employed are cadmium and lead chalcogenides (CdS [28], CdSe [44], CdTe [45], PbS [46], PbSe [47]), although a few reports dealing with InP [9] and InAs [48] can also be found. Core/shell QDs, i.e. those having more than one semiconductor material in a well-defined "onion-like" structure, like CdSe/ZnS [49], CdS/CdSe [50], or ZnSe/CdS [51], have also been tapped as sensitizers. Lastly, alloyed QDs would encompass those composed by different semiconductor materials homogeneously dispersed within the nanocrystals, such as CdSeS [52], PbSeS [46], $CuInS_2$ [53], or $ZnCuInS_2$ [54]. Among these materials, CdSe is the most widely studied, mainly due to its easy synthesis, stability, and good performance in sensitized devices. Very recently, another kind of QD outside this classification, the so-called lead iodide perovskites, has burst in the field of QDSSCs attaining astonishingly high conversion efficiencies in liquid junction devices (6.5 %) [55]. This material, though far from being a "colloidal" QD (it is grown in situ on the oxide substrate), opens up a promising new avenue for sensitizing metal oxides.

Three main approaches for preparing the narrow size-dispersed QD colloidal solutions have been reported: hot injection method, solvothermal route, and aqueous synthesis. The hot injection technique entails rapidly mixing highly reactive precursors at a high temperature [56]. In such a way a "burst nucleation" process occurs at the early stages, bringing about relatively monodisperse nuclei that would grow in a homogeneous way subsequently. For example, CdSe can be prepared by rapid injection of a mixture of $Cd(CH_3)_3$ and TOP-Se (trioctylphosphine selenide) in TOPO (triocylphosphine oxide) at ~300 °C under an Ar atmosphere. Subsequently, the temperature is lowered to allow the as-generated CdSe nuclei to grow in a controlled way [57]. In this case, TOPO acts not only as a solvent but also as a coordinating agent or surfactant governing the growth of nanocrystals. Note that in subsequent adaptations of this procedure, different precursors, surfactants, and the utilization of non-coordinating solvents have been tested [58]. This technique has also proven to be very efficient for the preparation of highly crystalline and monodisperse QDs such as CdTe [45], PbS [33], PbSe [47], InAs [48], and InP [9], among others.

The solvothermal synthesis is carried out by using an organic solvent as reaction medium in a sealed vessel [59]. The foremost advantage of this system arises from the fact that heating promotes the reaction and enhances the solubility of reagents. By choosing an appropriate organic solvent and a temperature higher than the normal solvent boiling point, raw materials like elemental S, Se, or Te, apparently poorly reactive, can be employed. As an example, Wang et al. proposed a solvothermal route to prepare highly crystalline CdSe QDs [60]. In this case, elemental selenium and cadmium myristate were mixed in toluene in the presence of oleic acid and TOP as surfactants and heated in a sealed vessel for several hours

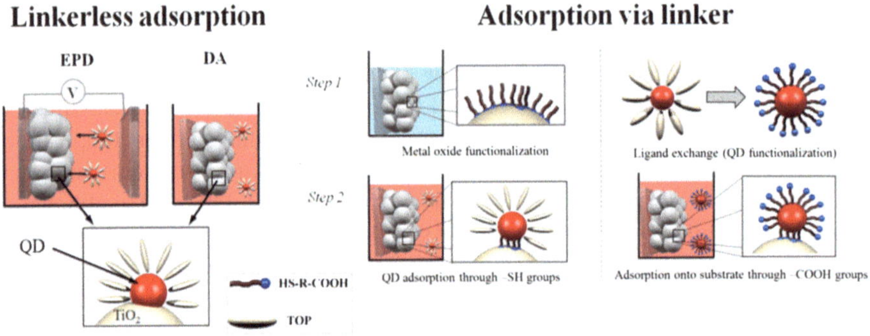

Fig. 1.5 Main sensitization methodologies: linkerless adsorption by either electrophoresis (EPD) or direct adsorption (DA) (*left*), and two-step sequential adsorption via a linker (*right*)

at 180 °C. This strategy has been widely used for preparing CdTe [60] and, more recently, ternary alloys such as $CuInS_2$ [61].

Finally, semiconductor QDs, especially cadmium chalcogenides, can also be synthesized in aqueous media. Concretely, semiconductor nanocrystals are obtained via a precipitation reaction between Cd^{2+} ions and a highly reactive source of chalcogen, in the presence of thiols as stabilizing agents [62]. The thiols are crucial in the process, since they strongly adsorb on the nanocrystal surface preventing uncontrollable growth. Concretely, Rogach et al. described the synthesis of highly luminescent CdTe QDs by the in situ reaction of $Cd(ClO_4)_2$ with H_2Te or NaHTe in water under a N_2 atmosphere using mercaptopropionic acid (MPA) as thiol stabilizer [63]. This procedure has also been adapted for the synthesis of CdS [28] and CdSe [64]. In this case and, in contrast to other procedures described above, where the nanocrystal is capped with inert organic molecules, the resulting QD can be obtained with a functionalized shell according to the moieties of the thiol capping molecules.

Once the semiconductor QDs are synthesized, two main routes have been developed to anchor them to the electron acceptor matrix: directly or via a linker (Fig. 1.5). One of the easiest ways is the electrophoretic deposition method [46, 65]. Commonly, the conductive substrate coated with the electron acceptor is soaked in the organic dispersion of QDs in front of a counter electrode. The application of a high dc voltage (in the range of kV) for several hours drives nanocrystals toward the electrode. Given the absence of charge in the nanocrystals, the electric field would solely interact with induced dipoles in the QDs, which account for the required high dc voltages. On the other hand, the direct adsorption of QDs on TiO_2 or ZnO (or even carbon nanostructures) can be attained by just immersing the bare electrode into the QD colloidal dispersion [9, 48, 83]. Guijarro et al. [66] recently demonstrated that the choice of the solvent where the QDs are dispersed is crucial to promote the direct adsorption of QDs. In general, direct adsorption suffers from the lack of a specific interaction between the QDs and the substrate, which indeed favors agglomeration of QDs over the attainment of high coverages. Adsorption through a linker helps solve this problem, by including bifunctional molecules that bind to both materials. Commonly, linkers have a carboxylic group (–COOH) that interacts specifically with TiO_2 (and other oxides) and a thiol group (–SH) that anchors to the QD surface (e.g., MPA, thioglycolic acid, and cysteine) [3]. Two main routes (consisting of two steps) to tether the QDs by means of these linkers or molecular wires have been reported. On the one hand, the bare TiO_2 electrode can be modified with the linker, by simple immersion in a solution containing the linker. This gives rise to a functionalized TiO_2 with free thiol groups pointing outward. Subsequently, the modified TiO_2 is immersed into the QD colloidal dispersion and the QDs are attached via the thiol group to the oxide [44]. In this case, a maximum QD coverage tantamount to that attained by direct adsorption has been reported [66]. On the other hand, the QDs can be modified with the linker prior to their attachment to the oxide resulting in functionalized QDs with a shell of external unsaturated carboxylic groups. For this purpose, ligand exchange of QDs must be carried out, to replace the native organic surfactans (TOP or oleic acid) by bifunctional ligands. It is worth noting that originally QDs are hydrophobic, but the ligand exchange gives rise to water-soluble QDs because of the favorable interaction between carboxylic groups and water. Commonly, the ligand exchange can be performed by heating at mild conditions a mixture of the QD colloidal dispersion with the linker [20, 49]. Likewise, some authors performed the ligand exchange by vigorously stirring under ambient conditions the biphasic mixture made with the linker dissolved in water and the QDs dispersed in an organic solvent. As a result, the QDs exchange the capping agent and migrate to the aqueous phase [10]. Once the QDs dispersed in the aqueous medium, the pH is adjusted (to around 7) to facilitate their adsorption on TiO_2 through the carboxylate groups of the QD capping agent [67]. This strategy has proven to lead to a QD coverage of ~34 %, higher than those typically obtained by direct adsorption from an organic phase. It is worth noting that the aqueous syntheses described above already release QDs capped with bifunctional molecules like MPA and consequently, sensitization can be performed in one step [28]. In this context, very recently Zhang et al. showed that the performance of QDSSCs sensitized from aqueous medium using MPA as a linker can be significantly enhanced by including a short annealing treatment [10]. This presumably removes the linker, allowing for direct contact between oxide and QDs, thus circumventing the deleterious aggregation of direct adsorption methodologies.

1.2.3 Electrolytes and Counter Electrodes

Not many different electrolytes have been proposed for colloidal QDSSCs. Aqueous polysulfide electrolytes (S^{2-}/S_n^{2-}) based on sodium sulfide and sulfur are the most commonly used [19]. Generally when this redox shuttle is employed, the concentration of the oxidized species is always lower than that of the reduced one in order to prevent recombination with electrons in the oxide. It is noteworthy that sometimes a mixed solution of methanol and water is used as a solvent [10, 68]. Methanol is a nonregenerative hole scavenger and could give rise to erroneous power conversion efficiencies.

Trying to take advantage of the enormous background gained from DSSCs during the last decades and considering the similarities between dye- and quantum dot-sensitized solar cells, organic electrolytes containing the I_3^-/I^- redox couple have also been tested for QDSSCs [9]. These electrolytes have given the best results for DSSCs. However, most QDs suffer from (photo)degradation when used in combination with this redox pair, unless they are protected with a TiO_2 coating [69].

Apart from these, other electrolytes have been investigated, although to a lesser extent. This is the case of the cobalt(III/II)-based systems in organic media [48, 70]. Recently, a thiourea-based organic redox shuttle has also been studied [71].

Regarding counter electrodes, platinized conductive glass has been studied with all the aforementioned electrolytes [9, 19, 48, 70, 71]. Pt counter electrodes are the most widespread in DSSCs, because they show low charge transfer resistance for organic I_3^-/I^- redox electrolytes. Nevertheless, in contact with aqueous polysulfide electrolytes, Pt electrodes get poisoned. Thus, alternative counter-electrode materials, such as Au [19], carbon [68] and metal sulfides like Cu_2S [10, 19], have been employed. The best efficiencies reported so far have been obtained for polysulfide liquid junction QDSSCs (5.4 %) employing Cu_2S on brass or Cu_2S/graphene oxide as counter electrodes [10, 72]. In the last case in situ deposited QDs were used instead of colloidal ones.

1.3 Solid-State Solar Cells with QD-Sensitized Photoanodes

The stability of the liquid junction QDSSCs is limited by the use of a liquid electrolyte as hole conductor, which is responsible not only for the possible leakage of the cells but also for the degradation of the light absorbers due to its commonly aggressive nature. In order to solve these constraints, liquid-junction solar cells are evolving toward solid state designs where the liquid electrolyte is substituted by an HTM. In most cases, the studies dealing with this kind of configuration involved the use of successive ionic layer adsorption and reaction (SILAR) [73] or chemical bath deposition (CBD) [74, 75] as methods for QD preparation. In fact, one of the highest conversion efficiencies obtained so far for a semiconductor-sensitized solar cell (6.18 %) employs CBD for growing in situ Sb_2S_3 on the surface of a mesoporous layer of TiO_2 and PCPDTBT (poly(2,6-(4,4-bis-(2-ethylhexyl)-4H-cyclopenta[2,1-b;3,4-b′]dithiophene)-alt-4,7(2,1,3-benzothiadiazole))) as HTM [76]. However, it is important to take into consideration that, in this and other studies that use CBD, there is a massive deposition of sensitizer. This leads to a configuration close to an ETA (extremely thin absorber) solar cell and to a loss of the quantum confinement typical of the QDs.

As mentioned in the introduction, the HTMs usually proposed for QDSSCs are mainly divided into molecular and polymeric hole transporters. In addition, there are some studies employing ionic liquids, whose properties are between those of typical liquid- and solid-phase HTMs [54, 77]. One of the main limitations of solid-state QDSSCs is the difficulty of the HTM to penetrate through the pores of the oxide semiconductor matrix, which restricts the thickness of the latter. In principle, the use of a molecular hole transporting material instead of a polymeric one could facilitate the complete filling of the mesoporous oxide. However, no large differences have been observed when these materials are compared [78].

There is very little work done on solid-state colloidal QDSSCs. The first study dates back from 2003. Zeng et al. [79] employed InP QDs as light harvesters and spiro-OMeTAD as HTM; low short circuit photocurrents (j_{sc}) of 0.035 mA cm^{-2} and open circuit voltages (V_{oc}) of 0.4 V were obtained under 58 mW cm^{-2} simulated solar light. Günes and coworkers [80] proposed a hybrid solar cell combining HgTe QDs and poly(3-hexylthiophene) (P3HT) as HTM; values of V_{oc} of 0.35 V, j_{sc} of 0.8 mA cm^{-2} and a fill factor (FF) of 0.4 were reported. In 2009, Kniprath et al. presented a cell constructed by layer-by-layer deposition of the electron conductor (TiO_2), sensitizer (CdSe QDs), and HTM (the conjugated polymer poly(9,9-dioctyl-fluorene-*co*-*N*-(4-butylenphenyl)diphenylamine)). The typical values obtained with this configuration were $j_{sc} = 0.05$ mA cm^{-2} and $V_{oc} = 1.0$ V, with a fill factor up to 0.31 and a maximum power efficiency of 0.020 % [81]. One year later, a cell consisting of ZnO nanocolumns sensitized with CdSe QDs through an aromatic linker and P3HT as HTM was reported by Zeng and coworkers. Short circuit currents of 0.14 mA cm^{-2} and open circuit voltages around 0.55 V were reached with very low fill factors [82]. More recently, Snaith et al. showed a hybrid design composed of mesoporous SnO_2 sensitized with PbS and infiltrated with spiro-OMeTAD as organic hole transporter. Higher short circuit currents (around 4.2 mA cm^{-2}), lower open circuit voltages (<0.45 V), and efficiencies over 0.5 % were observed [78]. Barceló et al. have presented an all solid-state solar cell fully processed in air incorporating CdSe QDs and 3,3‴-didodecyl-quaterthiophene (QT12) as HTM. In this work, different modes of attachment of the QDs to the oxide matrix were compared. Interestingly, the methods using colloidal QDs gave the best results (see Fig. 1.6). Low photocurrents, remarkable open circuit voltages (between 0.9 and 1 V), and a maximum power conversion efficiency of 0.34 % were obtained [83]. Very recently, Kim et al. have reported on an efficient and stable solid-state QDSSC displaying an impressive conversion efficiency of 3.3 %. The cells were based on co-capped colloidal PbS QDs linked to TiO_2 by MPA and P3HT as HTM [84].

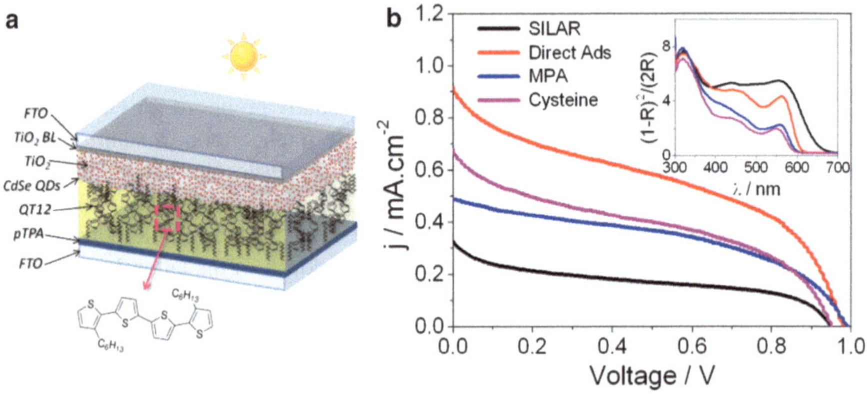

Fig. 1.6 (**a**) Schematic structure of the solid-state QDSSC tested by Barceló et al. (**b**) Current density–voltage curves of the cell with different modes of attachment. The inset presents the UV–vis diffuse reflectance spectra (transformed according to the Kubelka–Munk function) obtained for each photoanode. Adapted from [83]—reproduced by permission of the PCCP Owner Societies

1.4 Solar Cells with QD-Sensitized Photocathodes

Most frequently, the photoactive electrode in QDSSCs is the anode. An alternative consists in the use of a QD-sensitized photocathode together with a counter electrode. In such a case the operation mode is the inverse of that shown in Fig. 1.2 (Fig. 1.7). Upon light absorption and electron excitation, charge separation occurs by electron injection from the valence band of the p-type semiconductor to the excited sensitizer (henceforth, hole injection from the quantum dot into the semiconductor). Subsequently, the quantum dot is regenerated to its ground state by electron transfer to the oxidized form of the redox mediator. The resulting reduced species diffuses to the counter electrode (in this case the anode) and transfers one electron. Finally, the electron enters the external circuit and reaches the photocathode.

The possibility of using a p-type semiconductor photocathode in a DSSC was first demonstrated in 1999 [85]. Ever since, some relevant works studying this configuration have been published [86–92]. Nevertheless, as far as we know, there are only two reports exploring the feasibility of using quantum dots as light harvesters for p-type QDSSCs. In 2011, Chang et al. reported on a CdS-sensitized mesoscopic NiO photocathode operating in a polysulfide electrolyte [93]. The same year, Kang and coworkers showed that, compared to dye-sensitized photocathodes, CdS-sensitized NiO photocathodes exhibited two orders of magnitude faster hole transport (attributable to the passivation of surface states) [94]. It should be mentioned that both results have been obtained using QDs grown in situ by SILAR or CBD. We should emphasize that as far as we know, there are no reports on photocathodes sensitized by colloidal QDs. However, the work of Wu and Yeow should be underlined in this context [95]. Using spectroscopic measurements, these

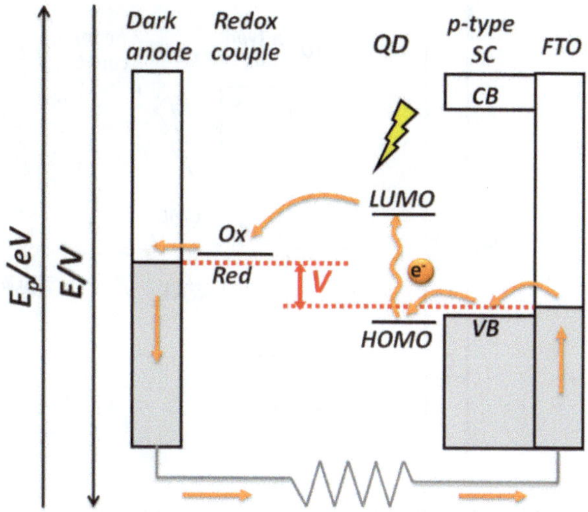

Fig. 1.7 Working scheme of a solar cell with a quantum dot-sensitized photocathode. SC denotes semiconductor

authors demonstrated that electron transfer occurs from NiO to colloidal CdSe–ZnS core–shell QDs.

In the following, the different materials proposed for QDSSCs based on a photocathode and the desirable characteristics for each of its components are reviewed.

Although several n-type semiconductors (TiO_2, ZnO, SnO_2, etc.) have been explored as electron conductors for QDSSCs (see Sect. 1.2.1 of this chapter), only NiO has been studied as a material accepting holes from the sensitizer and transporting them to the contact. This is probably because NiO is employed in a variety of applications, especially as electrochromic material for smart windows [96]. Nickel oxide can be easily synthesized directly as thin films following different routes such as hydrothermal synthesis [97], electrodeposition [98], and spray pyrolisis [93]. In addition, there also exist commercial nanopowders [86].

The valence band position of the hole transporting material is a significant parameter to be taken into account, as it determines the maximum V_{OC} attainable by the solar cell (see Fig. 1.7). In the case of NiO, the VB is located at 0.54 V vs NHE at pH 7 [99], which makes it difficult to obtain large open circuit voltages for the typically employed redox couples.

On the other hand, nickel oxide and in general p-type semiconductor oxides present lower carrier mobilities than n-type ones [89, 100]. This should favor recombination of holes in the VB of the oxide with (1) the excited electrons in the sensitizer and (2) the reduced species of the redox mediator. To build efficient solar cells it would be desirable to use p-type semiconductors with high hole mobilities and possibly with one-dimensional optimized structures.

The redox electrolytes employed in combination with sensitized photocathodes are the same used for sensitized photoanodes, that is, I_3^-/I^- for DSSCs and polysulfide electrolyte for QDSSCs. However, more recently, a cobalt

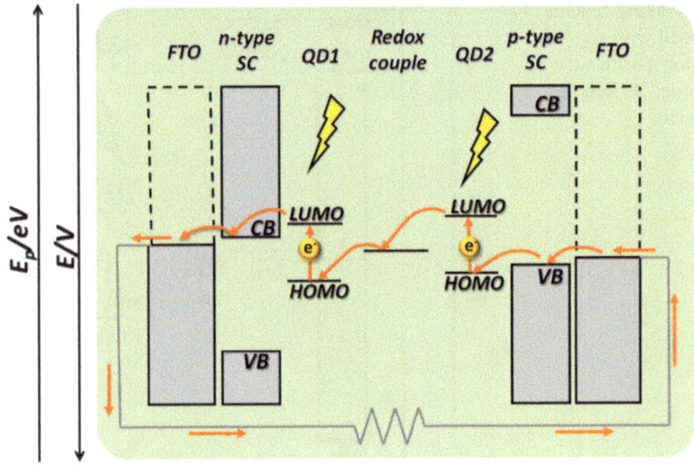

Fig. 1.8 Schematic representation of the working principle of a tandem QDSSC. SC denotes semiconductor

polypyridyl-based electrolyte has also been tested with remarkable results [88, 94, 101]. In any case, the photocurrents achieved are very low, which stems from:

- Low sensitizer loading.
- High recombination of the photoinjected holes in the VB of the NiO with the reduced light harvesters, favored by the low carrier mobility of this p-type semiconductor.
- The use of light absorbers that are normally designed for n-type semiconductors, with inappropriate LUMO position, very close to the VB edge of the oxide, facilitating recombination.

In this context, it is noteworthy that recombination could be hindered by implementing strategies already used for dye-sensitized photocathodes. For example, it has been demonstrated that the deposition of a thin layer of alumina over NiO prior sensitization prevents recombination between photoinjected holes in the oxide and the reduced light harvesters, increasing the carrier collection efficiency [91, 102].

The main interest of quantum dot-sensitized photocathodes is that they can be combined with photoactive anodes in a tandem configuration (see Fig. 1.8). This design presents some potential advantages:

- Its maximum theoretically power conversion efficiency is 42 % since two semiconductors are employed, being just 31 % for single junction configurations.
- The photovoltage in this kind of photovoltaic devices will be determined by the sum of the photopotentials of each electrode.
- It allows harvesting a broader window of the solar spectrum since two sensitizers, one in each electrode, are employed.

As a drawback, the photocurrent for this kind of configuration will be that of the least efficient electrode. This is the main reason explaining that the results obtained until now for this configuration in DSSCs are worse than those achieved with standard DSSCs [88].

1.5 Rationalization of the QDSSC Efficiency

Most recent studies on QDSSCs (and particularly those based on non-colloidal QDs) focus on the net conversion efficiency of the final devices. In order to devise strategies to improve the cell performance, the correlation between the efficiency and the different fundamental electron transfer/transport steps is required. In this respect, a number of experimental spectroscopic and theoretical tools have been developed to study the primary photochemical and photophysical events involving charge carriers. The corresponding studies complement the electrochemical and electrical measurements in the dark and under illumination [103].

1.5.1 The Use of Time-Resolved Techniques and Theoretical Methods

In the last decade, dozens of papers of theoretical nature have dealt with aspects such as dye-semiconductor binding [104–106], electronic structure [107, 108], or simulation of electron injection [106, 109]. On the other hand, theoretical approaches have been extensively applied to model the electronic structure and optical properties of QDs [110, 111] but, particularly, to calculate the efficiency of MEG [112–115]. Although it has been unambiguously proved that these methods provide reliable results, explain experimental data, and predict novel properties, they have been scarcely applied in the QDSSC field. Probably the development of these theoretical studies has been limited by the high computational cost and the uncertainty related to the chemical composition and morphology of the QDs and the interface QD/metal oxide.

Dahnovsky [116] has proposed a theoretical scheme to calculate the photocurrent in QDSSCs. The expression for the photocurrent is obtained by considering the quantum dynamics of the electron–electron and electron–phonon interactions in the QD, the interaction of QDs with light, the electron–phonon interaction in the metal oxide, and the interaction between the electronic states in the QD and those in the metal oxide. The final expression for the photocurrent could be solved by different computational schemes depending on the level of approximation.

Other authors have developed atomistic models to study the titanium dioxide–QD interface. Giustino et al. [117] have modelized a defect-free TiO_2/Sb_2S_3 QD interface and calculated the energy-level alignment based on a generalized gradient approximation to the density-functional theory (DFT). Using this type of approach, Qi and coworkers have also examined the geometry and electronic coupling between a CdS QD and a TiO_2 nanotube [118]. In this case, a detailed discussion of the interfacial electron transfer from the sulfur atoms of the CdS QD to the conduction band of TiO_2 upon visible illumination is presented. The electron transport along the TiO_2 nanotube is also simulated, and it is concluded that the one-dimensional directional pathway toward the substrate through titanium $d_{x^2-y^2}$ and d_{z^2} orbitals is favored (Sect. 1.2.1 of this chapter). Long and Prezhdo [119] have focused on the simulation of the ultrafast photoinduced electron transfer from a PbSe QD into rutile through a facet (110). In this case, the ab initio nonadiabatic molecular dynamics simulation employed indicates that the electron transfer successfully competes with energy losses due to electron–phonon relaxation.

The use of time-domain simulations for electron injection is required because the electron transfer from the QDs to the metal oxide is extremely fast, making difficult to invoke traditional electron transfer models such as the Marcus theory, which requires a slow transfer dynamics to allow for redistribution of vibrational energy. Probably, in the next years, the number of theoretical works in the QDSSC field will increase as these methods can extend our ability to understand the electronic and optical aspects of the photovoltaic conversion process at atomic-scale resolution, complementing phenomenological models and macroscopic measurements.

On the other hand, there exist a number of spectroscopic techniques that allow for the experimental detection of the chemical species generated upon QD photo-excitation. Most of these techniques are time resolved, making possible to study the kinetics of photoinduced elemental processes. Among these spectroscopic techniques, the most popular are:

- Femtosecond pump–probe spectroscopy, also called transient absorption spectroscopy, is based on changes in the absorption spectrum upon photoexcitation or charge transfer. In a typical experiment, QDs are photoexcited by means of a monochromatic pulse (or pump), while a weak probe (i.e., light of low intensity) is sent through the sample. Results are presented as difference absorption spectra, i.e., the absorption spectrum of the excited sample minus the absorption spectrum of the sample in the ground state (ΔA) after a variable time delay between the pump and the probe. Therefore, such spectrum gives information on the nature of the generated, while the time evolution of ΔA contains information on the photogenerated species dynamics. The transient absorption spectrum for a metal oxide film sensitized with QDs most often consists of a negative signal followed by a positive response. The negative signal is attributed to the ground state bleaching, while the positive featureless broad band extending at longer wavelengths is most likely due to the presence of free and trapped electrons in the metal oxide or localized holes in the QDs. In the case of a QD-sensitized metal oxide in contact with a HTM, the signal coming from the generation of oxidized HTM is also present.

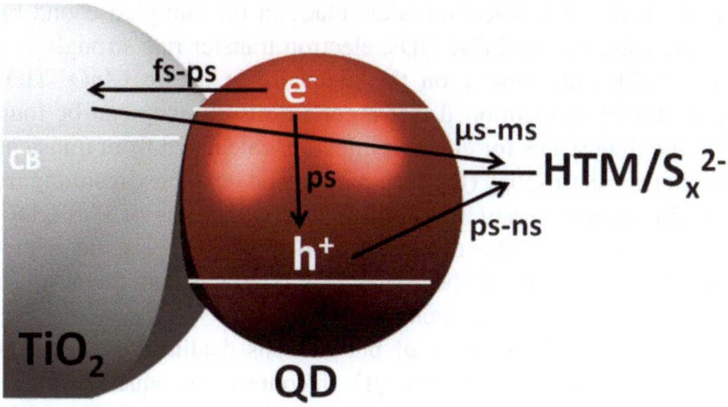

Fig. 1.9 Representation of a QD adsorbed on TiO_2 and the approximate time scales for electron and hole transfer processes

- Time-correlated single photon counting (sometimes is also referred to as emission lifetime). Here, the fluorescence of the sample is monitored as a function of time after excitation by a flash of light (a NanoLED excitation source with a pulse width in the ns range). For QD–metal oxide interfaces, light emission is due to the photoexcitation of QDs and their corresponding radiative transitions.
- Terahertz time-domain spectroscopy in which the processes are probed with short pulses of terahertz radiation. The configuration also uses a pump–probe scheme, as described for transient absorption spectroscopy [120]. The probing THz field between 0 and 8 meV provides a method to measure the photoconductivity in the oxide after selective photoexcitation of the QDs by the pump pulse. Thus, in contrast to the time-resolved spectroscopic techniques mentioned above, this technique allows to monitor the electron transfer from the excited QDs toward the metal oxide in QDSSCs unambiguously [47, 121].
- Finally, the lens-free heterodyne-detection transient-grating (LF-HD-TG) technique should be mentioned, although it is not a spectroscopic technique. The TG response monitors the change as a function of time in the refractive index of the sample upon photoexcitation. Assuming Drude's model, this change can be expressed as a linear function of the concentration of free photogenerated carriers. It should be noted that this optical technique is not carrier-specific, and that the relative contribution of each carrier to the TG signal is inversely proportional to its corresponding effective mass. However, in the case of the often-studied TiO_2/CdSe system, the effective mass of CB TiO_2 electrons is much larger than that of the carriers in CdSe, making their contribution to the TG signal negligible. In addition, the effective mass of CdSe holes is more than three times that of the CdSe electrons, being the TG signal dominated by the response of free electrons in CdSe [122].

Combining these techniques it has been possible to acquire a complete picture of the charge dynamics at the QD–metal oxide interface upon illumination (Fig. 1.9). Not only ultrafast transient absorption but also time-resolved terahertz spectroscopic measurements reveal that injection takes place in the sub-picosecond time scale [47, 123]. Interestingly, for CdSe QDs, electron transfer rate strongly depends on the QD size [124] and weakly on the metal oxide nature (ZnO, TiO_2, SnO_2) [33, 125]. It should be mentioned that a certain discrepancy can be found in the literature in the time scales for electron injection calculated from transient absorption and terahertz spectroscopy (femto to picosecond) and those resulting from time-resolved luminescence measurements (ns). This is the case of PbS QDs on TiO_2 nanoparticles [126, 127]. The discrepancy in the results can be related to the presence of defect centers at the oxide–QD interface that favor non-radiative recombination (luminescence quenching) [47].

On the other hand, the number of publications dealing with hole transfer in QDSSCs is scarce. In the case of CdSe QDs in contact with aqueous sulfide, it takes place on the nanosecond scale [128]. In contrast, the hole transfer from PbS toward spiro-OMeTAD seems to be much faster (ps) [129]. Finally, recombination, which takes place mainly by electron transfer from the metal oxide toward the oxidized hole conducting material, occurs on the microsecond–millisecond time scale [33, 129, 130]. It is worth noting that most often this process is studied by using electrochemical measurements such as photovoltage decay or impedance spectroscopy, rather than via spectroscopic techniques.

1.5.2 The Role of Capping Agents

The surface of colloidal QDs is covered with organic molecules, called capping agents. Apart from their synthetic utility, they are used to remove unsaturated valences, i.e. dangling bonds on the QD surface. This affects not only the electronic properties of the QDs [131], particularly the radiative/nonradiative carrier dynamics, but also the interaction with the metal oxide surface and charge transfer.

As mentioned in Sect. 1.2.2, the synthetic methods for colloidal QDs can be classified into two large groups depending on the use of either organic or aqueous solvents. The solvent employed determines the capping ligand nature. Long-chain molecules such as alkylphosphines, alkylphosphine oxides, and alkylamines are employed to protect the QDs in organic solvents. These molecules are dielectric barriers that may hinder interfacial charge transfer. On the other hand, colloidal aqueous QDs are stabilized with short-chain thiol-stabilizers, typically bifunctional molecules such as thioglycolic acid, mercaptopropionic acid, and cysteine. Their short chain length is highly favorable for efficient charge transfer.

However, most of the studies on colloidal QDSSCs have been performed using organic QD colloids. In such a case, it has been clearly demonstrated that electron injection toward the metal oxide depends on the capping agent. Sykora et al. [132] showed an enhancement in electron transfer for *n*-butylamine capped QDs adsorbed on TiO_2 with respect to QDs covered by tri-*n*-octylphosphine oxide (TOPO). This improvement was attributed mainly to two factors: an increased electronic coupling between the QDs and TiO_2 ascribed to the reduced length of the ligand and to an increased driving force for electron transfer resulting from the distortion in the QD electronic structure following the adsorption on TiO_2 [7, 129]. In this sense, selecting redox couples [71] or in general vein, hole transporting materials [83, 133] that efficiently interact with the QD capping agents may open a way to boost the efficiency of QDSSCs.

On the other hand, the capping molecules also play an important role protecting the QDs and enhancing the device stability under ambient conditions (avoiding encapsulation) [84].

Finally, some authors have tried to use dyes as ligands which can act as regenerating agents of the QD sensitizers and promote charge separation in the dye/QD/TiO_2 system [134, 135]. Depending on the QD and the dye employed, it is also possible to indirectly excite the dye by funneling harvested light via Förster resonance energy transfer [136, 137].

1.5.3 The Role of the Linker and Direct Adsorption

Guijarro et al. [66] have carefully studied the sensitization of TiO_2 nanoporous electrodes by incorporating presynthesized TOP-capped CdSe QDs. Covering the nanoporous TiO_2 films with QDs has been achieved by direct adsorption and via a molecular linker (MPA). Using the same QDs, the effect of the anchoring mode and coverage degree was tested. As mentioned above, in contrast to MPA-mediated attachment, for long adsorption times, direct adsorption leads to a high degree of QD aggregation, particularly in the outermost part of the TiO_2 nanoporous layer. This effect induces pore blockage and therefore, a diminution of the interfacial active area of the electrode, as revealed by the diminution of the charge accumulation region of the cyclic voltammograms (see Fig. 1.10). Importantly, at saturation, only 14 % of the real surface area of the TiO_2 layer was covered for both attachment modes. This result points to the necessity of porous oxide films with a more open structure, such as TiO_2 nanotubes. In such a case, the blockage of the pores should be partially hindered. Interestingly, the cyclic voltammograms for directly adsorbed colloidal CdSe TiO_2 nanotubes do not show a clear diminution of the active interfacial area even after 14 sensitization hours (Fig. 1.10). Instead, a pair of quasi-reversible peaks is observed, which can be ascribed to redox processes of interfacial QDs as there is a good correlation of its associated charge density and the amount of deposited QDs.

Regarding the QD coverage, for MPA attachment, the IPCE increases with loading, whereas a maximum (close to 40 % at the QD excitonic peak for P25 TiO_2 electrodes) is defined for intermediate coverages in the case of QD direct adsorption (see Fig. 1.11). In addition, for equivalent QD loading, the IPCE values are larger in the case of direct adsorption. This suggests that a direct contact (or close proximity) between the QDs and the oxide nanoparticles is beneficial for the efficiency of the photoanode. However, QD aggregation leads, for high QD coverages, to a severe drop in the IPCE. This decrease is probably linked to the

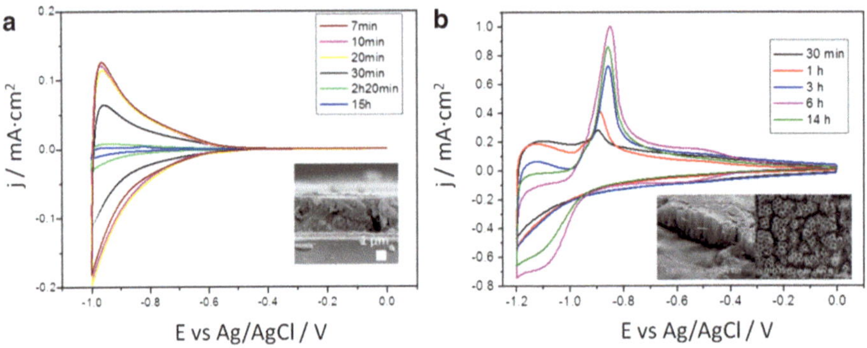

Fig. 1.10 Cyclic voltammograms for TiO_2 nanoparticulate (**a**) and TiO_2 nanotubes electrodes (**b**) after sensitization through direct adsorption of TOP-capped CdSe QDs for different adsorption times. Electrolyte: N_2-purged 0.5 M Na_2SO_3. Scan rate 20 mV s^{-1}

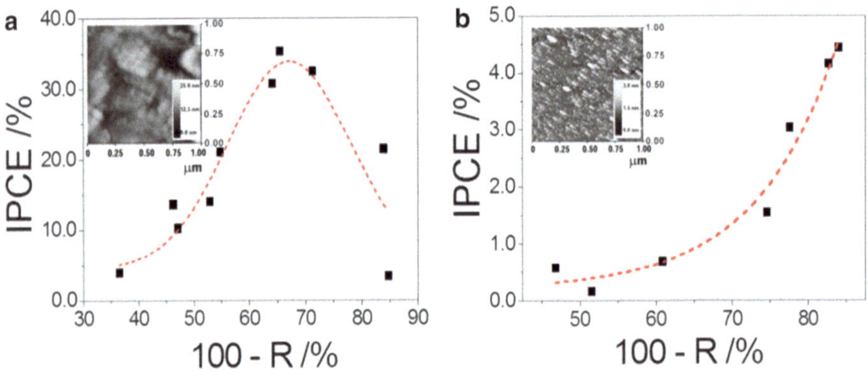

Fig. 1.11 IPCE at the excitonic peak vs 100-Reflectance % at the excitonic peak for direct (**a**) and MPA-mediated (**b**) adsorption. The inset shows the tapping-mode AFM images of an Nb-doped (1 1 0) rutile single crystal sensitized in the same way. The images clearly show the generation of aggregates for direct adsorption when a long adsorption time is employed. Adapted with permission from [66]

existence of QD–QD contacts, and more importantly to a blockage of the TiO_2 film nanochannels by accumulation of QD aggregates. Although these results were obtained in a three-electrode configuration, the final efficiency of two-electrode solar cells using both a polysulfide electrolyte [19] and a solid HTM [83] shows the same trend.

On the other hand, the photoanode efficiency strongly depends on the nature of the linker [44]. Not only the amount of adsorbed QDs is strongly dependent on the linker nature but also electron injection and recombination. A study of the subpicosecond-resolved dynamics for photogenerated carriers has revealed that direct adsorption yields faster electron injection than linker-assisted anchorage (Fig. 1.12a). In addition, when comparing the injection rates for samples using

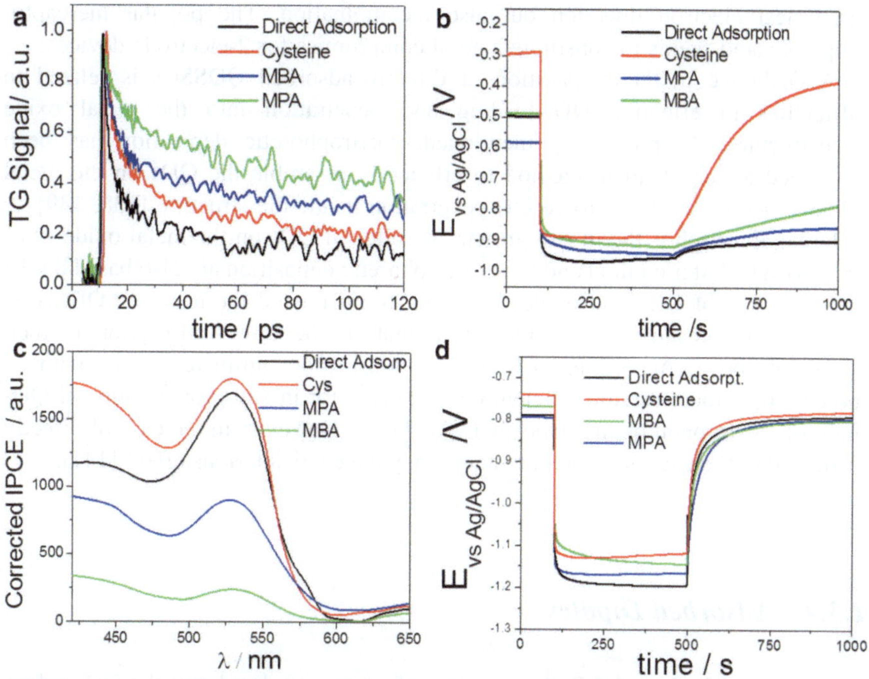

Fig. 1.12 (**a**) Transient grating responses for CdSe-sensitized TiO_2 electrodes using direct adsorption and cysteine, mercaptopropionic acid (MPA) and mercaptobenzoic acid (MBA) as linkers. All the curves are normalized to the maximum value. In this case a faster decay corresponds to a faster electron injection. (**c**) Corrected IPCE spectra, considering the light absorption, for TiO_2/CdSe QDs in N_2-purged aqueous 0.5 M Na_2SO_3. (**b** and **d**) Photopotential measurements obtained for transient illumination of electrodes sensitized with colloidal QDs in the absence (**b**, N_2-purged aqueous 0.5 M Na_2SO_3) and in the presence of an electron scavenger (**c**, 0.1 M S + 1 M Na_2S + 1 M NaOH). In this case, a larger photopotential value corresponds to a larger electron accumulation, while a faster decay corresponds to a faster electron recombination rate. Adapted with permission from [44]

linkers, even for small molecules such as cysteine or mercaptopropionic acid, important variations were found. Importantly, these measurements correlate well with the IPCE values measured in the absence of an electron scavenger (Fig. 1.12c) [44].

The presence of linker molecules tethering the QD may induce the generation of recombination centers, leading to electron lifetimes shorter than those obtained with direct adsorption (not only in the presence but also in the absence of an electron acceptor in the electrolyte) (Fig. 1.12b, d). This is not unexpected, as these linkers can efficiently trap photogenerated holes through generation of thyil radicals that could eventually recombine with electrons injected into TiO_2. According to these fundamental studies, and as shown experimentally, the direct adsorption method improves the final efficiency of the devices as long as QD aggregation is avoided [66, 83]. Regarding the linkers, cysteine shows not only the fastest electron injection but also recombination. The popular mercapto-propionic acid seems to constitute a good compromise for 2-electrode devices.

A challenge in the preparation of directly adsorbed QDSSCs is related to difficulties in effective QD loading and penetration into the metal oxide nanostructure. As previously mentioned, electrophoretic deposition has been envisaged as an adequate method to efficiently assemble the QDs on the metal oxide structure, as it can force QD penetration inside the structure [138, 139]. A different approach is the in situ growth of colloidal QDs on the metal oxide films [52, 64]. The latter method and the electrophoretic deposition are also based on the direct contact of QDs and oxide. However, the nature of the resulting QD/oxide interfaces is presumably different from that of the direct adsorption method described above. According to Rogach et al., a more intimate contact with an increased photoconversion efficiency is expected for in situ grown colloidal QDs than for electrophoretically deposited QDs [140]. However, in the case of directly grown QDs, the excess of capping agent may have a deleterious effect [141].

1.5.4 *Adsorbed Dipoles*

As mentioned in Sect. 1.5.2, the presence of organic molecules at the QD surface clearly influences the photoanode behavior. In this respect, it has been reported that not only the energy levels of CdS but also those of CdSe QDs can be shifted in a systematic fashion with respect to the TiO_2 bands using molecular dipoles (benzenethiol derivatives) as adsorbates on the QDs [142, 143]. More negative dipole moments are correlated with higher values of IPCE as a consequence of the upward shift of the QD conduction band. These results have been tested not only in liquid electrolytes [142, 143] but also in solid-state QDSSCs where the possibility of surface passivation by dipoles should also be considered [144]. Very recently it has been proposed that dipoles such as *p*-mercaptophenol may also act as hole mediators, capturing rapidly the holes before recombination inside the QD and funneling them toward a solid hole transporting material or liquid redox electrolyte [145]. In a similar way, it has been observed that a modification of directly adsorbed colloidal PbSe QDs (prepared according to [146]) on TiO_2 with *p*-mercaptophenol dramatically increases the resulting photocurrent (see Fig. 1.13). This molecule may act as a dipole, pushing up the QD conduction band, but in addition, it replaces the insulating shell of capping agents from the QD surface, favoring hole extraction and acting as a hole mediator.

Finally, the presence of adsorbed ions from the aqueous polysulfide redox electrolyte can also determine whether electron injection from the absorber occurs, as they are able to shift the QD bands. In this regard, it has been described the electron injection from CdSe (CBD) covered by CdS (SILAR) towards ZrO_2 [147], but also the band-edge tuning of colloidal Bi_2S_3 QDs on TiO_2 [148].

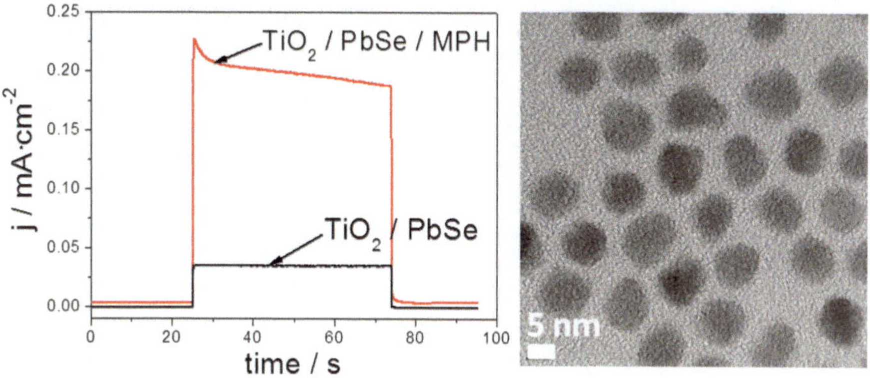

Fig. 1.13 Photocurrent transients recorded for PbSe-sensitized electrodes before and after modification with p-mercaptophenol (MPH). Measurements were done in a three-electrode cell using an Ag/AgCl/KCl (sat.) reference electrode and a Pt wire as a counter-electrode. An N_2-purged aqueous 0.5 M Na_2SO_3 was employed as electrolyte. Illumination was allowed only in the 380–1,100 nm range, with an irradiance of 50 mW cm^{-2}. A TEM image of the PbSe QDs is also included

1.5.5 Treatments with ZnS and Al_2O_3

One of the most frequent post-treatments of the QD-sensitized photoanodes is that resulting from applying 2 ZnS SILAR cycles. It was proposed initially for TiO_2 photoanodes sensitized by chemical bath deposition of CdSe, showing impressive improvements of the conversion efficiency. This method has also been applied to colloidal QDSSCs. Already in 2009, Giménez et al. applied the ZnS treatment to QDSSCs prepared with colloidal QDs [19]. An important improvement of IPCEs and photocurrents was observed. In addition, electrochemical impedance spectroscopy (EIS) measurements revealed that the post-treatment led to an increase in the charge transfer resistance of the sensitized electrode, which indicated a diminution of the recombination of TiO_2 with the electrolyte. Salant et al. [139] have demonstrated that the same method can be successfully applied to TiO_2 photoanodes sensitized with colloidal CdSe QDs by electrophoresis. No detailed mechanistic studies were performed, but important increases in the IPCE (and short-circuit photocurrent) and fill factor were observed. However, no significant improvement of the open circuit voltage was observed. In any case, the conversion efficiency increased by a factor of 4 in some of the cells. This post-treatment was also applied to CdSe-sensitized TiO_2 electrodes prepared either by chemical bath deposition or by adsorption of colloidal CdSe QDs, being the morphology of the ZnS layer addressed by means of HR-TEM [149]. Independently of the nature of the QDs, ZnS appears as a 2–5 nm thick amorphous layer surrounding both the CdSe NCs and the TiO_2 crystals. On the other hand, Agren and coworkers introduced the ZnS layer on TiO_2 electrodes modified with ZnSe/CdS and CdS/ZnSe core/shell QDs working with a thiourea-based organic redox couple [71]. No large effects were observed in the photocurrent, but the stability of the QDs improved. In this

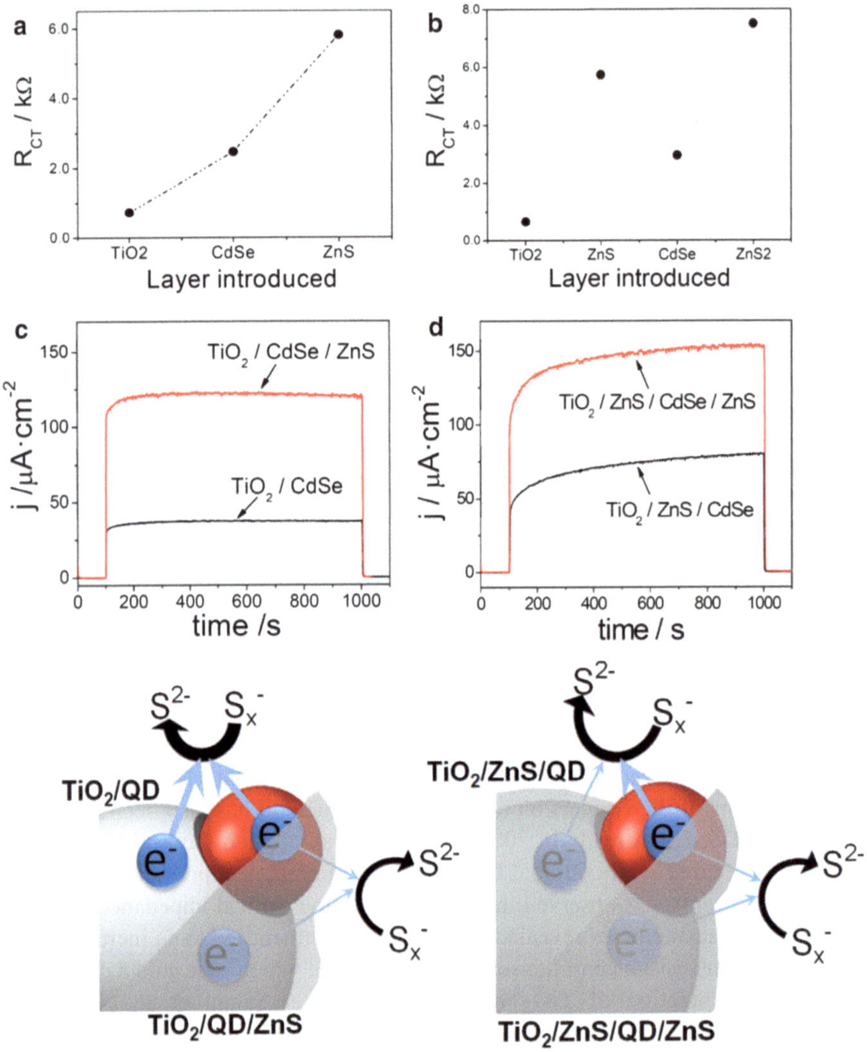

Fig. 1.14 Charge transfer resistances in contact with a polysulfide solution after the successive introduction of the different layers forming TiO_2/CdSe/ZnS (**a**) and TiO_2/ZnS/CdSe/ZnS (**b**) electrodes. Photocurrent transients for TiO_2/CdSe and TiO_2/CdSe/ZnS (**c**) and TiO_2/ZnS/CdSe and TiO_2/ZnS/CdSe/ZnS photoanodes (**d**). Drawings summarizing the effects of the introduction of the ZnS layers are depicted. The thickness of the arrows is proportional to the rate of the corresponding process. Adapted from [151]—reproduced with permission of the PCCP Owner Societies

case, the role of the ZnS layer would be similar to that of the amorphous TiO_2 layer as presented by Zaban and coworkers [150]. More recently, in a comprehensive study, Guijarro et al. demonstrated by means of spectroscopic, voltammetric, photocurrent, and EIS measurements that the role of the ZnS layer is dual (Fig. 1.14) [151]. On the one hand, it passivates the QDs and also the TiO_2 surface.

Both surface modifications seem to reduce recombination with the electrolyte, although the first one seems to be more important. On the other hand, TG measurements with subpicosecond resolution provided information on the carrier dynamics and revealed a second role of the ZnS as an enhancer of electron injection into the TiO_2.

Post-treatments based on other inorganic layers have been proposed. The performance of QDSSCs based on CdS/CdSe QDs prepared by SILAR is greatly increased by the use of a SiO_2 layer [152]. Again, the IPCE (and the photocurrent) is the most sensitive parameter. This procedure was later employed in the preparation of solid-state solar cells based on colloidal CdSe QDs and quaterthiophene as hole conducting material, although a detailed study on the effect of this layer was not carried out [83]. As far as we know, alumina has not been employed as protecting layer for QDSSCs although it has been employed in the context of dye-sensitized solar cells [153]. Interestingly, very recent results indicate an improvement in the conversion efficiency and, particularly, stability to air exposure of Schottky-type PbSe solar cells by the deposition of a conformal Al_2O_3 layer [154]. The extension of this concept to QDSSCs has not been demonstrated yet.

1.6 Conclusions

In spite of the fact that a great majority of the research on QDSSCs has been done with QDs prepared either by chemical bath deposition or by SILAR, the results already available with colloidal QDs suggest that the corresponding solar cells have the potential to achieve efficiencies comparable or even better than those obtained with in situ generated QDs. In fact, this has been corroborated by literature results appeared very recently showing high efficiencies in both liquid- and solid-state devices. In fact, Zhang et al. have reported on liquid-junction QDSSCs based on colloidal CdSe QDs with an efficiency as high as 5.4 %, one of the best performances reported so far regardless the nature of the QDs [10]. These results were attained by using colloidal hydrophilic QDs (QDs capped with MPA) together with a thermal annealing step. Even in this case, only a 34 % of the TiO_2 surface area was covered by the QDs, which indicates that the achievement of larger coverages may lead to even higher conversion efficiencies. On the other hand, significant results have been obtained very recently with solid-state QDSSCs prepared with colloidal QDs. Efficiencies around 3.3 % have been reported for devices based on TiO_2 sensitized with colloidal PbS QDs and P3HT as HTM [84].

There are several reasons suggesting that devices based on colloidal QDs may have better efficiencies than those based on in situ grown QDs:

1. The fact that the QD colloidal samples are close to monodisperse facilitates the formation of compact, ordered monolayers, difficult to obtain in the case of in situ QD growth. However, the use of colloidal QDs requires an appropriate design of the oxide template. Otherwise there could be regions of the oxide not attainable by the relatively voluminous colloidal particles.

2. Regarding carrier injection into the oxide, the existence of QD/QD interfacial traps and the possible formation of disordered multilayers should slow down the process. These disadvantages should be less important in the case of colloidal QDs as the formation of traps is, at least in part, avoided by the presence of the capping agent molecules.
3. In the case of devices prepared with in situ grown QDs, the existence of QD/QD interfacial traps, as well as traps at the interface between the oxide and the QD should lead to a substantial enhancement of recombination processes and therefore to a lower performance than in colloidal QDSSCs.
4. Taking advantage of the MEG is one of the reasons making attractive the research on QDSSCs. The fact that only defect-free, idealized systems have been reported to show the phenomenon (colloidal PbS QDs deposited on rutile single crystal electrodes [14]) is a first indication that SILAR and CBD QDs should not be prone to display this effect, probably due to the high concentration of defects characterizing these samples, which could act as carrier traps.

It is therefore advisable to keep on studying colloidal QDSSCs. Apart from the reasons presented above, colloids are adequate substrates for performing spectroscopic studies on the basis of their more defined emission and absorption spectra. In addition, the greater degree of control over the structure of the sensitized electrodes ensures a higher degree of reproducibility. It should also be mentioned that the possibility of varying at will the attachment mode offers the unique possibility of studying in a detailed way how simple, isolated factors affect charge injection.

On the other hand, a precise control of the size is considered crucial to adjust the resulting band gap and the position of the band edges. This can be achieved for colloidal QDs. The in situ growth methods do not allow decoupling size and coverage, limiting in this way the optimization of the devices. In addition, modifications triggered by the ligand exchange, adsorption of dipoles, etc., can be done in a more complete way in the case of colloids and, at best partially, in the case of in situ grown QDs. Finally, sensitizing with core/shell nanoparticles, alloyed or doped QDs, or QDs of ternary and quaternary materials, can only be achieved in a controlled way with presynthesized colloidal samples.

We believe that there is still a lot of room for improvement and many variables that can be played with to achieve tailor-made QDs suitable for different electron and hole transporting phases. There are still many ideas to be tried and many ways to proceed with these enticing materials. Recent developments seem to indicate this, as in the last year rather efficient liquid-electrolyte and solid-state QDSSCs based on colloids have been fabricated. As a result, efficiencies of colloidal QDSSCs currently equal or even outperform their counterparts based on in situ grown QDs. This is very encouraging, especially if one takes into account that the studies on the latter clearly outnumber those on the former.

Acknowledgments Financial support from the Spanish Ministry of Economy and Competitiveness through projects HOPE CSD2007-00007 (CONSOLIDER-INGENIO 2010), MAT2009-14004 (FONDOS FEDER), and PRI-PIBIN-2011-0816 is gratefully acknowledged. N.G. thanks the Spanish Ministry of Education for the award of an FPU grant.

References

1. Burda, C., Chen, X., Narayanan, R., El-Sayed, M.A.: Chemistry and properties of nanocrystals of different shapes. Chem. Rev. **105**, 1025–1102 (2005)
2. Nozik, A.J.: Quantum dot solar cell. Physica E **14**, 115–120 (2002)
3. Rühle, S., Shalom, M., Zaban, A.: Quantum-dot-sensitized solar cells. Chemphyschem **11**, 2290–2304 (2010)
4. Emin, S., Singh, S.P., Han, L., Satoh, N., Islam, A.: Colloidal quantum dot solar cells. Solar Energy **85**, 1264–1282 (2011)
5. Luther, J.M., Law, M., Beard, M.C., Song, Q., Reese, M.O., Ellingson, R.J., Nozik, A.J.: Schottky solar cells based on colloidal nanocrystal films. Nano Lett. **8**, 3488–3492 (2008)
6. Pattantyus-Abraham, A.G., Kramer, I.J., Barkhouse, A.R., Wang, X., Konstantatos, G., Debnath, R., Levina, L., Raabe, I., Nazeeruddin, M.K., Grätzel, M., Sargent, E.H.: Depleted-heterojunction colloidal quantum dot solar cells. ACS Nano **4**, 3374–3380 (2010)
7. Greenham, N.C., Peng, X., Alivisatos, A.P.: Charge separation and transport in conjugated-polymer/semiconductor-nanocrystal composites studied by photoluminescence quenching and photoconductivity. Phys. Rev. B **54**, 17628–17637 (1996)
8. O'Regan, B., Grätzel, M.: A low cost, high-efficiency solar based on dye-sensitized colloidal TiO_2 films. Nature **353**, 737–740 (1991)
9. Zaban, A., Micic, O.I., Gregg, B.A., Nozik, A.J.: Photosensitization of nanoporous TiO_2 electrodes with InP quantum dots. Langmuir **14**, 3153–3156 (1998)
10. Zhang, H., Cheng, K., Hou, Y.M., Fang, Z., Pan, Z.X., Wu, W.J., Hua, J.L., Zhong, X.H.: Efficient CdSe quantum dot-sensitized solar cells prepared by a postsynthesis assembly approach. Chem. Commun. **48**, 11235–11237 (2012)
11. Grätzel, M.: Solar energy conversion by dye-sensitized photovoltaic cells. Inorg. Chem. **44**, 6841–6851 (2005)
12. Guijarro, N., Lana-Villarreal, T., Shen, Q., Toyoda, T., Gómez, R.: Sensitization of titanium dioxide photoanodes with cadmium selenide quantum dots prepared by SILAR: photoelectrochemical and carrier dynamics studies. J. Phys. Chem. C **114**, 21928–21937 (2010)
13. Nozik, A.J.: Multiple exciton generation in semiconductor quantum dots. Chem. Phys. Lett. **457**, 3–11 (2008)
14. Parkinson, B.A.: Multiple exciton collection *in a* sensitized photovoltaic system. Science **330**, 63 (2010)
15. Nozik, A.J.: Peak external photocurrent quantum efficiency exceeding 100% via MEG in a quantum dot solar cell. Science **334**, 1530–1533 (2011)
16. Shah, A., Torres, P., Tscharner, R., Wyrsch, N., Keppner, H.: Photovoltaic technology: the case for thin-film solar cells. Science **285**, 692–698 (1999)
17. Spanhel, L., Weller, H., Henglein, A.: Photochemistry of semiconductor colloids. 22. Electron injection from illuminated CdS into attached TiO_2 and ZnO particles. J. Am. Chem. Soc. **109**, 6632–6635 (1987)
18. Hotchandani, S., Kamat, P.V.: Charge-transfer processes in coupled semiconductor systems. Photochemistry and photoelectrochemistry of the colloidal CdS–ZnO system. J. Phys. Chem. **96**, 6834–6839 (1992)
19. Giménez, S., Mora-Seró, I., Macor, L., Guijarro, N., Lana-Villarreal, T., Gómez, R., Diguna, L.J., Shen, Q., Toyoda, T., Bisquert, J.: Improving the performance of colloidal quantum-dot-sensitized solar cells. Nanotechnology **20**, 295204 (2009)
20. Lee, W., Kang, S.H., Min, S.K., Sung, Y.-E., Han, S.-H.: Co-sensitization of vertically aligned TiO_2 nanotubes with two different sizes of CdSe quantum dots for broad spectrum. Electrochem. Commun. **10**, 1579–1582 (2008)
21. Samadpour, M., Iraji zad, A., Taghavinia, N., Molaei, M.: A new structure to increase the photostability of CdTe quantum dot sensitized solar cells. J. Phys. D Appl. Phys. **44**, 045103 (2011)

22. Kongkanand, A., Tvrdy, K., Takechi, K., Kuno, M., Kamat, P.V.: Quantum dot solar cells. Tuning photoresponse through size and shape control of CdSe–TiO_2 architecture. J. Am. Chem. Soc. **130**, 4007–4015 (2008)
23. Toyoda, T., Shen, Q.: Quantum-dot-sensitized solar cells: effect of nanostructured TiO_2 morphologies on photovoltaic properties. J. Phys. Chem. Lett. **3**, 1885–1893 (2012)
24. Samadpour, M., Giménez, S., Iraji zad, A., Taghavinia, N., Mora-Seró, I.: Easily manufactured TiO_2 hollow fibers for quantum dot sensitized solar cells. Phys. Chem. Chem. Phys. **14**, 522–528 (2012)
25. Samadpour, M., Giménez, S., Boix, P.P., Shen, Q., Calvo, M.E., Taghavinia, N., Iraji zad, A., Toyoda, T., Míguez, H., Mora-Seró, I.: Effect of nanostructured electrode architecture and semiconductor deposition strategy on the photovoltaic performance of quantum dot sensitized solar cells. Electrochim. Acta **75**, 139–147 (2012)
26. Barceló, I., Lana-Villarreal, T., Gómez, R.: Efficient sensitization of ZnO nanoporous films with CdSe QDs grown by Successive Ionic Layer Adsorption and Reaction (SILAR). J. Photochem. Photobiol. A **220**, 47–53 (2011)
27. Sun, X.W., Chen, J., Song, J.L., Zhao, D.W., Deng, W.Q., Lei, W.: Ligand capping effect for dye solar cells with CdSe quantum dot sensitized ZnO nanorod photoanode. Opt. Express **18**, 1296–1301 (2010)
28. Luan, C., Vaneski, A., Susha, A.S., Xu, X., Wang, H.-E., Chen, X., Xu, J., Zhang, W., Lee, C.-S., Rogach, A.L., Zapien, J.A.: Facile solution growth of vertically aligned ZnO nanorods sensitized with aqueous CdS and CdSe quantum dots for photovoltaic applications. Nanoscale Res. Lett. **6**, 340 (2011)
29. Chen, J., Li, C., Song, J.L., Sun, X.W., Lei, W., Deng, W.Q.: Bilayer ZnO nanostructure fabricated by chemical bath and its application in quantum dot sensitized solar cell. Appl. Surf. Sci. **255**, 7508–7511 (2009)
30. Sudhagar, P., Song, T., Lee, D.H., Mora-Seró, I., Bisquert, J., Laudenslager, M., Sigmund, W. M., Park, W.I., Paik, U., Kang, Y.S.: High open circuit voltage quantum dot sensitized solar cells manufactured with ZnO nanowire arrays and Si/ZnO branched hierarchical structures. J. Phys. Chem. Lett. **2**, 1984–1990 (2011)
31. Vogel, R., Hoyer, P., Weller, H.: Quantum-sized PbS, CdS, Ag_2S, Sb_2S_3 and Bi_2S_3 particles as sensitizers for various nanoporous wide-bandgap semiconductors. J. Phys. Chem. **98**, 3183–3188 (1994)
32. Hossain, M.A., Jennings, J.R., Koh, Z.Y., Wang, Q.: Carrier generation and collection in CdS/CdSe-sensitized SnO_2 solar cells exhibiting unprecedented photocurrent densities. ACS Nano **5**, 3172–3181 (2011)
33. Leventis, H.C., O'Mahony, F., Akhtar, J., Afzaal, M., O'Brien, P., Haque, S.A.: Transient optical studies of interfacial charge transfer at nanostructures metal oxide/PbS quantum dot/organic hole conductor heterojuntion. J. Am. Chem. Soc. **132**, 2743–2750 (2010)
34. Hossain, M.A., Yang, G., Parameswaran, M., Jennings, J.R., Wang, Q.: Mesoporous SnO_2 spheres synthesized by electrochemical anodization and their application in CdSe-sensitized solar cells. J. Phys. Chem. C **114**, 21878–21884 (2010)
35. Shinde, D.V., Lim, I., Kim, C.S., Lee, J.K., Mane, R.S., Han, S.-H.: Upright-standing SnO_2 nanowalls: fabrication, dual-photosensitization and photovoltaic properties. Chem. Phys. Lett. **542**, 66–69 (2012)
36. Sheeney-Haj-Ichia, L., Basnar, B., Willner, I.: Efficient generation of photocurrents by using CdS/carbon nanotube assemblies on electrodes. Angew. Chem. Int. Ed. **44**, 78–83 (2005)
37. Farrow, B., Kamat, P.V.: CSe quantum dot sensitized solar cells. Shuttling electrons through stacked carbon nanocups. J. Am. Chem. Soc. **131**, 11124–11131 (2009)
38. Guo, C.X., Yang, H.B., Sheng, Z.M., Lu, Z.S., Song, Q.L., Li, C.M.: Layered graphene/ quantum dots for photovoltaic devices. Angew. Chem. Int. Ed. **49**, 3014–3017 (2010)
39. Becerril, H.A., Mao, J., Liu, Z., Stoltenberg, R.M., Bao, Z., Chen, Y.: Evaluation of solution-processed reduced graphene oxide films as transparent conductors. ACS Nano **2**, 463–470 (2008)

40. Chen, J., Xu, F., Wu, J., Qasim, K., Zhou, Y., Lei, W., Sun, L.-T., Zhang, Y.: Flexible photovoltaic cell based on a graphene-CdSe quantum dot nanocomposite. Nanoscale **4**, 441–443 (2012)
41. Lightcap, I.V., Kamat, P.V.: Fortification of CdSe quantum dots with graphene oxide. Excited state interactions and light energy conversion. J. Am. Chem. Soc. **134**, 7109–7116 (2012)
42. Brown, P., Kamat, P.V.: Quantum dot solar cells. Electrophoretic deposition of CdSe–C_{60} composites films and capture of photogenerated electrons with nC_{60} cluster shell. J. Am. Chem. Soc. **130**, 8890–8891 (2008)
43. Shibu, E.S., Sonoda, A., Tao, Z., Feng, Q., Furube, A., Masuo, S., Wang, L., Tamai, N., Ishikawa, M., Biju, V.: Photofabrication of fullerene-shelled quantum dots supramolecular nanoparticles for solar energy harvesting. ACS Nano **6**, 1601–1608 (2012)
44. Guijarro, N., Shen, Q., Giménez, S., Mora-Seró, I., Bisquert, J., Lana-Villarreal, T., Toyoda, T., Gómez, R.: Direct correlation between ultrafast injection and photoanode performance in quantum dot sensitized solar cells. J. Phys. Chem. C **114**, 22352–22360 (2010)
45. Bang, J.H., Kamat, P.V.: Quantum dot sensitized solar cells. A tale of two semiconductor nanocrystals: CdSe and CdTe. ACS Nano **3**, 1467–1476 (2009)
46. Benehkohal, N.P., González-Pedro, V., Boix, P.P., Chavhan, S., Tena-Zaera, R., Demopoulos, G.P., Mora-Seró, I.: Colloidal PbS and PbSeS quantum dot sensitized solar cells prepared by electrophoretic deposition. J. Phys. Chem. C **116**, 16391–16397 (2012)
47. Pijpers, J.J.H., Koole, R., Evers, W.H., Houtepen, A.J., Boehme, S., de Mello Donega, C., Vanmaekelberg, D., Bonn, M.: Spectroscopic studies of electron injection in quantum dot sensitized mesoporous oxide film. J. Phys. Chem. C **114**, 18866–18873 (2010)
48. Yu, P., Zhu, K., Normal, A.G., Ferrere, S., Frank, A.J., Nozik, A.J.: Nanocrystalline TiO_2 solar cells sensitized with InAs quantum dots. J. Phys. Chem. B **110**, 25451–25454 (2006)
49. Sambur, J.B., Parkinson, B.A.: CdSe/ZnS core/shell quantum dot sensitization of low index TiO_2 single crystal surfaces. J. Am. Chem. Soc. **132**, 2130–2131 (2010)
50. Pan, Z., Zhang, H., Cheng, K., Hou, Y., Hua, J., Zhong, Z.: Highly efficient inverted type-I CdS/CdSe core/shell structure QD-sensitized solar cells. ACS Nano **6**, 3982–3991 (2012)
51. Ning, Z., Tian, H., Yuan, C., Fu, Y., Qin, H., Sun, L., Agren, H.: Solar cells sensitized with type-II ZnSe–CdS core/shell colloidal quantum dots. Chem. Commun. **47**, 1536–1538 (2011)
52. Song, X., Wang, M., Shi, Y., Deng, J., Yang, Z., Yao, X.: In situ hydrothermal growth of CdSe(S) nanocrystals on mesoporous TiO_2 films for quantum dot-sensitized solar cells. Electrochim. Acta **81**, 260–267 (2012)
53. Hu, Z., Zhang, Q., Huang, Z., Li, D., Luo, Y., Meng, Q.: Aqueous colloidal $CuInS_2$ for quantum dot sensitized solar cells. J. Mater. Chem. **21**, 15903–15905 (2011)
54. Kuo, K.-T., Liu, D.-M., Chen, S.-Y., Lin, C.-C.: Core-shell $CuInS_2$/ZnS quantum dots assembled on short ZnO nanowires with enhanced photo-conversion efficiency. J. Mater. Chem. **19**, 6780–6788 (2009)
55. Im, J.-H., Lee, C.-R., Lee, J.-W., Park, S.-W., Park, N.-G.: 6.5% efficient perovskite quantum-dot-sensitized solar cell. Nanoscale **3**, 4088–4093 (2011)
56. de Mello Donegá, C., Liljeroth, P., Vanmaekelbergh, D.: Physicochemical evaluation of the hot-injection method, a synthesis route for monodisperse nanocrystals. Small **1**, 1152–1162 (2005)
57. Murray, C.B., Norris, D.J., Bawendi, M.G.: Synthesis and characterization of nearly monodisperse CdE (E = S, Se, Te) semiconductor nanocrystallites. J. Am. Chem. Soc. **115**, 8706–8715 (1993)
58. Peng, Z.A., Peng, X.: Formation of high-quality CdTe, CdSe. CdS nanocrystals using CdO as precursor. J. Am. Chem. Soc. **123**, 183–184 (2001)
59. Zhuang, Z., Peng, Q., Li, Y.: Controlled synthesis of semiconductor nanostructures in the liquid phase. Chem. Soc. Rev. **40**, 5492–5513 (2011)
60. Wang, Q., Pan, D., Jiang, S., Ji, X., An, L., Jiang, B.: A solvothermal route to size- and shape-controlled CdSe and CdTe nanocrystals. J. Cryst. Growth **286**, 83–90 (2006)

61. Li, T.-L., Teng, H.: Solution synthesis of high-quality $CuInS_2$ quantum dots as sensitizers for TiO_2 photoelectrodes. J. Mater. Chem. **20**, 3656–3664 (2010)
62. Li, Y., Jing, L., Qiao, R., Gao, M.: Aqueous synthesis of CdTe nanocrystals: progresses and perspectives. Chem. Commun. **47**, 9293–9311 (2011)
63. Rogach, A.L., Franzi, T., Klar, T.A., Feldmann, J., Gaponik, N., Lesnyak, V., Shavel, A., Eychmüller, A., Rakovich, Y.P., Donegan, J.F.: Aqueous synthesis of thiol-capped CdTe nanocrystals: state-of-the-art. J. Phys. Chem. C **111**, 14628–14637 (2007)
64. Wang, H., Luan, C., Xu, X., Kershaw, S.V., Rogach, A.L.: In situ versus ex situ assembly of aqueous-based thioacid capped CdSe nanocrystals within mesoporous TiO_2 films for quantum dot sensitized solar cells. J. Phys. Chem. C **116**, 484–489 (2012)
65. Salant, A., Shalom, M., Hod, I., Faust, A., Zaban, A., Banin, U.: Quantum dot sensitized solar cells with improved efficiency prepared using electrophoretic deposition. ACS Nano **4**, 5962–5968 (2010)
66. Guijarro, N., Lana-Villarreal, T., Mora-Seró, I., Bisquert, J., Gómez, R.: CdSe quantum dot-sensitized TiO_2 electrodes: effect of quantum dot coverage and mode of attachment. J. Phys. Chem. C **113**, 4208–4214 (2009)
67. Chen, J., Zhao, D.W., Song, J.L., Sun, X.W., Deng, W.Q., Liu, X.W., Lei, W.: Directly assembled CdSe quantum dots on TiO_2 in aqueous solution by adjusting pH value for quantum dot sensitized solar cells. Electrochem. Commun. **11**, 2265–2267 (2009)
68. Fan, S.-Q., Fang, B., Kim, J.H., Kim, J.-J., Yu, J.-S., Ko, J.: Hierarchical nanostructured spherical carbon with hollow core/mesoporous shell as a highly efficient counter electrode in CdSe quantum-dot-sensitized solar cells. Appl. Phys. Lett. **96**, 063501 (2010)
69. Shalom, M., Dor, S., Rühle, S., Grinis, L., Zaban, A.: Core/CdS quantum dot/shell mesoporous solar cell with improved stability and efficiency using an amorphous TiO2 coating. J. Phys. Chem. C **113**, 3895–3898 (2009)
70. Lee, H.J., Yum, J.-H., Leventis, H.C., Zakeeruddin, S.M., Haque, S.A., Chen, P., Seok, S.I., Grätzel, M., Nazeeruddin, M.K.: CdSe quantum dot-sensitized solar cells exceeding efficiency 1% at full-sun intensity. J. Phys. Chem. C **112**, 11600–11608 (2008)
71. Ning, Z., Yuan, C., Tian, H., Fu, Y., Li, L., Sun, L., Ågren, H.: Type-II colloidal quantum dot sensitized solar cells with a thiourea based organic redox couple. J. Mater. Chem. **22**, 6032–6037 (2012)
72. Santra, P.K., Kamat, P.V.: Mn-doped quantum dot sensitized solar cells: a strategy to boost efficiency over 5%. J. Am. Chem. Soc. **134**, 2508–2511 (2012)
73. Lee, H., Wang, M., Chen, P., Gamelin, D.R., Zakeeruddin, S.M., Grätzel, M.: Efficient CdSe quantum dot-sensitized solar cells prepared by an improved successive ionic layer adsorption and reaction process. Nano Lett. **9**, 4221–4227 (2009)
74. Chang, J.A., Rhee, J.H., Im, S.H., Lee, Y.H., Kim, H., Seok, S.I., Nazeeruddin, M.K., Gratzel, M.: High-performance nanostructured inorganic–organic heterojunction solar cells. Nano Lett. **10**, 2609–2612 (2010)
75. Moon, S.-J., Itzhaik, Y., Yum, J.-H., Zakeeruddin, S.M., Hodes, G., Grätzel, M.: Sb_2S_3-based mesoscopic solar cell using an organic hole conductor. J. Phys. Chem. Lett. **1**, 1524–1527 (2010)
76. Im, S.H., Lim, C.-S., Chang, J.A., Lee, Y.H., Maiti, N., Kim, H.-J., Nazeeruddin, M.K., Grätzel, M., Seok, S.I.: Toward interaction of sensitizer and functional moieties in hole-transporting materials for efficient semiconductor-sensitized solar cells. Nano Lett. **11**, 4789–4793 (2011)
77. Jovanovski, V., González-Pedro, V., Giménez, S., Azaceta, E., Cabañero, G., Grande, H., Bisquert, J.: A sulfide/polysulfide-based ionic liquid electrolyte for quantum dot-sensitized solar cells. J. Am. Chem. Soc. **133**, 20156–20159 (2011)
78. Snaith, H.J., Stavrinadis, A., Docampo, P., Watt, A.A.R.: Lead-sulphide quantum-dot sensitization of tin oxide based hybrid solar cells. Solar Energy **85**, 1283–1290 (2011)
79. Zeng, T., Gladwin, E., Claus, R.O.: Self-assembled InP quantum dot – TiO_2 solid Grätzel solar cell. MRS Symp. Proc. **764**, 171–175 (2003)

80. Günes, S., Neugebauer, H., Sariciftci, N.S., Roither, J., Kovalenko, M., Pillwein, G., Heiss, W.: Hybrid solar cells using HgTe nanocrystals and nanoporous TiO_2 electrodes. Adv. Funct. Mater. **16**, 1095–1099 (2006)
81. Kniprath, R., Rabe, J.P., McLeskey Jr., J.T., Wang, D., Kirstein, S.: Hybrid photovoltaic cells with II–VI quantum dot sensitizers fabricated by layer-by-layer deposition of water-soluble components. Thin Solid Films **518**, 295–298 (2009)
82. Zeng, T.-W., Liu, I.-S., Hsu, F.-C., Huang, K.-T., Liao, H.-C., Su, W.-F.: Effects of bifunctional linker on the performance of P3HT/CdSe quantum dot-linker-ZnO nanocolumn photovoltaic device. Opt. Express **18**, A357–A365 (2010)
83. Barceló, I., Campiña, J.M., Lana-Villarreal, T., Gómez, R.: A solid-state CdSe quantum dot sensitized solar cell based on a quaterthiophene as a hole transporting material. Phys. Chem. Chem. Phys. **14**, 5801–5807 (2012)
84. Kim, S., Im, S.H., Kang, M., Heo, J.H., Seok, S.I., Kim, S.-W., Mora-Seró, I., Bisquert, J.: Air-stable and efficient inorganic–organic heterojunction solar cells using PbS colloidal quantum dots co-capped by 1-dodecanethiol and oleic acid. Phys. Chem. Chem. Phys. **14**, 14999–15002 (2012)
85. He, J., Lindström, H., Hagfeldt, A., Lindquist, S.-E.: Dye-sensitized nanostructured p-type nickel oxide film as a photocathode for a solar cell. J. Phys. Chem. B **103**, 8940–8943 (1999)
86. Nattestad, A., Ferguson, M., Kerr, R., Cheng, Y.-B., Bach, U.: Dye-sensitized nickel(II)oxide photocathodes for tandem solar cell applications. Nanotechnology **19**, 295304 (2008)
87. Qin, P., Linder, M., Brinck, T., Boschloo, G., Hagfeldt, A., Sun, L.: High incident photon-to-current conversion efficiency of p-type dye-sensitized solar cells based on NiO and organic chromophores. Adv. Mat. **21**, 2993–2996 (2009)
88. Gibson, E.A., Smeigh, A.L., Le Pleux, L., Fortage, J., Boschloo, G., Blart, E., Pellegrin, Y., Odobel, F., Hagfeldt, A., Hammarström, L.: A p-type NiO-based dye-sensitized solar cell with an open-circuit voltage of 0.35 V. Angew. Chem. Int. Ed. **48**, 4402–4405 (2009)
89. Li, L., Gibson, E.A., Qin, P., Boschloo, G., Gorlov, M., Hagfeldt, A., Sun, L.: Double-layered NiO photocathodes for p-type DSSCs with record IPCE. Adv. Mater. **22**, 1759–1762 (2010)
90. Nattestad, A., Mozer, A.J., Fischer, M.K.R., Cheng, Y.-B., Mishra, A., Bauerle, P., Bach, U.: Highly efficient photocathodes for dye-sensitized tandem solar cells. Nat. Mater. **9**, 31–35 (2010)
91. Uehara, S., Sumikura, S., Suzuki, E., Mori, S.: Retardation of electron injection at NiO/dye/electrolyte interface by aluminium alkoxide treatment. Energy Environ. Sci. **3**, 641–644 (2010)
92. Odobel, F., Le Pleux, L., Pellegrin, Y., Blart, E.: New photovoltaic devices based on the sensitization of p-type semiconductors: challenges and opportunities. Acc. Chem. Res. **43**, 1063–1071 (2010)
93. Chan, X.-H., Jennings, J.R., Hossain, M.A., Yu, K.K.Z., Wang, Q.: Characteristics of p-NiO thin films prepared by spray pyrolysis and their application in CdS-sensitized photocathodes. J. Electrochem. Soc. **158**, H733–H740 (2011)
94. Kang, S.H., Zhu, K., Neale, N.R., Frank, A.J.: Hole transport in sensitized CdS–NiO nanoparticle photocathodes. Chem. Commun. **47**, 10419–10421 (2011)
95. Wu, X., Yeow, E.K.L.: Charge-transfer processes in single CdSe/ZnS quantum dots with p-type NiO nanoparticles. Chem. Commun. **46**, 4390–4392 (2010)
96. Lampert, C.M.: Electrochromic materials and devices for energy efficient windows. Solar Energy Mater. **11**, 1–27 (1984)
97. Kuang, D.-B., Lei, B.-X., Pan, Y.-P., Yu, X.-Y., Su, C.-Y.: Fabrication of novel hierarchical β-$Ni(OH)_2$ and NiO microspheres via an easy hydrothermal process. J. Phys. Chem. C **113**, 5508–5513 (2009)
98. Vera, F., Schrebler, R., Muñoz, E., Suarez, C., Cury, P., Gómez, H., Córdova, R., Marotti, R. E., Dalchiele, E.A.: Preparation and characterization of Eosin B- and Erythrosin J-sensitized nanostructured NiO thin film photocathodes. Thin Solid Films **490**, 182–188 (2005)

99. He, J., Lindström, H., Hagfeldt, A., Lindquist, S.-E.: Dye-sensitized nanostructured tandem cell-first demonstrated cell with a dye-sensitized photocathode. Sol. Energy Mater. Sol. Cells **62**, 265–273 (2000)
100. Mori, S., Fukuda, S., Sumikura, S., Takeda, Y., Tamaki, Y., Suzuki, E., Abe, T.: Charge-transfer processes in dye-sensitized NiO solar cells. J. Phys. Chem. C **112**, 16134–16139 (2008)
101. Gibson, E.A., Smeigh, A.L., Le Pleux, L., Hammarström, L., Odobel, F., Boschloo, G., Hagfeldt, A.: Cobalt polypyridyl-based electrolytes for p-type dye-sensitized solar cells. J. Phys. Chem. C **115**, 9772–9779 (2011)
102. Natu, G., Huang, Z., Ji, Z., Wu, Y.: The effect of an atomically deposited layer of alumina on NiO in p-type dye-sensitized solar cells. Langmuir **28**, 950–956 (2011)
103. Hodes, G.: Photoelectrochemical cell measurements: getting the basics right. J. Phys. Chem. Lett. **3**, 1208–1213 (2012)
104. De Angelis, F., Fantacci, S., Selloni, A., Nazeeruddin, M.K., Grätzel, M.: First-principles modeling of the adsorption geometry and electronic structure of Ru(II) dyes on extended TiO_2 substrates for dye-sensitized solar cell applications. J. Phys. Chem. C **114**, 6054–6061 (2010)
105. Xiao, D.Q., Martini, L.A., Snoeberger, R.C., Crabtree, R.H., Batista, V.S.: Inverse design and synthesis of acac-Coumarin anchors for robust TiO_2 sensitization. J. Am. Chem. Soc. **133**, 9014–9022 (2011)
106. Martsinovich, N., Ambrosio, F., Troisi, A.: Adsorption and electron injection of the N3 metal-organic dye on the TiO_2 rutile (110) surface. Phys. Chem. Chem. Phys. **14**, 16668–16676 (2012)
107. Le Bahers, T., Labat, F., Pauporté, T., Lainé, P.P., Ciofini, I.: Theoretical procedure for optimizing dye-sensitized solar cells: from electronic structure to photovoltaic efficiency. J. Am. Chem. Soc. **133**, 8005–8013 (2011)
108. Maggio, E., Martsinovich, N., Troisi, A.: Evaluating charge recombination rate in dye-sensitized solar cells from electronic structure calculations. J. Phys. Chem. C **116**, 7638–7649 (2012)
109. Martsinovich, N., Troisi, A.: How TiO_2 crystallographic surfaces influence charge injection rates from a chemisorbed dye sensitiser. Phys. Chem. Chem. Phys. **14**, 13392–13401 (2012)
110. Califano, M., Franceschetti, A., Zunger, A.: Temperature dependence of excitonic radiative decay in CdSe quantum dots: the role of surface hole traps. Nano Lett. **5**, 260–2364 (2005)
111. Efros, A.I.L., Rosen, M.: The electronic structure of semiconductor nanocrystals. Annu. Rev. Mater. Sci. **30**, 475–521 (2000)
112. Califano, M., Zunger, A., Franceschetti, A.: Efficient inverse Auger recombination at threshold in CdSe nanocrystals. Nano Lett. **4**, 525–531 (2004)
113. Rabani, E., Baer, R.: Theory of multiexciton generation in semiconductor nanocrystals. Chem. Phys. Lett. **496**, 227–235 (2010)
114. Rabani, E., Baer, R.: Distribution of multiexciton generation rates in CdSe and InAs nanocrystals. Nano Lett. **8**, 4488–4492 (2008)
115. Ellingson, R.J., Beard, M.C., Johnson, J.C., Yu, P., Micic, O.I., Nozik, A.J., Shabaev, A., Efros, A.L.: Highly efficient multiple exciton generation in colloidal PbSe and PbS quantum dots. Nano Lett. **5**, 865–871 (2005)
116. Dahnovsky, Y.: Quantum correlated electron dynamics in a quantum-dot sensitized solar cells: Keldysh function approach. Phys. Rev. B **83**, 165306 (2011)
117. Patrick, C.E., Giustino, F.: Structural and electronic properties of semiconductor-sensitized solar-cell interfaces. Adv. Funct. Mater. **21**, 4663–4667 (2011)
118. Dong, C., Li, X., Qi, J.: First-principles investigation on electronic properties of quantum dot-sensitized solar cells based on anatase TiO_2 nanotubes. J. Phys. Chem. C **115**, 20307–20315 (2011)
119. Long, R., Prezhdo, O.: Ab initio nonadiabatic molecular dynamics of the ultrafast electron injection from a PbSe quantum dot into the TiO_2 surface. J. Am. Chem. Soc. **133**, 19240–19249 (2011)

120. Ulbricht, R., Hendry, E., Shan, J., Heinz, T.F., Bonn, M.: Carrier dynamics in semiconductors studied with time-resolved terahertz spectroscopy. Rev. Mod. Phys. **83**, 543–586 (2011)
121. Cánovas, E., Moll, P., Jensen, S.A., Gao, Y., Houtepen, A.J., Siebbeles, L.D.A., Kinge, S., Bonn, M.: Size-dependent electron transfer from PbSe quantum dots to SnO_2 monitored by picosecond terahertz spectroscopy. Nano Lett. **11**, 5234–5239 (2011)
122. Shen, Q., Katayama, K., Yamaguchi, M., Sawada, T., Toyoda, T.: Study of ultrafast carrier dynamics of nanostructured TiO_2 films with and without CdSe quantum dot deposition using lens-free heterodyne detection transient grating technique. Thin Solid Films **486**, 15–19 (2005)
123. Zídek, K., Zheng, K., Ponseca, C.S., Messing, M.E., Wallenberg, L.R., Chábera, P., Abdellah, M., Sundström, V., Pullerits, T.: Electron transfer in quantum-dot-sensitized ZnO nanowires: ultrafast time-resolved absorption and terahertz study. J. Am. Chem. Soc. **134**, 12110–12117 (2012)
124. Robel, I., Kuno, M., Kamat, P.V.: Size-dependent electron injection from excited CdSe quantum dots into TiO_2 nanoparticles. J. Am. Chem. Soc. **129**, 4136–4137 (2007)
125. Tvrdy, K., Frantsuzov, P.A., Kamat, P.V.: Photoinduced electron transfer from semiconductor quantum dots to metal oxide nanoparticles. Proc. Natl. Acad. Sci. U.S.A. **108**, 29–34 (2011)
126. Hyun, B.-R., Zhong, Y.-W., Bartnik, A.C., Sun, L., Abruña, H.D., Wise, F.W., Goodreau, J.D., Matthews, J.R., Leslie, T.M., Borrelli, N.F.: Electron injection from colloidal PbS quantum dots into titanium dioxide nanoparticles. ACS Nano **11**, 2206–2212 (2008)
127. Yang, Y., Rodríguez-Córdoba, W., Xiang, X., Lian, T.: Strong electronic coupling and ultrafast electron transfer between PbS quantum dots and TiO_2 nanocrystalline films. Nano Lett. **12**, 303–309 (2012)
128. Chakrapani, V., Baker, D., Kamat, P.V.: Understanding the role of the sulfide redox couple (S^{2-}/S_n^{2-}) in quantum dot-sensitized solar cells. J. Am. Chem. Soc. **133**, 9607–9615 (2011)
129. Plass, R., Pelet, S., Krueger, J., Grätzel, M., Bach, U.: Quantum dot sensitization of organic–inorganic hybrid solar cells. J. Phys. Chem. B **106**, 7578–7580 (2002)
130. Guijarro, N., Lutz, T., Lana-Villarreal, T., O'Mahony, F., Gómez, R., Haque, S.A.: Toward antimony selenide sensitized solar cells: efficient charge photogeneration at spiro-OMeTAD/ Sb_2Se_3/metal oxide heterojunctions. J. Phys. Chem. Lett. **3**, 1351–1356 (2012)
131. Kilina, S., Ivanov, S., Tretiak, S.: Effect of surface ligands on optical and electronic spectra of semiconductor nanoclusters. J. Am. Chem. Soc. **131**, 7717–7726 (2009)
132. Szymanski, P., Fuke, N., Koposov, A.Y., Manner, V.W., Hoch, L.B., Sykora, M.: Effect of organic passivation on photoinduced electron transfer across the quantum dot/TiO_2 interface. Chem. Commun. **47**, 6437–6439 (2011)
133. Lana-Villarreal, T., Font-Sanchis, E., Sastre-Santos, A., Fernández-Lázaro, F., Gómez, R.: Characterization and polymerization of thienylphenyl and selenylphenyl amines and their interaction with CdSe quantum dots. Chemphyschem **12**, 1155–1164 (2011)
134. So, S., Fan, S.-Q., Choi, H., Kim, C., Cho, N., Song, K., Ko, J.: Stepwise cosensitization through chemically bonding organic dye to CdS quantum-dot-sensitized TiO_2 electrode. Appl. Phys. Lett. **97**, 263506 (2010)
135. Mora-Seró, I., Likodimos, V., Giménez, S., Martínez-Ferrero, E., Albero, J., Palomares, E., Kontos, A.G., Falaras, P., Bisquert, J.: Fast regeneration of CdSe quantum dots by Ru dye in sensitized TiO_2 electrodes. J. Phys. Chem. C **114**, 6755–6761 (2010)
136. Sarkar, S., Makhal, A., Lakshman, K., Bora, T., Dutta, J., Pal, S.K.: Dual-sensitization via electron and energy harvesting in CdTe quantum dots decorated ZnO nanorod-based dye-sensitized solar cells. J. Phys. Chem. C **116**, 14248–14256 (2012)
137. Buhbut, S., Itzhakov, S., Tauber, E., Shalom, M., Hod, I., Geiger, T., Garini, Y., Oron, D., Zaban, A.: Built-in quantum dot antennas in dye-sensitized solar cells. ACS Nano **4**, 1293–1298 (2010)
138. Poulose, A.C., Veeranarayanan, S., Varghese, S.H., Yoshida, Y., Maekawa, T., Kumar, D.S.: Functionalized electrophoretic deposition of CdSe quantum dots onto TiO_2 electrode for photovoltaic application. Chem. Phys. Lett. **539–540**, 197–203 (2012)

139. Salant, A., Shalom, M., Tachan, Z., Buhbut, S., Zaban, A., Banin, U.: Quantum rod-sensitized solar cell: nanocrystal shape effect on the photovoltaic properties. Nano Lett. **12**, 2095–2100 (2012)
140. Yang, H., Fan, W., Vaneski, A., Susha, A.S., Teoh, W.Y., Rogach, A.L.: Heterojunction engineering of CdTe and CdSe quantum dots on TiO_2 nanotube arrays: intricate effects of size-dependency and interfacial contact on photoconversion efficiencies. Adv. Funct. Mater. **22**, 2821–2829 (2012)
141. King, L.A., Riley, D.J.: Importance of QD purification procedure on surface adsorbance of QDs and performance of QD sensitized photoanodes. J. Phys. Chem. C **116**, 3349–3355 (2012)
142. Shalom, M., Ruhle, S., Hod, I., Yahav, S., Zaban, A.: Energy level alignment in CdS quantum dot sensitized solar cells using molecular dipoles. J. Am. Chem. Soc. **131**, 9876–9877 (2009)
143. Barea, E.M., Shalom, M., Giménez, S., Hod, I., Mora-Seró, I., Zaban, A., Bisquert, J.: Design of injection and recombination in quantum dot sensitized solar cells. J. Am. Chem. Soc. **132**, 6834–6839 (2010)
144. Chi, C.-F., Chen, P., Lee, Y.-L., Liu, I.-P., Chou, S.-C., Zhang, X.-L., Bach, U.: Surface modifications of CdS/CdSe co-sensitized TiO_2 photoelectrodes for solid-state quantum-dot-sensitized solar cells. J. Mater. Chem. **21**, 17534–17540 (2011)
145. Guijarro, N., Lana-Villarreal, T., Lutz, T., Haque, S.A., Gómez, R.: Sensitization of TiO_2 with PbSe quantum dots by SILAR: how mercaptophenol improves charge separation. J. Phys. Chem. Lett. **3**, 3367–3372 (2012)
146. Yu, W.W., Falkner, J.C., Shih, B.S., Colvin, V.L.: Preparation and characterization of monodisperse PbSe semiconductor nanocrystals in a noncoordinating solvent. Chem. Mater. **16**, 3318–3322 (2004)
147. Greenwald, S., Rühle, S., Shalom, M., Yahav, S., Zaban, A.: Unpredicted electron in CdS/CdSe quantum dot sensitized ZrO_2 solar cells. Phys. Chem. Chem. Phys. **13**, 19302–19306 (2011)
148. Peter, L.M., Wijayantha, K.G.U., Riley, D.J., Wagget, P.: Band-edge tuning in self-assembled layers of Bi_2S_3 nanoparticles used to photosensitize nanocrystalline TiO_2. J. Phys. Chem. B **107**, 8378–8381 (2003)
149. Giménez, S., Lana-Villarreal, T., Gómez, R., Agouram, S., Muñoz-Sanjosé, V., Mora-Seró, I.: Determination of limiting factors in photovoltaic efficiency in quantum dot sensitized solar cells: correlation between cell performance and structural properties. J. Appl. Phys. **108**, 064310 (2010)
150. Shalom, M., Dor, S., Rühle, S., Grinis, L., Zaban, A.: Core/CdS quantum dot/Shell mesoporous solar cells with improved stability and efficiency using an amorphous TiO2 coating. J. Phys. Chem. C **113**, 3895–3898 (2009)
151. Guijarro, N., Campiña, J.M., Shen, Q., Toyoda, T., Lana-Vilarreal, T., Gómez, R.: Uncovering the role of the ZnS treatment in the performance of quantum dot sensitized solar cells. Phys. Chem. Chem. Phys. **13**, 12024–12032 (2011)
152. Liu, Z., Miyauchi, M., Uemura, Y., Cui, Y., Hara, K., Zhao, Z., Sunahara, K., Furube, A.: Enhancing the performance of quantum dots sensitized solar cells by SiO_2 surface coating. Appl. Phys. Lett. **96**, 233107 (2010)
153. Palomares, E., Clifford, J.N., Haque, S.A., Lutz, T., Durrant, J.R.: Slow charge recombination in dye-sensitized solar cells (DSSC) using A_2O_3 coated nanoporous TiO_2 films. Chem. Commun. 1464–1465 (2002)
154. Liu, Y., Gibbs, M., Perkins, C.L., Tolentino, J., Zarghami, M.H., Bustamante, J., Law, M.: Robust, functional nanocrystal solids by infilling with atomic layer deposition. Nano Lett. **11**, 5349–5355 (2011)

Chapter 2
Hierarchically Nanostructured Photoelectrodes for Quantum-Dot-Sensitized Solar Cells

Eui-Hyun Kong, Yong-June Chang, and Hyun Myung Jang

Abstract Quantum-dot-sensitized solar cells (QDSSCs) recently have attracted a great deal of attention owing to their advantages that include high molar extinction coefficient of quantum dots, tunable energy gaps, and multiple exciton generation. However, various architectures have not yet been proposed as alternative working electrodes for QDSSCs. In this article, two hierarchical nanostructures will be presented as next-generation photoelectrodes for highly-efficient QDSSCs: (a) the tertiary-hierarchically-structured mesoporous spherical (MS) TiO_2 and (b) the sea urchin TiO_2 (SU TiO_2) particles composed of radially aligned rutile TiO_2 nanowires. The MS TiO_2 offers a high surface area, a high internal reflectance in the visible region, and good pore accessibility. A conversion efficiency of 1.9 % was achieved by CdS QDSSCs made with the MS TiO_2 photoelectrode, which corresponds to ~58 % improvement as compared with the values obtained from the conventional devices made with 20-nm-sized nanocrystalline TiO_2 film. Secondly, SU TiO_2 was incorporated into the TiO_2 nanoparticle (NP) network to construct the SU–NP composite film and applied to the CdS/CdSe/ZnS QDSSCs. A conversion efficiency of 4.2 % was achieved, which corresponds to ~20 % improvement as compared with the values obtained from the reference cell made of the NP film. We attribute this extraordinary result to the light scattering effect and efficient charge collection. Thus, the SU TiO_2 and MS TiO_2 can be promising materials for the photoanodes of QDSSCs.

Z.M. Wang, A. Neogi (eds.), *Nanoscale Photonics and Optoelectronics*, Lecture Notes in Nanoscale Science and Technology 9, DOI 10.1007/978-1-4419-7587-4_2,

E.-H. Kong • Y.-J. Chang • H.M. Jang (✉)
Department of Materials Science and Engineering, and Division of Advanced Materials Science, Pohang University of Science and Technology (POSTECH), Pohang 790-784, Republic of Korea
e-mail: hmjang@postech.ac.kr

J. Wu and Z.M. Wang (eds.), *Quantum Dot Solar Cells*, Lecture Notes in Nanoscale Science and Technology 15, DOI 10.1007/978-1-4614-8148-5_2,

2.1 Introduction

Dye-sensitized solar cells (DSSCs) consisting of a wide band-gap semiconductor film, a dye, and an electrolyte have been regarded as a promising alternative to silicon-based photovoltaic cells [1]. Instead of using a dye, the sensitization of a photoanode can be achieved through modification of the oxide surface with a narrow band-gap semiconductor quantum-dot (QD). Quantum-dot-sensitized solar cells (QDSSCs) recently have attracted a great deal of attention owing to their advantages over DSSCs. The advantages include (a) higher molar extinction coefficient [2] of QDs than ruthenium complexes [3, 4], (b) tunable energy gaps [5], and (c) multiple exciton generation (MEG) [6] which may potentially lead to a theoretical maximum efficiency [7] higher than that of DSSCs.

Recently, hierarchically structured oxide materials such as ZnO aggregates [8], nano-embossed hollow spherical TiO_2 [9], and mesoporous TiO_2 beads [10–12] have been reported as functional photoelectrode materials in the sensitized photovoltaic devices. These porous spherical systems consist of nano-sized crystallites that are aggregated to create micrometer- or submicrometer-sized secondary particles and thereby can function as light scattering materials with maintaining a large internal surface area for sufficient sensitizer-uptake. Thus, these bi-functional materials having 0–3 hierarchy seem to be more appealing than traditional nanocrystalline TiO_2. However, slow-trap-limited electron transport remains a problem in the charge collection kinetics.

Herein, we will introduce two hierarchical nanostructures: tertiary hierarchically structured TiO_2 and sea urchin TiO_2.

2.2 Theoretical Backgrounds

2.2.1 *Quantum Confinement Effect*

Quantum dots are semiconductor nanocrystals that have unique optical properties because of a combination of their band-gap energy and quantum-well phenomena. The dots of a few nanometers behave similarly to three-dimensional quantum wells. When an electron in the quantum dot is excited by a photon, it behaves as a particle confined in an infinite potential well. In the quantum dot, the density of states is discrete, and the band gap is determined by the size of the quantum dot. Through the theoretical calculations, the size dependence is plotted as a function of the inverse radius squared in Fig. 2.1a [13]. The experimental result also confirms that the absorption curves become blue-shifted with decreasing particle size as demonstrated in Fig. 2.1b [14].

In quantum mechanics, a proper approach to the behavior of an exciton is the 3-D model of a particle in a box [15] because the behavior of a particle is fundamentally described by the wavefunction Ψ which is the solution of the

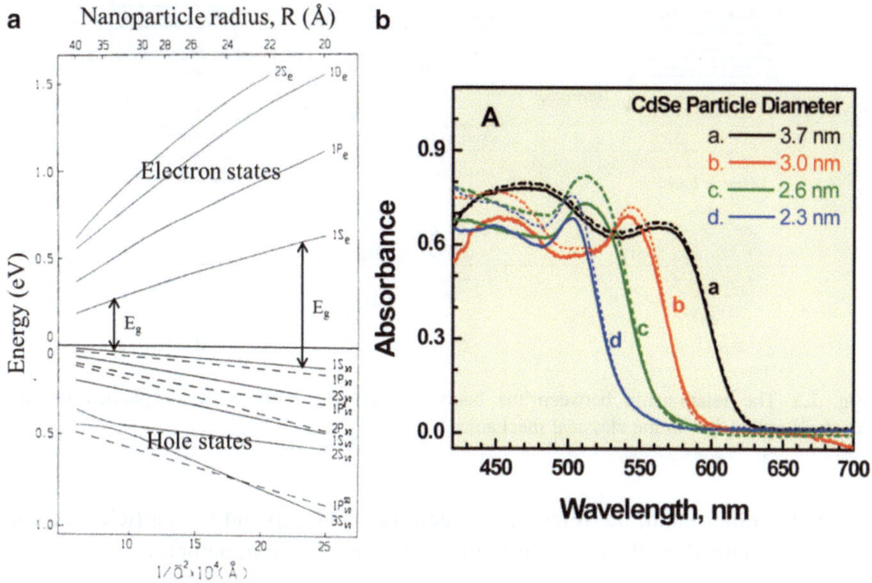

Fig. 2.1 (**a**) Theoretical size dependence of the electron and hole levels in CdSe nanocrystals. (**b**) Absorption spectra of 3.7, 3.0, 2.6, and 2.3 nm diameter CdSe quantum dots anchored on nanostructured TiO_2 films

time-independent Schrödinger equation. For a three-dimensional box, the wavefunctions and energies are given by the following equations:

$$\Psi_{n_x,n_y,n_z} = \sqrt{\frac{8}{L_x L_y L_z}} \sin\left(\frac{n_x \pi x}{L_x}\right) \sin\left(\frac{n_y \pi y}{L_y}\right) \sin\left(\frac{n_z \pi z}{L_z}\right) \tag{2.1}$$

$$E_{n_x n_y n_z} = \frac{h^2 \pi^2}{2m}\left[\left(\frac{n_x}{L_x}\right)^2 + \left(\frac{n_y}{L_y}\right)^2 + \left(\frac{n_z}{L_z}\right)^2\right] \tag{2.2}$$

where the three-dimensional wavevector is

$$\mathbf{k}_{n_x,n_y,n_z} = k_{n_x}\mathbf{x} + k_{n_y}\mathbf{y} + k_{n_z}\mathbf{z} = \frac{n_x \pi}{L_x}\mathbf{x} + \frac{n_y \pi}{L_y}\mathbf{y} + \frac{n_z \pi}{L_z}\mathbf{z} \tag{2.3}$$

where L is the length of the box and **x**, for example, denotes a unit vector along x-direction. If an exciton is a spherical particle, the relationship between energy and radius of a particle is given by

$$E^{\mathrm{e,h}} \propto \frac{1}{R^2} \tag{2.4}$$

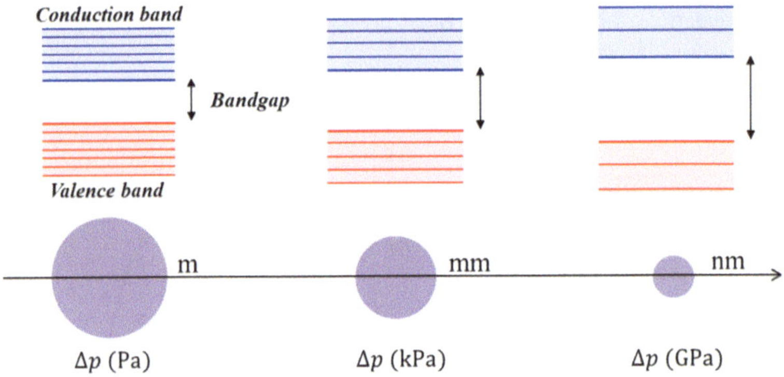

Fig. 2.2 The relationship between the band gap and the particle size explained through Young–Laplace law in the classical mechanics

On the other hand, the relation between the band gap and the particle size can also be explained by the classical Laplace–Young equation, namely,

$$\Delta p = 2\gamma H = \gamma\left(\frac{1}{R_1} + \frac{1}{R_2}\right) = \frac{2\gamma}{R} \propto \frac{1}{R} \quad \text{when } R_1 = R_2 = R \tag{2.5}$$

where Δp is the pressure difference across the fluid interface, γ is the surface tension, H is the mean curvature, and R_1 and R_2 are the two principal radii of curvature. The increase in pressure at the nanoscale results in strong forces toward the interior of the particle. This variation is responsible for the changes in inter-atomic interactions and the band gap [16, 17], as schematically represented in Fig. 2.2. Therefore, panchromatically-light-absorbing quantum-dot-sensitized solar cells can be designed by suitably exploiting the band gap engineering of the quantum-dots.

2.2.2 *Multiple Exciton Generation (=Carrier Multiplication)*

MEG involves the generation of electron–hole pairs per single photon. MEG may considerably enhance the energy conversion efficiency of QDSSCs. The energy gap between electron states is too small to be ignored because the bulk semiconductor has many atoms in total volume. Thus, excited carriers are rapidly relaxed with low phonon energy. That is, the photon energy in excess of the band gap is lost as heat through phonon emission following electron–phonon scattering when photon energy higher than band gap stimulates semiconductor materials (in Fig. 2.3a) [18–20]. The higher threshold photon energy is required for impact ionization, the process in a material by which one energetic charge carrier can lose energy by

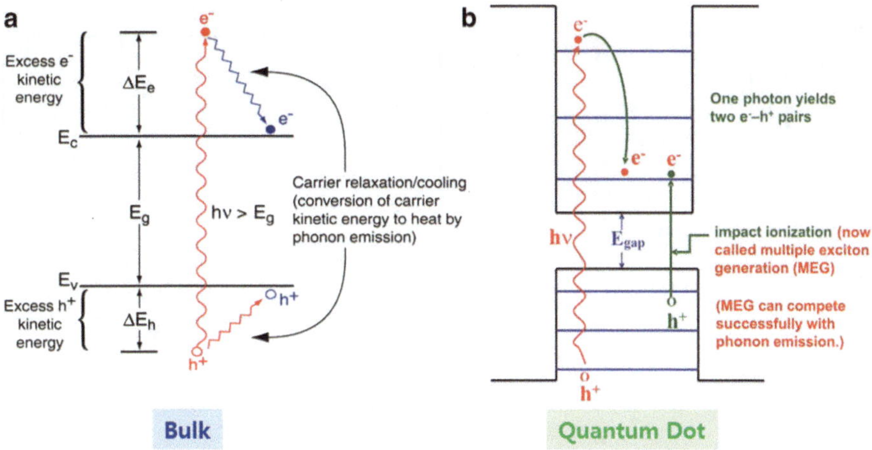

Fig. 2.3 (**a**) Hot carrier relaxation/cooling in bulk semiconductors. (**b**) Multiple electron–hole pair (exciton) generation (MEG) in quantum dots

the creation of other charge carriers, because crystal momentum (**k**) has to be conserved. This results in a low maximum quantum yield [21, 22].

On the other hand, a quantum-dot (QD) has 100–10,000 atoms in total volume. It is known that the energy separation between quantized levels in QDs can be many times higher than the phonon energy. In the relaxation process, several phonons should be emitted via electron–phonon scattering to meet the energy conservation, which accompanies multi-particle scattering events. As a result, the rate of excited electron relaxation can be retarded by the existence of discrete quantized energy levels. The slowed relaxation of hot excitons via phonon emission allows other channels for relaxation, such as MEG, to become competitive and even dominant in QDs (in Fig. 2.3b) [23]. Furthermore, the crystal momentum need not be conserved because the momentum is not a good quantum number for three-dimensionally confined carriers (from the Heisenberg Uncertainty Principle, the well-defined location of the electrons and holes in the nanocrystal makes the momentum uncertain). This results that the maximum quantum yield is very high because the lower threshold photon energy than bulk semiconductor is required for the impact ionization [24].

The difference in carrier multiplication (CM) thresholds observed in various quantum dots can be understood easily by using effective masses in simple bulk semiconductors without bringing up an exact structure of quantized states in these materials [25]. If one assumes that the energy required for an additional exciton is given by either the electron or the hole of an original exciton, then the CM threshold ($\hbar\omega_{CM}$) can be determined from the condition $\Delta E_{e(h)} = E_g$, where $\Delta E_{e(h)}$ is the energy of the electron (hole) excited by the photon in excess of the energy gap. This condition yields the following equation:

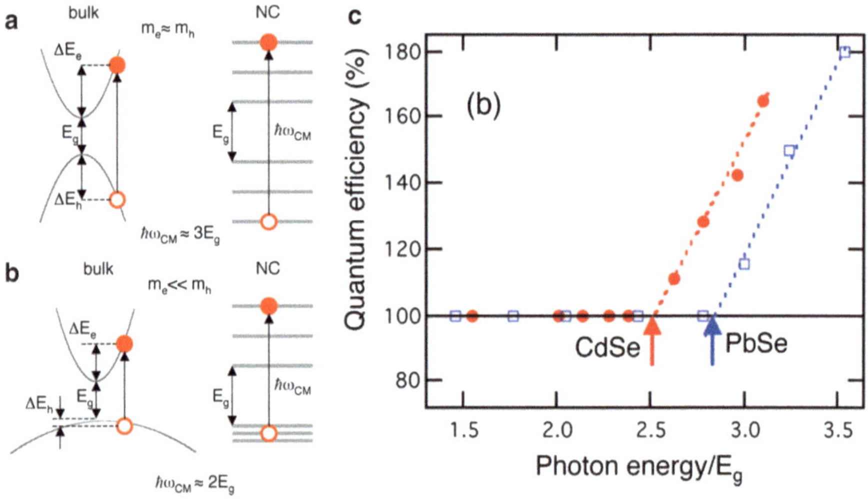

Fig. 2.4 "Bulk-like," carrier effective-mass considerations indicate that in the case of similar electron and hole masses ($m_e \approx m_h$) the onset of CM is 3.0 E_g (**a**), whereas it approaches 2.0 E_g (**b**) in the case for which the electron and hole masses are very dissimilar ($m_e \ll m_h$). (**c**) CM quantum efficiencies for CdSe (*solid circles*) and PbSe (*open squares*) nanocrystals as a function of excitation photon energy divided by E_g. The *two arrows* mark the onsets of CM

$$\hbar\omega_{CM} = \left(2 + \frac{m_e}{m_h}\right) E_g \tag{2.6}$$

In Eq. (2.6), it is assumed that $m_e \leq m_h$. In the lead salts, $m_e \approx m_h$ and hence $\hbar\omega_{CM} = 3E_g$ (Fig. 2.4a), which is consistent with the ~$3E_g$ threshold observed experimentally for PbSe nanocrystals (Fig. 2.4c). Because holes are heavier than electrons in CdSe, $\hbar\omega_{CM}$ should be smaller than $3E_g$ (Fig. 2.4b), which is again consistent with the experimental results that indicate a CM threshold of approximately $2.5E_g$ (Fig. 2.4c) [25].

2.2.3 Molar Extinction Coefficient (=Molar Absorptivity)

The concept of molar absorptivity is useful in analyzing optical data of sensitized solar cells. Let us first define the transmittance (T) to introduce this quantity. The ratio of radiant power transmitted (I) by a sample to the radiant power incident (I_0) on the sample is called the transmittance T:

$$T = I/I_0 \tag{2.7}$$

On the other hand, the absorbance (A) is defined by

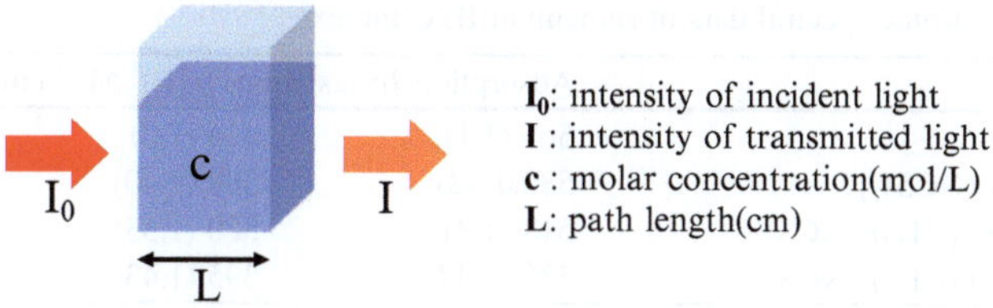

Fig. 2.5 Diagram of Beer–Lambert absorption

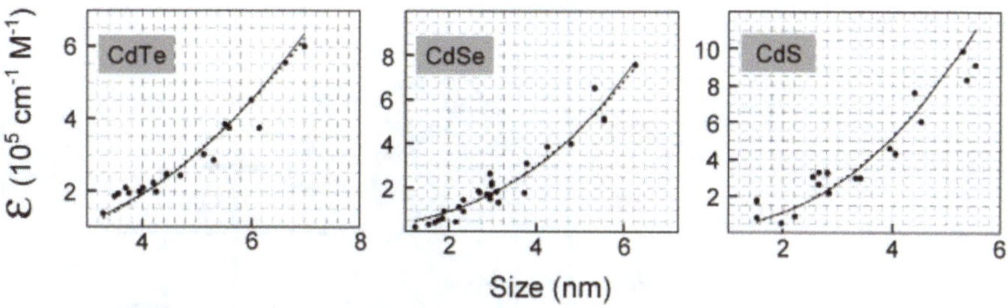

Fig. 2.6 Size-dependent extinction coefficient per mole of nanocrystals at the first excitonic absorption, ε, of CdTe, CdSe, and CdS nanocrystals

$$A = -\log T = \log(1/T) \qquad (2.8)$$

The transmittance and absorbance depend on the molar concentration (c), light path length in centimeters (L), and molar absorptivity ($\boldsymbol{\varepsilon}$) for the dissolved substance [26].

$$T = 10^{-\varepsilon cL} \quad \text{or} \quad A_\lambda = \varepsilon cL \qquad (2.9)$$

Beer's Law states that the molar absorptivity is constant and the absorbance is proportional to the concentration for a given substance dissolved in a given solute and measured at a given wavelength [27]. Since transmittance and absorbance are unitless, the units for molar absorptivity must cancel with units of measure in concentration and light path. Accordingly, the unit for molar absorptivity has $M^{-1}\ cm^{-1}$ or $L\ mol^{-1}\ cm^{-1}$. Most of laboratory spectrophotometers are fitted for use with 1 cm width cuvettes. Thus, molar absorptivity is expressed as

$$A_\lambda = \varepsilon cL = \varepsilon c \quad \text{when } L = 1\ \text{cm} \qquad (2.10)$$

The molar extinction coefficient values of quantum dots (such as CdS, CdSe, and CdTe) are ~10^5 cm^{-1} M^{-1} (Fig. 2.6) [2], an order of magnitude higher than ~10^4 cm^{-1} M^{-1} (Table 2.1) of ruthenium(II) complexes (such as N3 and N719) [4]. This implies that QDSSCs can absorb more light than DSSCs do at similar film thickness.

Table 2.1 Electronic spectral data of ruthenium(II) complexes

Complex	Adsorption λ(max) (nm) ($\epsilon/10^4$ M^{-1} cm^{-1})		
$[Ru(dcbpyH_2)_2(cl)_2]$	565 (1.1)	414 (1.0)	316 (3.45)
$[Ru(dcbpyH_2)_2(NCS)_2]$	538(1.42)	398 (1.40)	314 (4.82)
$(Bu_4N)_4[Ru(dcbpyH_2)_2(NCS)_2]$	518 (1.3)	380 (1.33)	308 (4.59)
$(Bu_4N)_2[Ru(dcbpyH_2)_2(NCS)_2]$	535 (1.47)	395 (1.43)	312 (4.91)

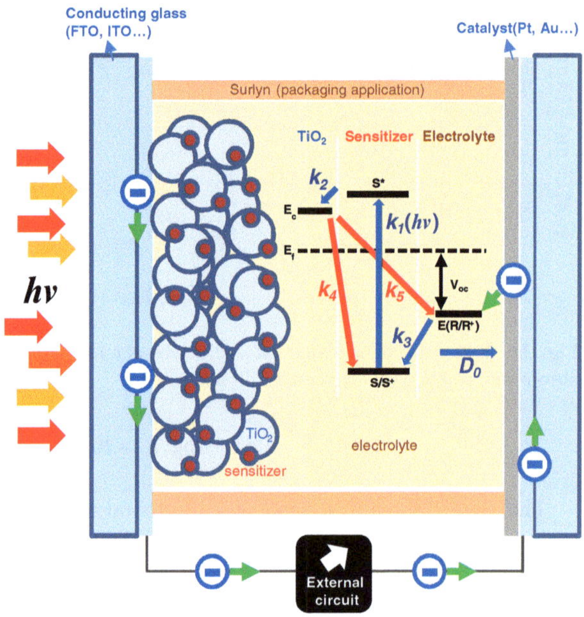

Fig. 2.7 The major kinetic processes in the operation of a QDSSC device

2.2.4 *Kinetic Processes Occurring in a QDSSC*

The energetics and key kinetic processes occurring in a QDSSC are illustrated in Fig. 2.7. Light is absorbed by quantum dots attached to the surface of a TiO_2 nanoparticle (k_1). The excited dye injects fast an electron into the TiO_2 conduction band (k_2) which is the source of the photocurrent. A redox shuttle in solution is required for the regeneration of the dye (k_3); diffusion of the oxidized form of the redox shuttle (D_o) to be reduced at the counter electrode completes the circuit. These processes compete with several reactions that inhibit the efficient operation of a QDSSC including decay (either radiative or nonradiative) of the chromophore's excited state as well as back-electron transfer from the TiO_2 nanoparticle to the oxidized dye (k_4) or the oxidized form of the redox shuttle (k_5). For a QD-sensitized DSSC, the decay of the QDs in an I^-/I_3^- redox system causes a rapid decrease of the photocurrent. For the QDSSCs using cadmium chalcogenides

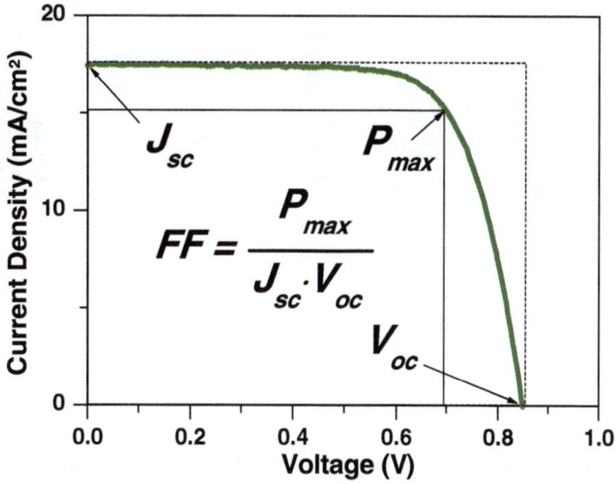

Fig. 2.8 Photocurrent density vs. voltage curve and the equation for fill factor

(S, Se or Te) as sensitizers, redox couple of polysulfide (S_2^-/S_x^{2-}) is a suitable system for stabilizing the QDs [28].

The power conversion efficiency of a solar cell, η, is usually expressed as

$$\eta = \frac{P_{\text{max}}}{P_{\text{in}}} = \frac{J_{\text{sc}} \cdot V_{\text{oc}} \cdot \text{FF}}{P_{\text{in}}} \tag{2.11}$$

where P_{max} and P_{in} are the maximum electrical power output and the incident solar power, respectively. The short circuit photocurrent density (J_{sc}) is the cell current measured at an applied potential of 0 V. J_{sc} is a function of the illumination intensity. The open-circuit photovoltage (V_{oc}) is the difference between the Nernstian potential of the solution, $E(\text{R/R}^+)$, and the quasi-Fermi level, E_{f} (i.e., potential of the semiconductor), and the fill factor (FF) is a term introduced to relate P_{max} to the parameters J_{sc} and V_{oc}, corresponding to the ratio of the area under a J–V curve defined by P_{max} to the area defined by J_{sc} and V_{oc} (Fig. 2.8).

2.3 Experimental Backgrounds

2.3.1 Electrochemical Impedance Spectroscopy (EIS)

EIS is used to analyze internal resistance in QDSSCs. Resulting impedance spectra can be fitted to an appropriate equivalent circuit, as shown in Fig. 2.9a [29–31]. In general, at least three internal resistances are found. Figure 2.9b shows a Nyquist plot. The three semicircular shapes are assigned to impedances related to charge transport at the Pt counter electrode (Z_{Pt}) in the high-frequency region (10^3–10^5 Hz), at the TiO_2/QD/electrolyte interface (Z_{ct}) in the middle-frequency region (10^0–10^3 Hz), and in the Warburg diffusion within the electrolyte (Z_{d}) in the

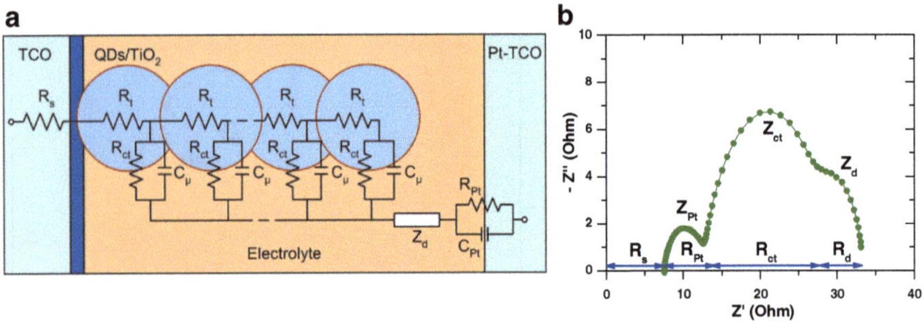

Fig. 2.9 (**a**) Equivalent circuit used to fit the EIS spectra. (**b**) A Nyquist plot taken showing each resistive element

low-frequency region (10^{-2}–10^{0} Hz), respectively. The resistance elements R_{Pt}, R_{ct}, and R_d are described as the real parts of Z_{Pt}, Z_{ct}, and Z_d, respectively [32]. R_s is the series resistance of the cell including the TCO sheet resistance and R_T is the electron transport resistance through the nanoparticle TiO_2 film.

The frequency (f) at maximum imaginary components of each semicircle in a Z''–Z' plot is the reciprocal of time constant (τ) for an equivalent circuit. Thus, the recombination lifetime (τ_r) between the electrons in the TiO_2 and the oxidized form of the redox couples in the solution can be calculated by adopting the following equation [33].

$$\tau_r = \frac{1}{\omega_{max}} = \frac{1}{2\pi f} \tag{2.12}$$

where ω_{max} refers to the peak frequency of the second semi-arc related to R_{ct}.

2.3.2 Intensity-Modulated Photocurrent/Photovoltage Spectroscopy (IMPS/IMVS)

One of the parameters that determine the photocurrent density is the charge-collection efficiency (η_{cc}). η_{cc} can be determined by the IMPS/IMVS. In brief, the IMPS measures the periodic current response to the intensity-modulated light, providing kinetic information on the charge transport under short-circuit conditions. The IMVS experiment is conducted with the same perturbed light but under open-circuit conditions, which offers the recombination lifetime. The electron transport time (τ_t) can be estimated using the equation $\tau_t = 1/(2\pi f_t)$, where f_t is the characteristic frequency minimum of the IMPS imaginary component. Similarly, the recombination lifetime (τ_r) can be determined using the relation of $\tau_r = 1/(2\pi f_r)$, where f_r is the characteristic frequency minimum of the IMVS imaginary component.

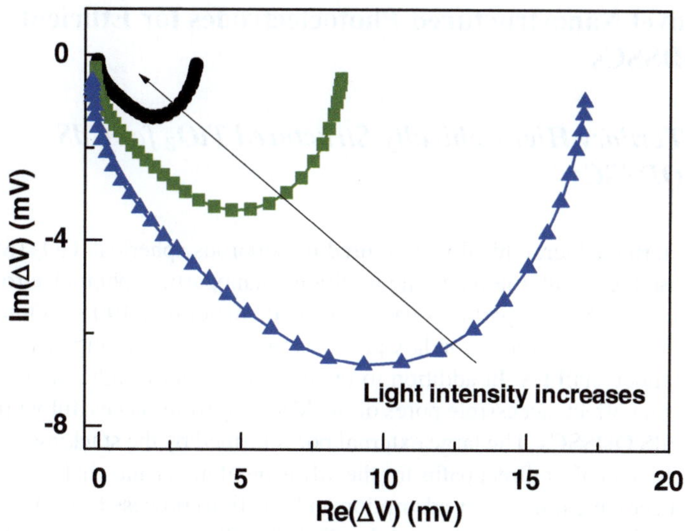

Fig. 2.10 IMVS measurements performed at different DC light intensities

As shown in Fig. 2.10, the semi-circle radius decreases with increasing light intensity. The frequency corresponding to the minimum in the complex plane increases as the light intensity increases, which indicates that the electron lifetime decreases. Therefore, the electron recombination is more pronounced at high illumination intensities. The IMPS data are similar to the data obtained in the IMVS measurements. At high frequencies the modulated photocurrent approaches zero, indicating that the modulation frequency is faster than the relaxation of the charge carrier density by transport to the contacts and back reaction.

The charge-collection efficiency (η_{cc}), electron diffusion coefficient (D_n), and effective electron diffusion length (L_n) can be calculated by the two values obtained from the above IMPS and IMVS analysis via the following equations [34, 35]:

$$\eta_{cc} = 1 - \frac{\tau_t}{\tau_r} \tag{2.13}$$

$$D_n = \frac{L^2}{(2.35\tau_t)} \tag{2.14}$$

where L is film thickness.

$$L_n = (D_n \times \tau_r)^{1/2} \tag{2.15}$$

2.4 Novel Nanostructured Photoelectrodes for Efficient QDSSCs

2.4.1 *Tertiary Hierarchically Structured TiO_2 for CdS QDSSCs*

Recently, tertiary-hierarchically-structured mesoporous spherical TiO_2 (MS TiO_2) was proposed as an alternative to the traditional nanoparticle photoelectrode in the QDSSCs [12]. MS TiO_2 particles have a mean diameter of 1,190 $\pm$ 60-nm, which consists of ~100-nm-sized secondary particles formed by clustering of 15-nm-sized primary nanocrystallites. In addition to the dual function of high loading capacity and scattering effect, accessible pores of the MS TiO_2 were successfully utilized for efficient CdS QDSSCs. The large external pores formed by the stacking of micrometer particles can offer direct paths for the diffusion of metal and chalcogen ions in a SILAR (successive ionic layer adsorption and reaction) process [36]. A conversion efficiency of 1.9 % was achieved by CdS QDSSCs composed of MS TiO_2 photoanode, noticeably higher than 1.2 % for conventional QDSSCs made of 20-nm-sized nanocrystalline TiO_2 (Solaronix, T20) and those of other reported CdS QDSSCs: 1.15 % with nanocrystalline TiO_2 film [28], 1.2 % with TiO_2 nanospheroidal electrodes [37], 1.46 % with ZnO nanotubes [38].

Figure 2.11a shows a morphological feature of the calcined MS TiO_2 particles. The high-magnification FE-SEM image indicates that the MS TiO_2 particles possess a tertiary hierarchical structure: the particles are composed of ~100-nm-sized secondary particles which, in turn, are made up of ~15 nm-sized primary nanocrystallites. It is observed that the CdS QDs are more thickly deposited with better surface coverage, forming agglomerates on the MS TiO_2 (Fig. 2.11b). In comparison, SEM images of the commercial nanocrystalline T20 film before and after the deposition of the CdS QDs are shown in Fig. 2.11c, d. As shown in the insets of Fig. 2.11, the MS TiO_2 has large external pores due to the stacking of micrometer-sized spherical colloids. On the other hand, T20 particles seem to be densely packed, thereby leaving only tiny mesopores. The following STEM analysis reveals more detailed information on the deposited CdS QDs grown after 12 SILAR cycles, where the SILAR (successive ionic layer adsorption and reaction) technique [36] is known as a modified version of chemical bath deposition.

Figure 2.12 shows the STEM images of the CdS QD-sensitized T20 (a, b) and MS TiO_2 (c, d) particles. The crystal size of the QDs is about 3–5 nm for both MS TiO_2 and T2 samples. However, the MS TiO_2 sample seemingly forms larger CdS agglomerates with better coverage on the TiO_2 surface. In contrast, most of the QDs seem to be individual on the T20 surface with less agglomeration of QDs, and partially uncovered surface is also found. This implies that the surface coverage of the QDs is relatively superior at the MS TiO_2 surface. As a result, the thickness of the CdS layer is estimated to be about 6.5–9.5 nm for the MS TiO_2 sample, whereas the thickness is nearly a half for the T20 (~3–6 nm). Electron energy loss

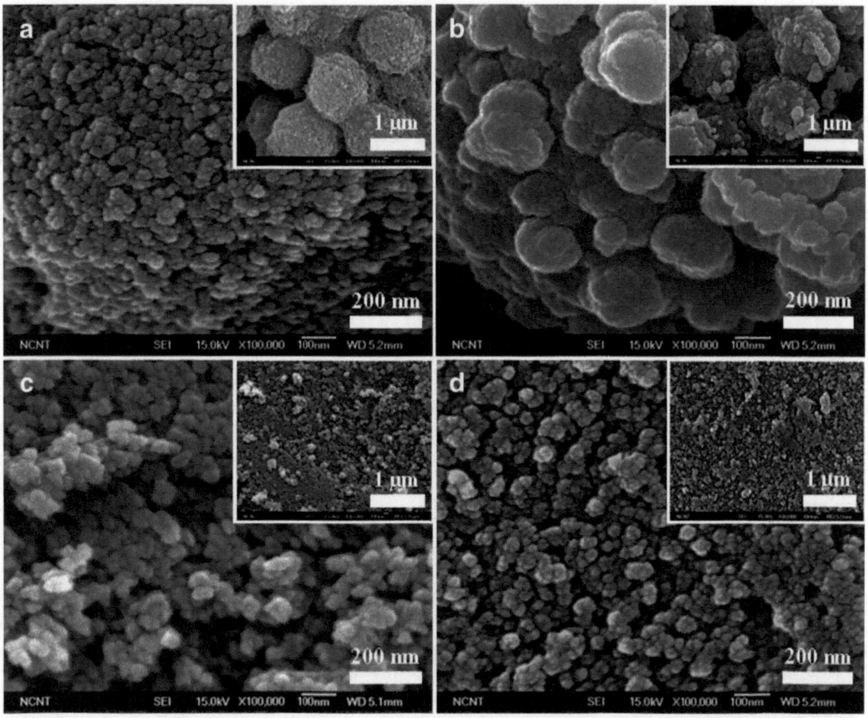

Fig. 2.11 *Top-view* images of the calcined MS TiO_2 film (**a**) before and (**b**) after CdS QD deposition, in comparison with those of the T20 film (**c**) before and (**d**) after CdS QD deposition

spectroscopy (EELS) was conducted to analyze the spatially resolved composition of the sensitized MS TiO_2 sample at a high resolution. A magnified TEM image of a single MS TiO_2 particle and corresponding EELS data are shown in Fig. 2.12e, f. It is clear that the CdS QDs are evenly distributed within an MS TiO_2 particle [12].

The X-ray diffraction patterns of the calcined TiO_2 films without and with SILAR process are presented in Fig. 2.13. After solvothermal treatment and subsequent calcination, well-resolved diffraction peaks appear as shown in Fig. 2.13a, which indicates the crystallization to anatase TiO_2 (JCPDS card No. 21-1272). From the XRD pattern presented in Fig. 2.13c, it can be concluded that the sintered T20 particles also exist in the anatase form. The average crystallite sizes of the calcined MS TiO_2 and T20 were 14.71 and 19.99 nm, respectively, as estimated from the full-width at half-maximum (FWHM) of the anatase (1 0 1) peak using the Scherrer equation. Figure 2.13b, d presents the XRD patterns taken after the CdS QD deposition with 12 SILAR cycles on the MS TiO_2 and the T20 films, respectively. It is confirmed that the CdS QDs are formed with a cubic structure according to the JCPDS card no. 80-0019. The crystal size was calculated from the FWHM of the CdS (1 1 1) peak. The diameter of the CdS QDs is estimated to be ~2.6 nm for both MS TiO_2 and T20.

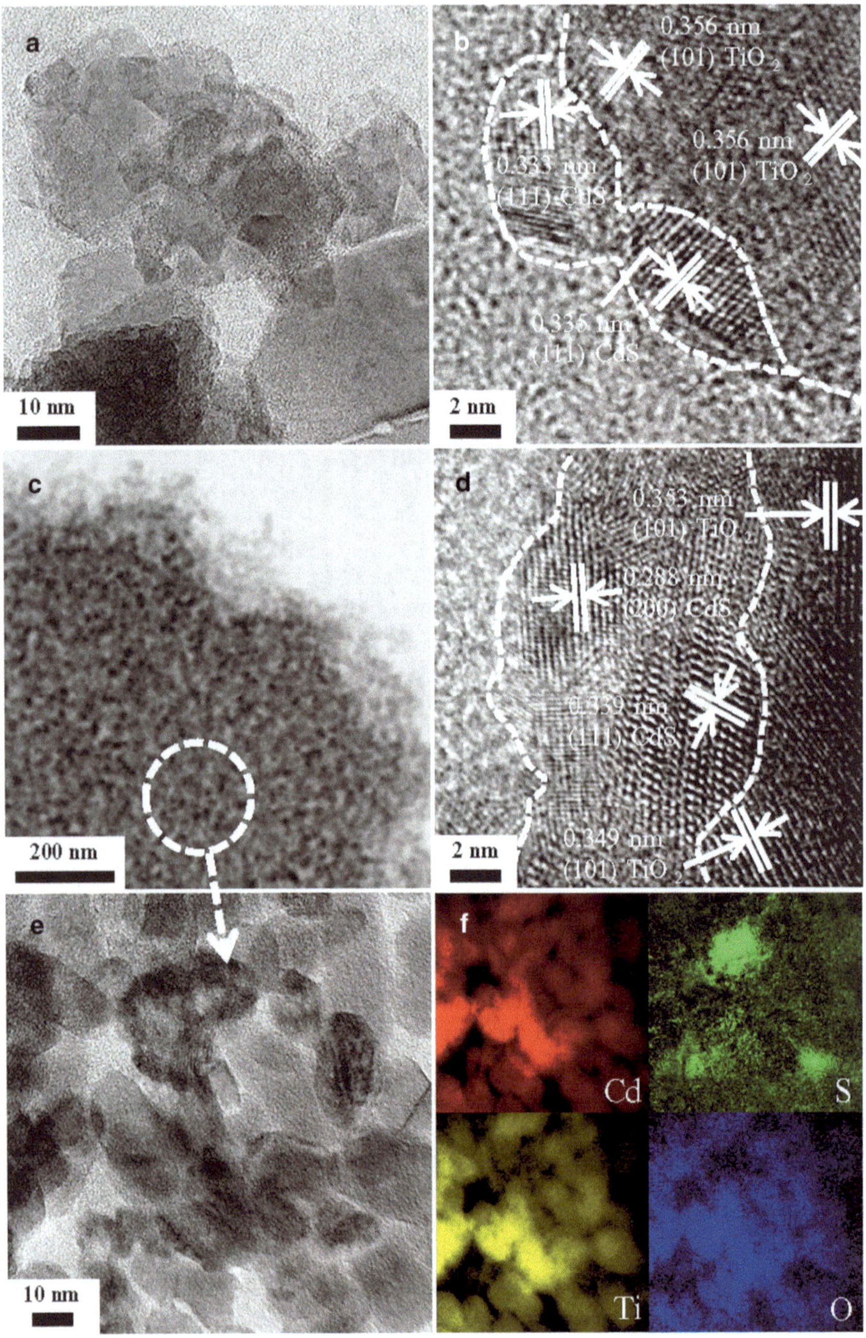

Fig. 2.12 STEM images of the CdS QD-sensitized TiO_2 particles. (**a**, **b**) T20 particles after SILAR process (**c**, **d**) A sliced MS TiO_2 particle after SILAR process (**e**, **f**) A high resolution TEM image of the sensitized MS TiO_2 particle, and corresponding EELS data

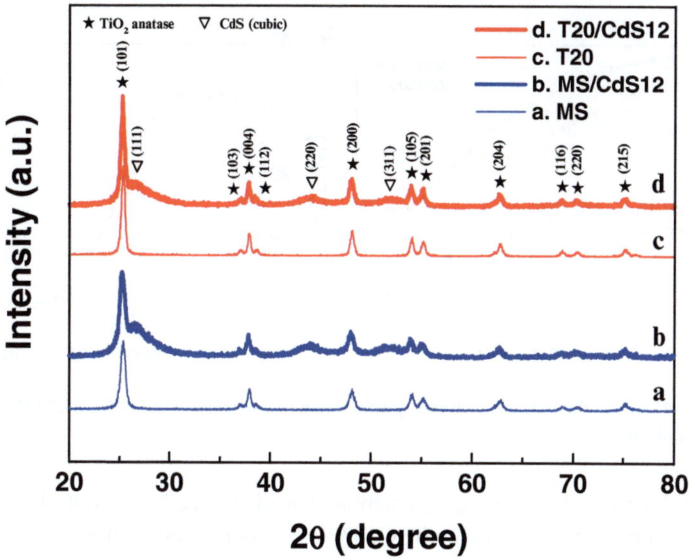

Fig. 2.13 XRD patterns of the calcined MS TiO_2 without (**a**) and with (**b**) SILAR process, and the calcined T20 without (**c**) and with (**d**) SILAR process

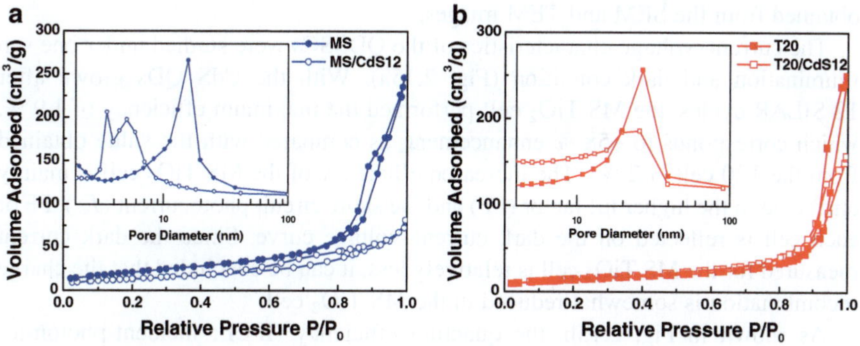

Fig. 2.14 Nitrogen sorption isotherms of the calcined MS TiO_2 (**a**) and the T20 (**b**) without and with SILAR process. The *inset* shows the corresponding pore-size distribution for both samples, as estimated by BJH analysis

The specific surface area and the pore-size distribution were characterized by measuring nitrogen-gas-sorption isotherms (Fig. 2.14a, b). For both MS TiO_2 and T20, H1-type hysteresis loops [39] and type IV isotherms with sharp capillary-condensation steps at a high relative pressure were observed. However, the estimated average pore size of each particle is markedly different: 15.9 nm for the MS TiO_2 and 26.9 nm for the T20. The specific surface area of the MS TiO_2 is estimated to be 76.02 $m^2\ g^{-1}$ which is about 1.5 times higher than that of the T20 (49.05 $m^2\ g^{-1}$). After the CdS QD deposition, the surface area of the MS TiO_2 is

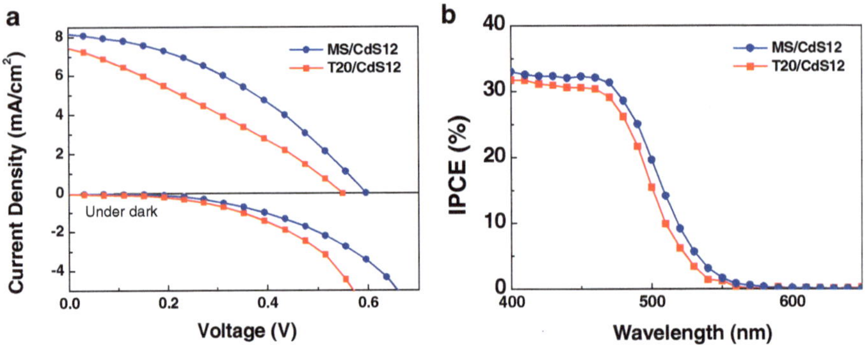

Fig. 2.15 (**a**) *I–V* and (**b**) IPCE (chopping frequency = 10 Hz) curves of CdS QDSSCs with the MS TiO_2 and the T20

greatly reduced to 48.52 $m^2\ g^{-1}$, whereas that of the T20 does not change much. Therefore, it can be inferred that the CdS QDs are deposited on the primary pores of the MS TiO_2 particles with good surface coverage, leading to a significant decrease in the surface area and the pore size. It is also confirmed that the primary pores still have a finite size, which can provide available sites for the interfacial reaction with the electrolyte. This assumption is in good agreement with the previous analysis obtained from the SEM and TEM images.

The current–voltage characteristics of the QDSSCs were studied under one sun illumination and dark condition (Fig. 2.15a). With the CdS QDs grown after 12 SILAR cycles, the MS TiO_2 cell performed the maximum efficiency of 1.9 %, which corresponds to ~58 % enhancement, as compared with the value obtained from the T20 cell (1.2 %). The increased efficiency of the MS TiO_2 cell is mainly attributed to the higher fill factor (FF) and the short-circuit photocurrent (J_{sc}). FF of each cell is reflected on the dark current–voltage curve. Since the dark current measured for the MS TiO_2 cell is relatively less, it can be concluded that the charge recombination is somewhat reduced in the MS TiO_2 cell.

As shown in Fig. 2.15b, the quantum efficiency (IPCE; incident-photon-to-current conversion efficiency) is also higher in the MS TiO_2 cell over the whole spectral range, as compared with that of the T20 cell. The improved photocurrent observed in QDSSCs with the tertiary-hierarchically-structured MS TiO_2 photoanode can be attributed to (a) the high surface area for QD loading and (b) the light scattering effect confirmed by the diffuse reflectance spectra presented in Fig. 2.16a. Owing to the presence of micrometer-sized tertiary colloidal particles, the MS TiO_2 is revealed to effectively scatter visible light (400–800 nm), which results in improved photovoltaic performances. The inset of Fig. 2.16a shows the difference in the transparency between the MS TiO_2 and T20. After the sensitization with the CdS QDs grown after 12 SILAR cycles, the reflectance spectra take on a new aspect (Fig. 2.16b): the reflectance of each sample is greatly reduced at the short wavelength range (<550 nm) due to the absorption of the incident light by the

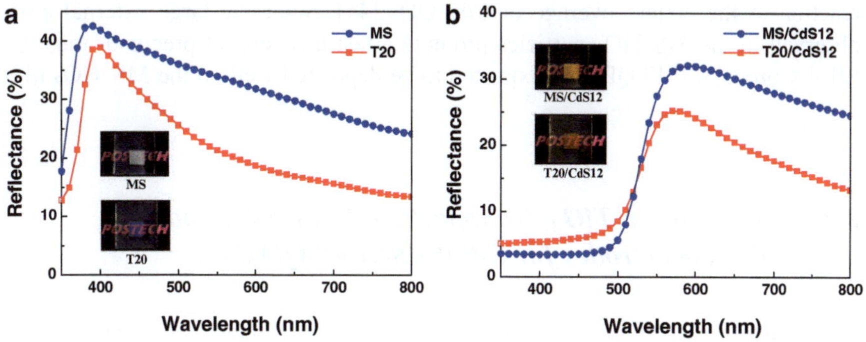

Fig. 2.16 Diffuse reflectance spectra of the CdS QD-sensitized MS TiO_2 and the T20 films (**a**) without and (**b**) with the SILAR process. The each *inset* compares *top-view* images of these two photoanodes

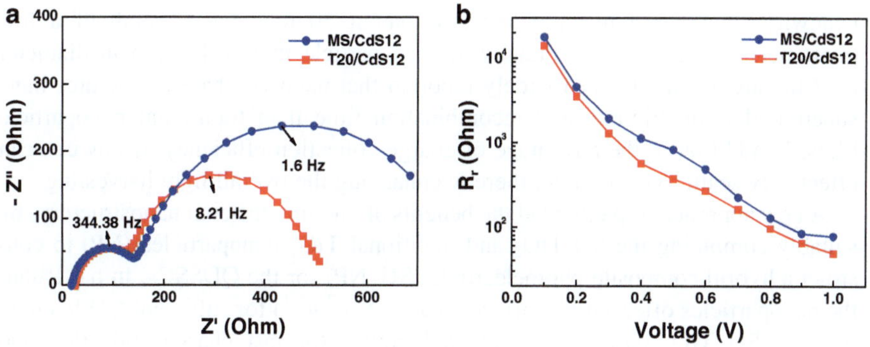

Fig. 2.17 (**a**) Nyquist plots of the EIS spectra and (**b**) R_{ct} for CdS QDSSCs with the MS TiO_2 and the T20 under dark

CdS QDs. As shown in the inset of Fig. 2.16b, the CdS QD-sensitized MS TiO_2 film is opaque with a tint of orange color while the sensitized T20 film is transparent [12].

The EIS was employed to examine the interfacial reactions in QDSSCs (Fig. 2.17). Typical Nyquist plots of both cells were obtained at the V_{oc} condition under dark using the equivalent circuit shown in the inset of Fig. 2.17a. τ_r is estimated to be 625 and 122 ms for the MS TiO_2 and the T20 cells, respectively. Figure 2.17b shows R_{ct} (charge recombination resistance) for both cells which decreases with increasing applied voltage [40]. It is shown that the MS TiO_2 cell has higher R_r than the T20 cell at various applied voltages. This result implies that the dark current arising from the recombination is greatly suppressed at the MS TiO_2/polysulphide electrolyte interface, which offers a higher V_{oc}, FF, and improved conversion efficiency. We attribute the reduced interfacial recombination reaction to the better coverage of CdS QDs [41]. Since the large external pores observed in the MS TiO_2 particles promote a fast diffusion of precursor ions in a SILAR process, CdS QDs are expected to be deposited well on the MS TiO_2 film.

2.4.2 Sea Urchin TiO_2–Nanoparticle Hybrid Composite Photoelectrodes for CdS/CdSe/ZnS QDSSCs

Hierarchically-structured materials are composed of nanocrystallites that are aggregated to form large secondary spheres, thereby can provide efficient light scattering with maintaining a large surface area for sufficient sensitizer-uptake. Although these 0–3 hierarchical materials can be a good alternative to the traditional nanocrystalline TiO_2, slow trap-limited charge transport remains a fundamental problem. To overcome this limit, a different type of hierarchical nanostructures was developed: sea urchin TiO_2 (SU TiO_2) formed by clustering nanowires that has a mean diameter of about 50 nm and a length of a few micrometers to construct a radially aligned particle that is 4–7 μm in diameter (1–3 hierarchy). It has been widely reported that nanowire-based structures have superior electron lifetime and recombination time than traditional nanoparticle films. In addition to the advantage of charge collection efficiency, nanowires can effectively scatter visible light, thereby enhancing the overall light harvesting.

A new approach implemented the benefits of one-dimensional nanostructures by suitably combining the SU TiO_2 and traditional TiO_2 nanoparticles (NP) to construct a hybrid composite photoelectrode (SU–NP) for the QDSSCs. In particular, the nanoparticles offer a high surface area (~75 $m^2\ g^{-1}$) for sufficient QD loading, whereas SU TiO_2 particles provide a highway for fast charge collection and multiple scattering centers within the photoelectrode (Fig. 2.18). CdS/CdSe/ZnS QDSSCs made of the SU–NP composite film exhibited remarkable improvement in power conversion efficiency: 4.2 versus 3.5 % for the reference cell made with the NP film [42].

Figure 2.19a–c shows morphological features of the calcined SU TiO_2 particles. The high-magnification FE-SEM image indicates that the SU TiO_2 particles are composed of nanowires that have a mean diameter of about 50 nm and a length of a few micrometers to construct a radially aligned particle that is 4–7 μm in diameter. In the inset of Fig. 2.19b, broken particles reveal the internal nanostructure of the SU TiO_2. It seems that many nanowires are densely packed within the spherical particle. The high-resolution image of a single nanowire constituting the SU TiO_2 (Fig. 2.19d) shows that the material is fully crystalline with the lattice spacing of 0.324 nm, which corresponds to the rutile (1 1 0) plane.

Time-dependent analysis provides more detailed information on the formation process of the SU TiO_2. Figure 2.20a shows that nanoparticles were initially created after 30 min solvothermal reaction. With the prolongation of reaction time (Fig. 2.20b, c: 1 and 1.5 h), the aggregates were formed which finally led to the

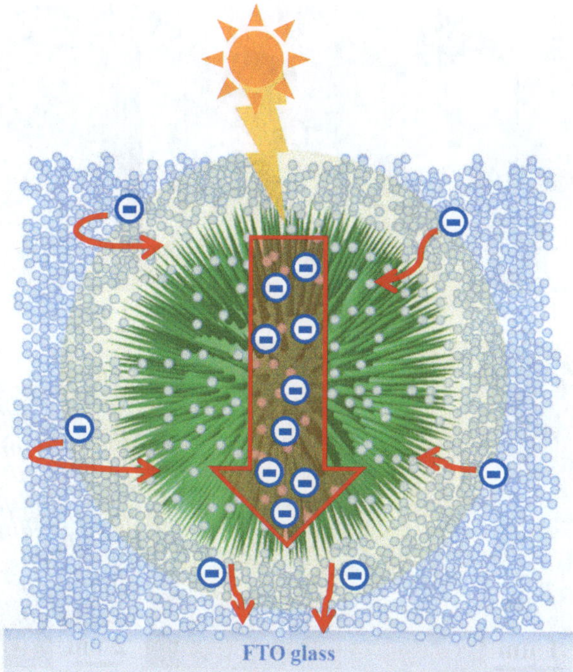

Fig. 2.18 A schematic diagram of a sea urchin TiO_2–nanoparticle composite (SU–NP) film that has fast charge collection and a light scattering effect

1–3 hierarchically-structured SU TiO_2 (Fig. 2.20d, 4 h). The anisotropic growth of nanowires on the spherical particle surface can be understood in terms of shape-controlled chemistry [43–45]. It is thought that Cl^- ions adsorb selectively onto the (1 1 0) crystal plane [46], which limit further growth of this plane, resulting in anisotropic growth and hence radially aligned TiO_2 nanowire formation.

The XRD patterns indicate that two distinct samples are in different crystalline phases (Fig. 2.21): SU TiO_2 in rutile and NP in anatase, respectively (JCPDS card Nos. 21-1276 and 21-1272).

Figure 2.22 shows the STEM images of the CdS/CdSe/ZnS QD-sensitized NP (a) and SU films (c). Most of the QDs seem to be individual on the NP film with less agglomeration of the QDs, and partially uncovered surface is also found. In contrast, the SU TiO_2 seemingly forms larger QDs agglomerates with better coverage on the surface of TiO_2 nanowires. This implies that the surface coverage of the QDs is superior at the SU TiO_2. EELS measurement was conducted to analyze the spatially resolved composition of the sensitized NP and SU TiO_2 samples at a high resolution. Magnified TEM images of NP and SU TiO_2 and the corresponding EELS data are shown in Fig. 2.22b, d.

The current–voltage characteristics of the QDSSCs were studied under one sun illumination (Fig. 2.23). The SU–NP cell shows V_{oc}, J_{sc}, fill factor (FF), and power conversion efficiency (η) of 531 mV, 18.2 mA cm^{-2}, 0.43, and 4.2 %, respectively; the reference NP cell performs 510 mV, 15.2 mA cm^{-2}, 0.45, and 3.5 %. Both J_{sc} and η of the SU–NP device are improved by around 20 %, compared to those of the

Fig. 2.19 SEM images of the SU TiO_2 particles at different magnifications (**a**, **b**, **c**). The *inset* of (**b**) shows a cross-sectional view of broken particles. A TEM image of a 1D nanocrystal in the SU TiO_2 particle (**d**). The *inset* of (**d**) shows a selected-area electron diffraction pattern

NP cell. The quantum efficiency (IPCE) provides detailed information on the light harvesting mechanism (inset of Fig. 2.23). It is observed that the overall IPCE values of the SU–NP cell are higher than the NP cell over the whole spectral range. This implies that efficiency improvement in the SU–NP composite QDSSC is mainly due to the enhanced current density.

It is well known that the photocurrent density (J_{sc}) can be calculated by the following expression [47]:

$$J_{sc} = q\eta_{lh}\eta_{inj}\eta_{cc}I_0 \tag{2.16}$$

where q is the elementary charge, η_{lh} is the light harvesting efficiency of a cell, η_{inj} is the charge-injection efficiency, η_{cc} is the charge-collection efficiency, and I_0 is the light flux. Here, η_{lh} is commonly determined by the amount of adsorbed QDs and light scattering of the films; η_{cc} is largely determined by the competition between the charge collection and recombination [34]. Since identical QDs (CdS/CdSe/ZnS) are applied to both the SU–NP and NP films, η_{inj} values of two QDSSCs are assumed to be same. Accordingly, it can be concluded that the photocurrent density in our experimental condition is determined by two parameters: the light harvesting efficiency (η_{lh}) and the charge-collection efficiency (η_{cc}).

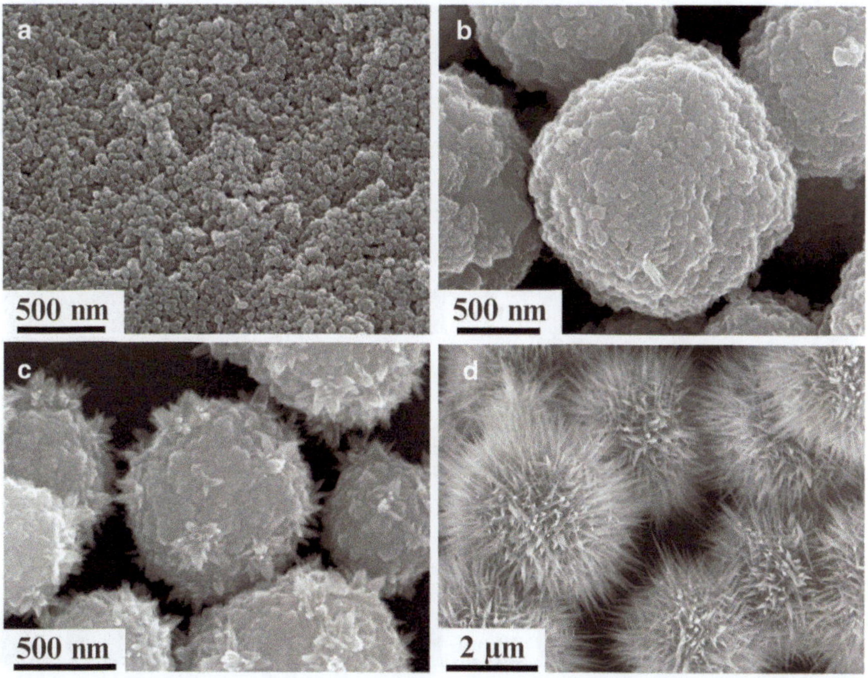

Fig. 2.20 SEM images of as-prepared SU TiO_2 at different reaction time: 30 min (**a**), 1 h (**b**), 1.5 h (**c**), and 4 h (**d**)

Fig. 2.21 XRD patterns of the SU TiO_2 particles and the TiO_2

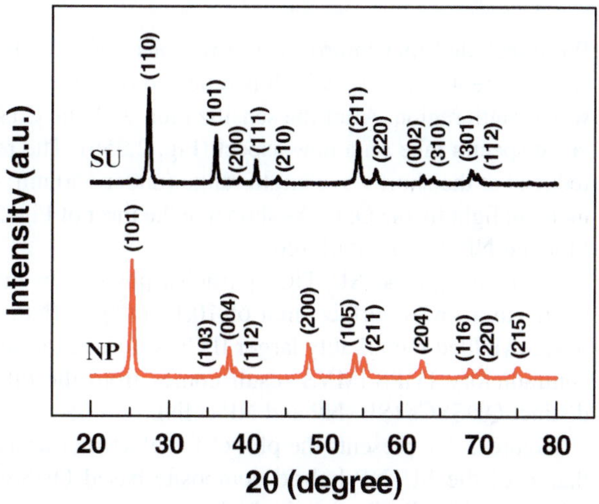

The η_{lh} value is explained by the reflectance spectra. Since the micrometer-sized SU TiO_2 particles can function as scattering centers (Fig. 2.24), the SU–NP composite film can effectively confine visible light (450–750 nm). The inset of Fig. 2.25a shows the difference in the transparency of the SU–NP and NP films.

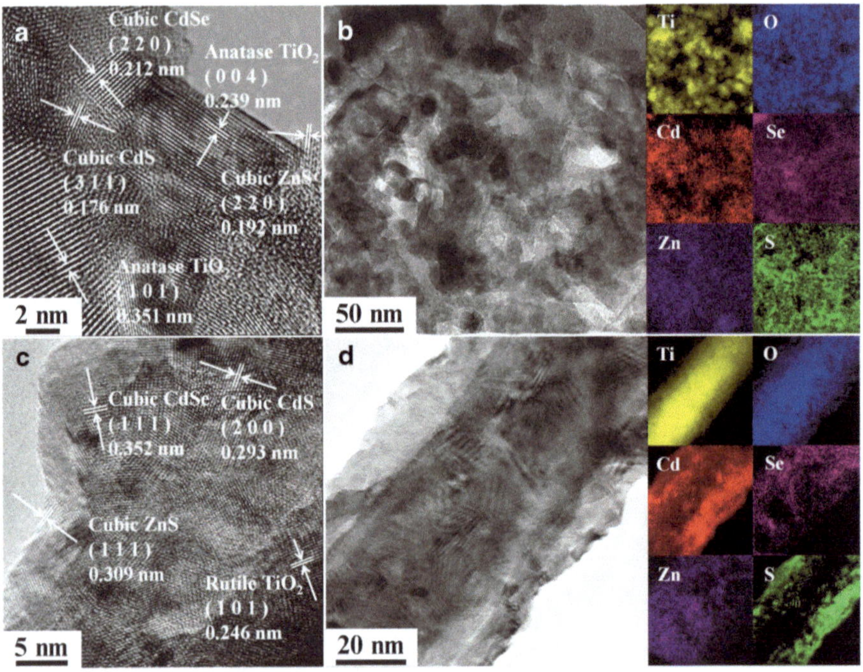

Fig. 2.22 STEM images of the CdS/CdSe/ZnS QD-sensitized TiO_2 nanocrystals. (**a, b**) The sensitized NP film and (**c, d**) SU film. Each *inset* compares EELS data of these two sensitized films

We concluded that improved red response of the NP–SU film contributes to the enhancement in the η_{lh}, which is reflected to the increased IPCE values at the long wavelength region. After the sensitization with the CdS/CdSe/ZnS QDs, the reflectance spectra take on a new aspect (Fig. 2.25b). The reflectance of each sample is reduced at the short wavelength range (under 650 nm) due to the absorption of the incident light by the QDs. As shown in the inset of Fig. 2.25b, the sensitized SU–NP film and NP film are dark red.

Even though the SU TiO_2 particles possess an inferior QD loading capacity owing to a lower surface area of 10.16 $m^2\ g^{-1}$ (NP: 73.29 $m^2\ g^{-1}$), the SU–NP device yielded remarkably larger IPCE values at the short wavelength region. This contradictory IMPS/IMVS result comes from the difference between η_{cc} in two distinct QDSSCs (SU–NP and NP cells).

Figure 2.26a presents the plot of the electron transport time (τ_{t}). It is noticeable that τ_{t} of the SU–NP hybrid composite-based QDSSC is larger than that of the reference NP cell. The relatively faster electron transport seems to be due to fewer interparticle junction within the SU–NP photoelectrode. The effective electron diffusion coefficient (D_n) of the SU–NP cell was 2.3 times larger than that of the NP cell under 100 W m^{-2} illumination (Fig. 2.26c).

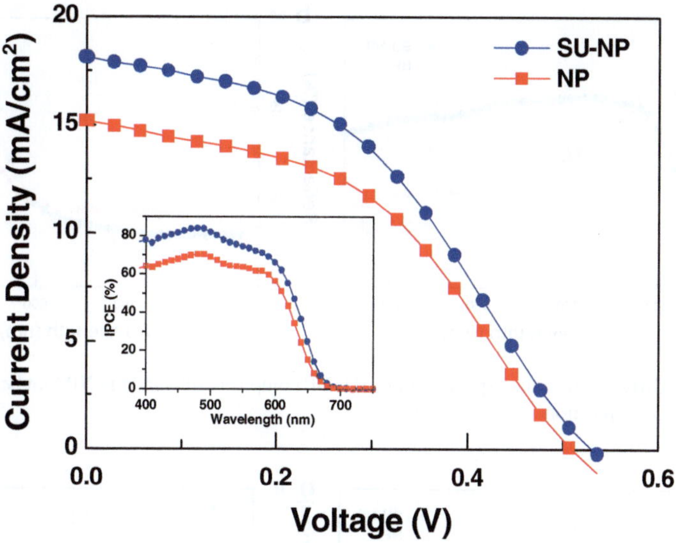

Fig. 2.23 *I–V* characteristics of the CdS/CdSe/ZnS QDSSCs with the SU–NP composite film and the reference NP film under illumination (AM 1.5, 100 mW cm^{-2}). The *inset* shows the corresponding IPCE spectra measured at chopping frequency of 3 Hz

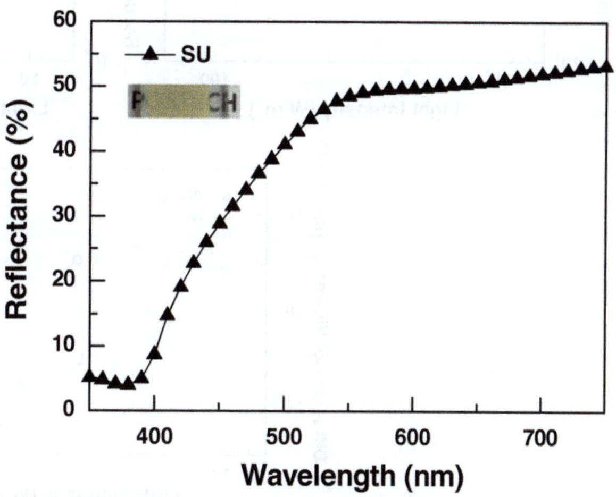

Fig. 2.24 Diffuse reflectance spectra of the SU film. The *inset* is the *top-view* image of the photoanode

Figure 2.26b shows that the recombination time (τ_r) of the SU–NP cell is longer than that of the NP device over the whole light-intensity range. The increased τ_r is attributed to the presence of the relatively defect-free single crystalline TiO_2 nanowires that have far fewer grain boundaries, which may result in the reduced electron trapping phenomena [48, 49]. Additionally, better surface coverage of the QDs at the SU–NP film can promote longer τ_r [42]. This assumption is in agreement

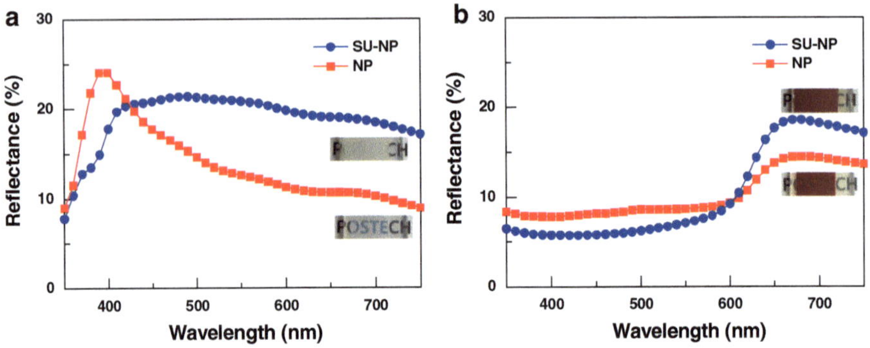

Fig. 2.25 Diffuse reflectance spectra of the SU–NP composite film and the NP film before (**a**) and after (**b**) QDs deposition

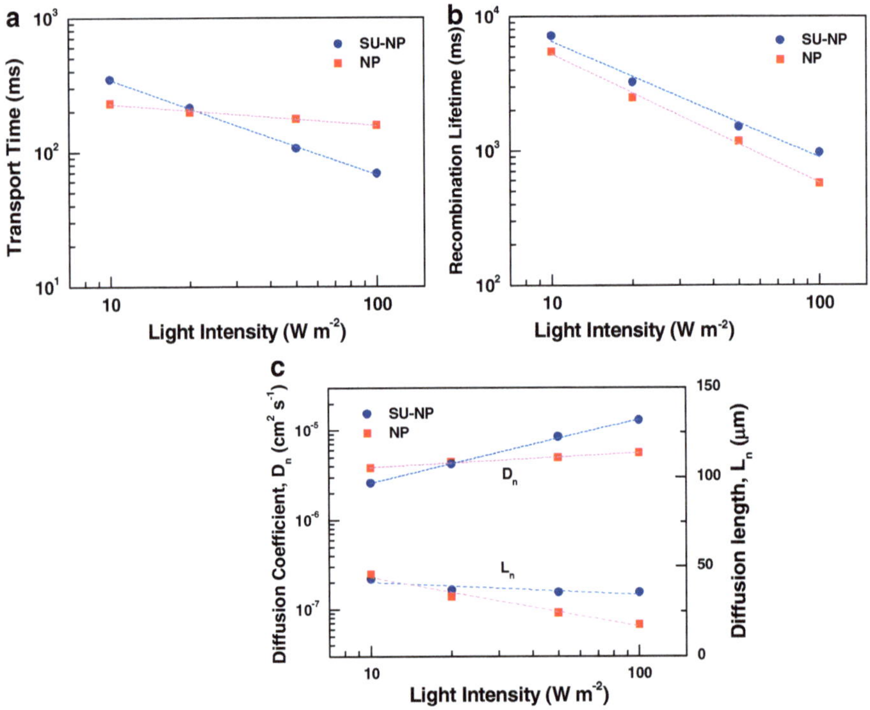

Fig. 2.26 Incident light-intensity-dependent transport time constant (**a**), recombination lifetime (**b**), and electron diffusion coefficient and effective diffusion length (**c**) for the QDSSCs based on the SU–NP composite film and the NP film measured under 625 nm LED illumination

with the previous analysis obtained from the TEM images. As shown in Fig. 2.26c, the effective electron diffusion length (L_n) of the SU–NP cell (35.69 μm) is around twice as thick as that of the NP cell (17.97 μm) under the light intensity of 100 W m^{-2}. On the basis of the increased τ_r, improved photoelectron densities n

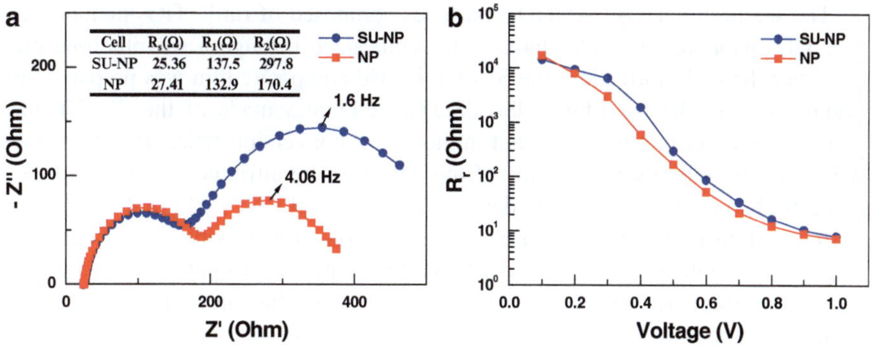

Fig. 2.27 (**a**) Nyquist plots of the EIS spectra and (**b**) R_{ct} for the CdS/CdSe/ZnS QDSSCs with the SU–NP composite and the NP films under dark. The *inset* of (**a**) is corresponding values of fitting parameters

($n \propto J_{sc}\tau_r$) [50] yield a higher V_{oc} [51] in the SU–NP cell. Consequently, η_{cc} of the SU–NP cell (92.9 %) was 20.6 % larger than that of the NP device.

Typical Nyquist plots of both cells under dark are presented in Fig. 2.27a. τ_r is estimated to be 625 and 246 ms for the SU–NP and the NP cells, respectively. Figure 2.27b shows R_{ct} for both cells which decreases with increasing applied voltage [40]. It is observed that the SU–NP cell has higher R_{ct} than the NP cell at various applied voltages. This result implies that the recombination is greatly suppressed at the SU TiO_2/polysulphide electrolyte interface, which offers a higher V_{oc}, J_{sc} and improved conversion efficiency. Therefore, it is reasonable to conclude that higher J_{sc} is mainly due to the higher η_{cc}, and partially due to the higher η_{lh} of the SU–NP hybrid composite QDSSCs.

2.5 Conclusions

In recent years, extensive research efforts have been made to implement high-performance QDSSCs. One of the research trends driving innovations is the material engineering of the oxide nanostrucutres for photoanodes. In this article, two representative approaches were introduced in the development of photoelectrodes for QDSSCs: tertiary-hierarchically-structured mesoporous spherical TiO_2 (MS TiO_2) and the sea urchin TiO_2 (SU TiO_2).

The tertiary-hierarchically-structured MS TiO_2 was used as a working electrode for CdS QDSSCs. A conversion efficiency of 1.9 % was achieved with the MS TiO_2, which is 58 % higher than the cells made with conventional nanocrystalline TiO_2 at AM 1.5 one sun illumination. The improved cell performance in the MS TiO_2 cell is attributed to its large surface area, high internal reflectance, and more accessible external pores.

The sea urchin TiO_2 (SU TiO_2) particles composed of rutile TiO_2 nanoneedles and TiO_2 nanoparticles (NP) have been synthesized through a solvothermal process. For the cell fabrication, the SU–NP hybrid composite film was prepared and compared with the NP film. CdS/CdSe/ZnS QDSSCs made of the SU–NP film achieved a noticeable improvement in the power conversion efficiency: 4.2 versus 3.5 % for the reference cell made of the NP film. We attribute this extraordinary result to the higher photocurrent density (J_{sc}) of the SU–NP cell (18.2 mA cm^{-2}), corresponding to about 20 % improvement in comparison with the NP cell. The origin of the enhanced J_{sc} was investigated in terms of two parameters: the light harvesting efficiency (η_{lh}) and the charge-collection efficiency (η_{cc}). The diffuse reflectance spectra and the intensity-modulated photocurrent/photovoltage spectra confirmed that both η_{lh} and η_{cc} were enhanced due to the internal light scattering and more efficient charge collection.

Therefore, it can be concluded that the MS TiO_2 and the SU–NP hybrid composite can be versatile building blocks for the QDSSCs.

Acknowledgments This work is financially supported by Pohang Steel Corporation (POSCO) through Steel Nano-Fusion Program (Project No. 2010Y110) and partly by the World Class University (WCU) program through the National Research Foundation (NRF) of Korea (Grant No. R31-2008-000-10059-0).

References

1. O'Regan, B., Grätzel, M.: A low-cost, high-efficiency solar cell based on dye-sensitized colloidal TiO_2 films. Nature **353**, 737–740 (1991)
2. Yu, W.W., Qu, L., Guo, W., Peng, X.: Experimental determination of the extinction coefficient of CdTe, CdSe, and CdS nanocrystals. Chem. Mater. **15**, 2854–2860 (2003)
3. Nazeeruddin, M.K., Kay, A., Rodicio, I., Humphry-Baker, R., Mueller, E., Liska, P., Vlachopoulos, N., Grätzel, M.: Conversion of light to electricity by cis-X2bis(2,2′-bipyridyl-4,4′-dicarboxylate)ruthenium(II) charge-transfer sensitizers (X = Cl-, Br-, I-, CN-, and SCN-) on nanocrystalline titanium dioxide electrodes. J. Am. Chem. Soc. **115**, 6382–6390 (1993)
4. Nazeeruddin, M.K., Zakeeruddin, S.M., Humphry-Baker, R., Jirousek, M., Liska, P., Vlachopoulos, N., Shklover, V., Fischer, C.-H., Grätzel, M.: Acid-base equilibria of (2,2-′-bipyridyl-4,4′-dicarboxylic acid)ruthenium(II) complexes and the effect of protonation on charge-transfer sensitization of nanocrystalline titania. Inorg. Chem. **38**, 6298–6305 (1999)
5. Gorer, S., Hodes, G.: Quantum size effects in the study of chemical solution deposition mechanisms of semiconductor films. J. Phys. Chem. **98**, 5338–5346 (1994)
6. Nozik, A.J.: Exciton multiplication and relaxation dynamics in quantum dots: applications to ultrahigh-efficiency solar photon conversion. Inorg. Chem. **44**, 6893–6899 (2005)
7. Hanna, M.C., Nozik, A.J.: Solar conversion efficiency of photovoltaic and photoelectrolysis cells with carrier multiplication absorbers. J. Appl. Phys. **100**, 074510 (2006)
8. Chou, T.P., Zhang, Q., Fryxell, G.E., Cao, G.Z.: Hierarchically structured ZnO film for dye-sensitized solar cells with enhanced energy conversion efficiency. Adv. Mater. **19**, 2588–2592 (2007)
9. Koo, H.-J., Kim, Y.J., Lee, Y.H., Lee, W.I., Kim, K., Park, N.-G.: Nano-embossed hollow spherical TiO_2 as bifunctional material for high-efficiency dye-sensitized solar cells. Adv. Mater. **20**, 195–199 (2008)

10. Chen, D., Huang, F., Cheng, Y.-B., Caruso, R.A.: Mesoporous anatase TiO_2 beads with high surface areas and controllable pore sizes: a superior candidate for high-performance dye-sensitized solar cells. Adv. Mater. **21**, 2206–2210 (2009)
11. Park, Y.C., Chang, Y.J., Kum, B.G., Kong, E.H., Son, J.Y., Kwon, Y.S., Park, T., Jang, H.M.: Size-tunable mesoporous spherical TiO_2 as a scattering overlayer in high-performance dye-sensitized solar cells. J. Mater. Chem. **21**, 9582–9586 (2011)
12. Park, Y.C., Kong, E.H., Chang, Y.J., Kum, B.G., Jang, H.M.: Tertiary hierarchically structured TiO_2 for CdS quantum-dot-sensitized solar cells. Electrochim. Acta **56**, 7371–7376 (2011)
13. Efros, A.L., Rosen, M.: The electronic structure of semiconductor nanocrystals. Annu. Rev. Mater. Sci. **30**, 475–521 (2000)
14. Kamat, P.V.: Quantum dot solar cells. Semiconductor nanocrystals as light harvesters. J. Phys. Chem. C **112**, 18737–18753 (2008)
15. Brus, L.E.: A simple model for the ionization potential, electron affinity, and aqueous redox potentials of small semiconductor crystallites. J. Chem. Phys. **79**, 5566–5571 (1983)
16. Kurisu, H., Tanaka, T., Karasawa, T., Komatsu, T.: Pressure induced quantum confined excitons in layered metal tri-iodide crystals. Jpn. J. Appl. Phys. **32**, 285–287 (1993)
17. Lee, C.-J., Mizel, A., Banin, U., Cohen, M.L., Paul Alivisatos, A.: Observation of pressure-induced direct-to-indirect band gap transition in InP nanocrystals. J. Chem. Phys. **113**, 2016–2020 (2000)
18. Nozik, A.J.: Spectroscopy and hot electron relaxation dynamics in semiconductor quantum wells and quantum dots. Annu. Rev. Phys. Chem. **52**, 193–231 (2001)
19. Williams, F., Nozik, A.J.: Irreversibilities in the mechanism of photoelectrolysis. Nature **271**, 137–139 (1978)
20. Boudreaux, D.S., Williams, F., Nozik, A.J.: Hot carrier injection at semiconductor-electrolyte junctions. J. Appl. Phys. **51**, 2158–2163 (1980)
21. Christensen, O.: Quantum efficiency of the internal photoelectric effect in silicon and germanium. J. Appl. Phys. **47**, 689–695 (1976)
22. Wolf, M., Brendel, R., Werner, J.H., Queisser, H.J.: Solar cell efficiency and carrier multiplication in $Si_{1-x}Ge_x$ alloys. J. Appl. Phys. **83**, 4213–4221 (1998)
23. Nozik, A.J.: Quantum dot solar cells. Physica E **14**, 115–120 (2002)
24. Nozik, A.J.: Multiple exciton generation in semiconductor quantum dots. Chem. Phys. Lett. **457**, 3–11 (2008)
25. Klimov, V.I.: Mechanisms for photogeneration and recombination of multiexcitons in semiconductor nanocrystals: implications for lasing and solar Energy. J. Phys. Chem. B **110**, 16827–16845 (2006)
26. Dean, J.A.: Lange's Handbook of Chemistry, 14th edn. McGraw-Hill, New York (1992)
27. Weast, R.C.: Handbook of Chemistry and Physics, 56th edn. CRC, Cleveland (1975)
28. Lee, Y.-L., Chang, C.-H.: Efficient polysulfide electrolyte for CdS quantum dot-sensitized solar cells. J. Power Sources **185**, 584–588 (2008)
29. Bisquert, J., Zaban, A., Greenshtein, M., Mora-Seró, I.: Determination of rate constants for charge transfer and the distribution of semiconductor and electrolyte electronic energy levels in dye-sensitized solar cells by open-circuit photovoltage decay method. J. Am. Chem. Soc. **126**, 13550–13559 (2004)
30. Ondersma, J.W., Hamann, T.W.: Impedance investigation of dye-sensitized solar cells employing outer-sphere redox shuttles. J. Phys. Chem. C **114**, 638–645 (2010)
31. Wang, Q., Ito, S., Grätzel, M., Fabregat-Sañtiago, F., Mora-Seró, I., Bisquert, J., Bessho, T., Imai, H.: Characteristics of high efficiency dye-sensitized solar cells. J. Phys. Chem. B **110**, 25210–25221 (2006)
32. Han, L., Koide, N., Chiba, Y., Mitate, T.: Modeling of an equivalent circuit for dye-sensitized solar cells. Appl. Phys. Lett. **84**, 2433–2435 (2004)
33. Kern, R., Sastrawan, R., Ferber, J., Stangl, R., Luther, J.: Modeling and interpretation of electrical impedance spectra of dye solar cells operated under open-circuit conditions. Electrochim. Acta **47**, 4213–4225 (2002)

34. Schlichthörl, G., Park, N.G., Frank, A.J.: Evaluation of the charge-collection efficiency of dye-sensitized nanocrystalline TiO_2 solar cells. J. Phys. Chem. B **103**, 782–791 (1999)
35. van de Lagemaat, J., Frank, A.J.: Nonthermalized electron transport in dye-sensitized nanocrystalline TiO_2 films: transient photocurrent and random-walk modeling studies. J. Phys. Chem. B **105**, 11194–11205 (2001)
36. Pathan, H.M., Lokhande, C.D.: Deposition of metal chalcogenide thin films by successive ionic layer adsorption and reaction (SILAR) method. Bull. Mater. Sci. **27**, 85–111 (2004)
37. Sudhagar, P., Jung, J.H., Park, S., Sathyamoorthy, R., Ahn, H., Kang, Y.S.: Self-assembled CdS quantum dots-sensitized TiO_2 nanospheroidal solar cells: structural and charge transport analysis. Electrochim. Acta **55**, 113–117 (2009)
38. Chen, J., Li, C., Zhao, D.W., Lei, W., Zhang, Y., Cole, M.T., Sun, X.W., Milne, W.I.: A quantum dot sensitized solar cell based on vertically aligned carbon nanotube templated ZnO arrays. Electrochem. Commun. **12**, 1432–1435 (2010)
39. Lowell, S., Shields, J.E.: Power Surface Area and Porosity. Chapman & Hall, London (1991)
40. Mora-Seró, I., Giménez, S., Fabregat-Sañtiago, F., Gómez, R., Shen, Q., Toyoda, T., Bisquert, J.: Recombination in quantum dot sensitized solar cells. Acc. Chem. Res. **42**, 1848–1857 (2009)
41. Guijarro, N., Lana-Villarreal, T., Mora-Seró, I., Bisquert, J., Gómez, R.: CdSe quantum dot-sensitized TiO_2 electrodes: effect of quantum dot coverage and mode of attachment. J. Phys. Chem. C **113**, 4208–4214 (2009)
42. Kong, E.-H., Chang, Y.-J., Park, Y.-C., Yoon, Y.-H., Park, H.-J., Jang, H.M.: Sea urchin TiO_2-nanoparticle hybrid composite photoelectrodes for CdS/CdSe/ZnS quantum-dot-sensitized solar cells. Phys. Chem. Chem. Phys. **14**, 4620–4625 (2012)
43. Jun, Y.W., Casula, M.F., Sim, J.H., Kim, S.Y., Cheon, J., Alivisatos, A.P.: Surfactant-assisted elimination of a high energy facet as a means of controlling the shapes of TiO_2 nanocrystals. J. Am. Chem. Soc. **125**, 15981–15985 (2003)
44. Peng, X.G., Manna, L., Yang, W.D., Wickham, J., Scher, E., Kadavanich, A., Alivisatos, A.P.: Shape control of CdSe nanocrystals. Nature **404**, 59–61 (2000)
45. Barnard, A.S., Curtiss, L.A.: Prediction of TiO_2 nanoparticle phase and shape transitions controlled by surface chemistry. Nano Lett. **5**, 1261–1266 (2005)
46. Adachi, M., Murata, Y., Takao, J., Jiu, J.T., Sakamoto, M., Wang, F.M.: Highly efficient dye-sensitized solar cells with a titania thin-film electrode composed of a network structure of single-crystal-like TiO_2 nanowires made by the "Oriented Attachment" mechanism. J. Am. Chem. Soc. **126**, 14943–14949 (2004)
47. Zhu, K., Neale, N.R., Miedaner, A., Frank, A.J.: Enhanced charge-collection efficiencies and light scattering in dye-sensitized solar cells using oriented TiO_2 nanotubes arrays. Nano Lett. **7**, 69–74 (2007)
48. Jiu, J.T., Isoda, S., Wang, F.M., Adachi, M.: Dye-sensitized solar cells based on a single-crystalline TiO_2 nanorod film. J. Phys. Chem. B **110**, 2087–2092 (2006)
49. Kang, S.H., Choi, S.H., Kang, M.S., Kim, J.Y., Kim, H.S., Hyeon, T., Sung, Y.E.: Nanorod-based dye-sensitized solar cells with improved charge collection efficiency. Adv. Mater. **20**, 54–58 (2008)
50. Schlichthörl, G., Huang, S.Y., Sprague, J., Frank, A.J.: Band edge movement and recombination kinetics in dye-sensitized nanocrystalline TiO_2 solar cells: a study by intensity modulated photovoltage spectroscopy. J. Phys. Chem. B **101**, 8141–8155 (1997)
51. Frank, A.J., Kopidakis, N., van de Lagemaat, J.: Electrons in nanostructured TiO_2 solar cells: transport, recombination and photovoltaic properties. Coord. Chem. Rev. **248**, 1165–1179 (2004)

Chapter 3
Hybrid Optoelectronic Devices with Colloidal Quantum Dots

Chien-Chung Lin

Abstract The progresses of colloidal quantum dots (CQDs) in the last decade have brought many developments in related optoelectronic devices. Genuine CQD optoelectronic devices, however, still suffer from low efficiency and short life span, which seriously hinder their practical use. A novel integration scheme is proposed and demonstrated to alleviate these problems. By combination of regular semiconductor devices and CQD thin film layers, we can achieve a much better performance and avoid the lifetime issue. The CQD layer can be applied in both photovoltaic reaction and general illumination via its strong absorption and emission characteristics. The design principles, fabrication technology, and device analysis will be discussed in detail.

3.1 Introduction

Since last decade, global warming and energy crisis have been haunting the future of human kind. Both problems are connected to high demands and exploitations of fossil-related fuel. How to reduce fossil fuel usage is thus an important issue when people are trying to alleviate these problems. Meanwhile if abolishing fossil fuel is a must, human kind still needs to find a way to generate energy and to disperse it wisely. This thought brought the rise of "green technologies," which shall be either energy-savers or alternative energy sources. Among them, there is the photovoltaic technology, which can convert the limitless sun power into the much desired electricity via different mechanisms. Since the invention of the solar cells in the 1950s, many methods and device architectures have been proposed and realized either in the lab or in the field, such as crystalline/amorphous Si-solar cell [1],

C.-C. Lin (✉)
Institute of Photonic System, College of Photonics, National Chiao Tung University, Tainan 71150, Taiwan
e-mail: chienchunglin@faculty.nctu.edu.tw

J. Wu and Z.M. Wang (eds.), *Quantum Dot Solar Cells*, Lecture Notes in Nanoscale Science and Technology 15, DOI 10.1007/978-1-4614-8148-5_3,

chalcopyrites [2, 3], dye-/quantum-dot-sensitized solar cells [4–6], and single-/multi-junction III–V solar cells [7–9]. While most of these solar cell technologies provide satisfactory results, it is III–V materials that claimed the highest power conversion efficiency (PCE) so far [10, 11] . Meanwhile, the detailed-balanced Shockley–Queisser limit (SQ limit) [12] puts an upper limit of 44 % for the cell with single band gap material, which is yet to be conquered.

Other than epitaxially growing semiconductor bulk materials for solar cells, another choice is to utilize other substances such as dye, polymer, or colloidal quantum dots (CQDs) (or nanocrystals) as the active layers in the devices. The benefits of adapting quantum dot material in the solar cell are mainly for low-cost, large area, flexible substrates, and potentially high efficiency. Many extraordinary works were reported for light detection [13], tandem solar cells [14], etc. However, most of the results were limited to 10 % PCE or less. How to combine both the versatility of the quantum dot material and the high PCE in semiconductor device becomes an interesting topic to research. One of the major limit in current solar cell is the lack of absorbing material at ultraviolet (UV) region. These high energy photons get absorbed quickly in the bulk material but the carriers are consumed near the surface by the traps and defects and thus cannot be collected. This portion of sunlight takes up to 7 % of the overall energy [15], and this percentage is even higher in the outer space. When the cost of a solar cell module is taken into account, every percent of PCE improvement can be important, not to mention a possible 10 % boost in efficiency. Therefore, efficient use of UV part of the solar spectrum for energy conversion is crucial. If we can harvest UV photons, the aforementioned S–Q limit can become closer to reality.

If the band gap of the device material can be tailored and the surface recombination velocity can be suppressed, the UV photons are actually just the same as visible ones. However, the generic semiconductor-based devices absorb visible photons very well, while they behave poorly in short wavelength regime. One possible solution is to utilize semiconductor-based quantum dots (QDs) such as CdS, CdSe, in the regular solar cell. These nano-scale materials have strong UV absorption and efficient visible down-conversion [16]. If this re-emitted photons can be properly collected by the host material underneath, the UV photons can then be counted as effective as other longer wavelength ones. Previous works [17–23] have demonstrated that it is possible to use CQDs or other substances (such as dye) to promote the efficiency of Si-based solar cells, either by anti-reflective (AR) effect due to the gradient refractive index introduced by the film or by the down-conversion transition of the CQDs' emission. Our work focuses on the novel combination of CQD with various different optoelectronic devices, such as Si nanopillar solar cell, GaAs-based solar cells, and UV light-emitting diode (LED).

In addition to photovoltaic technology, light generation is regarded as important as the electricity generation. After all, half of the daily life need some forms of light sources. Among all light-emitting mechanisms, radiative recombination of electrons and holes is a promising candidate with possible high efficiency. It was estimated that almost 20 % of electricity is used for general lighting [24]. The majority of indoor light bulbs today are still, however, in the same format from

100 years ago. If we could utilize the much more efficient radiative recombination in the material, we can save a lot of electricity and thus reduce the fossil fuel dependence and green house gas emission. From the research done by the United Nations, making the lighting more efficient is equivalent to taking about 50 million cars off the road [24]. The implication is clear enough.

Since the invention of light-emitting diodes (LEDs) in 1960 [25], scientists have made significant progress in the last 50 years. In the early days, due to the limitation on material band gap, we could only achieve red or orange LEDs with good efficiency. The breakthrough in GaN-based material growth led by Prof. Nakamura in 1990s completed the visible spectrum, and since then, an explosion of capital investment has taken off to manufacture super efficient LED for lighting. Among all these advances, obstacles such as high cost, low efficiency in green to yellow colors, efficiency droop, and defective GaN material still remain to be conquered. The inherent broad light emission spectrum of a generic LED can also hinder the potential display application. The CQD, on the other hand, possesses unique properties such as narrow emission linewidth, cheap manufacturing costs, and compliant to any substrates. These properties prompt CQD to become a good candidate for integration with regular semiconductor light-emitting device for performance enhancement.

Through these devices, the potential of CQD can be unleashed while the device performance and reliability can be maintained. Many aspects, however, are still under intensive study since the research activity of this field is just beginning.

Based on the aforementioned introduction, we will apply CQD in two important applications: photovoltaic and light emission. The introduction of this integration scheme is to solve the existing problems of traditional optoelectronic devices without getting into the trouble of pure CQD devices. In this chapter, we will cover the basic idea of our design and the details of implementation and applications. The chapter will be arranged in the following sequence: the hybrid platform design, the dispense method, the characterization, and finally the device demonstration. As we shall see in the next sections, the enhancement of hybrid CQD device can be a game-changer for the new generation of optoelectronic devices.

3.2 The Hybrid Design Platform

In the pure CQD devices, there are several different designs available such as metal–semiconductor junction, polymer–semiconductor, and semiconductor–semiconductor systems. shown in Fig. 3.1 [26]. However, all of them rely on the nanoparticles for light absorption or emission, and the charge transfer mechanism can be achieved by incorporating the surround medium such as TiO_2 or polymer [26]. The device usually suffers from the low efficiency, for example, sub-10 % PCE for solar cell. Also the lifetime and reliability issues for pure CQD devices are not easy to tackle with. Thus a new idea of implementing CQD materials needs to be found.

The basic design of our hybrid device is illustrated in Fig. 3.2. Both light absorption (solar cell) and emission (LED) can be implemented in this platform

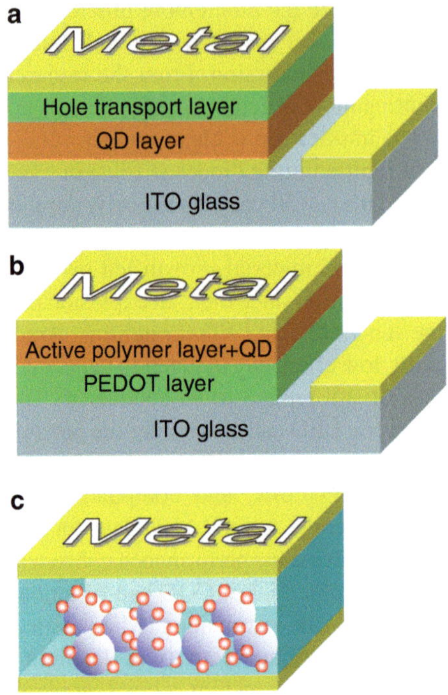

Fig. 3.1 Three types of pure colloidal quantum dot solar cells (**a**) metal–semiconductor junction, (**b**) polymer–semiconductor, and (**c**) semiconductor–semiconductor systems [26]

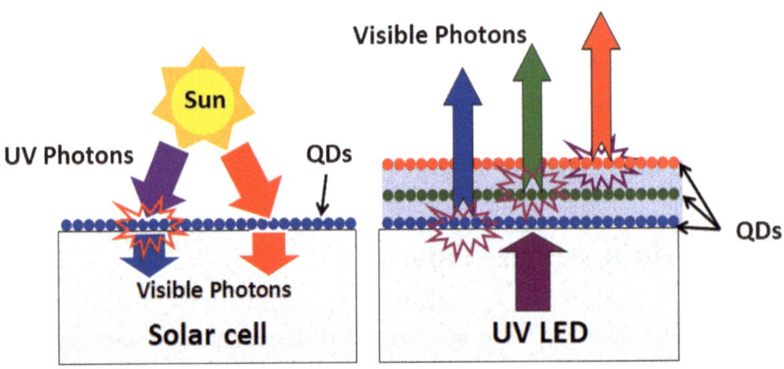

Fig. 3.2 Two basic optoelectronic devices adapting this hybrid integration of CQD layers. On the *left* is the solar cell, and on the *right* is the light-emitting diode

in which one or multiple layers of CQDs are laid onto the surface of the semiconductor devices and used for luminescent-down-shifting the incident photons or can be optically pumped for light emission. In the solar cell case, it is well known that the UV photons cannot be productive in terms of generating electron–hole pairs mainly due to short penetration depth and poor carrier collectivity. If a CQD layer is presented when the UV photons in the sunlight strike the device, the absorbed-then-emitted visible photons are expected from these nanocrystals, while other visible photons can pass through the CQD layer because of their smaller photon energies. These re-emitted visible photons from the CQD layer can be absorbed by the semiconductor layer underneath more efficiently compared to the UV counterparts. Thus we can enhance the overall utilization of the solar spectrum and boost our PCE.

To evaluate this luminescent down-shifting efficiency, however, is not an easy task. The nature of strong absorption of UV photons for both CQD layer and semiconductors complicates the problem. To simplify the situation, we can obtain a rough picture through the short-circuit current measurement. The short-circuit current is a direct evaluation of combined quantum efficiency and solar spectrum as it can be written as [27]:

$$J_{\mathrm{sc}} = \int j_{\mathrm{sc}}(\lambda)d\lambda = \frac{e}{hc}\int \lambda \times \mathrm{EQE}(\lambda) \times I_{\mathrm{AM1.5G}}(\lambda)d\lambda, \tag{3.1}$$

where J_{sc} is the short-circuit current, λ is the photon wavelength, $I_{\mathrm{AM1.5G}}$ is the intensity of the AM1.5G solar spectrum, and EQE(λ) is the wavelength-dependent external quantum efficiency (EQE) of the solar cell. So if the wavelength-dependent photocurrent can be measured, the enhancement factor of the photocurrent can be defined as:

$$\begin{aligned}
\text{Enhancement} &= \frac{\text{increase in the UV region of EQE}}{\text{overall increase in the EQE}} \\
&= \frac{\int_{\mathrm{UV}} j_{\mathrm{sc}}(\lambda)d\lambda\Big|_{\mathrm{CQD}} - \int_{\mathrm{UV}} j_{\mathrm{sc}}(\lambda)d\lambda\Big|_{\mathrm{no_CQD}}}{\int_{\mathrm{All}} j_{\mathrm{sc}}(\lambda)d\lambda\Big|_{\mathrm{CQD}} - \int_{\mathrm{All}} j_{\mathrm{sc}}(\lambda)d\lambda\Big|_{\mathrm{no_CQD}}} \\
&= \frac{\Delta\int_{\mathrm{UV}} \lambda\times\mathrm{EQE}(\lambda)\times I_{\mathrm{AM1.5G}}(\lambda)d\lambda}{\Delta\int_{\mathrm{All}} \lambda\times\mathrm{EQE}(\lambda)\times I_{\mathrm{AM1.5G}}(\lambda)d\lambda}
\end{aligned} \tag{3.2}$$

The assumption in Eq. (3.2) is that all the enhancement in UV region comes from the re-emitted photons of the CQD layer. This assumption can be true if the quantum yield of the dot is generally high, and all the UV photons are taken by the dots. If there are significant numbers of UV photons penetrate into GaAs, which might be the case if CQD is not efficient, the Eq. (3.2) will fail, and a more general discussion based on multiple layer structure with absorption and emission properties needs to be considered.

It is also important to choose the band gap of the CQD layer incorporated in our photovoltaic devices. The efficiency of the device could be influenced according to the type of nanocrystals on its surface. The quantum dot we pick for solar cell is CdS with 460 nm emission. The reason is as follows:

1. The absorption of the nanocrystals cannot mask the visible solar spectrum too much. This is due to a simple consideration: the conversion efficiency of nanocrystals is not as high as semiconductors, and the visible photons

(400–800 nm) usually generate most of electron–hole pairs in a solar cell. So the more we replace the functionality of semiconductor in the visible spectrum, the worse the overall PCE will be.

2. The emission wavelength of the nanocrystals is also important. If the emission wavelength is too short (like <400 nm), the re-emitted photons cannot be absorbed effectively due to the same reason to the UV solar photons: short penetration depth and poor carrier collection.

Based on these reasons, we believe using 460 nm CdS quantum dots should be a good choice because its absorption edge rises above 450 nm where a generic solar cell has lower quantum efficiency, and its emission of 460 nm photon can have about 30–40 % quantum efficiency in a regular solar cell.

In the light-emitter case, the CQD layers provide a wider choice for the color mixing with smaller footprint when compared to popular phosphors. The light emitted from CQD can have a narrower linewidth (meaning higher color purity and higher color rendering index [28]), and the color can be changed by the fabrication processes [29]. These features are not easy for phosphor particles to achieve. With a regular UV LED underneath, one or multiple layers of CQDs with different or the same sizes can be pumped simultaneously and the mixed color can be demonstrated. If the one-pass pumping is not good enough, a specialized mirror with high reflectivity in UV region and low in visible region can be designed and inserted into this scheme [30].

From this section, readers could realize that our idea is different in the way that the quantum dot layer plays a supporting role to augment the performances of the traditional optoelectronic devices. The traditional semiconductor devices enjoyed a long history of fruitful research and they are more reliable in many ways compared to the nanocrystals. However, as we mentioned previously, their inherent and fixed band gaps might be a barrier that hinders further improvement of device efficiency. Putting nanocrystal layers onto traditional devices can first expand the working bandwidth of the devices (due to the addition of different band gap materials) and second solve the lifetime or reliability issues of nanocrystals by simply replacing these removable layers whenever they are degraded. By this method, we can compensate for weaknesses of each material.

3.3 CQDs Preparation and Dispensing

Thanks to arduous research works in the past decades, the CQDs can be purchased readily from various vendors. The size of these nanocrystal can be modified by the fabrication processes and the product we used here is with an average size of 3.58 ± 0.22 nm in diameter for the 460 nm emission wavelength, 4.16 ± 0.21 nm for green and 5.41 ± 0.34 nm for red, respectively. The absorption and photoluminescence can be performed to characterize each types of the quantum dots as shown in Fig. 3.3. While three colors of red, green, and blue are all needed for

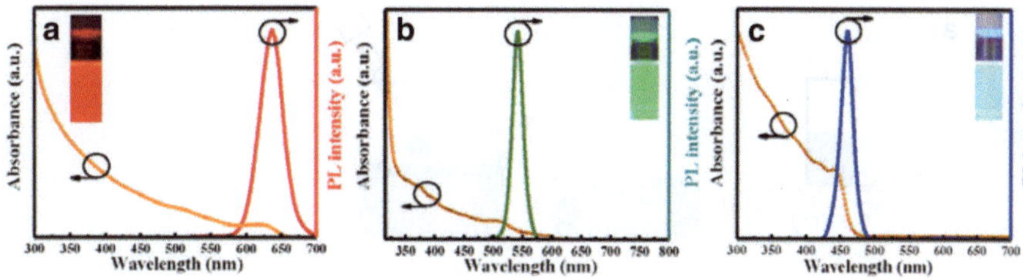

Fig. 3.3 Absorbance and photoluminescence intensity profiles for three different types of CQDs: (**a**) *red*, (**b**) *green*, and (**c**) *blue* [30]

light-emitting devices, their emission wavelength can be decisive for efficiency enhancement in the photovoltaic devices. In this section, we will discuss primarily the dispense method for CQD layers. The CQDs are usually dissolved in solutions like water or toluene. To use them in the solid form, certain dispense technologies needs to be developed.

The dispense of the quantum dot materials onto semiconductor surfaces can have two different categories: manual and automatic methods as shown in Fig. 3.4. Manual dispense is quick and easy to start with, and a precision scale is necessary to determine the amount of quantum dot materials about to be spread. Then a pipette for regular chemical disposal can be used for putting the materials. It is nature then to follow the regular drip-and-spin process in a spin-coater to finish the dispense. However extra cares need to be taken to obtain a uniform layer thickness, and different quantum dots might have different adhesion to the material surfaces.

The automatic method usually relies on the computerized machine to achieve the uniform dispense. Several technologies have been developed such as: mist coating, inkjet printing, reel-to-reel printing, and pulsed spray (PS) coating. [30–35]. Although promising results were obtained, issues such as cross-contamination between colors and the inefficient utilization of raw QD materials occur among various deposition technologies [31–34]. Because of the design of the deposition chamber in the mist coating method, the actual deposited volume of active material (such as quantum dots) is <10 % of the overall disposal, which leads to a waste of material [34]. Another inevitable disadvantage is the so-called coffee ring effect, which is caused by migration of the solutes toward the edge of the droplets during drying [33, 36]. The inkjet printing method has been demonstrated as a feasible method of QD device fabrication in the deposition of the substrate because of its simplicity and low cost; however, it produces nonuniform surfaces and inaccurate patterning [37, 38]. The separation of each pattern and the uniformity of RGB colors are critical to obtain successful CQD hybrid optoelectronic devices in industrial applications.

The optimal solution to these problems is the pulsed spray (PS) coating method, which can be used with an interval control through an air atomizing nozzle to control the thickness of the QD layer by layer [39]. The cross-contamination in red–green–blue (RGB) colors can be alleviated either using polydimethylsiloxane (PDMS) as the interface layer or use shadow mask. PDMS has a high degree of transparency, stability, and flexibility, which is suitable for the substrate of various

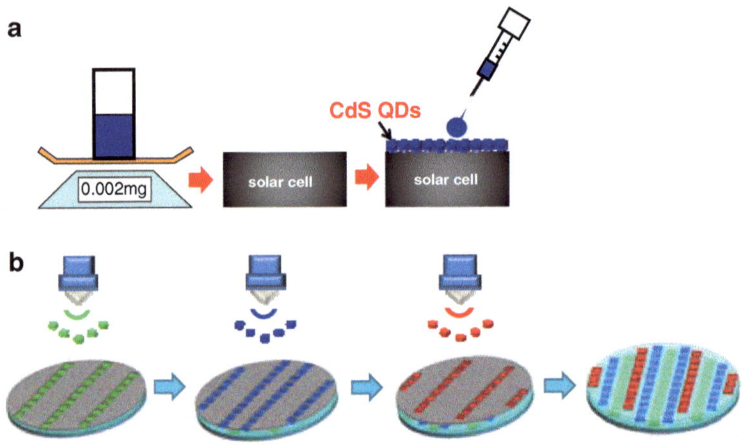

Fig. 3.4 Two different methods for CQD dispense on the device: (**a**) manual dispense by pipette (**b**) pulsed spray method

colors of CQDs and various applications. For the device mentioned in this chapter, we use pulsed spray coating primarily.

In traditional spraying techniques, the viscosity of the spraying mixtures can considerably affect the uniformity of the finished film. The interaction between the particles in the premixed solution can cause a gathering of the materials, and self-clustering can block the passage of the spray. The PS method can improve this issue by using two special designs: the air-injection mechanism in the nozzle and intermittent spraying frequency (5–10 Hz), and the constant stirring system. These methods can separate the target particles in the suspending solution more effectively than traditional methods. Thus, the atomized mixture of quantum dots and solvent can pass through a tiny nozzle to reach the desired surface, further reducing the chances of quantum-dot self-assembly. The actual operation was precisely controlled by a computer system, and the quantity of the spray can be monitored.

When the PS method is employed, many different procedures can be set up depending on applications. The blanket dispense without any mask is the easiest way which can generate a large area, uniform coating. In the general display application, however, an array of RGB pixels is more desirable. Figure 3.4b shows the process flowcharts of the pixelated RGB QD arrays on a 2-in. substrate. The concentrations of the RGB QD can be changed according to applications. The pixelated arrays were fabricated using the following steps: (1) host substrate preparation; (2) the mask was placed on the top of the substrate for alignment; and (3) the RGB QDs were sprayed onto the surface of the glass in the sequence of green, blue, and red. Similar steps can be followed for a 2-in. wafer white light source; however, no mask alignment is used.

The mask in our experiment was made of an aluminum plate, and an array of square (2 mm × 2 mm) holes were opened for patterning. The separation of the each pixel was 5.22 mm. The alignment between the various colors can be achieved by visual inspection and XY translational stage adjustment. In the pulsed spray procedure, the air pressure of the gun was approximately 0.5 MPa, and the pressures of the atomized air and swirl air were approximately 0.2 MPa.

The separation of the different color pattern formed by nanocrystals can be achieved either by the shadow mask in the horizontal plane or by the PDMS film vertically. The PDMS pre-polymer solution was poured over the glass substrate using the spin-coating method at 700 rpm for 60 s and baked in an oven at 70 °C. Subsequently, the PDMS film was ready for peeling. The thickness of the PDMS layer measured using SEM was approximately 80 μm. Because each PDMS layer was processed using the same parameters, we ensured that each layer in our structure had the same thickness. The thickness of the QD layer was difficult to measure because of the considerable thickness difference between the PDMS and the QD layer; only the TEM technique was used when they were manually dropped and dried. From the TEM image, we estimated the QD layer thickness at approximately 50–70 nm.

One advantage of the PS method is the uniformity of the CQD film. After PS QD deposition, we measured the AFM images of the surface showing the aggregation and clusters after layer-by-layer deposition. The AFM measurement of the blue, green, and red QD layer revealed a root-mean-square (RMS) roughness of 1.18, 0.816, and 0.788 nm, respectively. The RMS results indicate that uniform dispensing is possible with this technology.

3.3.1 General Characterization of a Hybrid CQD Optoelectronic Device

In this section, we will describe the basic procedure to characterize a hybrid device. Several measurements need to be performed to fully characterize a hybrid-type solar cell. The general *IV* under solar simulator and EQE measurement are necessary to find out the basic performance. For light-emitting device, the emission spectrum and intensity need to be taken, so an integrating sphere to collect photons is essential.

The power conversion efficiencies of solar cells were measured using a procedure complying with the international standard CEI IEC 60904-1. A 1,000 W/m^2, class-A Air Mass 1.5, Global (AM1.5G) light source, consisting of Newport 91192A, Newport 6271A (the Xeon lamp), and an AM1.5G filter (Newport 81088A) was used to characterize the devices [40], and the substrate temperature was kept at 25 ± 1 °C. In addition to the solar simulator, the measurement system also contains a power supply (Newport 69920), a probe stage, and a source meter with a four-wire mode (Keithley 2400). A calibration report by Newport Corporation shows that the temporal instability was 0.88 % and nonuniformity was 0.79 %. The spectrum of the solar simulator was measured using a calibrated spectrometer

(Soma S-2440) at wavelengths ranging from 300 to 1,100 nm. Before measurement, the intensity of the solar simulator was calibrated by a monocrystalline silicon reference cell with a 4-cm^2 illumination area (VLSI Standards, Inc.) [41].

For the EQE measurement, we employed a 300 W Xenon lamp (Newport 66984) and a monochromator (Newport 74112) as the light source. The beam spot at the sample was rectangular, and the spot size was roughly 3 mm^2. Calibration was conducted using a calibrated silicon photodetector with a known spectral response (Newport 818-UV). The EQE measurement was taken using a lock-in amplifier (Standard Research System, SR830), an optical chopper unit (SR540) operating at a 260-Hz chopping frequency, and a 1 Ω resistor in a shunt connection to convert the photocurrent to voltage. The temperature of the cells was actively controlled using the same method in the *IV* measurement [42, 43].

Photoluminescence (PL) and absorption spectra were used to understand the radiative characteristics of CdS QDs (Fig. 3.3). The PL emission spectrum of CdS QDs in toluene was measured using a 365-nm excited wavelength and showed two minor peaks at 460 and 600 nm. The absorbance spectrum shows an upward trend from 400 nm and beyond. The CdS QDs can absorb photons in the UV region and then re-emit the visible photons (in this case, they are blue). Each CdS QD has a diameter of approximately 5 nm and exhibits a uniform density distribution.

For the light-emitting devices, the spectrum of the emission, the analysis of the color temperature, and brightness of the luminance can be measured by a large integrating sphere. In our setup, the integrating sphere is coated OptoWhite (95 %)/ Zenith (97 %) to assure light scattering uniformly in the sphere. Therefore, we can assure that everywhere in the sphere should have the same power intensity. The total intensity could be calculated by integrating the intensity of whole area in the sphere. To conduct the electrical-optical measurement on LEDs, our samples can be located on a stage connects with a current/voltage source(Keithley 2400). When we put the sample into the sphere, the sphere will not be the ideal scattering sphere. Therefore, an auxiliary lamp is employed to better calibrate the overall measurement.

The characterization setup for the optoelectronic devices is important as we have to maintain both electrical and optical precision to obtain credible results. Many equipments need to be calibrate before the actual device are mounted and regular checkup is also necessary to keep the system for day-to-day operation.

3.3.2 CdS Sensitized Silicon Nanopillar Solar Cells

From this section on, we will introduce several examples of hybrid CQD devices which demonstrate the enhanced performance and the superiority of the design. First, we can use it on a silicon platform with nanopillar arrays (NPAs). To form these nano-scale structures, we use colloidal lithography method [44–48] to fabricate c-Si NPA solar cell. As shown in Fig. 3.5, the processes can be illustrated as follows: first, a monolayer of self-assembled 600 nm polystyrene (PS) nanospheres

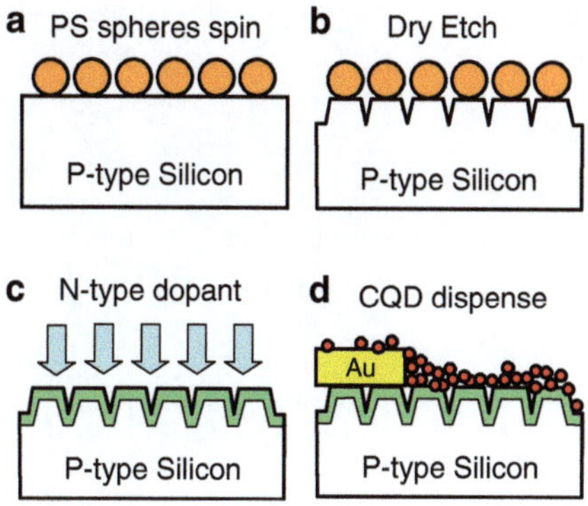

Fig. 3.5 The schematic fabrication flow for c-Si nanopillar arrays (NPA) solar cell with CdS quantum dots (QDs) layers: (**a**) spin-cast of nearly-close-packed polystyrene nanospheres monolayer, (**b**) resulting NPA after dry etching, (**c**) diffusion of an N-type layer. (**d**) SiN_x deposition and screen printing of the front and back electrodes, and then spin CdS QDs layers on the cell [18]

was spin-cast onto the surface of a p-type c-Si substrate. A nearly-close-packed 600 nm PS nanospheres monolayer was formed, and then the sample was etched by an inductively coupled plasma reactive ion etching (ICP-RIE) system, where a reactive gas Cl_2 with 1,000 W of RF power was introduced to dry etch the PS spheres and the silicon substrate under a 5 mTorr pressure at 60 °C. The sample was brushed and dipped into a pure acetone solution for 5 min to clean up the remaining PS spheres. Next, the NPA top 200 nm material was doped using phosphorus oxychloride ($POCl_3$) as the diffusion source to form an n-type layer. After the removal of the phosphosilicate glass (PSG) layer from the substrate surface with diluted hydrofluoric (HF) acid, an 80-nm-thick silicon nitride (SiN_x) layer was deposited to passivate the surface by using plasma-enhanced chemical vapor deposition (PECVD). Then the front and back metal contacts were screen-printed and a co-firing in a furnace system at 850 °C. Finally, a colloidal CdS QDs with a concentration of 5 mg/mL was spun on the surface of c-Si NPA solar cell, as shown in Fig. 3.5d. For comparison, the one without the CdS QDs layers was also fabricated simultaneously as the reference.

In Fig. 3.6a, we show the scanning electron microscopic (SEM) image of c-Si NPA in which the periodic close-packed pattern is accurately transferred onto the Si substrate after RIE. The height and shape of NPA can be controlled by varying the etching duration and the NPA in our process has the average height of 550 nm. The cross-section TEM was taken and the element of CdS QDs layers was confirmed by energy dispersive spectrometer (EDS).

The CQD type is CdS nanocrystal in this case, and the PL and absorption spectrum is the same as Fig. 3.3a. The absorbance spectrum has a sharp rising edge around 450 nm and then gradually increases from 375 nm towards 330 nm. It is obvious that CdS QDs exhibit photon down-conversion from ultraviolet to visible light (from ~330 to 550 nm). This converted light can provide more photons to be harvested in c-Si NPA solar cell. Figure 3.7 shows reflectance spectra by an

Fig. 3.6 A cross-sectional view of scanning electron microscopic (SEM) image of the c-Si nanopillar arrays [18]

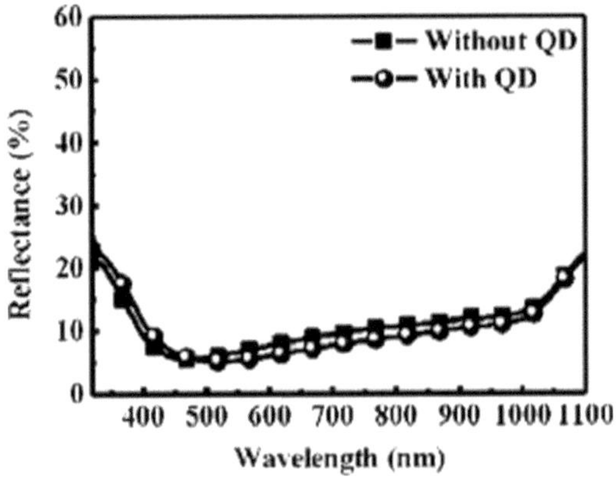

Fig. 3.7 The measured reflectance spectra for cells with c-Si nanopillar arrays (NPA) with and without CdS quantum dots (QDs) layers

integrating sphere at the normal incidence of light. The c-Si NPA with CdS QDs layers exhibited superior anti-reflective properties for the wavelength range of interest, especially in the range of wavelength from 450 to 1,100 nm, compared with the cell without CdS QDs layers. It is found that the CdS QDs layers provide a better antireflection layer for light harvest.

The photovoltaic *I–V* characteristics were measured under air mass 1.5 global illumination and room temperature conditions as mentioned in the previous section. Figure 3.8 shows the photovoltaic *I–V* characteristics of the c-Si NPA solar cell with and without CdS QDs layers. The device characteristics were summarized in Table 3.1. The short-circuit current density (J_{sc}) of the c-Si NPA solar cell with CdS QDs layers can reach 32.03 mA/cm^2, meanwhile the PCE is increased from 9.45 to 12.57 %, corresponding to a 33 % enhancement compared with the reference. It is also noticeable that open-circuit voltage (V_{oc}) shows negligible change. Additionally, the fill-factor (FF) is improved, which is an indicator of better series resistance of the device. To further investigate the reason behind this FF improvement, we measure the resistance between individual metal contacts with and without CdS QDs layers under a UV light irradiation of 355 nm wavelength. The change brought

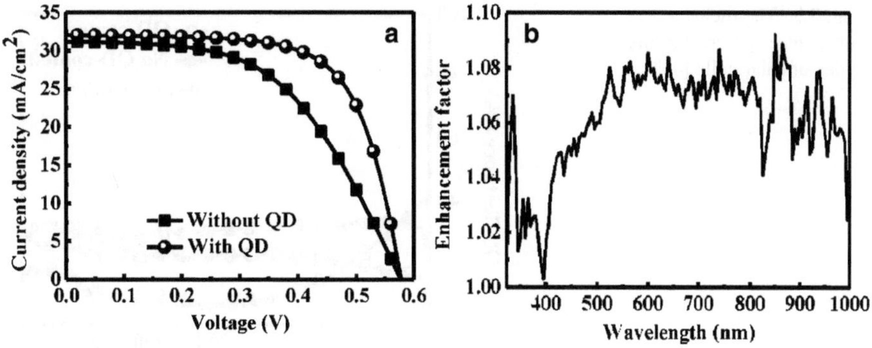

Fig. 3.8 (**a**) Photovoltaic *I–V* characteristics of the c-Si nanopillar arrays (NPA) solar cell with and without CdS quantum dots (QDs) layers. (**b**) The enhancement ratio of external quantum efficiency between QD device and no QD one

Table 3.1 Current–voltage characteristics of c-Si NPA solar cells with and without CdS QD layers

Type	V_{oc} (V)	J_{sc} (mA/cm^2)	FF (%)	η (%)
With CdS	0.57	32.03	67.9	12.57
Without CdS	0.57	31.17	52.6	9.45

by the QD coating is significant: the resistances dropped from 43.4 to 29.3 Ω. This 32 % reduction in electrical resistance should be the direct cause of our FF improvement. From the photovoltaic *IV* measurement, we can see the enhancement of the solar cell with QDs layers.

However, it is still unclear whether this improvement is from converted photons or from other factors. To investigate further, the spectrum of EQE was taken and the enhancement ratio was calculated between the QD device and no-QD device ($EQE_{QD}/EQE_{no\ QD}$). As shown in Fig. 3.8b, the enhancement is, to our disappointment, mainly from longer wavelength, which means the anti-reflective effect took over most of the improvement of performance. Nevertheless, the luminescent down-shifting effect did contribute about 1 % of the short circuit current increase and the spike in the UV wavelength of the enhancement ratio is encouraging.

3.4 CdS Sensitized GaAs Solar Cell

After the silicon solar cell experiment, several drawbacks of the design were discussed. As one could observe from the previous section, the actual enhancement in the UV region is quite minimal. We thought the LDS effect might be masked by the strong anti-reflection of NPAs. Also it is possible that the silicon is not very effective to take the re-emitted blue photons. To move forward, we choose a flat surface structure and a direct band gap material (GaAs) as our host. The same CQD

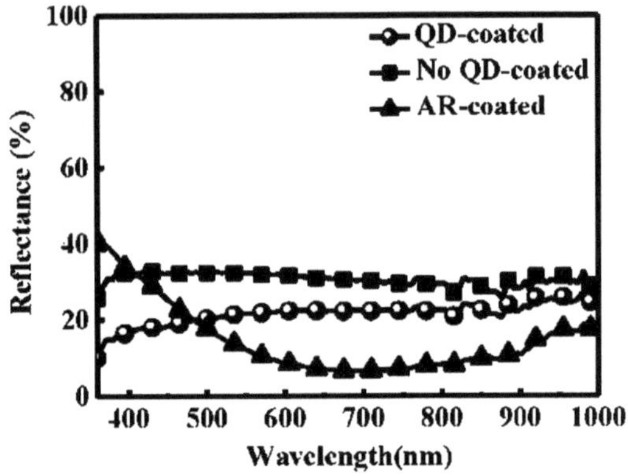

Fig. 3.9 The measured reflectance spectra for three types of solar cells [49]

with a major emission peak at 460 nm was applied in all experiment. In absorbance spectrum, a sharp rising edge was detected around 450 nm and a steady increase from 400 nm on. The PL and absorption of this CQD is described in Sect. 3 of this chapter.

Three groups of samples with different surface coatings were prepared: bare GaAs surface, anti-reflective coated surface, and CQD-coated surface. Taking reflectance spectrum among three groups, very distinctive features can be found across the whole spectrum range as shown in Fig. 3.9. When a CQD layer was applied to GaAs surface, even as thin as several tens of nanometer, due to its intermediate refractive index ($n \sim 2.4$), we could see an global reduction in reflectance spectrum compared to bare semiconductor surface (as in Fig. 3.9). The CQD layer is certainly not as good as the regular anti-reflective (AR) coating in longer wavelength but its low reflectance in shorter wavelength is impressive. Meanwhile the AR coating exhibits traditional feature of the minimal reflectance around the designated wavelength (700 nm in our case) and a larger reflection is expected in the UV region.

Next, we characterize the basic performance of these devices through photovoltaic *I–V* curves and EQE measurement by the same steps as in the previous section. The results of three different types of GaAs solar cells were shown in Fig. 3.10. The important device parameters are summarized in Table 3.2. The short-circuit current density (J_{sc}) of the CQD-coated sample can reach 26.99 mA/cm^2 and its PCE increases from 17.71 to 21.06 %, corresponding to a 18.9 % enhancement compared to the device without AR coating. When compared with the optimal AR-coated device, the CQD-coated sample still edges out by 4.1 % increase in J_{sc} and 2.7 % enhancement in overall PCE. Additionally, open-circuit voltages (V_{oc}) of these devices show no degradation, and the change in fill-factor (FF) is negligible.

To further understand the utilization efficiency of photons with different wavelengths, we measured the spectral response of the EQE, in which a xenon light source is used as the illumination source. Figure 3.11a shows the EQE as function of illumination wavelength for all three types of samples, and Fig. 3.11b shows the EQE enhancement factor of the CQD-coated sample, compared to bare

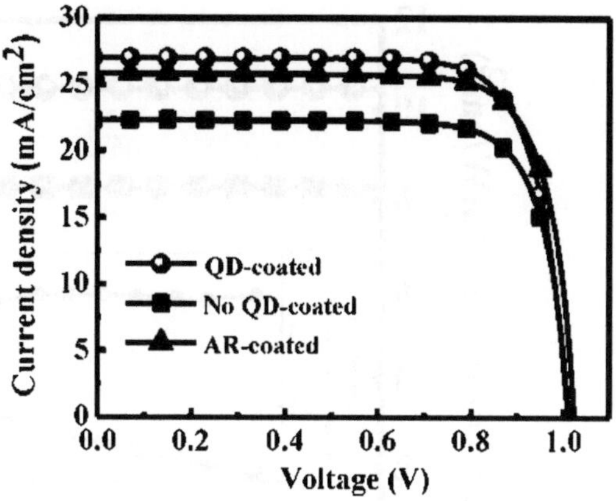

Fig. 3.10 Photovoltaic *I–V* characteristics of three types of solar cells [49]

Table 3.2 A summarized current–voltage characteristics of GaAs + CQD, GaAs only (No CQD), GaAs + AR coating solar cells

Type	V_{oc} (V)	J_{sc} (mA/cm^2)	FF (%)	η (%)
GaAs w/o AR + CdS QD	1.01	26.99	77.24	21.06
GaAs w/o AR	1.01	22.31	78.49	17.71
GaAs with AR	1.02	25.92	77.58	20.51

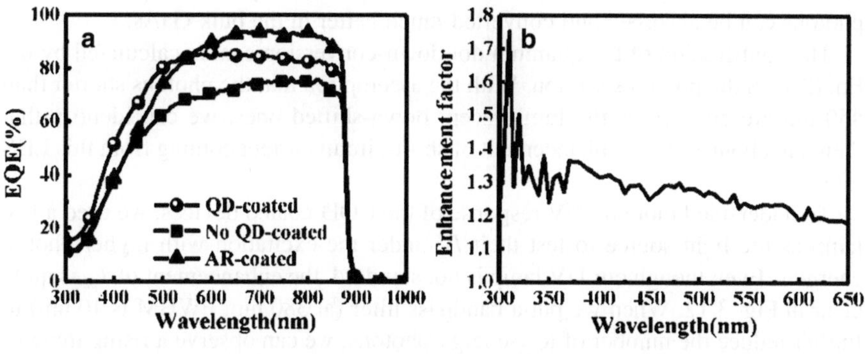

Fig. 3.11 (**a**) Measurement of external quantum efficiency of GaAs + CQD, GaAs only (No CQD), and GaAs + AR coating solar cells. (**b**) Enhancement of EQE between GaAs + CQD and GaAs only device. Peak at ~310 nm indicates photon down-conversion [49]

GaAs one. From EQE spectrum, we could see the universal enhancement of CQD-coated sample against the reference, and the curves of CQD-coated sample and AR-coated one crisscross at 554 nm. While the differences of EQE between CQD-coated and non-CQD-coated GaAs devices maximize at 520 nm range, the ratio grows rapidly once moved into ultraviolet regime ($\lambda < 350$ nm) and can be as high as 1.75 times. This result demonstrates an efficient luminescent down-shift of photons can subsidize the UV-generated carrier collection probability. In this

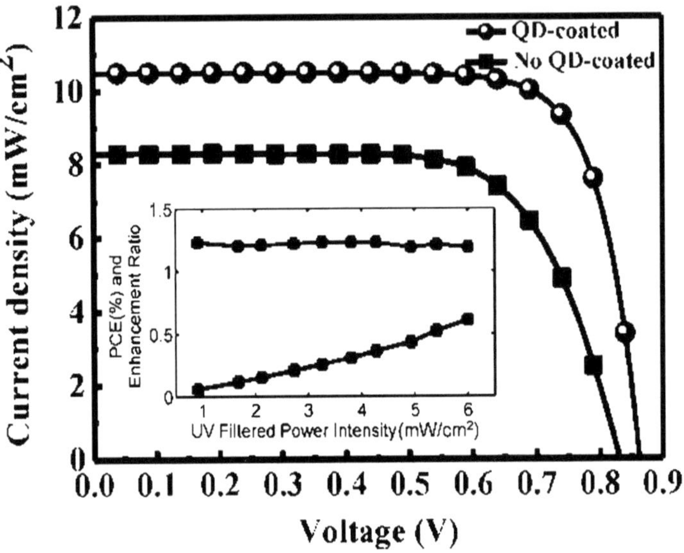

Fig. 3.12 *IV* measurement with UV light. The *inset* is the filtered UV PCE of the CQD/GaAs cell (*bottom line with a slope*) and the efficiency enhancement ($\eta_{CQD}/\eta_{no\ CQD}$, *top line*) [49]

scenario, the UV photons can be absorbed by CQD layer and re-emitted blue photons can be absorbed and converted much better in the bulk GaAs.

The contribution of the quantum dot down-conversion can be calculated by the Eq. (3.1) in the previous section. With the assumption that the photons shorter than 450 nm are counted as the luminescent down-shifted ones, we can identify that there are about 11 % of enhancement of short-circuit current coming from this LDS effect.

To understand more on UV response of the CQD-coated devices, we used a UV lamp as the light source to test their *IV* under the excitation with higher photon energies. Even though our UV lamp is not standard, the enhancement of J_{sc} is quite clear in Fig. 3.12. When we put a bandpass filter (at 380 nm, FWHM is 10 nm) to further reduce the number of low-energy photons, we can observe a rising trend of efficiency versus filtered power shown in Fig. 3.12, and a constant efficiency enhancement factor of 1.22 times to the no-CQD sample. Another interesting feature that is not seen in previous shown good device is the improvement of the fill factor (FF). For a device with a bad FF as in Fig. 3.12, adding CQD layer on its semiconductor surface modifies its photoconductivity dramatically [50] and the FF becomes much better. The increment can be as large as 30 % (from 57 to 74 %). The conductive atomic force microscopy (CAFM) can be used to verify this phenomenon. In the normal CAFM measurement, a fixed voltage (5 V) is applied on the tip of the probe, and the current is drawn from the surface atoms. We added an UV lamp during the measurement and with the lamp going on and off, we could observe the shift of surface current distribution as in Fig. 3.13. The increase of current due to UV excitation can be found as high as 76 %, if we compare the mean detected

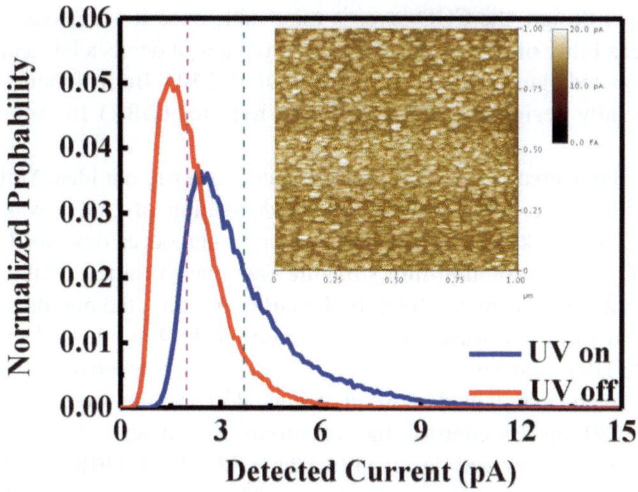

Fig. 3.13 The distribution of detected currents of QD on GaAs surface with and without UV lamp on. The *pink* and *blue vertical lines* indicate the average currents from the measurement. When UV lamp is off, the average current is 2.10 pA. If we turn on the UV lamp, the average jump up to 3.69 pA (an 76 % increase). The *inset* is the conductive AFM scanning plot [49]

current of each group. This is a clear indication of surface charges activation during the UV illumination.

From these experiments, we could see the introduction of CdS quantum dots on semiconductor surface brings several advantages: first is the aforementioned luminescent down-shifting and anti-reflection; second is the population increase of electronic carriers at the surface because of UV excitation. Both effects can greatly improve the EQE (optical) and fill factor (electrical) of our traditional GaAs-based solar cells. One of the benefits that we did not examine in this experiment is the removable feature of these quantum dot layers. When the quantum dots are degraded because of exposure to air, we can wash them out by proper solvent and re-coat the device with another fresh quantum dot material. With this way, the device can be relieved of the reliability concerns that usually accompany with these quantum dots.

3.5 Applications on LEDs

Other than solar cell, the CQD layer can also be employed for light-emitting devices. The narrow linewidth (<10 nm) can increase the color purity and variable band gap engineered by fabrication can be of great use for display purpose. As we mentioned before, the pure CQD LED still suffers from poor carrier injection and lifetime restriction. If a GaN-based UV LED is used for optical pumping, we can eliminate these two problems together: there won't be current through CQD layer and by proper design, the CQD layer is removable once it is decayed. To demonstrate the feasibility of this design, we have two types of devices fabricated: a large-area CQD-coated glass and an integrated CQD/UV LED. Both of them were loaded with a specially designed distributed Bragg reflector (DBR) for better pumping efficiency.

A simple large area sample was first fabricated to verify our idea. Multiple layers of PDMS film were made separately, and deposition of CQDs was performed between PDMS stacking layers by pulsed spray method as described before. An 11-pair HfO_2/SiO_2 DBR thin-film structure was placed on top of the sample for resonant cavity enhancement effect. In the DBR structure, the maximal reflectivity was designed for a UV wavelength of approximately 400 nm, and the refractive index of the HfO_2 and SiO_2 layers were 1.9 and 1.46, respectively. Over 90 % of reflectivity can be maintained between 365 and 430 nm; the wavelength of the UV source was 380 nm. By contrast, the reflectivity was lower in the visible spectral range. The mirror stack was designed to enable SiO_2/HfO_2 DBR to reflect most of the UV light and transmit the visible photons. Therefore, the reflected UV photons excited more RGB QDs and increased the efficiency of pumping.

Figure 3.14a shows the relative intensity of the large-area samples with DBR under a driving current of 350 mA in a UV pumping LED, compared with the reference without DBR. The result demonstrates that the large-area samples with DBR have higher intensity in red, blue, and green components than the reference without DBR. The reflection of light from the DBR structure can enhance the efficiency because the increased light path leads to a higher possibility of exciting the RGB QDs. The CIE color coordinates of the large-area samples with DBR, which provide a white light output, are (0.29, 0.29) shown in the inset of Fig. 3.14a. To further verify the pumping power dependence of each color, the driving current of the UV LED was varied from 100 to 400 mA. The enhancement of each color from the DBR sample over the non-DBR sample is shown in Fig. 3.14b. A saturation of this enhancement ratio was observed among red, green, and blue QD emissions beyond 250 mA, which indicates a possible quench in the quantum yield of the quantum dots. Additionally, the enhancement ratio of red emission was higher than that of the other colors, which can be attributed to the proximity of red QD layers to the UV source. Although the transparency in PDMS is high, slight UV light loss occurs in each PDMS layer. However, PDMS performs a crucial role in separating the RGB QDs in our structure and can further isolate the air from RGB QDs, which can stop the intensity degradation in RGB QDs. The green QDs exhibited a weaker enhancement in general. This may have been caused by the substantial aggregation of green QDs when they were sprayed and dried on the substrate. When dispersed and dried on a glass substrate, green QDs exhibited pronounced PL depression, whereas red and blue QDs retained comparable PL intensities, as shown in Table 3.3. This phenomenon indicates that green QDs experienced considerable aggregation during the solidification process, which is the main cause for the lower enhancement ratio observed for the LED. Further information can be acquired from the absorption spectrum. A smaller absorption of the green dot was observed from the direct measurement of reflection and

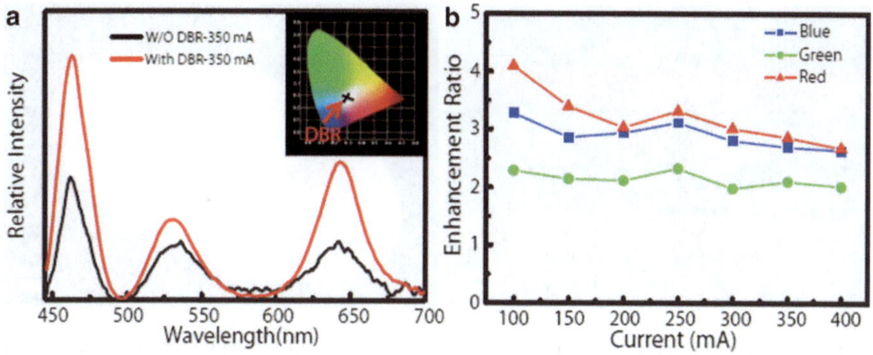

Fig. 3.14 (**a**) The relative intensity of the large-area samples with and without DBR operated under 350 mA. The *inset* shows the CIE color coordinates of the large-area samples with DBR. (**b**) Enhancement ratio of the intensity of large-area samples with and without DBR under various diode currents from 100 to 400 mA [30]

Table 3.3 The comparison between the luminous intensity of each RGB CQD colors versus the measured absorption percentage on glass

	Blue	Green	Red
Luminous intensity (a.u.)	100	11	60
Absorption at 380 nm (%)	8.8	2.3	4.7

transmission on solidified quantum dots on glass (Table 3.3), which also indicates a possible decrease in UV excitation efficiency.

The chromaticity coordinate of the large-area samples with DBR remained almost the same with the increase of the driving current, which indicates that the large-area sample with the DBR structure has high stability in color rendering, and this characteristic is crucial in solid-state lighting. With this pump-power-independent color mixing, our PDMS-layered structure can provide a tunable color-rendering scheme by adjusting the quantities of the RGB QDs of each layer. Additionally, the CIE coordinates of the large-area samples with DBR were compared to the National Television System Committee (NTSC) standard color triangle. The RGB color coordinates for the large-area samples with DBR were (0.69, 0.3), (0.19, 0.75), and (0.13, 0.05). Therefore, the area of the RGB triangle of the large-area samples with DBR was enhanced by 20 %, compared to the NTSC color gamut. This result can be attributed mainly to the narrow bandwidth in the RGB QDs, which have high color purities.

The images of the pixelated structure and 2-in. full-wafer sample under UV excitation are shown in Fig. 3.15. The size of each pixel was 2 mm × 2 mm, as shown in Fig. 3.15a, which can be scaled up and down according to the application. Figure 3.15b shows the full-size white light source under UV excitation.

After finishing the large area test, we further shrank the scale to an actual LED package. Figure 3.16a shows the schematic diagram of QDLED and a picture of a real QDLED in operation. Figure 3.16b shows the electroluminescence (EL) spectra

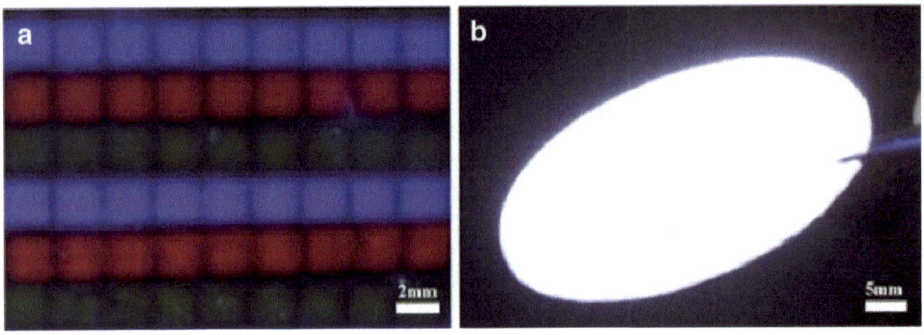

Fig. 3.15 Image of the (**a**) pixel pattern and (**b**) a 2-in. full wafer structure under UV excitation [30]

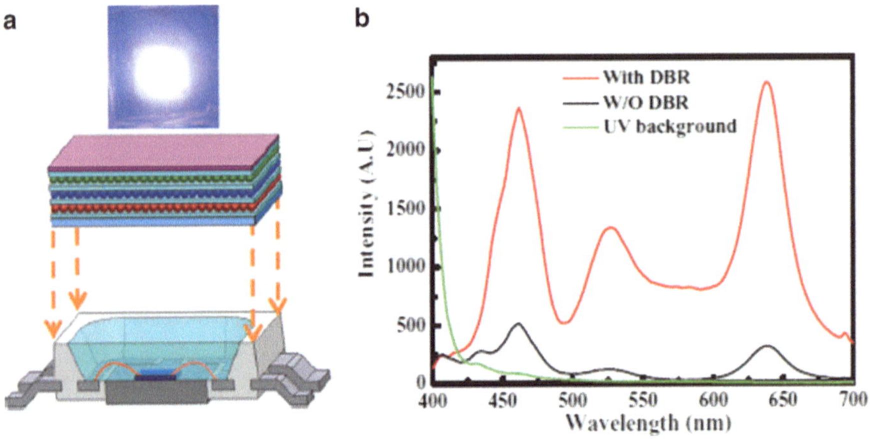

Fig. 3.16 (**a**) *Top*: the QDLED device in action; *bottom*: the device schematic diagram. (**b**) EL spectra with and without the DBR structure [30]

of our QDLED with and without the DBR structure. The background of pure UV LED emission is also shown in the figure. Three emission peaks occurred at 460 nm (blue band), 530 nm (green band), and 640 nm (red band), which were contributed by the RGB QDs. This indicated that EL with the DBR has stronger visible emission than that without the DBR structure.

The normalized EL spectra of UV chips with the DBR structure from 50 to 350 mA are shown in Fig. 3.17a. Higher RGB emission can be obtained by increasing the current, thereby achieving higher intensity. The chromaticity coordinates at various drive currents are shown in the inset of Fig. 3.17a and are marked by a red trace on the CIE map. Compared to our large-area installation, which exhibited almost no change in CCT, the integrated LED device demonstrated higher current dependence, which may have been caused by the thermal conduction problem inside the package. The luminous efficacy of radiation (LER) was calculated using the following formula [51, 52]:

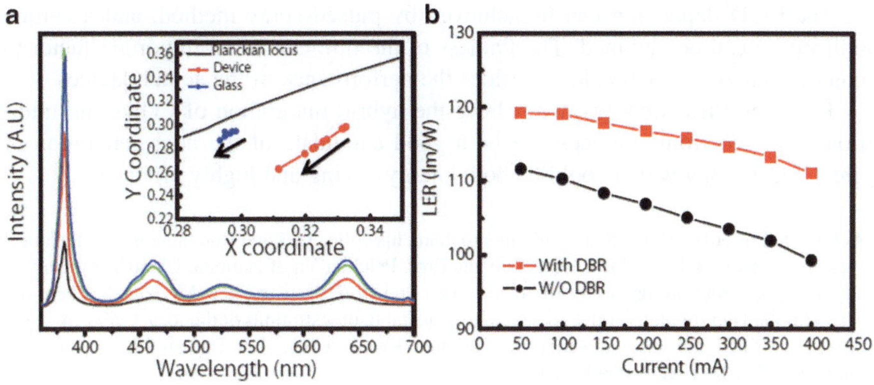

Fig. 3.17 (**a**) The relative emission spectrum of the QDLED with a DBR structure under various drive currents (50, 150, 250, and 350 mA, from *bottom* to *top curves*). The *inset* is the zoomed-in CIE1931 chromaticity diagram with traces of our QD material when dispensed on glass or packaged in the QDLED format under various pumping powers (for glass) or drive currents (for the QDLED device). (**b**) The luminous efficacy of optical radiation of UV chips with a DBR structure under currents between 50 and 400 mA [30]

$$\mathrm{LER} = 683 \frac{\mathrm{lm}}{\mathrm{W}} \frac{\int_\lambda P_{\mathrm{white}}(\lambda)V(\lambda)d\lambda}{\int_\lambda P_{\mathrm{white}}(\lambda)d\lambda} \tag{3.3}$$

where 683 lm/W is a normalization factor, $P_{\mathrm{white}}(\lambda)$ is the spectral power density of the light source, and $V(\lambda)$ is the human eye sensitivity function. The final results of the DBR sample and no-DBR sample are shown in Fig. 3.17b, which shows the superior performance of the DBR sample over all currents. Conversely, if the absolute intensity of emission is graphed against the driving current, the relative intensity rolls over beyond 350 mA because of the UV LED power saturation and the drop in quantum-dot efficiency with increasing temperature inside the package.

3.6 Conclusion

In conclusion, we applied the CQD layer onto traditional optoelectronic devices (including solar cells and integrated LEDs). All of them show significant improvement due to the introduction of CQD layers. In the solar cell case, the PCE can be as high as 33 %, and both silicon and GaAs materials respond to this effect. Further surface analysis indicates strong carrier activation under UV illumination, which confirms our assumption.

For the light-emitting device, large area CQD-coated glass and small footprint devices were fabricated. A good white light source can be obtained with good LER and small fluctuation of color temperature. With the help of UV DBR, the light emission can be greatly enhanced.

The CQD deposition can be achieved by pulsed spray method, and a certain uniformity can be obtained. The flatness of the film can be an important factor to commercialize this technology besides the performance of the actual devices.

Based on these findings, we believe the hybrid integration of CQDs and traditional optoelectronic devices can be a good candidate of the next generation of green technology which could be both energy-saving and highly efficient.

Acknowledgments The author would like to thank his colleagues and students for their technical supports, especially Prof. Hao-Chung Kuo and Prof. Peichen Yu of National Chiao Tung University for equipment sharing and fruitful discussions, and Dr. Hsin-Chu Chen, Mr. Kuo-Ju Chen, and Mr. Hau-Vei Han for their hard works on experimental demonstrations of the idea. C.C. Lin would also like to thank the financial support of National Science Council of Taiwan through the grant number: NSC101-2221-E-009-046-MY3.

References

1. Green MA (2007) Thin-film solar cells: review of materials, technologies and commercial status. J Mater Sci Mater Electron 18(S1):15–19
2. Siebentritt S (2002) Wide gap chalcopyrites: material properties and solar cells. Thin Solid Films 403–404:1–8
3. Klenk R, Klaer J, Scheer R, Lux-Steiner MC, Luck I, Meyer N, Ruhle U (2005) Solar cells based on $CuInS_2$—an overview. Thin Solid Films 480–481:509–514
4. Kongkanand A, Tvrdy K, Takechi K, Kuno M, Kamat PV (2008) Quantum dot solar cells. Tuning photoresponse through size and shape control of CdSe-TiO_2 architecture. J Am Chem Soc 130(12):4007–4015
5. Standridge SD, Schatz GC, Hupp JT (2009) Distance dependence of plasmon-enhanced photocurrent in dyesensitized solar cells. J Am Chem Soc 131(24):8407–8409
6. Zhang Q, Chou TP, Russo B, Jenekhe SA, Cao G (2008) Polydisperse aggregates of ZnO nanocrystallites: a method for energy-conversion-efficiency enhancement in dye-sensitized solar cells. Adv Funct Mater 18(11):1654–1660
7. Tanabe K (2009) A review of ultrahigh efficiency III-V semiconductor compound solar cells: multijunction tandem, lower dimensional, photonic up/down conversion and plasmonic nanometallic structures. Energies 2(3):504–530
8. Baur C, Bett A, Dimroth F, Siefer G, Meusel M, Bensch W, Kostler W, Strobl G (2007) Triple-junction III–V based concentrator solar cells: perspectives and challenges. J Sol Energy Eng 129(3):258–265
9. Takamoto T, Ikeda E, Kurita H, Ohmori M (1997) Over 30% efficient InGaP/GaAs tandem solar cells. Appl Phys Lett 70(3):381–383
10. Guter W, Schone J, Philipps SP, Steiner M, Siefer G, Wekkeli A, Welser E, Oliva E, Bett AW, Dimroth F (2009) Current-matched triple-junction solar cell reaching 41.1% conversion efficiency under concentrated sunlight. Appl Phys Lett 94(22):223504
11. Yu P, Chang CH, Chiu CH, Yang CS, Yu JC, Kuo HC, Hsu SH, Chang YC (2009) Efficiency enhancement of GaAs photovoltaics employing antireflective indium tin oxide nanocolumns. Adv Mater (Deerfield Beach, Fla) 21(16):1618–1621
12. Shockley W, Queisser HJ (1961) Detailed balance limit of efficiency of p-n junction solar cells. J Appl Phys 32(3):510–519
13. Geyer S, Porter VJ, Halpert JE, Mentzel TS, Kastner MA, Bawendi MG (2010) Charge transport in mixed CdSe and CdTe colloidal nanocrystal films. Phys Rev B 82(15):155201

14. Wang X, Koleilat GI, Tang J, Liu H, Kramer IJ, Debnath R, Brzozowski L, Barkhouse DAR, Levina L, Hoogland S, Sargent EH (2011) Tandem colloidal quantum dot solar cells employing a graded recombination layer. Nat Photonics 5(8):480–484
15. Caldwell MM (1979) Plant life and ultraviolet radiation: some perspective in the history of the earth's UV climate. Bioscience 29(9):520–525
16. Sun Q, Wang YA, Li LS, Wang D, Zhu T, Xu J, Yang C, Li Y (2007) Bright, multicoloured light-emitting diodes based on quantum dots. Nat Photonics 1(12):717–722
17. Trupke T, Green MA, Würfel P (2002) Improving solar cell efficiencies by down-conversion of high-energy photons. J Appl Phys 92(3):1668–1674
18. Chen H-C, Lin C-C, Han H-W, Tsai Y-L, Chang C-H, Wang H-W, Tsai M-A, Kuo H-C, Yu P (2011) Enhanced efficiency for c-Si solar cell with nanopillar array via quantum dots layers. Opt Express 19(S5 Suppl 5):A1141–A1147
19. Klampaftis E, Ross D, McIntosh KR, Richards BS (2009) Enhancing the performance of solar cells via luminescent down-shifting of the incident spectrum: a review. Sol Energy Mater Sol Cells 93(8):1182–1194
20. Klampaftis E, Richards BS (2011) Improvement in multi-crystalline silicon solar cell efficiency via addition of luminescent material to EVA encapsulation layer. Prog Photovolt Res Appl 19(3):345–351
21. Strümpel C, McCann M, Beaucarne G, Arkhipov V, Slaoui A, Švrcek VC, del Cañizo C, Tobias I (2007) Modifying the solar spectrum to enhance silicon solar cell efficiency—an overview of available materials. Sol Energy Mater Sol Cells 91(4):238–249
22. Pi X, Li Q, Li D, Yang D (2011) Spin-coating silicon-quantum-dot ink to improve solar cell efficiency. Sol Energy Mater Sol Cells 95(10):2941–2945
23. Mutlugun E, Soganci IM, Demir HV (2007) Nanocrystal hybridized scintillators for enhanced detection and imaging on Si platforms in UV. Opt Express 15(3):1128–1134
24. The rapid transition to energy efficient lighting: an integrated policy approach by United Nations Environment Programme. The enlighten initiative, 2012 Brochure EN0412, p. 5: Electricity for lighting is responsible for nearly 20% of total end use electrical consumption and almost 6% of global greenhouse gas (GHG) emissions
25. Holonyak N, Bevacqua SF (1962) Coherent (visible) light emission from Ga ($As_{1-x}P_x$) junctions. Appl Phys Lett 1(4):82–83
26. Kamat PV (2008) Quantum dot solar cells. semiconductor nanocrystals as light harvesters. J Phys Chem C 112:18737–18753
27. Sze SM (1981) Physics of Semiconductor Devices, 2nd edn. Wiley, New York, pp 804–805 (Chapter 14)
28. Tsao, J.Y., Coltrin, M.E., Crawford, M.H., Simmons, J.A.: Solid-state lighting: an integrated human factors, technology, and economic perspective. In: Proceedings of the IEEE, vol. 98, no. 7, pp. 1162–1179, July 2010
29. Dabbousi BO, Rodriguez-Viejo J, Mikulec FV, Heine JR, Mattoussi H, Ober R, Jensen KF, Bawendi MG (1997) (CdSe)ZnS core–shell quantum dots: synthesis and characterization of a size series of highly luminescent nanocrystallites. J Phys Chem B 101(46):9463–9475
30. Chen K-J, Chen H-C, Tsai K-A, Lin C-C, Tsai H-H, Chien S-H, Cheng B-S, Hsu Y-J, Shih M-H, Tsai C-H, Shih H-H, Kuo H-C (2012) Resonant-enhanced full-color emission of quantum-dot-based display technology using a pulsed spray method. Adv Funct Mater 22 (24):5138–5143. doi:10.1002/adfm.201200765
31. Zhao JL, Bardecker JA, Munro AM, Liu MS, Niu YH, Ding IK, Luo JD, Chen BQ, Jen AKY, Ginger DS (2006) Nano Lett 6:463
32. Zhu T, Shanmugasundaram K, Price SC, Ruzyllo J, Zhang F, Xu J, Mohney SE, Zhang Q, Wang AY (2008) Appl Phys Lett 92:023111
33. Wood V, Panzer MJ, Chen J, Bradley MS, Halpert JE, Bawendi MC, Bulovic V (2009) Adv Mater 21:2151
34. Haverinen HM, Myllyla RA, Jabbour GE (2009) Appl Phys Lett 94:073108
35. Sargent EH (2012) Colloidal quantum dot solar cells. Nat Photonics 6:133–135

36. Kshirsagar A, Pickering S, Xu J, Ruzyllo J (2011) ECS Trans 35:71
37. Cho KS, Lee EK, Joo WJ, Jang E, Kim TH, Lee SJ, Kwon SJ, Han JY, Kim BK, Choi BL, Kim JM (2009) Nat Photonics 3:341
38. Kim TH, Cho KS, Lee EK, Lee SJ, Chae J, Kim JW, Kim DH, Kwon JY, Amaratunga G, Lee SY, Choi BL, Kuk Y, Kim JM, Kim K (2011) Nat Photonics 5:176
39. Kuo HC, Hung CW, Chen HC, Chen KJ, Wang CH, Sher CW, Yeh CC, Lin CC, Chen CH, Cheng YJ (2011) Opt Express 19:A930
40. ASTMG 173-03, Standard Tables for Reference Solar Spectral Irradiances. ASTM International, West Conshohocken, Pennsylvania (2005)
41. Emery KA, Osterwald CR (1989) Solar cell calibration methods. Sol Cells 27:445–453
42. Smestad GP, Krebs FC, Lampet CM, Granqvist CG, Chopra KL, Mathew X, Takakura H (2008) Reporting solar cell efficiencies in solar energy materials and solar cells. Sol Energy Mater Sol Cells 92:371–373
43. Tseng PC, Yu P, Chen HC, Tsai YL, Han HW, Tsai MA, Chang CH, Kuo HC (2011) Angle-resolved characteristics of silicon photovoltaics with passivated conical-frustum nanostructures. Sol Energy Mater Sol Cells 95:2610–2615
44. Tsai MA, Yu PC, Chiu CH, Kuo HC, Lu TC, Lin SH (2010) Self-assembled two-dimensional surface structures for beam shaping of GaN-based vertical-injection light-emitting diodes. IEEE Photon Technol Lett 22(1):12–14
45. Li XH, Song RB, Ee YK, Kumnorkaew P, Gilchrist JF, Tansu N (2011) Light extraction efficiency and radiation patterns of III-nitride light-emitting diodes with colloidal microlens arrays with various aspect ratios. IEEE Photon J 3(3):489–499
46. Ee, Y.K., Kumnorkaew, P., Tong, H., Arif, R.A., Gilchrist, J.F., Tansu, N.: Enhancement of light extraction efficiency of InGa quantum wells light-emitting diodes with polydimethylsiloxane concave microstructures. In: Light-Emitting Diodes: Materials, Devices, and Applications for Solid State Lighting Xiii, vol. 7231 (2009)
47. Ee YK, Kumnorkaew P, Arif RA, Tong H, Zhao HP, Gilchrist JF, Tansu N (2009) Optimization of light extraction efficiency of III-nitride LEDs with self-assembled colloidal-based microlenses. IEEE J Sel Top Quantum Electron 15(4):1218–1225
48. Ee YK, Arif RA, Tansu N, Kumnorkaew P, Gilchrist JF (2007) Enhancement of light extraction efficiency of InGaN quantum wells light emitting diodes using SiO_2/polystyrene microlens arrays. Appl Phys Lett 91(22):221107
49. Lin CC, Chen HC, Tsai YL, Han HV, Shih HS, Chang YA, Kuo HC, Yu P (2012) Highly efficient CdS-quantum-dot-sensitized GaAs solar cells. Opt Express 20:A319–A326
50. Leatherdale CA, Kagan CR, Morgan NY, Empedocles SA, Kastner MA, Bawendi MG (2000) Photoconductivity in CdSe quantum dot solids. Phys Rev B Condens Matter Mater Phys 62:2669–2680
51. Nizamoglu S, Erdem T, Sun XW, Demir HV (2010) Opt Lett 35:3372
52. Erdem T, Nizamoglu S, Sun XW, Demir HV (2010) Opt Express 18:340

Chapter 4
Control of Photoinduced Charge Transfer in Semiconducting Quantum Dot-Based Hybrids

Zhihua Xu, Corey R. Hine, Mathew M. Maye, Qinpeng Meng, and Mircea Cotlet

Abstract Self-assembly of water-soluble colloidal core/shell CdSe/ZnS quantum dots with varying ZnS shell thickness with a conjugated polyelectrolyte results in a series of inorganic/organic hybrids with tunable charge transfer properties and with the shell acting as a tunneling barrier towards photoinduced hole transfer from quantum dot to polymer. The rate for hole transfer decreases exponentially with the increase of quantum dot shell thickness and the hole transfer process is assessed to be a dynamic quenching process. An increase in hole transfer rate is observed from solution to thin-film phase and is related to an improved interfacial morphology in quantum dot/polymer affecting the hole transfer rate, namely the donor–acceptor distance. Single particle spectroscopy experiments reveal fluctuating dynamics of hole transfer at the conjugated polymer/quantum dot interface, increased heterogeneity in the hole transfer rate with the increase of quantum dot's shell thickness.

4.1 Introduction

Hybrid semiconducting organic/inorganic composites are appealing materials for optoelectronic and biosensory applications. In such hybrid materials properties from each of the component materials are combined to create new properties that may not necessarily be present in individual components. For example, conjugated polymers have been blended with semiconductor quantum dots (QDs) and proposed for applications such as light-emitting diodes (LEDs) [1–4], photovoltaic solar cells

Z. Xu • Q. Meng • M. Cotlet (✉)
Center for Functional Nanomaterials, Brookhaven National Laboratory, 735 Brookhaven Avenue, Upton, NY 11973, USA
e-mail: cotlet@bnl.gov

C.R. Hine • M.M. Maye
Department of Chemistry, Syracuse University, 1-014 Center for Science and Technology, Syracuse, NY 13244, USA

J. Wu and Z.M. Wang (eds.), *Quantum Dot Solar Cells*, Lecture Notes in Nanoscale Science and Technology 15, DOI 10.1007/978-1-4614-8148-5_4,

(PVs) [5–8], or biosensing [9]. The interaction between the excited states of a conjugated polymer and a quantum dot, which determines the optoelectronic properties of the resulting hybrid material, needs to be fully understood. The highest occupied molecular orbital (HOMO) and the lowest unoccupied molecular orbital (LUMO) of the conjugated polymer and quantum dot in such hybrids is usually of type-II alignment, and this facilitates both charge separation and energy transfer between the components under exposure to light. Charge separation can be achieved by either electron transfer from the photoexcited polymer to the quantum dot or by hole transfer from the photoexcited quantum dot to the conjugated polymer. When both organic and inorganic components undergo photoexcitation, all above three types of interactions may be present and may compete with each other in a hybrid, thus leading to a complicated photoresponse. Both energy transfer and charge separation have been observed between CdSe quantum dots and their conjugated ligands [10, 11]. Energy transfer between conjugated polymers and quantum dots has been previously investigated and both Förster- and Dexter-type mechanisms have been proposed [12–16]. Photoinduced electron transfer from a conjugated polymer to a quantum dot has also been intensively investigated, along with the increased interest in the utilization of such organic/inorganic hybrids in photovoltaic applications [17–20]. Hole transfer from a photoexcited quantum dot to a conjugated polymer on the other hand remains little explored and understood. Hole transfer from quantum dots to conjugated polymers is an essential process in achieving high power conversion efficiency in photovoltaic solar cells, but at the same time it is a detrimental process for applications such as light emitting diodes (LEDs) and biosensing. Working photovoltaic solar cells based on quantum dot/conjugated polymer hybrids were demonstrated already and they prove indirectly the presence of hole transfer. Nevertheless, there exist just few reports on the hole transfer dynamics between quantum dots and conjugated polymers, in particular between InP core-only quantum dots and a poly(thiophene) derivative and only in solution phase [21]. Better design of new hybrids for optoelectronic and biosensory applications requires a systematic study of hole-transfer in conjugated polymer/QD hybrids. For example, the development of PV and LED devices based on conjugated polymer/QD thin films invokes more interest in the characterization of hole-transfer in solid films rather than in solution. Similarly, biosensors utilizing conjugated polymer/QD hybrids can benefit in sensitivity if one can understand how hole-transfer affects QD's emission properties.

Here, we report the photoinduced hole-transfer (HT) from a CdSe/ZnS core/shell QD with varying shell thickness (Fig. 4.1a) to a water-soluble conjugated polymer, poly(9,9′-bis(6-*N,N,N*-trimethylammoniumhexyl) fluorene-alt-phenylene22 (FHQ, MW ~ 15 kDa) (Fig. 4.1b) [22, 23]. For this particular donor/acceptor hybrid, the emission peaks of FHQ (450 nm) and QD (634–638 nm) are well separated (see Fig. 4.4a), allowing selective excitation of only the QD in a hybrid, for example by exciting with 590 nm light, so that HT from photoexcited QD to conjugated polymer can be probed exclusively by monitoring the photoluminescence (PL) from the QD. We first investigate photoinduced HT in FHQ/QD composite both in solution and in solid thin films by ensemble-based PL measurement, then, we use

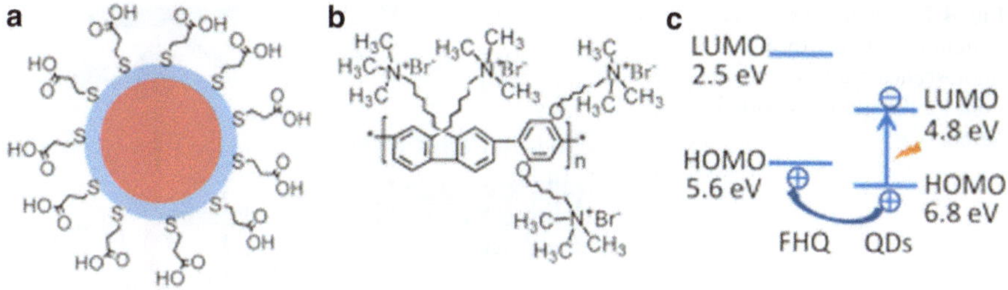

Fig. 4.1 (**a**) MPA-capped core/shell CdSe/ZnS quantum dots (QDs) and (**b**) cationic conjugated polymer FHQ. (**c**) Scheme for photoinduced hole transfer between QD and FHQ. Reproduced in part from [23]

single particle spectroscopy to probe the dynamics of HT and its heterogeneity at single conjugated polymer/QD interface and to probe the effect of HT on the photoluminescence blinking dynamics of QDs.

4.2 Materials and Methods

4.2.1 Synthesis of Core/Shell Nanocrystals

The CdSe QD cores (≈6 nm) were synthesized following traditional methods using trioctylphosphine oxide (TOPO) and hexadecylamine (HDA) as the capping agents [24–26]. The resulting TOPO/HDA capped QDs were first purified free of excess ligands by multiple methanol extractions, followed by precipitation in dry acetone. The TOPO/HDA-QDs were then redispersed and stored in dried chloroform before undergoing shell addition. The concentration of the CdSe QD core was then calculated using the size- dependent extinction coefficient [27–29]. Next, ZnS shells were deposited in a layer-by-layer fashion via the SILAR method [30–32] in order to tailor thickness [31, 32]. The sulfur precursor consisted of elemental sulfur dissolved in octadecene (ODE), and the zinc precursor was zinc acetate dissolved in oleylamine (OAm). These precursors were injected sequentially, allowing a minimum of 30 min between injections to allow for shell annealing. The process was repeated until a desired thickness was grown, 1.2, 1.7, 2.2, 2.7, and 4.2 nm. Shell growth was calculated based on desired increase in QD volume, as well as the QD concentration in the reaction medium. Upon reaching a desired shell thickness, an aliquot of the CdSe/ZnS QDs was collected from the reaction solution and purified as described above. Alternatively, CdSe cores were purchased from NN-labs, and ZnS shells were deposited as described above. Each CdSe/ZnS QD was then phase transferred to aqueous buffers using the recently reported modular phase transfer method [33]. Briefly, to a solution of QDs in chloroform, a 3,000-fold molar excess of L-histidine (His) in basic methanol was added, and the QDs were phase transferred to buffer via a single homogeneous step. After vortexing, the

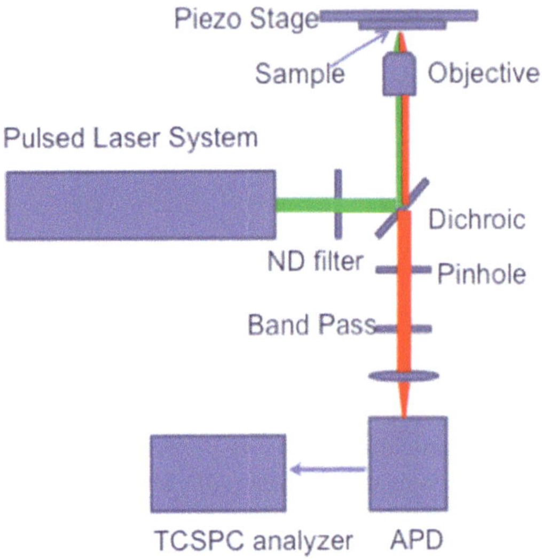

Fig. 4.2 Schematics of the scanning-stage confocal fluorescence microscope used for the studies reported herein

His-QDs were collected in a 10 mM borate buffer (pH 8.3), and excess TOPO, HDA, OAm was back-extracted via chloroform. Finally, excess His was removed using a 100 kDa molecular weight centrifugal filter (Millipore) and 3–5 mL of 10 mM borate buffer. Finally, the His-QDs were further functionalized with MPA in a homogenous ligand exchange step in which a 6,000-molar ratio of [MPA]:[QD] is added. During this process, the MPA and His undergo monolayer exchange, resulting in MPA-capped QDs. The MPA-QDs were then purified and quantified as described above.

Blends of FHQ/QD were prepared from a 200 nM borate buffered solution of QDs by adding FHQ at various molar ratios. For concentrated thin films, blended solutions were spin-casted on pre-cleaned cover-glasses. For single particle experiments, blended solutions were diluted (1,000×) and spin-casted on pre-cleaned cover-glasses to ensure a concentration of 1 QD/μm^2 or less. For diluted QD-only samples, QDs were spin-casted directly on pre-cleaned cover-glass. For both concentrated thin films and single particle experiments, samples were purged with nitrogen prior (20 min) and during the measurement.

4.2.2 Spectroscopic Characterization Methods

Ensemble PL lifetimes, fluorescence lifetime imaging microscopy (FLIM) and single particle fluorescence microscopy measurements were performed on a home-built confocal scanning stage microscope, schematically shown in Fig. 4.2, and based on an inverted microscope (Olympus IX81, 1.4 NA 100× oil objective) equipped with a piezo scanning stage (Physik Instruments, Germany) and coupled to a diode-pumped solid state laser system delivering 590 nm pulses of 250 fs width, at 8 MHz repetition rate (Spectra Physics, Newport). Experiments were performed with an average power at the sample of about 100 nW. The emitted photoluminescence was collected in epi-illumination format, spectrally separated from the excitation laser light by a dichroic mirror (Semrock, DiO-594) and by a band pass filter (Semrock FF01-628/40), and detected by a single photon counting avalanche photodiode (APD, MPD Pico-Quant) and a time-analyzer (PicoHarp 300, PicoQuant). Data acquisition and data analysis were performed with the commercial Symphotime analysis software (Picoquant). Dynamic light scattering experiments were performed on a commercial Malvern Zetasizer NS.

4.3 Results and Discussion

The MPA-capped CdSe/ZnS QDs were synthesized as described in the previous paragraph have average shell thicknesses of 1.2 nm (QD-1), 1.7 nm (QD-2), 2.2 nm (QD-3), 2.7 nm (QD-4), and 4.2 nm (QD-5), respectively. Their structural and photophysical parameters are given in Table 4.1. QD's core size and shell thickness were estimated based on transmission electron microscopy experiments by size distribution analysis. These are negatively charged water-soluble QDs and as such they can electrostatically bind cationic FHQ [22] to form a stable organic/inorganic nanocomposite [34]. This is demonstrated by the dynamic light scattering (DLS) measurements shown in Fig. 4.3 where a clear increase in hydrodynamic volume is

Table 4.1 Structural and photophysical parameters of core/shell QDs

CdSe/ZnS	Shell thickness (nm)	QY (%)	PL peak (nm)	τ_{QD} (ns)
QD-1	1.2	7.2	634	9
QD-2	1.7	9.4	635	12
QD-3	2.2	16	636	21
QD-4	2.7	22	637	32
QD-5	4.2	25	638	32

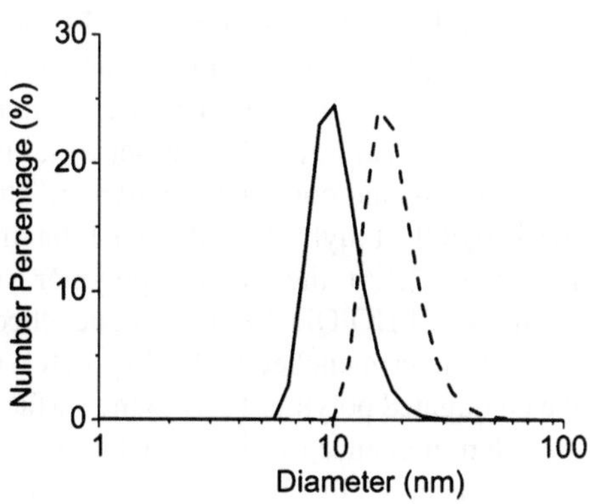

Fig. 4.3 Dynamic light scattering for QD-2 (*line*) and FHQ/QD-2 hybrids (*dash line*)

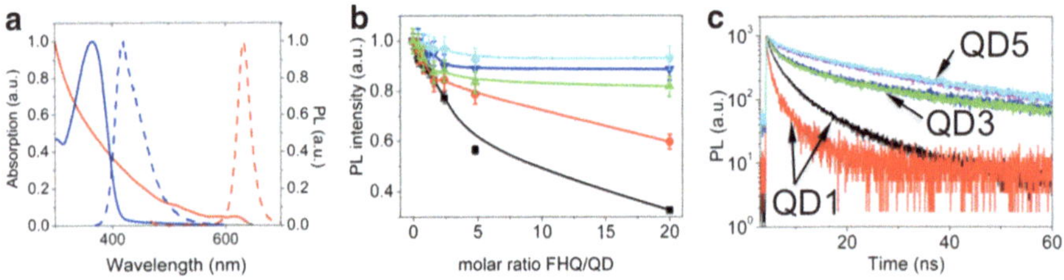

Fig. 4.4 (**a**) UV–vis (*solid*) and PL (*dash*) spectra of QD-1 (*red*) and FHQ (*blue*) in aqueous solution. (**e**) PL intensity vs FHQ/QD molar ratio, $R_{FHQ/QD}$, for hybrids based on QD-1 (*black square*), QD-2 (*red dots*), QD-3 (*green triangles*), QD-4 (*blue triangles*), and QD-5 (*magenta rhombs*). (**f**) PL decays of QD-1 (*black*) and FHQ/QD-1 ($R_{FHQ/QD} = 20$, *red*), of QD-2 (*green*) and FHQ/QD-2 ($R_{FHQ/QD} = 20$, *blue*) and of QD-3 (*cyan*) and FHQ/QD-3 ($R_{FHQ/QD} = 20$, *magenta*). Reproduced in part from [23]

observed from QDs-only to QDs blended with FHQ. Assuming that the ZnS shell acts as a tunneling barrier for hole-transfer from the photoexcited QD to FHQ, this series of QDs can be used to probe and at the same time to demonstrate distance-dependent hole transfer in conjugated polymer/QD hybrids. With increased shell thickness, the PL peak of QDs slightly shifts from 634 nm (QD-1, QD-2) to 635 nm (QD-3, QD-4) and to 638 nm (QD-5), and the PL quantum yield increases from 7 % for QD-1 up to 25 % for QD-5 and the PL lifetime follows a similar trend (see Table 4.1). This shell thickness dependency of both PL quantum yield and lifetime for QDs is consistent with previous reports suggesting increased leakage of the exciton wavefunction into the shell and improved passivation of surface defects with increasing shell thickness [35–37].

4.3.1 Hole-Transfer in Solution

We probed photoinduced hole-transfer in solution by blending QDs (200 nM in borate buffer) with FHQ at various FHQ/QD molar ratios (denoted $R_{FHQ/QD}$), from 0:1 to 20:1, by measuring the PL intensity and PL lifetime at the emission peak of the QD ($\lambda_{exc} = 590$ nm). These experimental results are shown in Fig. 4.4. The molar ratio $R_{FHQ/QD} \sim 20$ was set as an upper limit at which one expects that all QDs have bound conjugated polymer. The strongest PL quenching was observed for FHQ/QD-1 hybrid, with QD-1 having the thinnest ZnS shell (1.2 nm, see Fig. 4.4b–c). At the highest polymer concentration ($R_{FHQ/QD\text{-}1} = 20$), the PL intensity of FHQ/QD-1 hybrid is quenched 67 % (Fig. 4.4b) and the corresponding PL lifetime is quenched 62 % (Fig. 4.4c), suggests efficient electrostatic binding of the conjugated polymer FHQ to the surface of QD and dynamic quenching of QD's photoluminescence by hole-transfer to FHQ. The dynamic PL quenching of QDs by hole-transfer to FHQ is a result of the presence of the ZnS shell acting as an additional tunneling barrier for HT, and this is opposite to as it has been reported previously for hole transfer in solution in the case of InP quantum dot/P3HT hybrids [21]. For InP QD/P3HT hybrids, significant changes in the PL intensity of InP core-only QDs due to hole-transfer to P3HT were accompanied by very little changes in the related PL lifetime because QDs complexed with P3HT would experience complete PL quenching by efficient HT, that is, static quenching. CdSe/ZnS QDs with thicker shells also experience dynamic PL quenching, somewhat less dramatic, 40 % for FHQ/QD-2, 18 % for FHQ/QD-3, 11 % for FHQ/QD-4 and 7 % for FHQ/QD-5 for the same $R_{\mathrm{FHQ/QD}}$ molar ratio of 20:1 (Fig. 4.4b).

The PL lifetimes for QDs and FHQ/QD hybrids are given by:

$$\tau_{\mathrm{QD}} = 1/(k_{\mathrm{R}} + k_{\mathrm{NR}}), \qquad \tau_{\mathrm{FHQ/QD}} = 1/(k_{\mathrm{R}} + k_{\mathrm{NR}} + k_{\mathrm{HT}}) \tag{4.1}$$

with k_{R}, k_{NR} radiative and nonradiative rates for QD, respectively, and k_{HT} the rate for HT from photoexcited QD to FHQ. k_{HT} can then be calculated by

$$k_{\mathrm{HT}} = 1/\tau_{\mathrm{FHQ/QD}} - 1/\tau_{\mathrm{QD}} \tag{4.2}$$

Applying Eq. (4.2) to the PL lifetime data from Fig. 4.4c gives an estimated rate $k_{\mathrm{HT}} \sim 2.0 \times 10^8\ \mathrm{s}^{-1}$ for FHQ/QD-1 in solution for $R_{\mathrm{FHQ/QD\text{-}1}} = 20$ which is the hybrid exhibiting the strongest quenching. The rate for charge transfer can be written as

$$k_{\mathrm{HT}} \propto \frac{4\pi^2}{h}\langle H^2\rangle \frac{1}{(4\pi\lambda k_{\mathrm{B}}T)^{1/2}} \exp\left(-\frac{(\Delta G^0 + \lambda)^2}{4\pi\lambda k_{\mathrm{B}}T}\right) \tag{4.3}$$

With H electronic coupling, ΔG^0 the free energy and λ reorganization energy. ΔG^0 is determined by the HOMO and LUMO levels of the donor and acceptor components (Fig. 4.1c). H is determined by the overlap between the hole wavefunction of QD and the HOMO of FHQ, which is a function of donor–acceptor separation distance. Assuming the HOMO and LUMO of both components remain unchanged when changing only the ZnS shell thickness, t_{s}, the rate k_{HT} for FHQ/QD hybrids should be sensitive only to changes in donor–acceptor separation distance, in this case changes on the ZnS shell thickness t_{s}, by the following relationship:

$$k_{\mathrm{HT}} = K_0 \exp(-\beta t_{\mathrm{s}}) \tag{4.4}$$

where β is a damping factor. Figure 4.5 (black squares) shows the calculated HT rates (k_{HT}) for various ZnS shell thicknesses by using the PL lifetime data from Fig. 4.4c and a fit by Eq. (4.4) (black line) yielding a damping factor $\beta \sim 2.8\ \mathrm{nm}^{-1}$ that confirms the tunneling barrier character for the ZnS shell against the hole-transfer from QD to FHQ. The obtained β for FHQ/QD hybrids is comparable to that of CdSe/CdS core/shell nanorods adsorbed with small molecule hole acceptors like phenothiazine (PTZ) ($\beta \sim 4.0$) [38]. Extrapolation of the dependency k_{HT} vs ZnS shell thickness t_{s} from

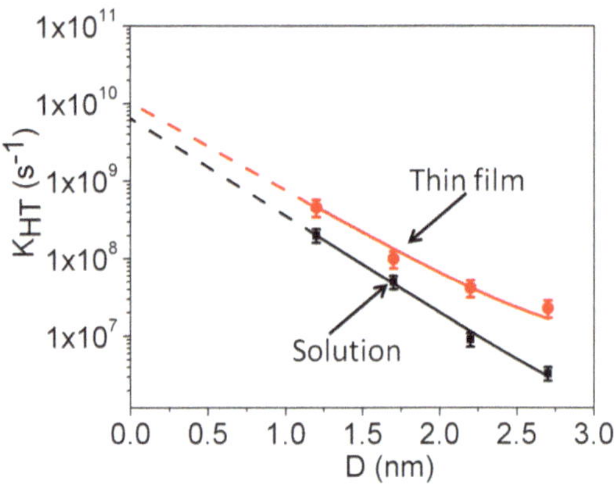

Fig. 4.5 ZnS shell thickness (t_s) vs hole transfer rate (k_{HT}) dependency for FHQ/QD hybrids in solutions (*black*) and in thin films (*red*). Fits according to Eq. (4.4) are shown as *solid lines*. Reproduced from [23]

Fig. 4.5 to zero shell thickness reveals a rate $k_{HT}(0) \sim 5.2 \times 10^9\ s^{-1}$ for the hole transfer between CdSe core-only and FHQ in solution, which is similar to that reported for CdSe core-only nanocrystals adsorbed with PTZ molecules [39].

4.3.2 Hole-Transfer in Concentrated Thin Films

We also investigated the steady-state and transient PL of FHQ/QD hybrids (for $R_{FHQ/QD} = 20$) in solid thin films and compared that with the bare QDs spin-casted on a glass surface. Figure 4.6 summarizes confocal FLIM experiments for thin-film samples of FHQ/QD-1, FHQ/QD-3, and FHQ/QD-5 hybrids and for their corresponding bare QDs. Both FHQ/QD hybrid thin film and bare QD samples were prepared from solutions of similar QD concentration (200 nM). Bare QDs on glass surface exhibit average PL lifetimes similar to those observed in solution as can be seen in Fig. 4.6a–c where PL decays are shown in left panels and FLIM images in the center panels. FHQ/QD hybrids exhibit quenched PL lifetimes, except for the FHQ/QD-5 that are hybrids based on QDs with the thickest ZnS shell (4.2 nm) (Fig. 4.6a–c with PL decays in left panels and FLIM images in right panels). From the PL decay data for FHQ/QD-1 from Fig. 4.6a we calculated an average rate for hole transfer of $k_{HT}(\text{FHQ/QD-1}) \sim 4.6 \times 10^8\ s^{-1}$, with FHQ/QD-1 being the hybrid with the strongest hole transfer. For hybrids in thin film, the average hole transfer rate k_{HT} scales exponential with the ZnS shell thickness (see Fig. 4.5, red dots), with a damping factor $\beta \sim 2.7\ nm^{-1}$, a similar behavior to that observed in solution. Extrapolation of the exponential dependency from Fig. 4.5 to zero shell thickness provides a limiting rate $k_{HT}(0) \sim 1.1 \times 10^{10}\ s^{-1}$ for the hole transfer between the CdSe core-only QD and the FHQ polymer for solid thin films. The increase in hole transfer rate for FHQ/QD hybrids in thin films compared to solution is clearly seen in the k_{HT} vs t_s dependencies from Fig. 4.5 and this increase in hole transfer in thin films might be attributed to a denser packing of the conjugated polymer chains on the QD's surface in the case of solid thin films, which in turn increases the electronic coupling between the donor and acceptor components.

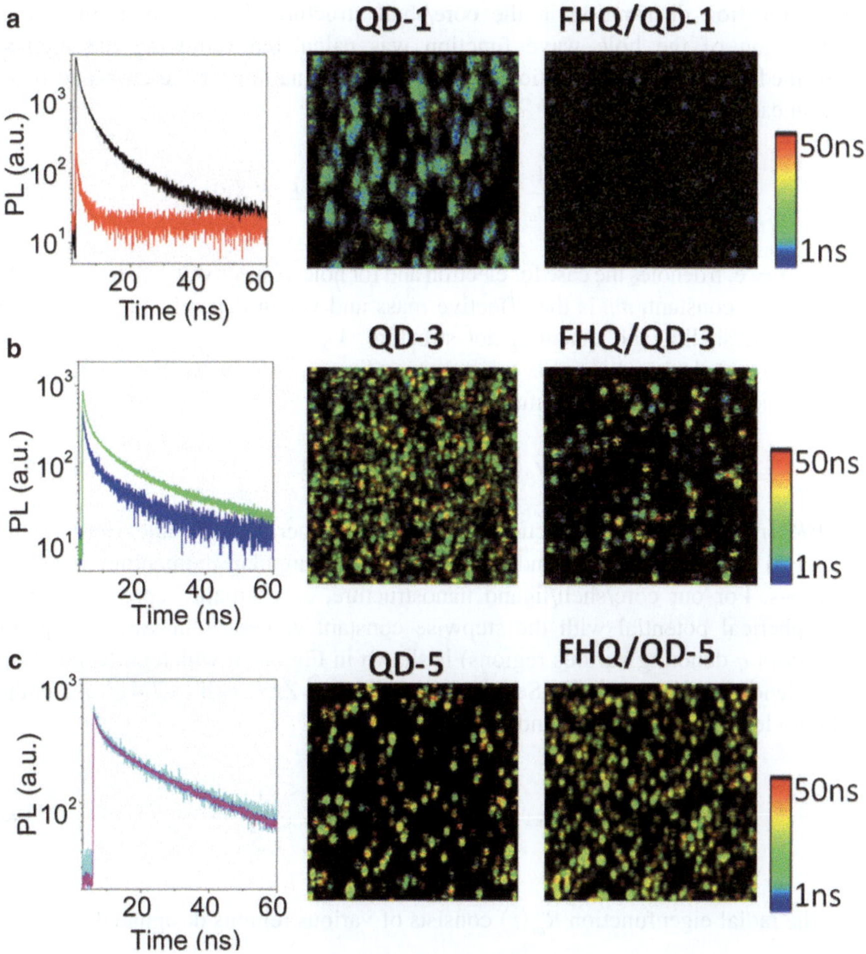

Fig. 4.6 Photoinduced hole transfer in concentrated FHQ/QD thin films. *Left*: average PL decays from (**a**) QD-1 on surface (*black*) and FHQ/QD-1 hybrid thin films (*red*), from (**b**) QD-3 on surface (*green*) and FHQ/QD-3 hybrid thin films (*blue*) and from (**c**) QD-5 on surface (*cyan*) and FHQ/QD-3 thin films (*magenta*). *Right*: FLIM images (15 μm × 15 μm) of QD on surface and of FHQ/QD hybrid thin films for (**a**) QD-1, (**b**) QD-3 and (**c**) QD-5. Reproduced from [23]

4.3.3 Theoretical Estimation of the Tunneling Barrier Effect of ZnS Shell

To further demonstrate the tunneling barrier character of the ZnS shell on the HT rate we used the effective mass theory to estimate the CdSe hole wave function density at the surface of ZnS shell in the QDs [36, 37]. In this model, QDs are treated as particles confined in spherical wells of finite depth. In these calculations, the Schrodinger equation for the QD system is solved numerically to obtain the hole wave function distribution in the core/shell structure. In particular, the radial distribution of the hole wave function was calculated following the method described in [40, 41]. The stationary Schrödinger equation for the envelope function in each band is given by:

$$\left[-\frac{\hbar}{2}\nabla\frac{1}{m_\alpha}\nabla+V_\alpha(\mathbf{r})\right]\Psi_\alpha(\mathbf{r})=E_\alpha\Psi_\alpha(\mathbf{r}) \tag{4.4}$$

where $\alpha = \mathrm{e}, \mathrm{h}$ denotes the case for electron and for hole, respectively, where $\hbar = h/2\pi$ is Planck's constant, m_α is the effective mass and which depends on the position in the core/shell/ligand quantum dot structure, $V_\alpha(\mathbf{r})$ is the potential, E_α is the energy eigenvalue, and $\psi_\alpha(\mathbf{r})$ is the corresponding eigenfunction. For a spherically symmetric potential, the eigenfunction is written as:

$$\psi_\alpha(\mathbf{r}) = R^\alpha_{nl}(r)Y_{lm}(\theta,\varphi) \tag{4.5}$$

with $R^\alpha_{nl}(r)$ the radial wave function and $Y_{lm}(\theta,\varphi)$ a spherical harmonic. Here n is the principal quantum number and l and m are the angular momentum quantum numbers. For our core/shell/ligand nanostructure, a schematic representation of the spherical potential with the stepwise constant values V_z in various regions (subscript q denoting various regions) is shown in Fig. 4.7a, which corresponds to the valence band edges for CdSe core (−6.8 eV) and ZnS shell (−7.4 eV), and the HOMO level of the MPA ligands (−6.7 eV).

$$V_\alpha(r) = \begin{cases} V_{\alpha,1} & \text{for} \quad 0 \le r < r_1 \\ V_{\alpha,2} & \text{for} \quad r_1 \le r < r_2 \\ V_{\alpha,3} & \text{for} \quad r_2 \le r < r_3 \end{cases} \tag{4.6}$$

The radial eigenfunction $R^\alpha_{nl}(r)$ consists of various regions designated $R^\alpha_{nl,q}$:

$$R^\alpha_{nl}(r) = \begin{cases} R^\alpha_{nl,1} & \text{for} \quad 0 \le r < r_1 \\ R^\alpha_{nl,2} & \text{for} \quad r_1 \le r < r_2 \\ R^\alpha_{nl,3} & \text{for} \quad r_2 \le r < r_3 \end{cases} \tag{4.7}$$

In regions where $E_\alpha > V_{\alpha,q}$, the solution is a linear combination of the spherical Bessel and Neumann function j_l and n_l:

$$R^\alpha_{nl,q} = A^\alpha_{nl,q}j_l(k_{nl,q}r) + B^\alpha_{nl,q}n_l(k_{nl,q}r) \tag{4.8}$$

with

$$k^\alpha_{nl,q} = \left[\frac{2m_\alpha(E_\alpha - V_{\alpha,q})}{\hbar^2}\right]^{\frac{1}{2}} \tag{4.9}$$

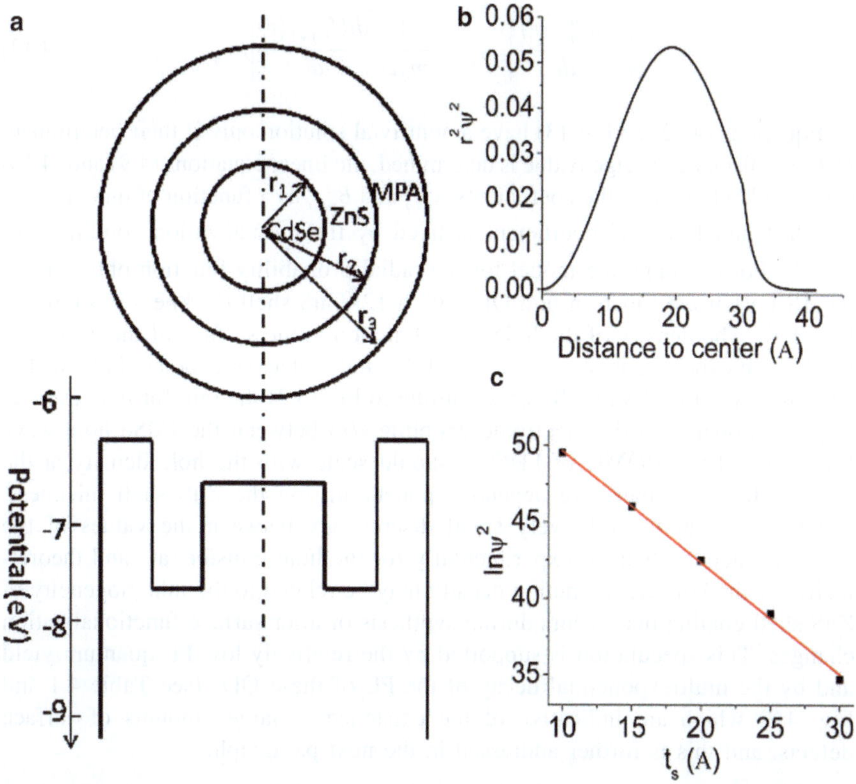

Fig. 4.7 (**a**) Schematic representation of the valence band alignment at the core/shell/ligand (CdSe/ZnS/MPA) quantum dot structure; (**b**) radial probability function of 1S hole wave function in CdSe/ZnS QD with shell thickness of 12 Å; (**c**) hole wave function density at ZnS surface vs ZnS shell thickness

In regions where $E_\alpha < V_{\alpha,q}$, the solution is a linear combination of the two Hankel function $h_l^+ h_l^+$ and h_l^-:

$$R^\alpha_{nl,q} = A^\alpha_{nl,q} h_l^+\left(i\kappa_{nl,q} r\right) + B^\alpha_{nl,q} h_l^-\left(i\kappa_{nl,q} r\right) \tag{4.10}$$

with

$$\kappa^\alpha_{nl,q} = \left[\frac{2m_\alpha\left(V_{\alpha,q} - E_\alpha\right)}{\hbar^2}\right]^{\frac{1}{2}} \tag{4.11}$$

The solution must satisfy the following fitting conditions:

$$R^\alpha_{nl,q}\left(r_q\right) = R^\alpha_{nl,q+1}\left(r_q\right) \tag{4.12}$$

$$\frac{1}{m_q}\frac{dR^{\alpha}_{nl,q}(r)}{dr}\bigg|_{r=r_q}. = \frac{1}{m_{q+1}}\frac{dR^{\alpha}_{nl,q+1}(r)}{dr}\bigg|_{r=r_q}. \tag{4.13}$$

Equations (4.12) and (4.13) have a nontrivial solution only if their determinant $D(E_\alpha) = 0$. Once the eigenvalue is determined, the linear equations (4.9) and (4.10) can be solved yielding the coefficients $A^{\alpha}_{nl,q}$ and $B^{\alpha}_{nl,q}$ as a function of one of them. The last undetermined coefficient is fixed by the normalization condition for $R^{\alpha}_{nl}(r)^2$. For example, the model for the radial probability function obtained for the 1S hole wave function in a QD with a 12ÅZnS shell thickness is shown in Fig. 4.7b. The density of the hole wave function at the surface of the ZnS shell vs shell thickness t_s is shown in Fig. 4.7c, and it decays exponentially with a damping factor of 7.3/nm. Since according to Eq. (4.3) the rate for hole transfer is proportional with the electronic coupling (H) between the CdSe hole wave function and the HOMO of FHQ, it should scale with the hole density at the shell surface and therefore depend exponentially on the ZnS shell thickness. We believe that the relatively small discrepancy between the values of the damping factors obtained experimentally for the hole transfer rate and theoretically for the hole wavefunction density may be related to the inhomogeneity of ZnS shell coating that occurs during synthesis or after surface functionalization changes. This speculation is supported by the relatively low PL quantum yield and by the multiexponential decay of the PL of these QDs (see Table 4.1 and Fig. 4.2) which are indicative of the existence of large amounts of surface defects, and this is further addressed in the next paragraph.

4.3.4 *Hole-Transfer in Isolated FHQ/QD Hybrids*

We performed single particle PL spectroscopy on diluted samples of isolated QDs and of FHQ/QD hybrids and typical single molecule PL intensity and PL lifetime trajectories recorded from such nanoentities are shown in Fig. 4.8. The effect of ZnS shell thickness on the HT can be also seen at the single FHQ/QD hybrid level: from thin (1.2 nm for QD-1, Fig. 4.8a) to intermediate (2.2 nm for QD-3, Fig. 4.8b) to thick shell (4.2 nm, QD-5, Fig. 4.8c), hole transfer quenches the PL intensity and lifetime in FHQ/QD hybrids from very efficient (FHQ/QD-1, Fig. 4.8a), to intermediate (FHQ/QD-3, Fig. 4.8b) to negligible (FHQ/QD-5, Fig. 4.8c). Due to the very efficient HT in the case of FHQ/QD-1 hybrid, very few isolated FHQ/QD-1 hybrids are "visible" by single particle PL spectroscopy and the PL signal from such hybrids is too weak to render any reliable analysis (see Fig. 4.8a, lower panel). Due to the negligible HT in FHQ/QD-5 hybrids, we noticed no significant difference between PL signals from isolated QD-5 and FHQ/QD-5 hybrids (Fig. 4.8c). Since only QD-2, QD-3, and QD-4 and their related hybrids displayed decent PL signals for reliable data analysis, they are considered for the discussion of HT at the single hybrid level.

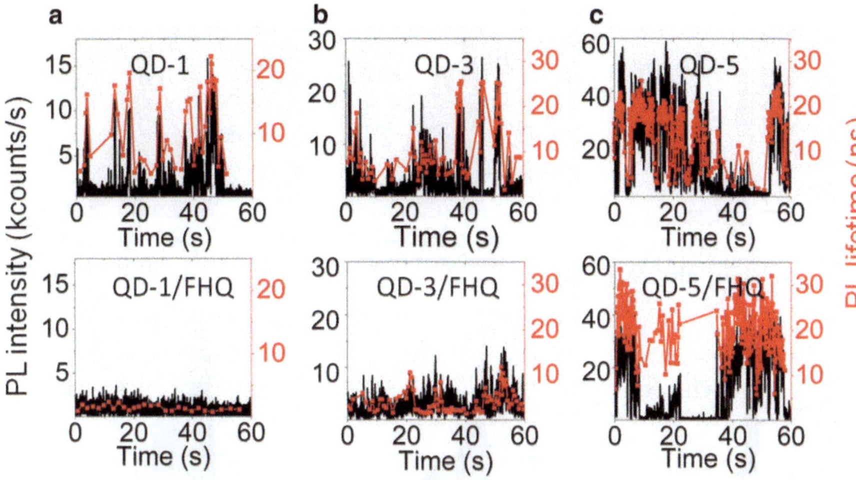

Fig. 4.8 Time trajectories of single molecule PL intensity (*black*) and PL lifetime (*red*) for isolated QDs (*upper panel*) and FHQ/QD-1 hybrids (*lower panel*): (**a**) QD-1 and FHQ/QD-1; (**b**) QD-3 and FHQ/QD-3; (**c**) QD-5 and FHQ/QD-5. Reproduced from [23]

The speculation of inhomogeneous coating of the CdSe core by the ZnS shell is supported by the dynamics of the PL intensity and lifetime observed in single particle experiments. For example, for isolated QD-3 (2.2 nm shell thickness), the PL intensity exhibits random fluctuations with a wide distribution of on-states (Fig. 4.8b, upper panel), different from the well-known "telegraph-like" blinking signal [42]. The PL lifetimes for an isolated QD-3 show correlated fluctuations with the PL intensity (Fig. 4.8b), suggesting fluctuating nonradiative recombination rate for photogenerated excitons. Similar "non-telegraphic"PL blinking was observed previously for isolated QDs and attributed to photoinduced electron trapping in QD's surface or in external defects [43, 44] and for QDs coupled to a fullerene acting as an external electron trap [45]. Therefore, the non-telegraphic PL blinking of isolated QD-3 could be related to inhomogeneous coating of the CdSe core by the ZnS layer. The same behavior could also be related to structural defects induced by large lattice mismatch between CdSe and ZnS, which can create nonradiative recombination sites [36, 46]. For isolated FHQ/QD hybrids, HT becomes an additional quenching pathway for the excited state of QD, since both the PL intensity and PL lifetime for the hybrid are quenched when compared to the isolated QD-3 (Fig. 4.8b). Furthermore, both the PL intensity and PL lifetime for isolated FHQ/QD hybrids exhibit quasi-continuous on-states, indicating fluctuating HT rate at single particle level, similar to previous single molecule reports on electron transfer between QDs and different acceptors [47–49] or other organic molecules [45, 50–53]. In order to characterize the degree of fluctuation of HT rate in single FHQ/QD hybrids, we constructed distributions of PL lifetimes from 50 randomly selected nanoparticles of each category, QD-2, QD-3, QD-4, and their FHQ/QD related hybrids and these data are shown in Fig. 4.9. The heterogeneity in PL lifetime and therefore in the rate for HT was assessed by the standard deviation σ,

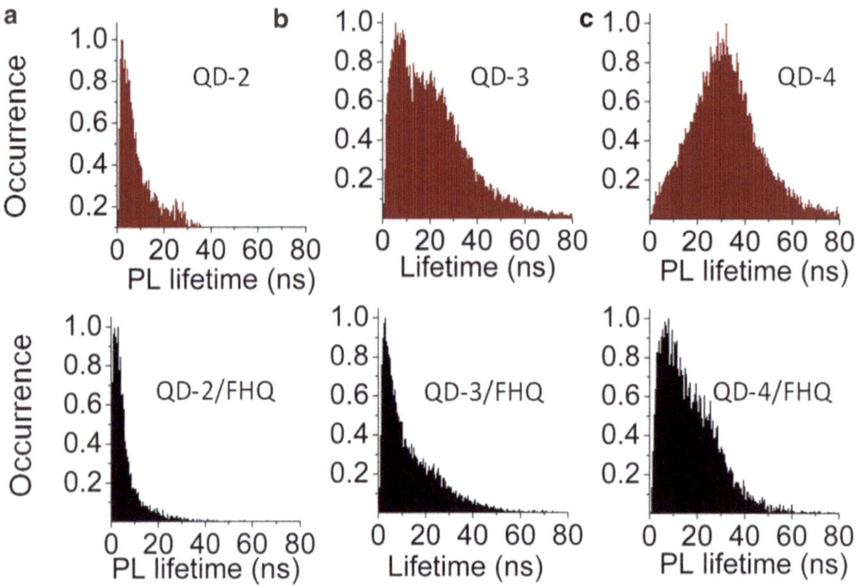

Fig. 4.9 Distributions of PL lifetimes for isolated QDs on surface (*red*, *upper panels*) and isolated FHQ/QD hybrids (*black*, *lower panels*) on glass surface: (**a**) QD-2 and FHQ/QD-2; (**b**) QD3 and FHQ/QD-3; (**c**) QD4 and FHQ/QD-4. Reproduced in part from [23]

$$\sigma(\tau) = \sqrt{\frac{1}{N}\sum_{i=1}^{N}(\tau_i - \tau_{\text{ave}})^2} \tag{4.14}$$

with N being the number of PL lifetime events contributing to a distribution, τ_i the PL lifetime of event i, and τ_{ave} the average lifetime calculated here as $\tau_{\text{ave}} = (1/N)\sum\tau_i$. With the increase in ZnS shell thickness, from QD-2 to QD-3 and to QD-4, PL lifetime distributions shift to larger values (Fig. 4.9a–c), consistent with the ensemble data (Figs. 4.4 and 4.5). At the same time, σ becomes smaller with the increase of ZnS shell thickness (Fig. 4.10a), indicating a reduced fluctuation in the nonradiative recombination rate for photogenerated excitons in QDs. It is known that an increase in the shell thickness for QDs leads to a reduced density of surface traps [35–37]. Therefore, the data from Fig. 4.10a for isolated QDs further support our previous speculation that fluctuations in the PL lifetime and PL intensity of isolated QDs are strongly dependent on the density of surface traps, here determined by the ZnS shell inhomogeneity. Coupling of a QD with the FHQ polymer significantly increases the heterogeneity in the measured PL lifetimes, as the standard deviation σ increases considerably for any given ZnS thickness (Fig. 4.10a). Consequently, HT induces additional heterogeneity (fluctuation) into the exciton decay rate of a QD when this is part of a FHQ/QD hybrid. The contribution of HT to the overall fluctuation present in an FHQ/QD hybrid can be seen by plotting the ratio R_σ of standard deviations for single FHQ/QD hybrids and for isolated QDs vs shell thickness (Fig. 4.10b).

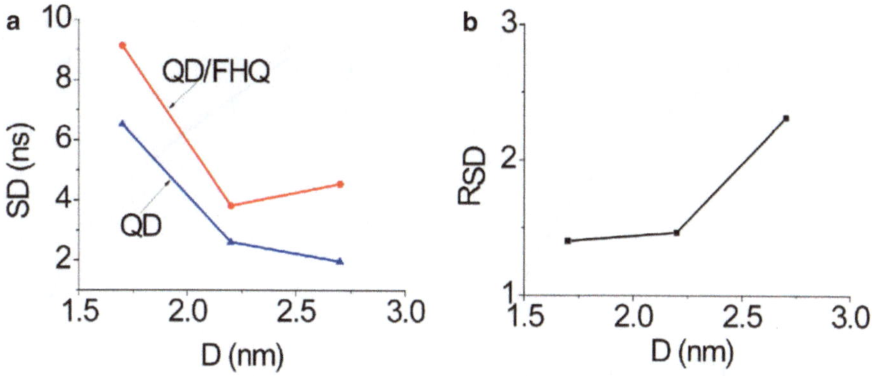

Fig. 4.10 (**a**) Standard deviation σ of PL lifetime vs ZnS shell thickness, t_s, for the distributions shown in Fig. 4.9a–c. (**b**) Ratio (R_σ) of standard deviations of PL lifetime for FHQ/QD hybrids and for QDs vs ZnS shell thickness, t_s. Reproduced in part from [23]

Surprisingly, R_σ increases with the increase of ZnS shell thickness, suggesting an increase in the degree of fluctuation in HT for thicker shells, contrary to the trend for HT rate. One possible explanation for such observed behavior is given below. An increase in QD shell thickness leads to a larger surface area that can complex with the FHQ polymer. In other words, a QD with thicker ZnS shell can couple more polymer chains and meanwhile it can possess a larger degree of variation in surface coupling, which indubitably induces a larger static and dynamic heterogeneity in the PL lifetime when HT is present.

We investigated the effect of hole transfer on the PL blinking dynamics of QDs. PL blinking of isolated QDs, or the switching of their emitted PL between bright (on) and dark (off) states has been related to photoinduced charging of QD core [42]: a QD will stay bright (on state) when the core is neutral and photogenerated electron-hole recombine radiatively or will become dark (off state) when the core is charged and efficient Auger recombination overwhelms the radiative pathway. This model has been widely used to explain PL blinking phenomena of quantum dots, although recent experimental observations and theoretical models have proposed the existence of additional blinking mechanisms [54–58]. PL blinking dynamics for isolated QDs is often characterized by the distributions of the on-times, $P(t_{on})$, and off-times, $P(t_{off})$, with t_{on} and t_{off} defined by a threshold value, here for example set at three-times the standard deviation over the average background signal [59]. $P(t_{on})$ and $P(t_{off})$ distributions corresponding to isolated QD and FHQ/QD hybrids are shown in Fig. 4.11. Each of these distributions contains data extracted from 50 randomly selected PL intensity traces of equal probing time (60 s). For isolated QD-2, QD-3 and QD-4, the $P(t_{off})$ and $P(t_{on})$ distributions are well fitted with a truncated power-law function [60–62].

$$P(t) = bt^{-m} \exp\left(-\frac{t}{\tau}\right) \tag{4.15}$$

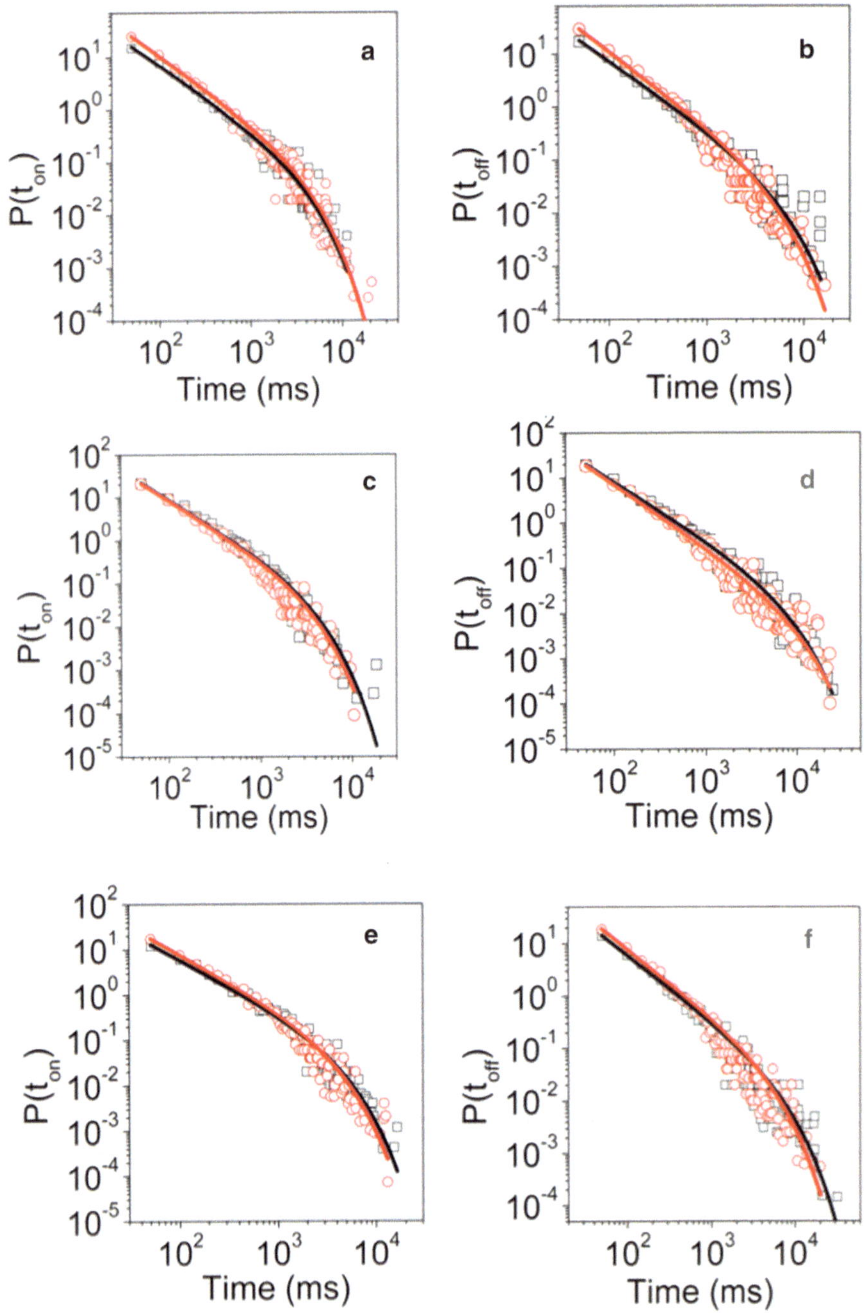

Fig. 4.11 Probability distributions $P(t_{on})$, (**a**, **c**, **e**) and $P(t_{off})$, (**b**, **d**, **f**) for QD-2 (**a**, **b**, *black squares*), FHQ/QD-2 hybrids (**a**, **b**, *red dots*), for QD-3 (**c**, **d**, *black squares*) and FHQ/QD-3 hybrids (**c**, **d**, *red dots*) and for QD-4 (**e**, **f**, *black squares*) and FHQ/QD-4 hybrids (**e**, **f**, *red dots*). Reproduced in part from [21]

with m, power law exponent and τ, crossover time from power-law to exponential decay behavior. Similar blinking behavior of isolated QDs has been observed previously [53, 60, 61] and associated with the presence of surface trap charges (defects) inducing additional nonradiative recombination and thus affecting QD blinking dynamics. $P(t_{off})$ and $P(t_{on})$ distributions for FHQ/QD hybrids can also be fitted well with Eq. (4.15) (see Fig. 4.11), and they do not differ significantly from those corresponding to isolated QDs, thus indicating a trivial effect of HT on QD's blinking dynamics at the timescale probed in our experiments (from 20 ms to 20 s). This is in contrast to the effect of electron transfer (ET) on QD's blinking dynamics observed in the case of QDs coupled with electron acceptors such as fullerenes [45, 47] or titanium oxide [48] where photoinduced electron transfer was found to reduce t_{on}. In the present case, when a QD forms a complex with FHQ, hole transfer clearly emerges as an additional pathway competing with radiative recombination, as it is demonstrated by the observed quenched PL intensity and PL lifetime (see Fig. 4.8), two observables that are associated with the on-states of the QD. The on-states for FHQ/QD-2 resemble the gray states of isolated QDs proposed in a few [63, 64]. On the other hand, hole transfer is expected to affect the dynamics of Auger recombination and of other nonradiative recombinations by regulating the charging and discharging processes of QD core, and hence affecting the PL blinking dynamics. However, we observed no significant effect of HT on the blinking dynamics of QDs at timescales longer than 20 ms, a binning time we were forced to use in order to obtain trustable data for blinking analysis. This suggests the necessity to probe blinking dynamics in the presence of HT at even shorter time scale and to compare this with the effect of electron transfer on the same process.

4.4 Conclusion

In summary, photoinduced hole transfer in hybrids based on CdSe/ZnS QD and a cationic conjugated polymer FHQ has been investigated in solution and solid film, and down to the single particle level. The HT rate in FHQ/QD hybrids exhibits clear exponential dependency on the thickness of ZnS shell, providing the tunneling barrier character of the shell towards HT. In addition, the rate is found to be larger in solid films than in solution, suggesting a significant effect of interfacial morphology on the HT. Single particle spectroscopy not only confirms the shell thickness dependency observed in ensemble measurements but also reveals the fluctuating dynamics of HT at the single FHQ/QD interface. From the analysis of distributions of PL lifetimes of isolated hybrids, the heterogeneity of HT is found to increase with the thickness of ZnS shell. Although HT quenches the intensity of the on-states of quantum dots, it shows little effect on their on-time and off-time distributions at long timescale as probed here.

Acknowledgments Part of this research was carried out at the Center for Functional Nanomaterials of Brookhaven National Laboratory in Upton, New York, under the Office of

Science US-DOE Contract No. DE-AC02-98CH10886. M.M.M. thanks the AFOSR (FA9550-10-1-0033) and ACS-PRF (51303-DNI10) for financial support. We thank Dr. H.L. Wang from Los Alamos National Laboratory in New Mexico and Dr. H.Y. Woo from Pusan National University in South Korea for providing the conjugated polymer.

References

1. Colvin, V.L., Schlamp, M.C., Alivisatos, A.P.: Light-emitting-diodes made from cadmium selenide nanocrystals and a semiconducting polymer. Nature **370**, 354–357 (1994)
2. Dabbousi, B.O., Bawendi, M.G., Onitsuka, O., Rubner, M.F.: Electroluminescence from CdSe quantum-dot polymer composites. Appl. Phys. Lett. **66**, 1316–1318 (1995)
3. Tessler, N., Medvedev, V., Kazes, M., Kan, S.H., Banin, U.: Efficient near-infrared polymer nanocrystat light-emitting diodes. Science **295**, 1506–1508 (2002)
4. Koktysh, D.S., Gaponik, N., Reufer, M., Crewett, J., Scherf, U., Eychmuller, A., Lupton, J.M., Rogach, A.L., Feldmann, J.: Near-infrared electroluminescence from HgTe nanocrystals. Chemphyschem **5**, 1435–1438 (2004)
5. Huynh, W.U., Dittmer, J.J., Alivisatos, A.P.: Hybrid nanorod-polymer solar cells. Science **295**, 2425–2427 (2002)
6. Campbell, I.H., Crone, B.K.: Quantum-dot/organic semiconductor composites for radiation detection. Adv. Mater. **18**, 77–79 (2006)
7. McDonald, S.A., Konstantatos, G., Zhang, S.G., Cyr, P.W., Klem, E.J.D., Levina, L., Sargent, E.H.: Solution-processed PbS quantum dot infrared photodetectors and photovoltaics. Nat. Mater. **4**, 138–U14 (2005)
8. Thelakkat, M., Maria, S., Susha, A.S., Sommer, M., Talapin, D.V., Rogach, A.L.: Semiconductor block copolymer nanocomposites with lamellar morphology via self-organization. Macromolecules **41**, 6081–6088 (2008)
9. Rogach, A.L., Jiang, G.X., Susha, A.S., Lutich, A.A., Stefani, F.D., Feldmann, J.: Cascaded FRET in conjugated polymer/quantum dot/dye-labeled DNA complexes for DNA hybridization detection. ACS Nano **3**, 4127–4131 (2009)
10. Early, K.T., Sudeep, P.K., Emrick, T., Barnes, M.D.: Polarization-driven stark shifts in quantum dot luminescence from single CdSe/oligo-PPV nanoparticles. Nano Lett. **10**, 1754–1758 (2010)
11. Odoi, M.Y., Hammer, N.I., Sill, K., Emrick, T., Barnes, M.D.: Observation of enhanced energy transfer in individual quantum dot-oligophenylene vinylene nanostructures. J. Am. Chem. Soc. **128**, 3506–3507 (2006)
12. Anni, M., Manna, L., Cingolani, R., Valerini, D., Creti, A., Lomascolo, M.: Forster energy transfer from blue-emitting polymers to colloidal CdSe/ZnS core shell quantum dots. Appl. Phys. Lett. **85**, 4169–4171 (2004)
13. Chang, T.W.F., Musikhin, S., Bakueva, L., Levina, L., Hines, M.A., Cyr, P.W., Sargent, E.H.: Efficient excitation transfer from polymer to nanocrystals. Appl. Phys. Lett. **84**, 4295–4297 (2004)
14. Mahrt, R.F., Kaufmann, S., Stoferle, T., Moll, N., Scherf, U., Tsami, A., Talapin, D.V., Murray, C.B.: Resonant energy transfer within a colloidal nanocrystal polymer host system. Appl. Phys. Lett. **90** (2007)
15. Mahrt, R.F., Stoferle, T., Scherf, U.: Energy transfer in hybrid organic/inorganic nanocomposites. Nano Lett. **9**, 453–456 (2009)
16. Stefani, F.D., Lutich, A.A., Jiang, G., Susha, A.S., Rogach, A.L., Feldmann, J.: Energy transfer versus charge separation in type-II hybrid organic–inorganic nanocomposites. Nano Lett. **9**, 2636–2640 (2009)

17. Greenham, N.C., Peng, X.G., Alivisatos, A.P.: Charge separation and transport in conjugated-polymer/semiconductor-nanocrystal composites studied by photoluminescence quenching and photoconductivity. Phys. Rev. B **54**, 17628–17637 (1996)
18. Ginger, D.S., Greenham, N.C.: Photoinduced electron transfer from conjugated polymers to CdSe nanocrystals. Phys. Rev. B **59**, 10622–10629 (1999)
19. Lin, Z.Q., Xu, J., Wang, J., Mitchell, M., Mukherjee, P., Jeffries-EL, M., Petrich, J.W.: Organic–inorganic nanocomposites via directly grafting conjugated polymers onto quantum dots. J. Am. Chem. Soc. **129**, 12828–12833 (2007)
20. Heinemann, M.D., von Maydell, K., Zutz, F., Kolny-Olesiak, J., Borchert, H., Riedel, I., Parisi, J.: Photo-induced charge transfer and relaxation of persistent charge carriers in polymer/nanocrystal composites for applications in hybrid solar cells. Adv. Funct. Mater. **19**, 3788–3795 (2009)
21. Selmarten, D., Jones, M., Rumbles, G., Yu, P.R., Nedeljkovic, J., Shaheen, S.: Quenching of semiconductor quantum dot photoluminescence by a pi-conjugated polymer. J. Phys. Chem. B **109**, 15927–15932 (2005)
22. Kang, M., Nag, O.K., Nayak, R.R., Hwang, S., Suh, H., Woo, H.Y.: Signal amplification by changing counterions in conjugated polyelectrolyte-based FRET DNA detection. Macromolecules **42**, 2708–2714 (2009)
23. Xu, Z., Hine, C.R., Maye, M.M., Meng, Q., Cotlet, M.: Shell thickness dependent photoinduced hole transfer in hybrid conjugated polymer/quantum dot nanocomposites: from ensemble to single hybrid level. ACS Nano **6**, 4984–4992 (2012)
24. Qu, L.H., Peng, X.G.: Control of photoluminescence properties of CdSe nanocrystals in growth. J. Am. Chem. Soc. **124**, 2049–2055 (2002)
25. Talapin, D.V., Mekis, I., Gotzinger, S., Kornowski, A., Benson, O., Weller, H.: CdSe/CdS/ZnS and CdSe/ZnSe/ZnS core-shell-shell nanocrystals. J. Phys. Chem. B **108**, 18826–18831 (2004)
26. Li, J.J., Wang, Y.A., Guo, W.Z., Keay, J.C., Mishima, T.D., Johnson, M.B., Peng, X.G.: Large-scale synthesis of nearly monodisperse CdSe/CdS core/shell nanocrystals using air-stable reagents via successive ion layer adsorption and reaction. J. Am. Chem. Soc. **125**, 12567–12575 (2003)
27. Yu, W.W., Qu, L.H., Guo, W.Z., Peng, X.G.: Experimental determination of the extinction coefficient of CdTe, CdSe, and CdS nanocrystals. Chem. Mater. **15**, 2854–2860 (2003)
28. Liu, Y., Kim, M., Wang, Y., Wang, Y.A., Peng, X.: Highly luminescent, stable, and water-soluble CdSe/CdS core-shell dendron nanocrystals with carboxylate anchoring groups. Langmuir **22**, 6341–6345 (2006)
29. Jasieniak, J., Smith, L., van Embden, J., Mulvaney, P., Califano, M.: Re-examination of the size-dependent absorption properties of CdSe quantum dots. J. Phys. Chem. C **113**, 19468–19474 (2009)
30. Xie, R.G., Kolb, U., Li, J.X., Basche, T., Mews, A.: Synthesis and characterization of highly luminescent CdSe-Core CdS/Zn0.5Cd0.5S/ZnS multishell nanocrystals. J. Am. Chem. Soc. **127**, 7480–7488 (2005)
31. Chen, Y., Vela, J., Htoon, H., Casson, J.L., Werder, D.J., Bussian, D.A., Klimov, V.I., Hollingsworth, J.A.: "Giant" multishell CdSe nanocrystal quantum dots with suppressed blinking. J. Am. Chem. Soc. **130**, 5026 (2008)
32. Htoon, H., Malko, A.V., Bussian, D., Vela, J., Chen, Y., Hollingsworth, J.A., Klimov, V.I.: Highly emissive multiexcitons in steady-state photoluminescence of individual "giant" CdSe/CdS core/shell nanocrystals. Nano Lett. **10**, 2401–2407 (2010)
33. Zylstra, J., Amey, J., Miska, N.J., Pang, L., Hine, C.R., Langer, J., Doyle, R.P., Maye, M.M.: A modular phase transfer and ligand exchange protocol for quantum dots. Langmuir **27**, 4371–4379 (2011)
34. Mayya, K.S., Schoeler, B., Caruso, F.: Preparation and organization of nanoscale polyelectrolyte-coated gold nanoparticles. Adv. Funct. Mater. **13**, 183–188 (2003)
35. Hines, M.A., Guyot-Sionnest, P.: Synthesis and characterization of strongly luminescing ZnS-Capped CdSe nanocrystals. J. Phys. Chem. **100**, 468–471 (1996)

36. Dabbousi, B.O., RodriguezViejo, J., Mikulec, F.V., Heine, J.R., Mattoussi, H., Ober, R., Jensen, K.F., Bawendi, M.G.: (CdSe) nS core-shell quantum dots: synthesis and characterization of a size series of highly luminescent nanocrystallites. J. Phys. Chem. B **101**, 9463–9475 (1997)
37. Zhu, H.M., Song, N.H., Lian, T.Q.: Controlling charge separation and recombination rates in CdSe/ZnS type I core-shell quantum dots by shell thicknesses. J. Am. Chem. Soc. **132**, 15038–15045 (2010)
38. Jiang, Z.J., Kelley, D.F., Leppert, V.: Static and dynamic emission quenching in core/shell nanorod quantum dots with hole acceptors. J. Phys. Chem. C **113**, 19161–19171 (2009)
39. Huang, J.E., Huang, Z.Q., Jin, S.Y., Lian, T.Q.: Exciton dissociation in CdSe quantum dots by hole transfer to phenothiazine. J. Phys. Chem. C **112**, 19734–19738 (2008)
40. Haus, J.W., Zhou, H.S., Honma, I., Komiyama, H.: Quantum confinement in semiconductor heterostructure nanometer-size particles. Phys. Rev. B **47**, 1359–1365 (1993)
41. Schooss, D., Mews, A., Eychmuller, A., Weller, H.: Quantum-dot quantum-well CDS/HGS/CDS - theory and experiment. Phys. Rev. B **49**, 17072–17078 (1994)
42. Efros, A.L., Rosen, M.: Random telegraph signal in the photoluminescence intensity of a single quantum dot. Phys. Rev. Lett. **78**, 1110–1113 (1997)
43. Schlegel, G., Bohnenberger, J., Potapova, I., Mews, A.: Fluorescence decay time of single semiconductor nanocrystals. Phys. Rev. Lett. **88** (2002)
44. Zhang, K., Chang, H.Y., Fu, A.H., Alivisatos, A.P., Yang, H.: Continuous distribution of emission states from single CdSe/ZnS quantum dots. Nano Lett. **6**, 843–847 (2006)
45. Xu, Z.H., Cotlet, M.: Quantum dot-bridge-fullerene heterodimers with controlled photoinduced electron transfer. Angew. Chem. Int. Ed. **50**, 6079–6083 (2011)
46. Baranov, A.V., Rakovich, Y.P., Donegan, J.F., Perova, T.S., Moore, R.A., Talapin, D.V., Rogach, A.L., Masumoto, Y., Nabiev, I.: Effect of ZnS shell thickness on the phonon spectra in CdSe quantum dots. Phys. Rev. B **68** (2003)
47. Song, N.H., Zhu, H.M., Jin, S.Y., Zhan, W., Lian, T.Q.: Poisson-distributed electron-transfer dynamics from single quantum dots to C60 molecules. ACS Nano **5**, 613–621 (2011)
48. Jin, S.Y., Lian, T.Q.: Electron transfer dynamics from single CdSe/ZnS quantum dots to TiO (2) nanoparticles. Nano Lett. **9**, 2448–2454 (2009)
49. Issac, A., Jin, S.Y., Lian, T.Q.: Intermittent electron transfer activity from single CdSe/ZnS quantum dots. J. Am. Chem. Soc. **130**, 11280 (2008)
50. Biju, V., Micic, M., Hu, D.H., Lu, H.P.: Intermittent single-molecule interfacial electron transfer dynamics. J. Am. Chem. Soc. **126**, 9374–9381 (2004)
51. Wang, Y.M., Wang, X.F., Ghosh, S.K., Lu, H.P.: Probing single-molecule interfacial electron transfer dynamics of porphyrin on TiO2 nanoparticles. J. Am. Chem. Soc. **131**, 1479–1487 (2009)
52. Cotlet, M., Masuo, S., Luo, G.B., Hofkens, J., Van der Auweraer, M., Verhoeven, J., Mullen, K., Xie, X.L.S., De Schryver, F.: Probing conformational dynamics in single donor-acceptor synthetic molecules by means of photoinduced reversible electron transfer. Proc. Natl. Acad. Sci. USA **101**, 14343–14348 (2004)
53. Bell, T.D.M., Stefan, A., Masuo, S., Vosch, T., Lor, M., Cotlet, M., Hofkens, J., Bernhardt, S., Mullen, K., van der Auweraer, M., Verhoeven, J.W., De Schryver, F.C.: Electron transfer at the single-molecule level in a triphenylamine-perylene imide molecule. Chemphyschem **6**, 942–948 (2005)
54. Jha, P.P., Guyot-Sionnest, P.: Trion decay in colloidal quantum dots. ACS Nano **3**, 1011–1015 (2009)
55. Zhao, J., Nair, G., Fisher, B.R., Bawendi, M.G.: Challenge to the charging model of semiconductor-nanocrystal fluorescence intermittency from off-state quantum yields and multiexciton blinking. Phys. Rev. Lett. **104**, 157403 (2010)
56. Rosen, S., Schwartz, O., Oron, D.: Transient fluorescence of the off state in blinking CdSe/CdS/ZnS semiconductor nanocrystals is not governed by Auger recombination. Phys. Rev. Lett. **104**, 157404 (2010)

57. Frantsuzov, P.A., Marcus, R.A.: Explanation of quantum dot blinking without the long-lived trap hypothesis. Phys. Rev. B **72**, 155321 (2005)
58. Galland, C., Ghosh, Y., Steinbruck, A., Sykora, M., Hollingsworth, J.A., Klimov, V.I., Htoon, H.: Two types of luminescence blinking revealed by spectroelectrochemistry of single quantum dots. Nature **479**, 203–U75 (2011)
59. Kuno, M., Fromm, D.P., Hamann, H.F., Gallagher, A., Nesbitt, D.J.: Nonexponential "blinking" kinetics of single CdSe quantum dots: a universal power law behavior. J. Chem. Phys. **2000**(112), 3117–3120 (2000)
60. Drndic, M., Wang, S., Querner, C., Emmons, T., Crouch, C.H.: Fluorescence blinking statistics from CdSe core and core/shell nanorods. J. Phys. Chem. B **110**, 23221–23227 (2006)
61. Peterson, J.J., Nesbitt, D.J.: Modified power law behavior in quantum dot blinking: a novel role for biexcitons and Auger ionization. Nano Lett. **9**, 338–345 (2009)
62. Xu, Z., Cotlet, M.: Photoluminescence blinking dynamics of colloidal quantum dots in the presence of controlled external electron traps. Small **8**, 253–258 (2012)
63. Spinicelli, P., Buil, S., Quelin, X., Mahler, B., Dubertret, B., Hermier, J.P.: Bright and grey states in CdSe-CdS nanocrystals exhibiting strongly reduced blinking. Phys. Rev. Lett. **102** (2009)
64. Kipp, T., Ma, X.D., Tan, H., Mews, A.: Fluorescence enhancement, blinking suppression, and gray states of individual semiconductor nanocrystals close to gold nanoparticles. Nano Lett. **10**, 4166–4174 (2010)

Chapter 5
Theory of Quantum Dot Arrays for Solar Cell Devices

Stanko Tomić

Abstract Vertically or laterally coupled semiconductor quantum dot (QD) arrays emerged recently as promising structures for the next generation of high-efficiency intermediate band solar cell (IBSC), due to their ability to facilitate the formation of mini-bands. The quantum coupling effect, that exists between states in QDs of an array, influences the electronic and optical properties of such structures. We present here a method based on multi-band $\mathbf{k} \cdot \mathbf{p}$ Hamiltonian combined with periodic boundary conditions, applied to predict the electronic and optical properties of InAs/GaAs QDs based vertical and lateral QD arrays. Formation of the intermediate band (IB) in all cases was achieved via delocalisation of the electron ground state ($e0$). By changing the geometry of the QD arrangement in arrays we have identified conditions for the IB to be separated by a pure zero density of states from the rest of states in the conduction band (CB). Due to symmetry of the QD array lattice and QD states itself, which are C_{2v} for the zinc blende QDs, the electronic and absorption structure needs to be obtained via sampling throughout the reciprocal space in the first Brillouin zone of QD arrays.

5.1 Introduction

The intermediate-band solar cell (IBSC) [1] is one of the promising candidates for next generation solar cells (SC) with an estimated detailed balance efficiency of 63.8 % under idealized conditions and maximum concentration [2]. The intermediate band (IB) material is characterised by the existence of an electronic energy band of allowed states within the conventional-energy band gap, E_g(CB,VB), of the host material, splitting it into two subgaps, E_g(CB,IB) and E_g(IB,VB). This band

S. Tomić (✉)
Joule Physics Laboratory, School of Computing, Science and Engineering,
University of Salford, Manchester M5 4WT, UK
e-mail: s.tomic@salford.ac.uk

J. Wu and Z.M. Wang (eds.), *Quantum Dot Solar Cells*, Lecture Notes in Nanoscale Science and Technology 15, DOI 10.1007/978-1-4614-8148-5_5,

allows the creation of additional electron–hole pairs from the absorption of two subband gap energy photons. Under this assumption, the first photon promotes an electron from the valence band (VB) to the IB, and a second photon simultaneously pumps another electron from the IB to the conduction band (CB). To this end, it is necessary that the IB is half filled with electrons so that it can supply electrons to the CB as well as receive them from the VB. The electron–hole pairs generated in this way add up to the conventionally generated ones by the absorption of a single photon, the third one, pumping an electron from the VB directly to the CB. Therefore, if quasi-Fermi level separations between bands are sustainable upon external excitation, the photocurrent of the solar cell, and ultimately its efficiency, is enhanced. According to this concept, increase in photocurrent in IBSC occurs without degradation of the output voltage of the cell. The output voltage is given by the split between the CB's electron and the VB's hole quasi-Fermi levels, which is still limited by the total band gap E_g of the barrier material.

There are currently two approaches pursued to achieve the operation of the IBSC: (1) bulk material doped with impurities that form the band in otherwise forbidden energy gap of the host material and (2) formation of bands via quantum coupling of discrete states in the neighbouring QDs of a periodic QD array matrix. In approach (1), to design IB in bulk material via doping with impurities, formation of the band is demonstrated theoretically by using the electronic structure methods with periodic boundary conditions [3–6], which introduce artificial periodicity between impurities. In reality though, existence of random impurities only exaggerate the unwanted effect of the nonradiative Shockley-Read-Hall (SRH) recombination, leading to strong deterioration of the efficiency [7]. Concept (2), formation of bands via quantum coupling of discrete states in the neighbouring QDs of a periodic array is nothing less ambiguous. Although the formation of bands in the solid state is supported by weakly bound *valence electrons* in the outermost shells of an atom, in the IBSC based on QD arrays, the requirement is for the electron ground state (that could be understood as deep core electrons in analogy to solid state) to form a band! In this concept two different types of transitions, intraband (between IB and CB) and interband (between VB and IB, and between VB and CB), either radiative or nonradiative, need to be tuned to be on a similar time scale, in order to facilitate formation of the relevant quasi-Fermi levels (for CB, VB and most critically for IB!) and to ensure that the operation of the device benefits from the existence of the IB.

Introduction of new bands increases chances for additional radiative and nonradiative recombination processes involving IBs. To overcome this problem, the IBSC requires solar concentration to ensure its optimal operation when the carrier photogeneration is outperforming recombination via IB. So far, the most effort has been devoted to fabricate SC devices based on semiconductor quantum dots (QDs) and to study its operation principle [8–15], with limited or no evidence of IB operation. Despite this significant effort, evaluation of carrier dynamics via IB for IBSC presents both fundamental and technical obstacles, limiting its application and penetration into the PV market.

In this chapter we review theoretical methods for the analysis of electronic and optical properties of semiconductor QD arrays. We also relate the absorption properties to radiative times that exist between various bands in the InAs/GaAs QD-based IBSC.

5.2 Electronic Structure

The quantum mechanical description of electrons in Quantum Dot Array Based Solar Cells requires detailed knowledge of their wavefunctions, $\psi_{n,\,\mathbf{k}}(\mathbf{r})$, which are found by solving the Schrödinger equation:

$$H_0\psi_{n,\mathbf{k}}(\mathbf{r}) = E_{n,\mathbf{k}}\psi_{n,\mathbf{k}}(\mathbf{r}). \tag{5.1}$$

The Hamiltonian in Eq. (5.1), $H_0 = p^2/2m_0 + V(\mathbf{r})$, is a function of the quantum mechanical momemtum operator, $\mathbf{p} = -i\hbar\nabla$, and the crystal potential experienced by electrons, $V(\mathbf{r}) = V(\mathbf{r} + \mathbf{R})$, which is a periodic function, with the periodicity of the crystal lattice, **R**. According to Bloch's theorem, the solutions to this Schrödinger equation can be written as: $\psi_{n,\,\mathbf{k}}(\mathbf{r}) = e^{i\mathbf{k}\cdot\mathbf{r}}u_{n,\,\mathbf{k}}(\mathbf{r})$, where **k** is the electron wavevector, n is the band index and $u_{n,\,\mathbf{k}}$ is the *cell-periodic* function, with the same periodicity as the crystal lattice. The cell-periodic function, $u_{n,\,\mathbf{k}}$, satisfies the equation:

$$H_{\mathbf{k}}u_{n,\mathbf{k}} = E_{n,\mathbf{k}}u_{n,\mathbf{k}} \tag{5.2}$$

where the Hamiltonian:

$$H_{\mathbf{k}} = H_0 + H'_{\mathbf{k}} = \frac{p^2}{2m_0} + V + \frac{\hbar^2k^2}{2m_0} + \frac{\hbar\mathbf{k}\cdot\mathbf{p}}{m_0} \tag{5.3}$$

is given as a sum of two terms: the unperturbed Hamiltonian, H_0, which in fact equals the exact Hamiltonian at $\mathbf{k} = 0$ (i.e. at the Γ point in the Brillouin zone), and the "perturbation", $H_{\mathbf{k}}'$. Equation (5.3) is called the $\mathbf{k}\cdot\mathbf{p}$ Hamiltonian [16–18]. The $\mathbf{k}\cdot\mathbf{p}$ perturbation theory is most accurate for small values of **k**. However, if enough terms are included in the perturbative expansion, then the theory can in fact be reasonably accurate for any value of **k** in the entire Brillouin zone [19–23]. The quantities required for the parametrisation of the $\mathbf{k}\cdot\mathbf{p}$ Hamiltonian: the band edge energies, $E_{n,\,0}$, and the optical matrix elements, $\langle u_{n,\,0} \mid \mathbf{p} \mid u_{n',\,0}\rangle$, are typically inferred from experimental data and detailed atomistic-based theories.

Particular strengths of the $\mathbf{k}\cdot\mathbf{p}$ theory are straightforward inclusion of the spin–orbit interaction, the strain effects on the band structure via deformation potential theory [24] and piezoelectricity due to anisotropy of the strain around the 3D shape of the QD. With all those effects included, the $\mathbf{k} \cdot \mathbf{p}$ Hamiltonian suitable for the analysis of QD arrays becomes:

$$\hat{H}_{\mathbf{k}\cdot\mathbf{p}} = \frac{p^2}{2m_0} + V + \frac{\hbar^2 k^2}{2m_0} + \frac{\hbar\mathbf{k}\cdot\mathbf{p}}{m_0} + \frac{1}{4m_0^2c^2}(\sigma \times \nabla V)\cdot(\mathbf{k}+\mathbf{p}) + \mathbf{D}^{\varepsilon}\cdot\varepsilon + V_{\mathrm{PZ}}(\varepsilon) \tag{5.4}$$

where the $\mathbf{D}^{\varepsilon}$ is the deformation potential operator and $\sigma = (\sigma_x, \sigma_y, \sigma_z)$ is a vector consisting of the three Pauli spin matrices, $\boldsymbol{\varepsilon}$ is the strain tensor and $V_{\mathrm{PZ}}(\boldsymbol{\varepsilon})$ is the piezoelectric potential [25].

The theoretical model of a QD array electronic structure, presented in this chapter, is based on the 8-band $\mathbf{k} \cdot \mathbf{p}$ implementation [26] of the Hamiltonian in Eq. (5.4) adapted for QD nanostructures [27, 28]. This Hamiltonian takes into account band mixing between the lower conduction band s-anti-bonding and top three valence band p-bonding states (all spin degenerate), in the underlying QD and barrier materials, and includes the spin–orbit interaction, strain and the first order piezoelectric field.

5.2.1 Plane Waves Implementation of QD Array Electronic Structure

The QD as a three-dimensional object breaks the translational symmetry of the bulk material along all three Cartesian directions implying operator replacement $k_\nu \rightarrow -i\,\partial/\partial\,\nu$ in Eq. (5.4), where $\nu = (x, y, z)$. To solve the multi-band system of Schrödinger equations, Eq. (5.4), the plane wave (PW) methodology is employed as an expansion method [29–32]. In the PW representation the eigenvalues (E_n) and coefficients ($A_{n,\mathbf{k}}$) of the nth-eigenvector, [$\psi_n(\mathbf{r}) = \sum_{\mathbf{k}} A_{n,\mathbf{k}} \mid \mathbf{k}\rangle$ and $\mid \mathbf{k}\rangle = e^{i\mathbf{kr}}$], are linked by the relation

$$\sum_{m,\mathbf{k}'} h_{m,n}(\mathbf{k}', \mathbf{k}) A_{n,\mathbf{k}} = E_n A_{n,\mathbf{k}} \tag{5.5}$$

where

$$h_{m,n}(\mathbf{k}', \mathbf{k}) = \frac{1}{\Omega}\langle \mathbf{k}' | H_{\mathbf{k}\cdot\mathbf{p}} | \mathbf{k}\rangle, \tag{5.6}$$

are the Fourier transform of the Hamiltonian matrix elements, and $m, n \in \{1, \ldots, 8\}$ are the band indexes of the 8-band $\mathbf{k}\cdot\mathbf{p}$ Hamiltonian. All the elements in the Hamiltonian matrix, Eq. (5.5), can be expressed as a linear combination of different kinetic and strain related terms and its convolution with the characteristic function of the actual QD shape, $\chi_{\mathrm{qd}}(\mathbf{k})$ [32, 33]. The whole $\mathbf{k}$-space is discretized by embedding the QD in a rectangular box of dimensions L_x, L_y and L_z and volume

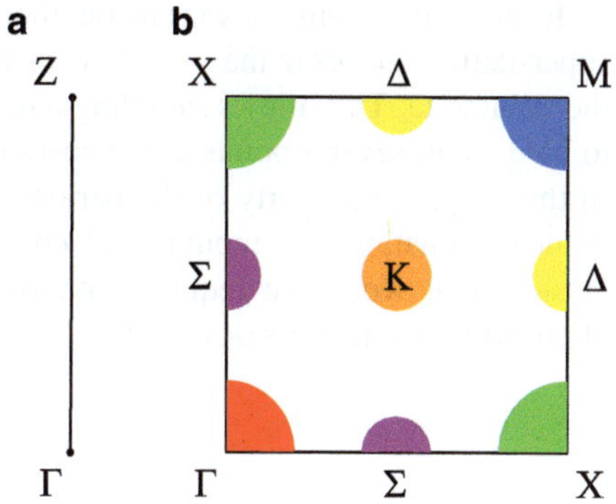

Fig. 5.1 Schematic of the first Brillouin zone along with few characteristic points of (**a**) 1D vertically aligned QD array and (**b**) 2D in-plane arranged QD array

$\Omega = L_x \times L_y \times L_z$ and choosing the **k**-vectors in the form of $\mathbf{k} = 2\pi(n_x/L_x, n_y/L_y, n_z/L_z)$, where n_x, n_y and n_z are integers whose change controls the convergence of the method.

Several characteristics have contributed to the popularity of the plane wave method: plane-wave representation of all operators in the envelope Hamiltonian is analytical, strain and piezo field distributions can be calculated analytically in Fourier space and a relatively small number of plane waves is sufficient for good accuracy [32].

However, there is one particular strength of the plane wave method when its application to periodic QD array structures is concerned: It inherently assumes *periodic Born-von Karman boundary conditions*. In such a way, it introduces translational symmetry of the system. This leads to physical interaction of a QD with its periodically replicated images through: (a) electronic coupling between states of neighbouring dots; (b) propagation of strain field of neighbouring dots and (c) Coulomb interaction between carriers in the dot with carriers in its images.

The electronic structure of a QD array is characterised by a Brillouin zone (BZ) determined by the QD array dimensions, Fig. 5.1 [34–41]. To calculate the electronic structure the only modification required to the basis set is to replace the reciprocal lattice vectors in the PW expansion with those shifted due to the QD-superlattice (QD-SL):

1. in the case of vertically aligned QDs in a columnar QD array:

$$|k_z\rangle = e^{ik_z z} \to |k_z + K_z\rangle = e^{i(k_z+K_z)z} \tag{5.7}$$

   with L_x and L_y chosen to be large enough to prevent the lateral coupling;
2. in the case of laterally aligned QDs in an in-plane QD matrix:

$$|\mathbf{k}_{||}\rangle = e^{i\mathbf{k}_{||}\mathbf{r}_{||}} \to |\mathbf{k}_{||} + \mathbf{K}_{||}\rangle = e^{i(\mathbf{k}_{||}+\mathbf{K}_{||})\mathbf{r}_{||}} \tag{5.8}$$

where $\mathbf{k}_{||} = k_x\hat{x} + k_y\hat{y}$, $\mathbf{K}_{||} = K_x\hat{x} + K_y\hat{y}$ and $\mathbf{r}_{||} = x\hat{x} + y\hat{y}$, and L_z is chosen to be large enough to prevent the vertical coupling.

In the expressions for expansion above: $0 \leq K_\nu \leq \pi/L_\nu$, where the L_ν are the super-lattice vectors in the $\nu = (x, y, z)$ directions, and correspond to those forming the volume Ω. This allows sampling along the **K** points of a QD-SL Brillouin zone to be done at several points at the cost of a single QD calculation at each point **K**. In this way, the property of the periodic boundary conditions can be exploited to reduce dramatically computational costs and to avoid calculations on large QD clusters that would be required otherwise. All the results presented here were obtained by using the kppw code [31, 32].

5.3 Absorption

The expression giving the absorption coefficient within the dipole approximation is,

$$\alpha(\hbar\omega) = \frac{\pi e^2}{c\epsilon_0 m_0^2 \bar{n}\omega\Omega} \sum_{\mathbf{K}} \sum_{i,f} |\langle i|\hat{\mathbf{e}} \cdot \mathbf{p}(\mathbf{K})|f\rangle|^2 \delta[E_f(\mathbf{K}) - E_i(\mathbf{K}) - \hbar\omega](f_i - f_f), \tag{5.9}$$

where e is the electron charge, $\bar{n}$ the refractive index, ε_0 the dielectric permittivity of vacuum, c the speed of light, Ω the volume of the structure, $\hat{\mathbf{e}}$ the unit vector of light polarization and $\mathbf{p(K)}$ is the momentum operator. The optical dipole matrix element is given by $\langle i|\hat{\mathbf{e}} \cdot \mathbf{p}(\mathbf{K})|f\rangle$ and varies inside the Brillouin zone (BZ) of an QD array [38]. The initial and final state energies are $E_i(\mathbf{K})$ and $E_f(\mathbf{K})$, which also varies throughout the QD array BZ, $\hbar\omega$ is the photon energy, and $f_{i(f)}$ is the initial (final) state Fermi–Dirac distribution function.

(i) in the case of vertically aligned QDs in an columnar QD array:

$$\sum_{\mathbf{K}} \to \sum_{K_z} \quad \textit{and} \quad \sum_{K_z} \to \frac{L_z}{2\pi} \int dK_z \tag{5.10}$$

and L_x and L_y are chosen to be large enough to prevent the lateral coupling;
(ii) in the case of laterally aligned QDs in an in-plane QD matrix:

$$\sum_{\mathbf{K}} \to \sum_{K_x} \sum_{K_y} \quad \textit{and} \quad \sum_{K_x} \to \frac{L_x}{2\pi} \int dK_x, \ \sum_{K_y} \to \frac{L_y}{2\pi} \int dK_y. \tag{5.11}$$

In order, to avoid integration throughout the whole BZ in laterally arranged QDs array, we sample the BZ at several characteristic points (Γ, X, M, Σ, Δ, K), see Fig. 5.1. In accordance with the symmetries of the QD (which is C_{2v}) and rectangular lattice of a QD array, the absorption coefficient can be approximated by:

$$\alpha \approx \frac{1}{16}[\alpha(\Gamma) + 2\alpha(X) + \alpha(M) + 4\alpha(\Sigma) + 4\alpha(\Delta) + 4\alpha(K)] \tag{5.12}$$

5.3.1 Band to Continuum Absorption

Absorption between the IB and CB can be successfully calculated using Eq. (5.9) for states in the CB closely spaced to the bottom of the CB edge of the barrier material [41]. For states higher up in energy, in the continuum of the CB, the expression Eq. (5.9) needs to be modified to take into account the specific shape of the wave functions in the continuum [42, 43]. We derive here the PW representation of the expression for the band-to-continuum absorption.

In the parabolic approximation for the energy dispersion in continuum, the normalised wave function of the corresponding state in continuum becomes a plane wave:

$$\psi_c(\mathbf{r}) = \frac{1}{\sqrt{\Omega}} e^{i\mathbf{k}'\mathbf{r}}, \tag{5.13}$$

with the wave vector $\mathbf{k}'$ related to the energy, E', in the continuum by: $E' = \hbar^2\mathbf{k}'^2/2\,m_b^*$. Combining with the normalised bound state

$$\psi_i(\mathbf{r}) = \frac{1}{\sqrt{\Omega}} \sum_{\mathbf{k}} A_{i,\mathbf{k}} e^{i\mathbf{k}\mathbf{r}} \tag{5.14}$$

the bound-continuum dipole that enters Eq. (5.9) becomes:

$$\langle \psi_i(\mathbf{r}) | \hat{\mathbf{e}} \cdot \ \mathbf{p} | \psi_c(\mathbf{r}) \rangle = \frac{1}{\Omega} \sum_{\mathbf{k},\mathbf{k}'} \delta(\mathbf{k}-\mathbf{k}') \int_{\Omega} A_{i,\mathbf{k}}^* \frac{\partial H_{\mathbf{k}\cdot\mathbf{p}}(\mathbf{k})}{\partial \mathbf{k}} e^{i(\mathbf{k}-\mathbf{k}')\mathbf{r}} d\mathbf{r} \tag{5.15}$$

or after realisation of the momentum conservation criteria, $\delta(\mathbf{k}-\mathbf{k}')$:

$$\langle \psi_i(\mathbf{r}) | \hat{\mathbf{e}} \cdot \ \mathbf{p} | \psi_c(\mathbf{r}) \rangle = \frac{1}{\Omega} \sum_{\mathbf{k}} A_{i,\mathbf{k}}^* \frac{\partial H_{\mathbf{k}\cdot\mathbf{p}}(\mathbf{k})}{\partial \mathbf{k}} \int_{\Omega} d\mathbf{r} = \sum_{\mathbf{k}} A_{i,\mathbf{k}}^* \frac{\partial H_{\mathbf{k}\cdot\mathbf{p}}(\mathbf{k})}{\partial \mathbf{k}} \tag{5.16}$$

The absorption on bound-free transitions is given by a similar expression to Eq. (5.9) in which the summation over discrete final states turns into an integral over the final-state wave vector $\mathbf{k}'$:

$$\sum_f \to \frac{\Omega}{(2\pi)^3} \int d\mathbf{k}'. \tag{5.17}$$

After substitution of Eqs. (5.16) and (5.17) into Eq. (5.9):

$$\alpha(\hbar\omega) = \frac{e^2}{8\pi^2 c\varepsilon_0 m_0^2 \bar{n}\omega} \sum_i^{N_b} \int_{\mathbf{k}'} \left| \sum_{\mathbf{k}} A_{i,\mathbf{k}}^* \frac{\partial H_{\mathbf{k}\cdot\mathbf{p}}(\mathbf{k})}{\partial \mathbf{k}} \right|^2 (f_i - f_{\mathbf{k}'}) \frac{\Gamma}{\Gamma^2 + \left(\frac{\hbar^2\mathbf{k}'^2}{2m_b^*} - E_i - \hbar\omega \right)^2} d\mathbf{k}' \tag{5.18}$$

where N_b is the number of bound states in the CB, and Γ the line broadening of the final state. The energy conservation criteria in Eq. (5.9) is relaxed by Lorentzian line shape.

5.4 Model QD Arrays

The model QD array considered here consists of InAs/GaAs QDs with truncated pyramidal shape. The size and shape of the QD are controlled by the pyramid base length, b, its height h, and truncation factor, t, defined as a ratio between length of the pyramid side at h and length of the pyramid base b and is set to 0.5, see Fig. 5.2a. The QDs are embedded in the tetragonal like unit cell, $\Omega = (L_x, L_y, L_z)$, Fig. 5.2a.

For vertically coupled QDs, Fig. 5.2b, the vertical periodicity of the QD array is controlled by $L_z = d_z + h + L_{wl}$, where d_z is the vertical separation of the QDs in subsequent layers, i.e. the distance between the top of the QDs and the bottom of the wetting layer (WL) in the subsequent layer. The d_z is varied with $L_x = L_y$ chosen to be large enough to prevent lateral electronic coupling.

For lateral, in-plane, quantum coupled QD arrays, QDs are arranged in a two-dimensional $x - y$ lattice, as shown in Fig. 5.2c. In the square lattice model, $L_x = L_y = b + d$, where $d = d_x = d_y$ is the lateral separation of the QDs, i.e. the side to side distance between two neighbouring QDs. The parameter d is varied with L_z chosen to be large enough to prevent vertical electronic coupling.

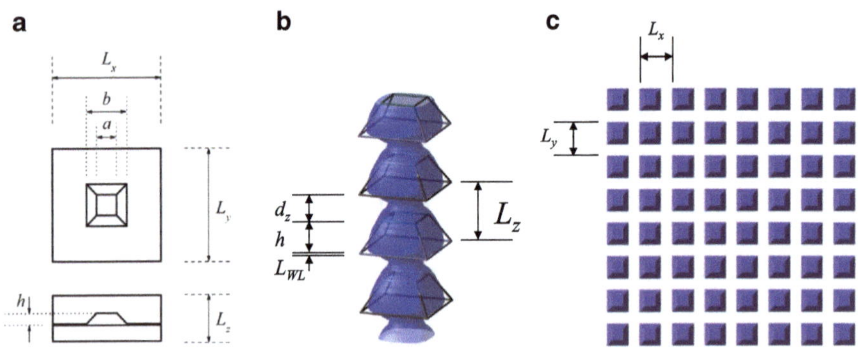

Fig. 5.2 Schematic of a single QD that is used to makes an array (**a**), QD truncation factor is defined as $t = a/b$, while all the other relevant parameters are defined in the main text. The vertically aligned 1D QD array (**b**) and 2D laterally aligned QD array (**c**)

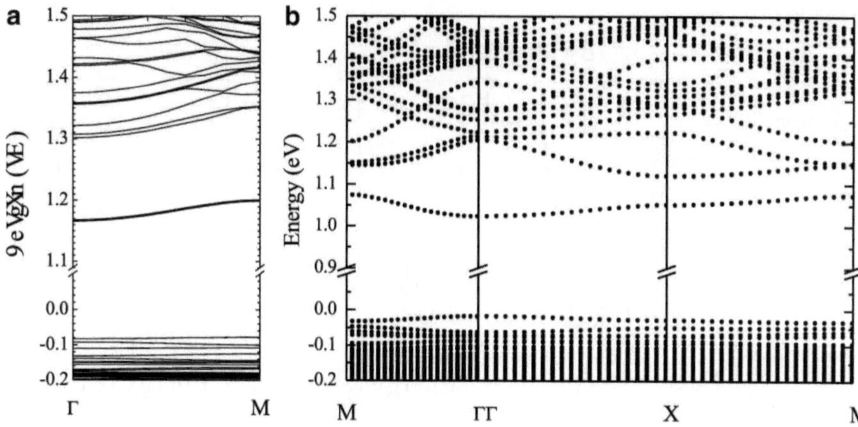

Fig. 5.3 Electronic structure of: (**a**) 1D vertically aligned QD array made of QDs $b = 6$ nm, $d_z = 4$ nm and (**b**) 2D in-plane arranged QD matrix made of QDs, $b = 10$ nm, $d_x = d_y = 1$ nm through the first Brillouin zone

5.5 Electronic Structure of QD Arrays

The formation of an IB through energy dispersion inside the first BZ is closely connected with the QD periodicity, defined via d_z in the case of a 1D vertically aligned QD array, or d_x and d_y in the case of a 2D laterally aligned QD array. As shown in Fig. 5.1, for a 1D BZ of vertical coupled QDs, two characteristic points, Γ and Z, and points along the path between them are used to quantify the energy dispersion for different K_z values, Fig. 5.3a.

Figure 5.3b shows the band dispersion of the laterally coupled 2D QD array along different characteristic directions inside the first BZ, following the sequence M-Γ-X-M of a square lattice. For the laterally coupled QD array, where the 2D BZ contains more characteristic points than in the 1D case, several different directions and characteristic points in the BZ need to be calculated, in order to locate the IB's minimum and maximum, i.e. IB bandwidths and gaps between them. The close proximity of the energy points at M and X, of the IB formed by $e0$ states, confirms this. Energy dispersion along different **K** values finally forms the electronic band structure of the QD array material.

5.6 Absorption of QD Arrays

5.6.1 *Vertical QD Arrays*

We first analysed the absorption spectra of three representative 1D vertical QD arrays, Fig. 5.4, made of: (a) small QD of $h = 3$ nm height, $b = 6$ nm base length and $L_x = 20$ nm; (b) medium QD of $h = 6$ nm, $b = 12$ nm and $L_x = 20$ nm and

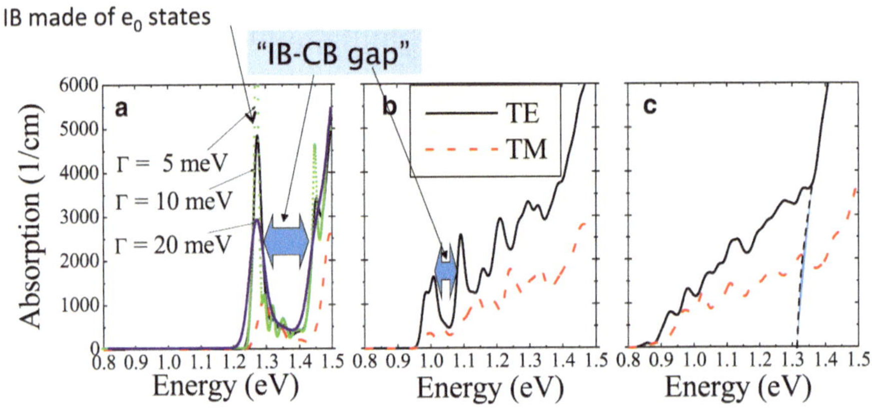

Fig. 5.4 Intraband transversal electric (TE) and transversal magnetic (TM) components of the absorption spectra of three representative QD arrays discussed in the main text, calculated with Γ=10 meV. TE spectra of structure (**a**) for $\Gamma = 5$, 10 , 20 meV (*dotted*, *solid*, and *dashed lines*, respectively) demonstrate the robustness of the IB with respect to broadening. (**a**) QD size in array is $b = 6$nm, (**b**) QD size in array is $b = 12$ nm and (**c**) QD size is $b = 20$ nm. In (**c**) *dotted line* is given for orientation and represents GaAs bulk-like absorption edge

(c) big QD of $h = 10$ nm, $b = 20$ nm and $L_x = 40$ nm. Comparison of the absorption characteristics, computed using Eq. (5.9), of the three representative, vertically aligned QD arrays given in Fig. 5.4, suggests that QD arrays with small to medium QD sizes ($b \sim 6 - 12$ nm) are the best candidates for the active region of a high efficiency IBSC. Those samples exhibit a well-defined absorption peak related to the IB that is separated from the rest of the absorption spectra by a very low density of states, even when line broadening of $\Gamma = 20$ meV is assumed. This is essential for providing not only a VB to IB absorption path but also for opening an energy gap between the IB and the rest of CB, thereby a second photon with an energy below the barrier material energy gap to be absorbed. Moving from medium to big QDs, this peak is reduced and becomes increasingly hybridized with the rest of the absorption spectra as the density of both CB and VB states increases, due to increased QD size. As a consequence the larger QD array ($b = 20$ nm) would behave simply as a bulk-like material with a redshifted absorption spectrum, extending bulk absorption capabilities towards higher wavelengths, but not providing energy separation for a third quasi-Fermi level within the IB. This statement is valid only if one assumes pure InAs for the QD region. If the amount of In is reduced, due to group III intermixing [44], or if In grading within the QD region is assumed [45], the design of an isolated IB may be possible even with large dots due to an effective reduction in the confinement both in the CB and VB of the QD region, which would facilitate formation of the IBs. The first four optical transitions have recently been observed in photoreflectance measurements on a QD array structure ($b = 10$ nm), for IBSC [46]. Our theoretical results for the similar structure (b) ($b = 12$ nm) are in very good agreement, taking into account differences in the QD sizes and the approximations assumed in the model presented here.

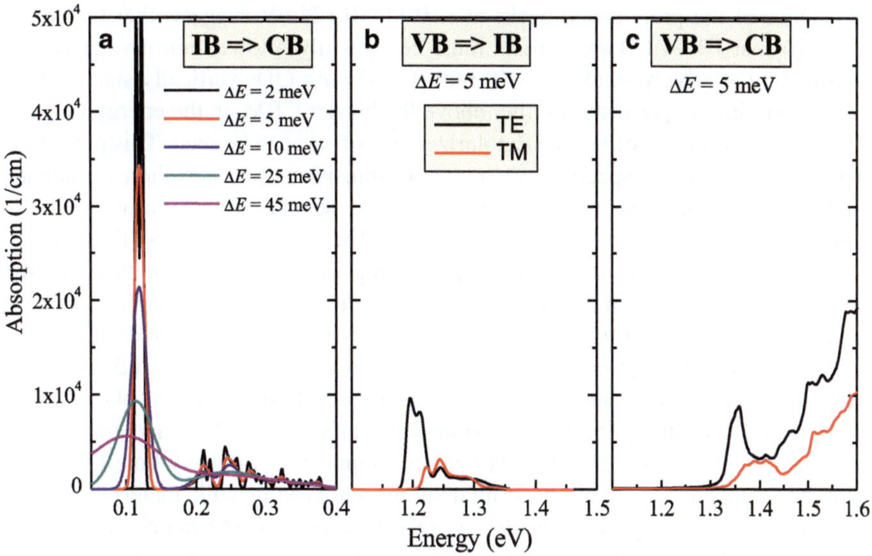

Fig. 5.5 Absorption spectra of QD array with vertical periodicity $d_z = 5$ nm. (**a**) IB → CB absorption, (**b**) VB → IB absorption and (**c**) VB → CB absorption. Both the TE (*solid line*) and TM (*dashed line*) polarizations are presented. To monitor the strength of the IB → CB absorption coefficient, the spectra were broadened by changing the line broadening from $\Delta E = 2$ meV to $\Delta E = 25$ meV

Under the assumption that the IB is of the finite width, and that carrier dynamics are such that formation of three distinct quasi-Fermi levels under external illumination is supported, the absorption spectra between CB and VB mini-bands given in Fig. 5.4 can be decomposed into three different absorption pathways: $\alpha^{\mathrm{VB}\rightarrow\mathrm{IB}}(\hbar\omega)$, $\alpha^{\mathrm{IB}\rightarrow\mathrm{CB}}(\hbar\omega)$ and $\alpha^{\mathrm{VB}\rightarrow\mathrm{CB}}(\hbar\omega)$.

Figure 5.5 shows all relevant absorption coefficients of the vertical QD array made of truncated pyramidal InAs/GaAs QDs with $b = 10$ nm, $h = 2.5$ nm and $t = 0.5$, and with vertical spacing between the dots in the array of $d_z = 5$ nm. First in Fig. 5.5a the transitions from the QD array CB ground state, e_0, miniband to all other electron minibands higher in energy are presented. A sharp double-peak that appears at 0.117 and 0.125 eV (FWHM=14 meV) corresponds to all of the momentum-allowed transitions between e_0 and e_1, e_2 minibands. This IB→CB absorption peak is completely TE polarized[1] due to S symmetry of the QD states forming the e_0 miniband and $P_{x,y}$ symmetry of the QD states forming the e_1 and e_2 minibands. It should be emphasized that the e_1 and e_2 minibands are already overlapped with the conduction band minima (CBM) of the barrier material.

[1] The TE means that the optical field is polarized along any of, $\hat{e}_x + \hat{e}_y$, directions that are "in-plane" of the structure, and is perpendicular to the QD array growth axis; TM means that the optical field is polarized along, $\hat{e}_z$, direction parallel to the QD array growth axes.

Existence of the pronounced peaks at $\sim$ 0.117–0.124 eV suggests that the condition (ii) that IB $\rightarrow$ CB excitations must be optically allowed and strong can be fulfilled by relatively small (b = 10 nm), InAs/GaAs QDs vertically stacked with d_z = 5 nm in the QD array. Further above the barrier CBM, in the energy range of 0.2–0.4 eV, a spectra of weak TE polarized transitions can be seen. This part of the IB $\rightarrow$ CB absorption spectra corresponds to transitions between the e_0 miniband and largely delocalized states between QDs. The second absorption process relevant for IBSC operation is associated with the VB $\rightarrow$ IB transitions. This absorption coefficient has been calculated again using Eq. (5.9), with the first electron miniband e_0 now taken to be final and unoccupied. As initial minibands we considered the first 1,000 hole minibands starting from the h_0 miniband. The lowest energy of the 1,000th hole miniband goes well below the valence band maximum (VBM) in the barrier material. The calculated absorption coefficient of VB $\rightarrow$ IB process is shown in Fig. 5.5b. The absorption peak is at $\sim$ 1.2 eV and again of TE polarization. This time the TE polarization originates from heavy hole QD states that have no admixture of the orbital of the underlying bulk material with P_z symmetry. Due to highly strained InAs QD material in GaAs matrix, the biaxial component of the compressive strain splits the heavy from the light hole states in the VB, while the hydrostatic component of the strain pushes heavy hole states upwards on the absolute energy scale, making the h_0 miniband of heavy hole character. The shoulder in the absorption spectra between 1.23 eV and 1.35 eV corresponds to transitions between deep VB minibands and the e_0 miniband. The third absorption process, VB $\rightarrow$ CB, is shown in Fig. 5.5c. This absorption spectra was obtained using the first 1,000 minibands in the VB as initial states and 100 empty minibands in the CB, starting from the e_1 miniband, as final states. The absorption edge for this process is at 1.33 eV and is determined by energy distance between the h_0 and e_1 minibands.

In the analysis above all absorption coefficients were calculated assuming line broadening in Eq. (5.9), of ΔE = 5 meV. While such a sharp line seems to be justifiable for the interband transitions, such as VB $\rightarrow$ IB and VB $\rightarrow$ CB, that occurs on a ns time scale, and for high quality QD samples, it might be questionable for the intraband transitions (IB $\rightarrow$ CB) that are dominated by nonradiative phonon-assisted scattering, which occurs on a very fast ps time scale [41, 47, 48]. To assess this effect we plot in Fig. 5.5a the absorption coefficient for several different broadenings. For the more realistic broadening for intraband transitions of ΔE = 45 meV, we estimate IB $\rightarrow$ CB absorption coefficient to drop to $\sim 5 \times 10^3$ cm^{-1}.

5.6.2 Lateral QD Arrays

Figure 5.6a, b illustrates the energy dispersion of the two characteristic laterally arranged QD arrays, analysed in this section, along the M-Γ- X - M path in the square lattice BZ. The array (a), of small closely coupled QDs, consists of small

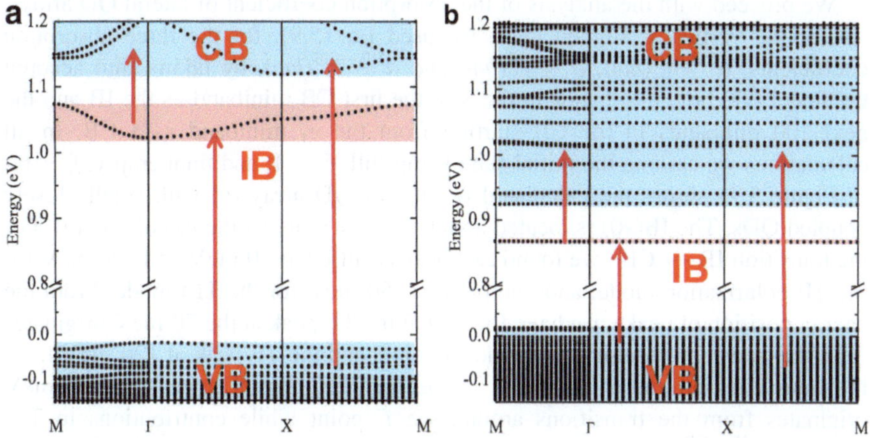

Fig. 5.6 Electronic structure of the laterally coupled 2D QD array throughout the first BZ: (**a**) array made of QDs with $b = 10$ nm and $d_x = d_y = 1$ nm and (**b**) array made of QD with $b = 20$ nm and $d_x = d_y = 5$ nm

QDs with lateral size $b = 10$ nm and inter-dot spacing, $d_x = d_y = 1$ nm, which corresponds to an areal density of 8.3×10^{11} cm^{-2}; the array (b), of big weakly coupled QDs, consists of big QDs with lateral size $b = 20$ nm and inter-dot spacing of $d_x = d_y = 5$ nm, which corresponds to an areal density of 1.6×10^{11} cm^{-2}. Comparing the position of the IB and CB states in Fig. 5.6a, b, it is observed that electrons in the array (b) of large QDs are more confined inside the QD's CB. Electrons in the array (a), with small closely coupled QDs, are less strongly bounded and more dispersed inside the first BZ, producing wider IBs in the CB. This indicates that a small strongly coupled QD array structure is a more suitable candidate to form an IB. The wave functions of QD array with small QDs are more likely to penetrate into the barrier region and, if this region is not that wide, to overlap with the wave functions of neighbouring QDs. It can also be seen from Fig. 5.6b that strong confinement of states in the QD array with large QDs causes almost dispersionless, discrete, electronic structure, with negligible IB band width. The estimated width of the IB formed from e_0 in QD array (a) is $\Delta E_{IB(e0)} = 51$ meV, separated from the rest of the CB with zero DOS of 46 meV. It is interesting to notice that the electronic structure of QD array (a) exhibits the *indirect gap nature*; i.e. while the VB maximum is at the point M, the CB minimum is at a different point, X. This emphasises the need for careful electronic structure calculations throughout the whole BZ, and not just at the Γ point, if one wants to estimate faithfully widths of the IB and gaps between them in laterally arranged QD arrays. In both arrays, due to much larger electron effective mass in the VB, the VB states are practically thermalised at room temperature, as we could not find states that are separated by more than 25 meV.

We proceed with the analysis of the absorption coefficient of lateral QD arrays. As for the vertically aligned QDs, we used Eq. (5.9), for the three absorption coefficients, $\alpha^{\mathrm{VB}\rightarrow\mathrm{IB}}(\hbar\omega)$, $\alpha^{\mathrm{IB}\rightarrow\mathrm{CB}}(\hbar\omega)$ and $\alpha^{\mathrm{VB}\rightarrow\mathrm{CB}}(\hbar\omega)$, by taking into account the first 1,000 hole minibands in the VB, the first CB miniband as the IB and the next 100 minbands in the CB starting from the e_1 miniband as a CB. In all calculations we assume the initial state being full, $f_i = 1$, and final empty, $f_f = 0$.

Figure 5.7a shows the calculated results for QD array (a) with small closely coupled QDs. The IB($e0$) is located about ~1.01 eV above the top of the VB. For the transition IB → CB , we found two peaks centred at 70 meV and 120 meV for the TE polarization mode, and one peak at 250 meV for the TM mode. From the energy position of peaks, we have found that the TE peak at the 70 meV originates from the transitions around the X and M points and the TE peak at 120 meV from the transitions around the Γ point in the first BZ. The TM peak at 250 meV originates from the transitions around the Γ point while contributions to TM polarized $\alpha^{\mathrm{IB}\rightarrow\mathrm{CB}}$ from other points in the BZ are negligible. By adopting more realistic line broadening of $\Delta E = 25$ meV for $\alpha^{\mathrm{IB}\rightarrow\mathrm{CB}}(\hbar\omega)$, the two TE peaks hybridise in one with much lower value of the absorption coefficient of $\sim 1.5 \times 10^4\ \mathrm{cm}^{-1}$.

For the transition VB → IB and VB → CB, the absorption coefficients are almost the same, $\sim 5 \times 10^3\ \mathrm{cm}^{-1}$. Another feature found in VB→IB absorption spectra is the multi-peak nature of the absorption profile. We ascribed this effect to the narrow dispersion of the states in VB, which is due to the much heavier effective mass of holes compared to the effective mass of electrons in the CB. This can be confirmed by almost the flat dispersion of the VB states as shown in Fig. 5.6a, b. Flat dispersion indicates that the coupling driven wavefunction overlap in VB is not as large as those occurred in the CB, thus the observed optical transition VB → IB is predominately governed by the discrete nature of the isolated QD. We also found that the absorption at energy above 1.3 eV is almost zero, indicating that the optical absorption is occurring between IB and the limited number of energy states around the top of VB, which is similar to the case of IB → CB. Absorption between VB → CB is in principle the same as for the bulk-like material and is known to have an approximate parabolic absorption profile. Our results showed almost the same parabolic profile, except that the large amount of subpeaks appeared at the low energy range. We claim that the appearance of multiple peaks is in principle the same phenomenon as observed in the VB → IB transition which was attributed to the flat band dispersion of VB.

In Fig. 5.7b we show the calculated results for the QD array (b) with large weakly coupled QDs. The main difference compared to the array (a) is that IB → CB and VB → IB absorption coefficients are significantly reduced, while VB → CB is largely similar. With increasing size of QD the overlap between wave functions and the momentum operator of the underlying bulk material [49], which is in fact the dipole matrix element, decreases. Since the absorption coefficients depend on those dipole matrix elements the absorption coefficients

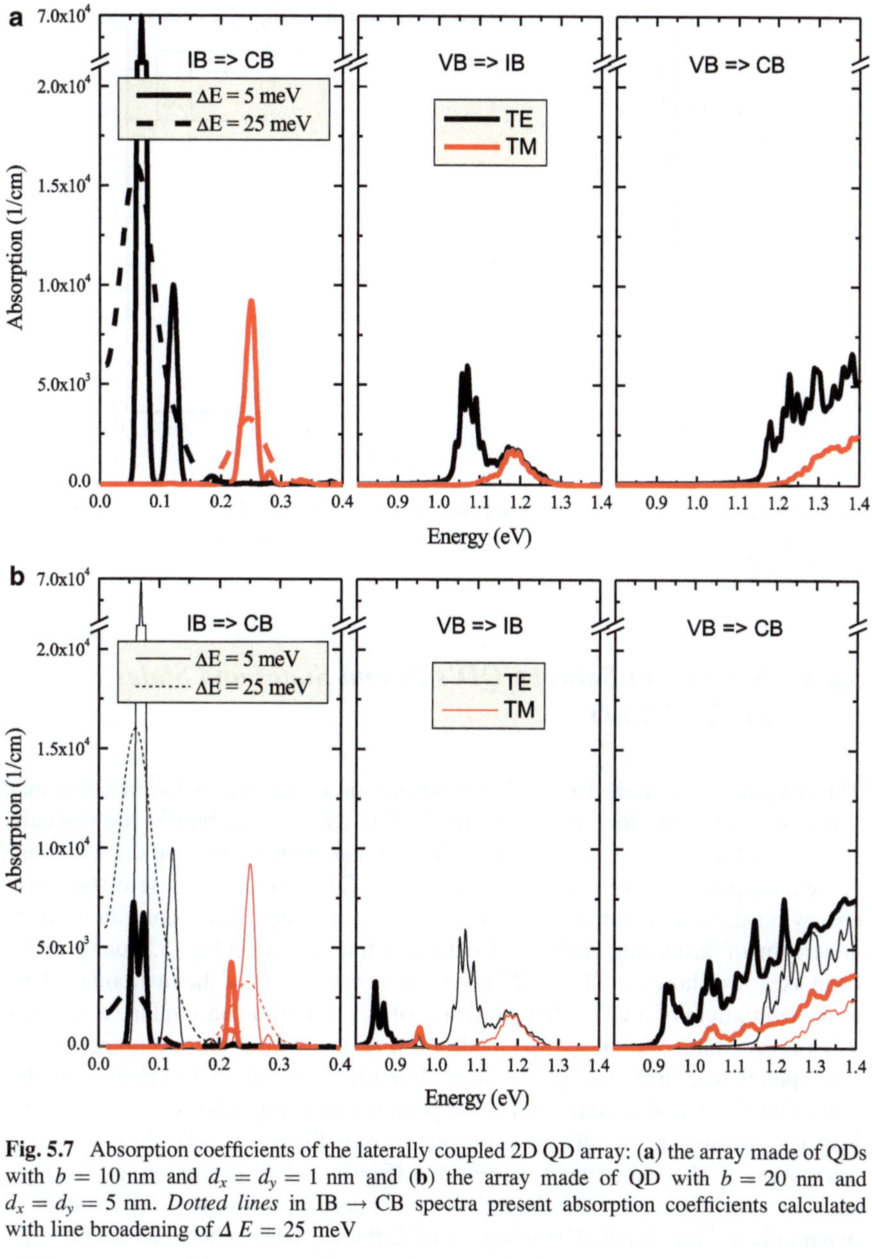

Fig. 5.7 Absorption coefficients of the laterally coupled 2D QD array: (**a**) the array made of QDs with $b = 10$ nm and $d_x = d_y = 1$ nm and (**b**) the array made of QD with $b = 20$ nm and $d_x = d_y = 5$ nm. *Dotted lines* in IB $\rightarrow$ CB spectra present absorption coefficients calculated with line broadening of $\Delta E = 25$ meV

that depends on confined states are reduced. On the other hand VB $\rightarrow$ CB absorption is, as already mentioned, of bulk-like nature and less dependent on the confined states, as it involves many confined as well as delocalised states above the VB and CB edge.

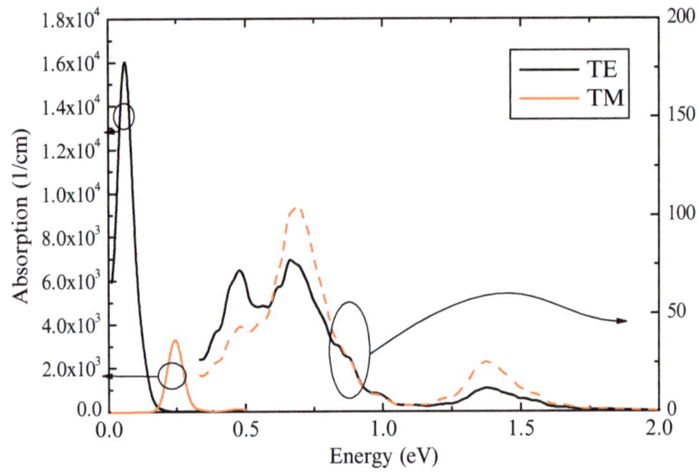

Fig. 5.8 Bound-to-bound spectra of 2D QD array with $b = 10$ nm and $d_x = d_y = 1$ nm, left axis, and bound-to-continuum right axis. All spectra are broaden with the line broadening of $\Delta E = 25$ meV

5.6.3 *Absorption Between QD's Bound States and States in Continuum*

The remaining question here is; if the absorption coefficient between QD bound states and those in close proximity to the CB edge in the barrier are optically allowed and finite (see Fig. 5.5a for $\hbar\omega$ in the range between 0.2 and 0.4 eV), what is the absorption between states confined in the QD region and those higher up in energy, in the continuum of the CB? Can those transitions also contribute to the generation of the current via IB → Continuum transitions? In Fig. 5.8, on the right-hand axes, we show the TE and TM absorption coefficient of the first bound states to the continuum of a QD with lateral size of $b = 10$ nm in an in-plane array with $d_x = d_y = 1$ nm. We have assumed all first six states are occupied, which corresponds to a high doping condition. Even if there are six electrons available in the QD, the bound to continuum absorption reaches only ~100 cm^{-1}. If there is just one electron in the QD the absorption drops further to ~10–30 cm^{-1}. For orientation, in the same figure we plot the IB($e0$) → CB absorption, on the right axes. It is obvious that the bound to continuum absorption is 2–3 orders of magnitude weaker than that between bound to bound states or between bound states and states in the proximity of the barrier CB edge, Fig. 5.5a, with very little ability to contribute to current generation.

5.7 Absorption Coefficient and Radiative Transition Times

The radiative recombination is the detailed balance counterpart of the corresponding rate of electron–hole pair generation by thermal radiation [50]. Therefore, if the transition strength is large the radiative recombination is strong and the radiative lifetime small. To compare the radiative recombination between different transitions, $|a\rangle \rightarrow |b\rangle$ and $|c\rangle \rightarrow |d\rangle$, not only the transition strength but also the transition energy needs to be taken into account. In this section we prove that radiative lifetime follows the trend of the optical dipole matrix element, and relate this to the absorption coefficient of the QD array.

According to the Fermi Golden rule, the transition time, $\tau_{i\,f}$, from an initial state $|i\rangle$ to a final state $|f\rangle$ due to the spontaneous emission of photons (radiative lifetime) of angular frequency ω is given by:

$$\frac{1}{\tau_{if}} = \frac{2\pi}{\hbar} \sum_{\mathbf{k}} |\langle i|H'_{\mathbf{k}}|f\rangle|^2 \delta(E_i - E_f - \hbar\omega) \tag{5.19}$$

where E_i, E_f, are the energies of the initial and final state, respectively. The Hamiltonian, $H_{\mathbf{k}}'$, of the electron interaction with electromagnetic field is obtained by replacing $\mathbf{k}$ with $\mathbf{k} + (q/\hbar)\mathbf{A}$ in the kinetic part of the Hamiltonian, Eq. (5.4), (where $\mathbf{A} = A_0\hat{\mathbf{e}}$ is the magnetic vector potential), i.e. $H_{\mathbf{k}}' = H_{\mathbf{k}}[\mathbf{k} + (q/\hbar)\mathbf{A}] - H_{\mathbf{k}}[\mathbf{k}]$. In the dipole approximation $\mathbf{A}$ is considered constant in space, and all the terms quadratic in $\mathbf{A}$ are neglected. By substituting the expressions for $H_{\mathbf{k}}'$ and $\mathbf{A}$ into Eq. (5.19), and after replacing summation over $\mathbf{k}$ vector with integration, i.e. $\sum_{\mathbf{k}} \rightarrow [\Omega/(2\pi)^3]\int d\mathbf{k}$ we arrived at

$$\frac{1}{\tau_{if}} = \frac{\Omega}{(2\pi)^3} \int d\mathbf{k} \frac{\pi}{\varepsilon\omega\Omega} \left[|\hat{e}_x \mathbf{p}_{if}|^2 + |\hat{e}_y \mathbf{p}_{if}|^2 + |\hat{e}_z \mathbf{p}_{if}|^2 \right] \delta(E_i - E_f - \hbar\omega), \tag{5.20}$$

where $\varepsilon = \varepsilon_0\varepsilon_r$. After integration ($d\mathbf{k} \rightarrow 4\pi\, k^2 d\, k$), using $\omega = ck/\bar{n}$ and $\bar{n}^2 = \varepsilon_r$ the final result reads:

$$\frac{1}{\tau_{if}} = \frac{4}{3} \frac{\bar{n}(E_i - E_f)}{\pi\hbar^2 c^3 \varepsilon_0} \left[|\hat{e}_x \cdot \mathbf{p}_{if}|^2 + |\hat{e}_y \cdot \mathbf{p}_{if}|^2 + |\hat{e}_z \cdot \mathbf{p}_{if}|^2 \right], \tag{5.21}$$

where the factor $(4/3)$ comes from averaging over all possible light directions and spin degenerate transitions between i and f. In Fig. 5.9 the momentum matrix elements, (a)–(c), the transition strength taking into account only the states at $K_z = 0$, (d)–(f), and transition strength summed over all K_z states, (g)–(i) are given. The momentum matrix elements corresponding to the CB $\rightarrow$ IB and IB $\rightarrow$ VB transitions, are $|\hat{e}_x \mathbf{p}_{e1,e0}|^2/P_0^2 = 0.044$ and $|\hat{e}_x \mathbf{p}_{e0,h0}|^2/P_0^2 = 0.223$ (in units of bulk interband momentum matrix element, equal to $P_0 = 9.37$ eVÅ for InAs). Those intensities correspond to transition energies $E_{e1} - E_{e0} = 117.3$ meV and $E_{e0} - E_{h0} = 1.192$ eV, respectively. As can be seen from Fig. 5.9a, b, for both transitions the $|\hat{e}_z \mathbf{p}_{if}|^2$ is negligible, and due to symmetry reasons $|\hat{e}_y \mathbf{p}_{if}|^2 = |\hat{e}_x \mathbf{p}_{if}|^2$.

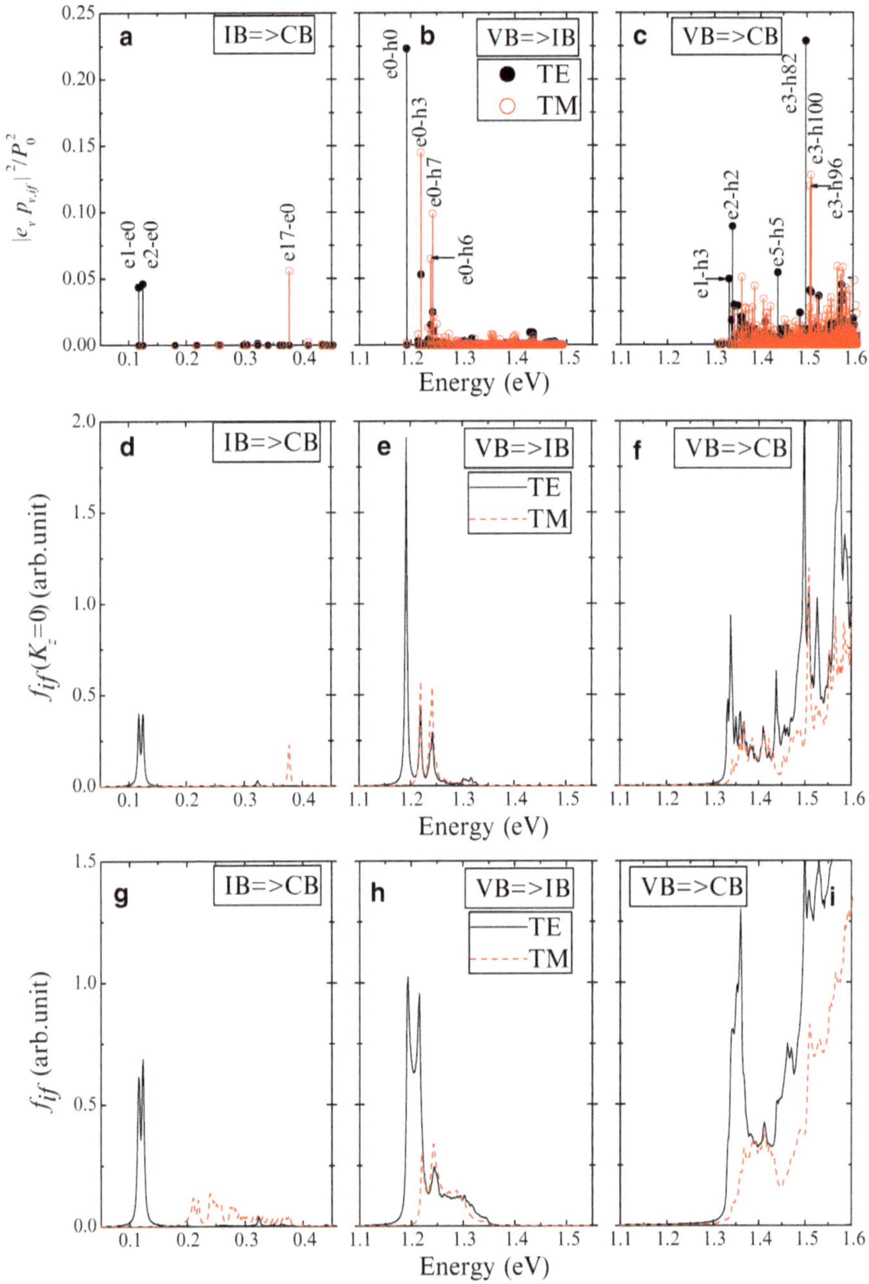

Fig. 5.9 Momentum matrix elements of the QD array with vertical periodicity $d_z = 5$ nm at $K_z = 0$: (**a**) IB → CB transitions, (**b**) VB → IB transitions, (**c**) VB → CB transitions; the transition strength for the same array at the band edge $K_z = 0$, (**d**) IB → CB transitions, (**e**) VB → IB transitions, (**f**) VB → CB transitions; and the transition strength for the same array summed over all K_z states, (**g**) IB → CB transitions, (**h**) VB → IB transitions and (**i**) VB → CB transitions. Both the TE (*solid symbol* or *solid line*) and TM (*open symbol* or *dashed line*) polarizations are shown

Next, we express the radiative transition times in terms of the value of transition strength at energy that corresponds to the transition energy between states i and f, i.e. $\hbar\omega = E_i - E_f$. We can write: $\bar{f}_{if}(\hbar\omega = E_i - E_f) = \int |\hat{\mathbf{e}}\cdot\mathbf{p}_{if}|^2\delta(E_i - E_f - \hbar\omega)d\hbar\omega = |\hat{\mathbf{e}}\cdot\mathbf{p}_{if}|^2\cdot 1$. The ratio between CB → IB and IB → VB radiative transition times, after noting from Fig. 5.9a, b that for these transitions $|\hat{e}_z\mathbf{p}_{if}|^2 \ll |\hat{e}_x\mathbf{p}_{if}|^2$, can be expressed in terms of the transition strength as:

$$\frac{\tau_{e1,e0}}{\tau_{e0,h0}} = \frac{E_{e0} - E_{h0}}{E_{e1} - E_{e0}}\left(\frac{\bar{f}_{e0,h0}}{\bar{f}_{e1,e0}}\right) \approx \frac{E_{e0} - E_{h0}}{E_{e1} - E_{e0}}\left(\frac{|\hat{e}_x\mathbf{p}_{e0,h0}|^2}{|\hat{e}_x\mathbf{p}_{e1,e0}|^2}\right) \approx 52. \tag{5.22}$$

In Fig. 5.9a, d the additional peak due to $|\hat{e}_z\mathbf{p}_{ij}|^2$ momentum matrix element (TM polarization) can be identified at energy ~ 0. 25 eV, above the double peak that corresponds to $e_1 \to e_0$ and $e_2 \to e_0$ transitions. All these peaks in the intraband spectra have been previously observed in single InAs/GaAs QD structures, both theoretically and experimentally [51–53]. It is interesting to note that, after summation over K_z in the first Brillouin zone of the QD array, this peak disappears, Fig. 5.9g. This suggests its very strong variation with K_z inside the miniband. The momentum matrix elements $|\hat{e}_x\mathbf{p}_{e1,e0}|^2$ and $|\hat{e}_x\mathbf{p}_{e2,e0}|^2$ are much less sensitive on K_z. Strong variation of the momentum matrix elements inside the QD array minibands has already been reported [38]. This indicates the importance of the integration over K_z in calculation of the absorption spectra of QD arrays.

The relation between the absorption coefficient, Eq. (5.9), and radiative transition time given by Eq. (5.21) can be established as:

$$\alpha(\hbar\omega_{if}) \propto \frac{1}{\omega_{if}\tau_{if}(E_i - E_f)}. \tag{5.23}$$

Using values at $K_z = 0$ as $\omega^{\mathrm{VB(h0)}\to\mathrm{IB(e0)}} \approx 10\times\omega^{\mathrm{IB(e0)}\to\mathrm{CB(e1)}}$, $|E_{\mathrm{VB(h0)}} - E_{\mathrm{IB(e0)}}| \approx 10\times|E_{\mathrm{IB(e0)}} - E_{\mathrm{CB(e1)}}|$ and $\tau_{\mathrm{CB(e1),IB(e0)}} \approx 50\times\tau_{\mathrm{IB(e0),VB(h0)}}$ it follows: $\alpha^{\mathrm{IB(e0)}\to\mathrm{CB(e1)}} \approx 2\times\alpha^{\mathrm{VB(h0)}\to\mathrm{IB(e0)}}$. As can be seen comparing Fig. 5.9d, e, g, h, the relative change of the ratio between transition strengths, $f_{\mathrm{IB(e0),CB(e1)}}/f_{\mathrm{VB(h0),IB(e0)}}$, at $K_z = 0$ and after summation over all K_z values, another factor of ~ 2 can be extracted. Finally one gets $\alpha^{\mathrm{IB(e0)}\to\mathrm{CB(e1)}} \approx 4\times\alpha^{\mathrm{VB(h0)}\to\mathrm{IB(e0)}}$, assuming integration over all K_z states, which correspond to results shown in Fig. 5.5.

Following the detailed balance drift-diffusion transport model developed for IBSCs proposed by Strandberg and Reenaas [54], and using information on the relevant energy gaps E_g(VB,IB)=1.2 eV, E_g(IB,IC)=0.12 eV and E_g(VB,CB)= 1.32 eV, absorption and various radiative times between them presented here, we have estimated that the radiative limit efficacy of InAs/GaAs QD based QD array vary between 24 % and 39 % for the light concentrations of 1 and 1,000, respectively, Fig. 5.10.

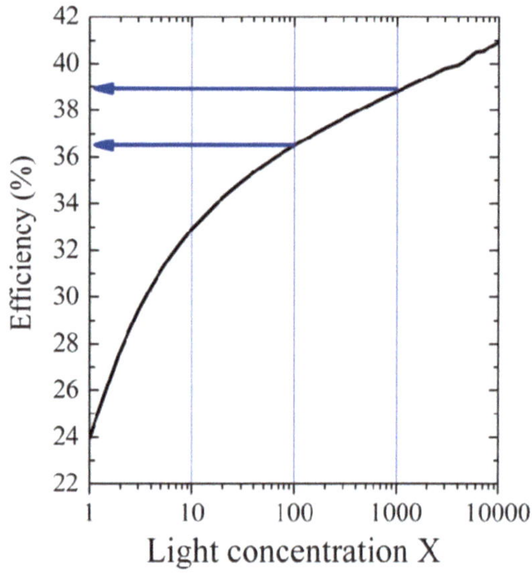

Fig. 5.10 Power conversion efficiency versus light concentration factor of a vertically aligned 1D QD array made of QDs with $b = 10$ nm and $d_z = 5$ nm

5.8 Conclusions

In this chapter we have presented a comprehensive theoretical model for the design and modelling of electronic structure, absorption and radiative times of IBSC based on InAs/GaAs QDs arranged in 1D vertical or 2D lateral arrays. In vertically arranged QD arrays, the IB bandwidths and gaps are determined by the energy values at the edges of the Brillouin zone, i.e. $E_i(K_z = 0)$ and $E_i(K_z = \pi/L_z)$. In laterally arranged QD arrays realistic prediction of the band structure, IB bandwidths, gaps between them, as well as absorption properties require detailed examination of the whole first Brillouin zone, not just at single Γ point. For the predicted anergy gaps of E_g(VB,IB) = 1. 2 eV, E_g(IB,IC) = 0. 12 eV and E_g(VB, CB) = 1. 32 eV that realistic InAs/GaAs QD array can offer and calculated absorption spectra, by using the drift-diffusion model adapted for IBSC, we have estimated radiative limit efficiency in the range of 24% to 39 % for light concentration of 1 to 1,000, respectively.

Acknowledgements The author wishes to thank Y. Okada, T. Sogabe and N. Vukmirović for many useful discussions and suggestions. The author is grateful to the New Energy and Industrial Technology Development Organization (NEDO), Japan, for financial support under grant: "Research and Development on Innovative Solar Cells: Post-Silicon solar cells for ultra-high efficiencies". The author also wishes to thank the STFC e-Science, UK, for providing the computational resources.

References

1. Wolf, M.: Proc. IRE **48**, 1246 (1960)
2. Luque, A., Martí, A.: Phys. Rev. Lett. **78**, 5014 (1997)
3. Wahnón, P., Tablero, C.: Phys. Rev. B **65**, 165115 (2002)
4. Palacios, P., Aguilera, I., Wahnon, P., Conesa, J.C.: J. Phys. Chem. C **112**, 9525 (2008)
5. Palacios, P., Sanchez, K., Conesa, J., Fernandez, J., Wahnon, P.: Thin Solid Films **515**, 6280 (2007)
6. Bailey, C.L., Liborio, L., Mallia, G., Tomić, S., Harrison, N.M.: Phys. Rev. B **81**, 205214 (2010)
7. López, N., Reichertz, L.A., Yu, K.M., Campman, K., Walukiewicz, W.: Phys. Rev. Lett. **106**, 028701 (2011)
8. Guimard, D., Morihara, R., Bordel, D., Tanabe, K., Wakayama, Y., Nishioka, M., Arakawa, Y.: Appl. Phys. Lett. **96**, 203507 (2010)
9. Oshima, R., Takata, A., Okada, Y.: Appl. Phys. Lett. **93**, 083111 (2008)
10. Martí, A., Antolín, E., Stanley, C.R., Farmer, C.D., López, N., Díaz, P., Cánovas, E., Linares, P.G., Luque, A.: Phys. Rev. Lett. **97**, 247701 (2006)
11. Hubbard, S.M., Cress, C.D., Bailey, C.G., Raffaelle, R.P., Bailey, S.G., Wilt, D.M.: Appl. Phys. Lett. **92**, 123512 (2008)
12. Blokhin, S., Sakharov, A., Nadtochy, A., Pauysov, A., Maximov, M., Ledentsov, N., Kovsh, A., Mikhrin, S., Lantratov, V., Mintairov, S., Kaluzhniy, N., Shvarts, M.: Semiconductors **43**, 514 (2009)
13. Sablon, K.A., Little, J.W., Mitin, V., Sergeev, A., Vagidov, N., Reinhardt, K.: Nano Lett. **11**, 2311 (2011).
14. Bailey, C.G., Forbes, D.V., Raffaelle, R.P., Hubbard, S.M.: Appl. Phys. Lett. **98**, 163105 (2011)
15. Okada, Y., Morioka, T., Yoshida, K., Oshima, R., Shoji, Y., Inoue, T., Kita, T.: J. Appl. Phys. **109**, 024301 (2011)
16. Luttinger, J.M., Kohn, W.: Phys. Rev. **97**, 869 (1955)
17. Kane, E.O.: J. Phys. Chem. Solid. **1**, 249 (1957)
18. Kane, E.O.: Semiconduct. Semimetal. **1**, 75 (1966)
19. Cardona, M., Pollak, F.H.: Phys. Rev. **142**, 530 (1966)
20. Richard, S., Aniel, F., Fishman, G.: Phys. Rev. B **70**, 235204 (2004)
21. Rideau, D., Feraille, M., Ciampolini, L., Minondo, M., Tavernier, C., Jaouen, H., Ghetti, A.: Phys. Rev. B **74**, 195208 (2006)
22. Foreman, B.A.: Phys. Rev. B **76**, 045327 (2007)
23. Saïdi, I., Radhia, S.B., Boujdaria, K.: J. Appl. Phys. **107**, 043701 (2010)
24. Bir, G.L., Pikus, G.E.: Symmetry and Strain-Induced Effects in Semiconductors. Wiley, New York (1974)
25. Loehr, J.P.: Physics of Strained Quantum Well Lasers. Kluwer, Boston (1998)
26. Pidgeon, C.R., Brown, R.N.: Phys. Rev. **146**, 575 (1966)
27. Pryor, C.: Phys. Rev. B **57**, 7190 (1998)
28. Stier, O., Grundmann, M., Bimberg, D.: Phys. Rev. B **59**, 5688 (1999)
29. Cusack, M.A., Briddon, P.R., Jaros, M.: Phys. Rev. B **54**, R2300 (1996)
30. Vukmirović, N., Indjin, D., Jovanović, V.D., Ikonić, Z., Harrison, P.: Phys. Rev. B **72**, 075356 (2005)
31. Tomić, S., Sunderland, A.G., Bush, I.J.: J. Mater. Chem. **16**, 1963 (2006)
32. Vukmirović, N., Tomić, S.: J. Appl. Phys. **103**, 103718 (2008)
33. Andreev, A.D., Downes, J.R., Faux, D.A., O'Reilly, E.P.: J. Appl. Phys. **86**, 297 (1999)
34. Li, S.-S., Xia, J.-B., Yuan, Z.L., Xu, Z.Y., Ge, W., Wang, X.R., Wang, Y., Wang, J., Chang, L. L.: Phys. Rev. B **54**, 11575 (1996)
35. Andreev, A.: In: Proceedings of SPIE, vol. 3284, pp. 151–161 (1998); Conference on In-Plane Semiconductor Lasers: From Ultraviolet to Mid-Infrared II, San Jose, CA, 26–28 Jan 1998

36. Lazarenkova, O.L., Balandin, A.A.: J. Appl. Phys. **89**, 5509 (2001)
37. Shao, Q., Balandin, A.A., Fedoseyev, A.I., Turowski, M.: Appl. Phys. Lett. **91**, 163503 (2007)
38. Tomić, S., Jones, T.S., Harrison, N.M.: Appl. Phys. Lett. **93**, 263105 (2008)
39. Klos, J.W., Krawczyk, M.: J. Appl. Phys. **106**, 093703 (2009)
40. Bragar, I., Machnikowski, P.: J. Appl. Phys. **112**, 124318 (2012)
41. Tomić, S.: Phys. Rev. B **82**, 195321 (2010)
42. Todorovic, G., Milanovic, V., Ikonic, Z., Indjin, D.: Solid State Comm. **110**, 103 (1999)
43. Tadić, M., Ikonić, Z.: Phys. Rev. B **52**, 8266 (1995)
44. Joyce, P.B., Krzyzewski, T.J., Bell, G.R., Joyce, B.A., Jones, T.S.: Phys. Rev. B **58**, R15981 (1998)
45. Tomić, S., Howe, P., Harrison, N.M., Jones, T.S.: J. Appl. Phys. **99**, 093522 (2006)
46. Canovas, E., Marti, A., Lopez, N., Antolin, E., Linares, P., Farmer, C., Stanley, C., Luque, A.: Thin Solid Films **516**, 6943 (2008)
47. Zibik, E.A., Wilson, L.R., Green, R.P., Bastard, G., Ferreira, R., Phillips, P.G., Carder, D.A., Wells, J.-P.R., Cockburn, J.W., Skolnick, M.S., Steer, M.J., Hopkinson, M.: Phys. Rev. B **70**, 161305 (2004)
48. Sauvage, S., Boucaud, P., Lobo, R.P.S.M., Bras, F., Fishman, G., Prazeres, R., Glotin, F., Ortega, J.M., Gérard, J.-M.: Phys. Rev. Lett. **88**, 177402 (2002)
49. Andreev, A.D., O'Reilly, E.P.: Appl. Phys. Lett. **87**, 213106 (2005)
50. van Roosbroeck, W., Shockley, W.: Phys. Rev. **94**, 1558 (1954)
51. Sauvage, S., Boucaud, P., Julien, F.H., Gérard, J.-M., Thierry-Mieg, V.: Appl. Phys. Lett. **71**, 2785 (1997)
52. Aleshkin, V., et al.: Light absorption and emission in InAs/GaAs quantum dots and stepped quantum wells. In: Alferov Z.I., Esaki L. (eds.) 10TH International Symposium on Nanostructures: Physics and Technology. Proceedings of The Society of Photo-Optical Instrumentation Engineers (SPIE), vol. 5023, pp. 209–212, 2003. 10th International Symposium on Nanostructures, ST Petersburg, Russia, 17–21 June 2002
53. Vorobjev, L.E., Firsov, D.A., Shalygin, V.A., Fedosov, N.K., Panevin, V.Y., Andreev, A., Ustinov, V.M., Cirlin, G.E., Egorov, V.A., Tonkikh, A.A., Fossard, F., Tchernycheva, M., Moumanis, K., Julien, F.H., Hanna, S., Seilmeier, A., Sigg, H.: Semicond. Sci. Tech. **21**, 1341 (2006)
54. Strandberg, R., Reenaas, T.W.: J. Appl. Phys. **105**, 124512 (2009)

Chapter 6
Material Selection for the Quantum Dot Intermediate Band Solar Cell

Steven Jenks and Robert Gilmore

Abstract The main limitation of the conventional solar conversion device is that low energy photons cannot excite charge carriers to the conduction band and therefore do not contribute to the device's current. Another limitation is that high energy photons are not efficiently used due to a poor match of the solar spectrum to the energy gap. However, when intermediate bands are introduced into the energy gap of a conventional device, low energy photons can be used to promote charge carriers in a stepwise manner to the conduction band and photons are better matched to the energy transitions between bands. Solar cells with intermediate bands can have conversion efficiencies that exceed thermodynamic limits of the conventional solar cells. A device based on the confined electron levels of quantum dots, called the quantum dot intermediate band solar cell, is a physical realization of the intermediate band solar cell. We discuss the design criteria for selecting materials for the quantum dot intermediate band solar cells. With the aid of the finite element method, we perform numerical simulations on two types of quantum dot geometry and identify optimal material systems that are considered candidates for the quantum dot intermediate band solar cell with efficiencies greater than 46 % for unconcentrated light and greater than 62 % for fully concentrated light. Materials considered in this work are the technologically important III–V semiconductors and their alloys.

6.1 Introduction

In this section, we review some useful conversion efficiencies with the view of not just establishing the upper limit but estimating the potential for current design improvement and uncovering the fundamental quantities that determine this upper limit, as these become design criteria for real world devices. In a limited

S. Jenks (✉) • R. Gilmore
Department of Physics, Drexel University, 3141 Chestnut Street,
Philadelphia, PA 19104-2875, USA
e-mail: sej52@drexel.edu; robert.gilmore@drexel.edu

J. Wu and Z.M. Wang (eds.), *Quantum Dot Solar Cells*, Lecture Notes in Nanoscale Science and Technology 15, DOI 10.1007/978-1-4614-8148-5_6,

sense this is analogous to Carnot's determination of the maximum efficiency that can be attained in the conversion of heat energy to useful work, which provides a quantitative measure of the degree to which the output efficiency of a real heat engine could still be improved. The merit of the heat engine can be appraised in terms of the limit set out by the second law of thermodynamics.

In 1961, Shockley and Queisser published a landmark paper aimed at determining a theoretical justifiable upper efficiency limit [1]. Before then, the treatment of photovoltaic efficiency was based on empirical values for constants describing the characteristics of the conversion device. In general, they were all fairly consistent with observed values and accepted as theoretical limits. They realized that there exists a limit based on a consequence of the nature of atomic processes required by the basic laws of physics and referred to this upper efficiency limit as the detailed balance limit. This limit is calculated in accordance with the list of assumptions [1, 2]:

1. The photovoltaic device absorbs blackbody radiation at a temperature of $T_s =$ 6,000 K and ambient $T_a = 300$ K and emits blackbody radiation at ambient $T_a = 300$ K;
2. Only radiative transitions occur between the bands;
3. All photons above the lowest energy gap are absorbed;
4. Carrier mobility is infinite and as a consequence, the quasi-Fermi energy levels are constant throughout the cell so $\mu = qV$;
5. Only one electron–hole pair is created per photon;
6. A perfect mirror is located on the back of the device so that radiation makes a double pass through the cell and can only escape through the front area of illumination;
7. The net photon flux (number of incident minus number of emitted photons) is equal to the number of charge carrier pairs collected at the contacts (detailed balance assumption).

It is important to note that as photons with energy greater than the band gap are absorbed to create an electron–hole pair, any excess energy beyond that of the band gap will be lost due to thermalization and the carriers will relax to the band edges before circuit extraction or recombination. This makes an absorbed photon with $\hbar\omega > E_g$ have the same effect as an absorbed photon with energy $\hbar\,\omega = E_g$ and is the reason why the number of absorbed photons, rather than absorbed energy, is necessary to make these calculations.

The sun is modeled as a blackbody with a surface temperature of $T_s = 6{,}000$ K. The photon and emitted energy flux density are both derived from Planck's law over the energy range from E_1 and E_2 [3]

$$\dot{N} = \frac{2\pi F(\theta)}{h^3 c^2} \int_{E_1}^{E_2} \frac{E^2}{e^{E/k_B T} - 1} dE \tag{6.1}$$

$$\dot{E} = \frac{2\pi F(\theta)}{h^3 c^2} \int_{E_1}^{E_2} \frac{E^3}{e^{E/k_B T} - 1} dE \tag{6.2}$$

where T is the temperature, k_B is Boltzmann's constant, and $F(\theta) = \sin^2\theta$ is a geometrical factor that arises from integrating over the relevant angular range, where θ, $0 \leq \theta \leq \pi/2$, is defined by the angle to the device surface normal. The half angle subtended by the sun is $\theta = 0.26°$ such that $F_{sun}(0.26°) = 2.16 \times 10^{-5}$ [3]. Integrating over all E gives the total emitted power density $F(\theta)\sigma_s T^4$, where σ_s is Stefan's constant.

$$\sigma_s = \frac{2\pi^5 k_B^4}{15 h^3 c^2}$$

At the top of Earth's atmosphere, using the reduced geometric factor F_{sun}, the incoming power density falling on a photovoltaic device with planar geometry is $P_s = 1{,}584$ Watts per meter squared (W/m^2). If the incoming light intensity is increased beyond this value, light is concentrated and the incoming power density will be increased by a concentration factor, X. For example, a concentration factor of $X = 10$ would increase the power density to $P_s = 15{,}840$ W/m^2. Theoretically, a system would achieve the full concentration when $X = 1/F_{sun} = 46{,}198$. Mathematically, increasing concentration is equivalent to increasing the half angle subtended by the sun. Practical systems include parabolic reflectors or Fresnel lenses that place the photovoltaic device at the focus.

Since only radiative events are considered, generation and recombination events are signaled by photon absorption or emission. In addition, the device is assumed to have perfect charge carrier collection meaning that photogenerated charge carriers surviving radiative recombination will be collected by the external circuit. Therefore, the net photon flux will be equal to the number of charge carrier pairs collected at the contacts and be equal to charge carrier flux. When the charge carrier flux is multiplied by the electric charge q, the current density of the device is found

$$\begin{aligned} J(E_g, T_s, T_a, X, V) = q[&XF_{sun}N(E_g, \infty, T_s, 0) + (1 - XF_{sun})N(E_g, \infty, T_a, 0) \\ &- F(\pi/2)N(E_g, \infty, T_a, qV)] \end{aligned} \tag{6.3}$$

where we have made use of the function

$$N(E_1, E_2, T, \mu) = \frac{2\pi}{h^3 c^2} \int_{E_1}^{E_2} \frac{E^2}{e^{(E-\mu)/k_B T} - 1} dE \tag{6.4}$$

describing the photon flux, the geometric factor $F(\theta)$, and the concentration factor X. The first term on the right-hand side of Eq. (6.3) represents the current density generated from radiation the device absorbs from the sun at the characteristic temperature T_s over the angular range $0 < \theta < 0.26°$, the second term in the equation represents the current density generated from radiation the device absorbs from ambient[1] at the characteristic temperature T_a over the angular range $0.26^\circ < \theta < \pi/2$, and the last term in the equation represents the recombination of carriers through photon emission at the characteristic temperature T_a and uniform potential $q\,V$ that do not contribute to the current density over the angular range $0 < \theta < \pi/2$ [3–5]. It should be noted that within the photovoltaic device, photons are continuously absorbed and emitted but only when one leaves there will be a net electronic transition toward lower energies. Transitions are governed by a first order differential equation using *Fermi's Golden rule* and quasi-thermal probability distribution functions [6]. Under the boundary condition that the device is assumed thick enough to achieve full absorption of photons, the solution leads to the emitted photon flux $N(E_g, \infty, T_a, q\,V)$.

For each value of E_g, there exists a voltage V_m that maximizes the power density delivered by the device such that maximum efficiency of the device is

$$\eta(E_g, T_s, T_a, X) = \frac{J(E_g, T_s, T_a, X, V_m) \cdot V_m}{X \cdot 1584}. \tag{6.5}$$

Following the prescription of finding the detailed balance limiting efficiency, we have found the maximum efficiency as a function of the band gap E_g for unconcentrated light $\eta(E_g, T_s = 6{,}000\,K, T_a = 300\,K, X = 1)$ and fully concentrated light $\eta(E_g, T_s = 6{,}000\,K, T_a = 300\,K, X = 1/F_{sun})$ (see Fig. 6.1). Fully concentrated light has a maximum efficiency of 41 % when there is a band gap of 1. 1 eV, while unconcentrated light has a maximum efficiency of 31 % when there is a band gap of 1. 31 eV.

Efficiency of the single energy gap solar conversion device seems to drop off for small energy gaps, $E_g < 1$ eV, and for larger energy gaps, $E_g > 2$ eV. Intuitively this can be understood in terms of the power density $P_d = J \cdot V$ that is delivered by the device. For smaller band gaps, most photons would be absorbed thereby increasing current density but a good portion of the their energy would be wasted through thermalization. In addition, since the device operates at a potential difference proportional to the band gap, each extracted carrier's potential energy will be small. For larger band gaps, a good portion of the incident photons would not be absorbed and current density would decrease. In both cases, efficiency drops off due to the current density being inversely proportional to the operating voltage.

As a consequence of the detailed balance limiting efficiency calculations, the device's band gap becomes a fundamental quantity influencing efficiency. Further, photons energetically located near the band gap are converted most efficiently. If the solar resource were monochromatic, conversion would be optimal because the device's bandgap would be tuned to match the light's energy. Unfortunately the

[1] Photon absorption includes a contribution from thermal photons that is assumed to behave like a blackbody over the rest of the hemisphere. Although the contribution to the current is negligible, it is a standard assumption in efficiency calculations.

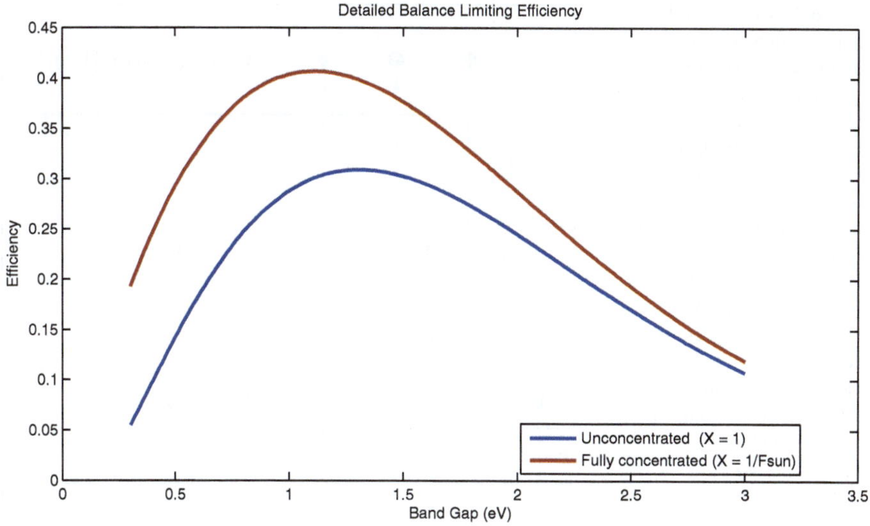

Fig. 6.1 Detailed balance limiting efficiency of a photovoltaic device as a function of the band gap for unconcentrated light $X = 1$ and fully concentrated light $X = 1/F_{\text{sun}}$. The efficiency was calculated assuming the sun is modeled as a blackbody with characteristic temperature 6, 000 K. For fully concentrated light, the maximum efficiency of about 41% occurs approximately at a band gap of $E_g = 1.\ 1$ eV, while unconcentrated light leads to a maximum efficiency of about 31 % with a band gap of $E_g = 1.\ 31$ eV

solarre source is not monochromatic and with a single band gap, low energy photons are not absorbed and higher energy photons lose energy through thermalization. As such the best conversion based on a single band gap is 41 %.

6.2 Intermediate Band Solar Cell

The main limitations of the photovoltaic conversion device are that low energy photons cannot excite charge carriers to the conduction band, therefore do not contribute to the devices's current, and high energy photons are not efficiently used due to a poor match to the energy gap. However, if intermediate levels are introduced into the energy gap of a conventional solar cell, then low energy photons can be used to promote charge carriers in a stepwise manner to the conduction band. In addition, the photons would be better matched with energy transitions between bands. Figure 6.2 illustrates this type of structure. In this case, there are two intermediate bands between the valence and conduction bands, allowing for a total of six upward electronic transitions. Luque et al. describe this type of device as the intermediate band solar cell (IBSC) [4]. This is a multi-step or ladder approach to increase efficiency. It will be shown that the maximum efficiency of a photovoltaic conversion device using one or two intermediate bands is greater than the single band gap conventional device.

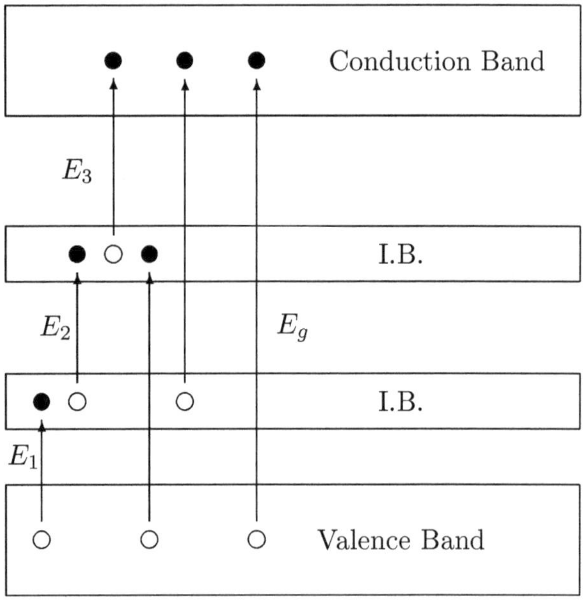

Fig. 6.2 This 4-Band diagram depicts two intermediate bands and shows the possible electronic transitions (reverse transitions are not shown but do occur). There is a total of six upward transitions, with E_1, E_2, and E_3 making up the three independent ones. The large energy gap, E_g, is the normal gap between the conduction and valence bands. For the purposes of this chapter, we make the assumption that $E_1 \leq E_2 \leq E_3$

In an IBSC where there is one intermediate band located between the conduction and valence band, an electron from the valence band can be excited to either the intermediate or conduction band. Additionally, an electron located in the intermediate band can be excited to the conduction band. In total, there are three upward energy transitions in this device: E_1, E_2, and E_g. E_1 represents valence to intermediate band, E_2 represents intermediate to conduction band, and E_g represents the conventional band gap between the valence and conduction band. The two intermediate transitions E_1 and E_2 are independent of each other, while the band gap transition E_g is a function of the two intermediate ones: $E_g = E_1 + E_2$. Proper operation of the IBSC requires that no charge carriers are to be extracted from the intermediate band [4]. This is important for two reasons. First, this would deplete the carrier density available for excitation to higher bands, including the conduction band. Second, charge carriers are extracted from intermediate bands at lower potential differences than from the conduction band. This would reduce the power output of the IBSC and thus the conversion efficiency. Since carriers would only be extracted from the valence and conduction bands, the IBSC would operate similarly to the conventional solar cell in the sense that the operating voltage would be proportional to E_g.

When the conventional solar cell is exposed to light, carrier populations increase beyond equilibrium and cause the Fermi energy level to split into two chemical potentials. These chemical potentials represent steady state solutions that allow treatment of the carrier dynamics in each band separately. More specifically, they are used to determine the probability that an electron will be in any state in the conduction band and a hole will be in any state in the valence band. When the IBSC is exposed to light, the Fermi energy level must split into the number of total bands. In the case with one intermediate band, the Fermi energy level will split into three chemical potentials μ_c, μ_i, and μ_v that represent steady state solutions that allow treatment of the carrier dynamics in the conduction, intermediate, and valence band, respectively. Each band in the IBSC must possess its own chemical potential. If the intermediate band is thermally coupled to the conduction or valence band, excited electrons will lose kinetic energy through phonon emission before the necessary next event occurs for proper IBSC operation: (1) electrons are not extracted from conduction band or (2) electrons are not promoted from the intermediate to conduction band. As a result, the Fermi energy level will split into two chemical potentials and carrier dynamics will be dominated by these two chemical potentials as in the conventional solar conversion device.

Using performance characteristics and assumptions of the IBSC, limiting efficiency calculations are performed similar to the detailed balance calculation developed by Shockley and Queisser. The following list of assumptions is used to carry out limiting efficiency calculations for the IBSC [4]:

1. The solar cell absorbs blackbody radiation at a temperature of $T_s = 6{,}000$ K and ambient $T_a = 300$ K and emits blackbody radiation at ambient $T_a = 300$ K;
2. Only radiative transitions occur between the bands;
3. All photons above the lowest energy gap are absorbed and no high energy photon is used in a low energy process;
4. Carrier mobility is infinite and as a consequence, the quasi-Fermi energy levels are constant throughout the cell;
5. Only one electron–hole pair is created per photon;
6. A perfect mirror is located on the back of the cell so that radiation makes a double pass through the cell and can only escape through the front area of illumination;
7. No carriers are extracted from the intermediate band(s);
8. The net photon flux (number of incident minus number of emitted photons) is equal to the number of charge carrier pairs collected at the contacts (detailed balance assumption).

To find the limiting efficiency, our task lies in finding the characteristic IBSC parameters that maximize the delivered power density.

For the IBSC with one intermediate band, photons with energy $E_1 \leq \hbar\omega < E_2$ are absorbed to promote an electron to the intermediate band and create a hole in the valence band, any excess energy beyond that of the energy transition E_1 will be lost due to thermalization and carriers will relax to the band edges before another radiative event occurs. This makes an absorbed photon with $E_1 < \hbar\omega < E_2$ have the same effect as an absorbed photon with energy $\hbar\,\omega = E_1$. Photons with energy $E_2 \leq \hbar\omega < E_g$ are absorbed to promote an electron from the intermediate band to the conduction band, any excess energy beyond that of the energy transition E_2 will be lost due to thermalization and carriers will relax to the band edges before circuit extraction or recombination. This makes an absorbed photon with $E_2 < \hbar\omega < E_g$ have the same effect as an absorbed photon with energy $\hbar\omega = E_2$. Photons with energy $\hbar\omega \geq E_g$ are absorbed to promote an electron to the conduction band and create a hole in the valence band, any excess energy beyond that of the energy transition E_g will be lost due to thermalization and carriers will relax to the band edges before circuit extraction or recombination. This makes an absorbed photon with $\hbar\omega > E_g$ have the same effect as an absorbed photon with energy $\hbar\omega = E_g$. The net photon flux is equal to the number of charge carrier pairs collected at the contacts so that the current density is just

$$\begin{aligned}J(E_1, E_2, E_g, T_s, T_a, X, V) = q[&XF_{\text{sun}}N(E_g, \infty, T_s, 0) + (1 - XF_{\text{sun}})N(E_g, \infty, T_a, 0)\\&- F(\pi/2)N(E_g, \infty, T_a, qV)] + q[XF_{\text{sun}}N(E_2, E_g, T_s, 0)\\&+ (1 - XF_{\text{sun}})N(E_2, E_g, T_a, 0)\\&- F(\pi/2)N(E_2, E_g, T_a, \mu_{cI})]\end{aligned} \tag{6.6}$$

where we have made use of Eq. (6.4), the geometric factor $F(\theta)$, and the concentration factor X. The terms in the first bracket on the right-hand side of the equation represent the current density generated from the promotion of electrons from the valence band to the conduction band less recombination events from the reverse transition, while the terms in the second bracket represent the current density generated from the promotion of electrons from the intermediate band to the conduction band less recombination events from the reverse transition. In both bracketed terms, the IBSC absorbs radiation from the sun at the characteristic temperature T_s over the angular range $0 < \theta < 0.26^\circ$ and from ambient at the characteristic temperature T_a over the angular range $0.26^\circ < \theta < \pi/2$, while the IBSC emits radiation at the characteristic temperature T_a and characteristic uniform chemical potential over the angular range $0 < \theta < \pi/2$.

For each energy configuration (E_1, E_2, E_g), there exists a voltage V_m that maximizes the IBSC power density output $V_m \cdot J_m(E_1, E_2, E_g, T_s, T_a, X, V)$. However, the current density equation contains an additional chemical potential μ_{cI} and we must assign a value if we are going to determine Eq. (6.6). Proper operation of the IBSC requires that there is no current extracted from the intermediate band(s), i.e. the current entering the intermediate must equal the current leaving the intermediate band. This sets up the constraint

$$\begin{aligned}q[XF_{\text{sun}}N(E_2, E_g, T_s, 0) + (1 - XF_{\text{sun}})N(E_2, E_g, T_a, 0) - F(\pi/2)N(E_2, E_q, T_a, \mu_{cI})]\\= q[XF_{\text{sun}}N(E_1, E_2, T_s, 0) + (1 - XF_{\text{sun}})N(E_1, E_2, T_a, 0)\\- F(\pi/2)N(E_1, E_2, T_a, \mu_{Iv})]\end{aligned} \tag{6.7}$$

and when considered in conjunction with

$$qV = \mu_{cv} = \mu_{cI} + \mu_{Iv} \tag{6.8}$$

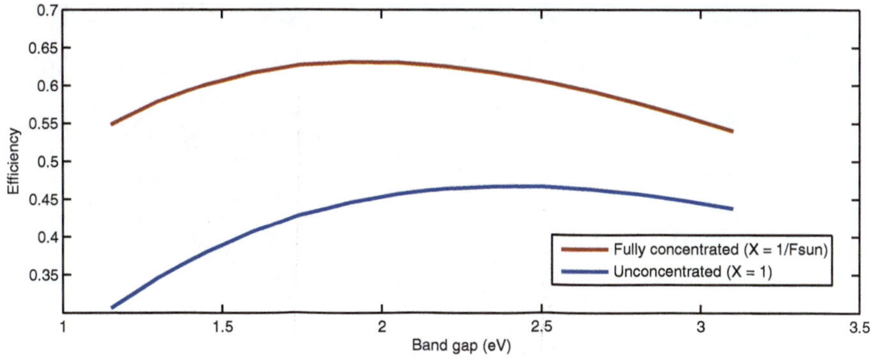

Fig. 6.3 Limiting efficiency of the IBSC with one intermediate band as a function of the largest band gap E_g for unconcentrated light $X = 1$ and fully concentrated light $X = 1/F_{sun}$. The efficiency was calculated assuming the sun is modeled as a blackbody with characteristic temperature 6, 000 K. For fully concentrated light, the maximum efficiency of about 63. 2 % occurs approximately at a band gap of $E_g = 1.93$ eV, while unconcentrated light leads to a maximum efficiency of about 46. 8 % with a band gap of $E_g = 2.40$ eV

$$\mu_{cI} = \mu_c - \mu_I, \quad \mu_{Iv} = \mu_I - \mu_v \tag{6.9}$$

for a given value V, all the chemical potentials are determined and the current density for the IBSC is solved.

Following the prescription outlined above, the limiting efficiency of the IBSC is found as a function of the band gap E_g by varying energy transition E_1 above the valence band for unconcentrated $X = 1$ and fully concentrated $X = 1/F_{sun}$ light (see Fig. 6.3). Fully concentrated light has a maximum efficiency of 63.2 % when the largest band gap is $E_g = 1.93$ eV, while unconcentrated light has a maximum efficiency of 46.8 % when the largest band gap is $E_g = 2.40$ eV. As the concentration factor increases, so does the limiting efficiency up to the maximum of 63.2 %, a result that is similar to the conventional solar conversion device. It should be noted that two patterns have emerged when considering limiting efficiency calculations for both the IBSC and conventional solar conversion device: (1) as light concentration increases, efficiency increases and (2) the large band gap E_g decreases as light concentration increases.

As the performance of the one intermediate band IBSC model is compared to the conventional solar conversion model, it is evident that conversion efficiency has significantly improved. For unconcentrated light, performance has increased from 31 to 46.8 %. For fully concentrated light, performance has increased from 41 to 63.2 %. This increase is dramatic and the motivating factor for further research in this type of structure.

After calculating the significant efficiency improvements of the IBSC, the interest surrounding the sensitivity of efficiency as a function of the two independent energy transitions E_1 and E_2 should be investigated. Presumably, a physical device that employs the characteristics of an IBSC will not exactly match the

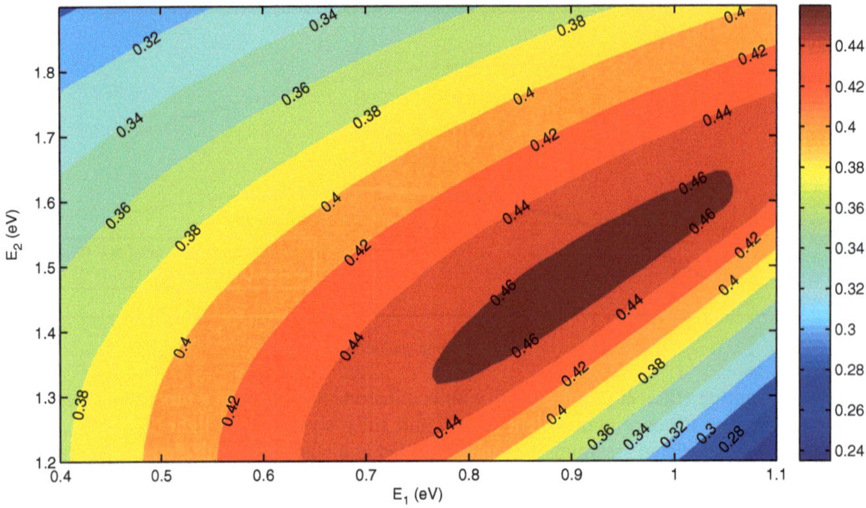

Fig. 6.4 Limiting efficiency of the IBSC with one intermediate band as a function of the two energy transitions E_1 and E_2 for unconcentrated light $X = 1$. The *color bar* on the right-hand side of the contour plot summarizes what efficiencies the various colors represent

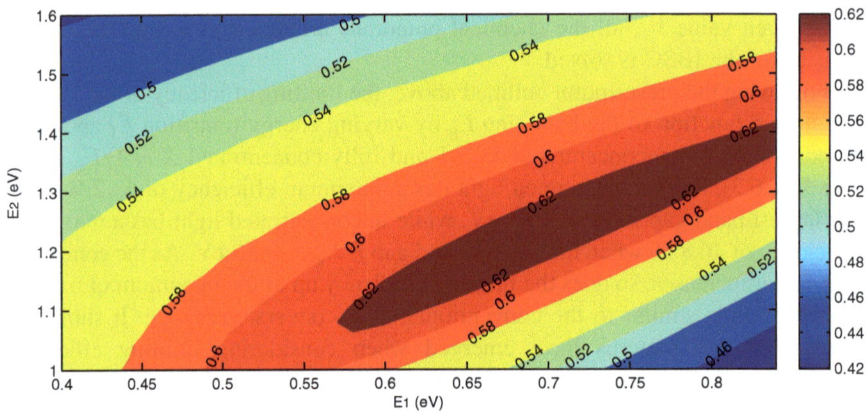

Fig. 6.5 Limiting efficiency of the IBSC with one intermediate band as a function of the two energy transitions E_1 and E_2 for fully concentrated light $X = 1/F_{\text{sun}}$. The *color bar* on the right-hand side of the contour plot summarizes what efficiencies the various colors represent

energy band configuration that leads to the theoretical maximum efficiency. Therefore, it would be prudent to calculate the sensitivity of these energy levels to see how efficiency responds. Does the limiting efficiency drop off suddenly if energy levels are slightly removed from the optimized position? To answer this question, we calculated the limiting efficiency of the IBSC as a function of the two energy transitions E_1 and E_2 for both unconcentrated and fully concentrated light (see Figs. 6.4 and 6.5).

For unconcentrated light, the maximum efficiency of 46. 8 % occurs at the energy band transitions $E_1 = 0.92$ eV and $E_2 = 1.48$ eV, but Fig. 6.4 shows that an efficiency of $\geq$ 46 % can occur at various band configurations. In comparison with the conventional conversion device, most energy band configurations of the IBSC have a greater limiting efficiency than 31 %. Additionally, it does not seem that efficiency drops off suddenly for band configurations outside the optimized position.

For fully concentrated light, the maximum efficiency of 63.2 % occurs at the energy band transitions $E_1 = 0.70$ eV and $E_2 = 1.23$ eV but Fig. 6.5 shows that an efficiency of $\geq$ 62 % can occur at various band configurations. Specifically at band configurations ranging in the intervals 0. 57 eV $\leq E_1 \leq$ 0. 88 eV and 1.06 eV $\leq E_2 \leq$ 1.43 eV. In comparison with the conventional conversion device, most energy band configurations of the IBSC have a greater limiting efficiency than 41 %. Again, it does not seem that efficiency drops off suddenly for band configurations outside the optimized position.

In an IBSC where there are two intermediate bands located between the conduction and valence band (see Fig. 6.2), there is a total of six upward energy transitions in this device E_1, E_2, E_3, E_4, E_5, and E_g: E_1 represents valence to first intermediate band, E_2 represents first intermediate to second intermediate band, E_3 represents second intermediate to conduction band, E_4 represents valence to second intermediate band, E_5 represents first intermediate band to conduction band, and E_g represents the conventional band gap between the valence and conduction band. The three intermediate transitions E_1, E_2, and E_3 are independent of each other, while the three other transitions are a function of the intermediate transitions.

$$E_4 = E_1 + E_2 \tag{6.10}$$

$$E_5 = E_2 + E_3 \tag{6.11}$$

$$E_g = E_1 + E_2 + E_3 \tag{6.12}$$

For each energy configuration (E_1, E_2, E_3), there exists a voltage V_m that maximizes the IBSC power density output $V_m \cdot J_m(E_1, E_2, E_3, T_s, T_a, X, V)$. However, as with the one intermediate band IBSC, the current density contains chemical potentials that need to be solved for. To find all the chemical potentials for a particular energy configuration, the operation of the IBSC demands the following constraints:

1. The current entering intermediate band I_1 must equal the current leaving the intermediate band I_1 ($J_{vI_1} = J_{I_1I_2} + J_{I_1c}$) and the current entering intermediate band I_2 must equal the current leaving the intermediate band I_2 ($J_{I_1I_2} + J_{vI_2} = J_{I_2c}$).
2. The chemical potential μ_{I_2v} must equal the sum of the chemical potentials μ_{I_1v} $+\mu_{I_2I_1}$. The chemical potential μ_{cI_1} must equal the sum of the chemical potentials $\mu_{I_2I_1} + \mu_{cI_2}$. Finally and as before, the chemical potential $\mu_{cv} = qV$ must equal the sum of the quasi-Fermi levels $\mu_{I_1v} + \mu_{I_2I_1} + \mu_{cI_2}$

where we have simplified the current density notation such that J_{vI_1} is equal to

$$q[XF_{\text{sun}}N(E_1, E_2, T_s, 0) + (1 - XF_{\text{sun}})N(E_1, E_2, T_a, 0) - F(\pi/2)N(E_1, E_2, T_a, \mu_{vI_1})]$$

and so on. The same general approach to find all the chemical potentials (total of 6) is used.

Following the prescription above, the limiting efficiency of the IBSC is found by varying the energy transitions E_1 and E_2 above the valence band for unconcentrated $X = 1$ and fully concentrated $X = 1/F_{\text{sun}}$ light. Fully concentrated light has a maximum efficiency of 72.4 % for the energy transitions of 0.59 eV, 0.93 eV, and 1.05 eV for a band gap $E_g = 2.57$ eV, while unconcentrated light has a maximum efficiency of 52. 1 % for the energy transitions of 0.85 eV, 1.20 eV, and 1.43 eV for a band gap $E_g = 3.48$ eV. As the concentration factor increases, so does the limiting efficiency up to the maximum of 72.4%, a result that is directionally similar to both the conventional solar conversion device and IBSC with one intermediate band.

As the performance of the two intermediate band IBSC model is compared to the conventional solar conversion model, it is evident that conversion efficiency has again significantly improved. For unconcentrated light, performance has increased from 31 to 52.1%. For fully concentrated light, performance has increased from 41 to 72.4 %. When the performance of the two intermediate band IBSC is compared to the one intermediate band IBSC, conversion efficiency improves. For unconcentrated light, performance has increased from 46.8 to 52.1 %. For fully concentrated light, performance has increased from 63.2 to 72.4 %.

6.3 Quantum Dot Intermediate Band Solar Cell

With the advance in microfabrication technology, it is possible to coherently grow two or more dissimilar semiconductors. These types of structures are called heterostructures [7]. Of real interest in these structures are the effects caused by the different band gaps and relative energy alignment with respect to each other. As an example, consider a heterostructure consisting of compound semiconductors aluminum arsenide and gallium arsenide (AlAs/GaAs) material that alternates along the z-direction (see Fig. 6.6). The band gap of the pure compound semiconductor AlAs is larger than the band gap of the pure compound semiconductor GaAs. In addition to the size of the band gaps, the relative band alignment between the two permits the GaAs conduction band to sit below the conduction band of AlAs and the GaAs valence band to sit above the valence band of AlAs. As a result, there is an energy separation between the GaAs and AlAs valence bands. This is called the valence band offset (see Fig. 6.6). The conduction band offset is defined similarly. The alternating materials and offsets create one-dimensional potential wells. Within each well, additional GaAs energy levels below its valence band could lie

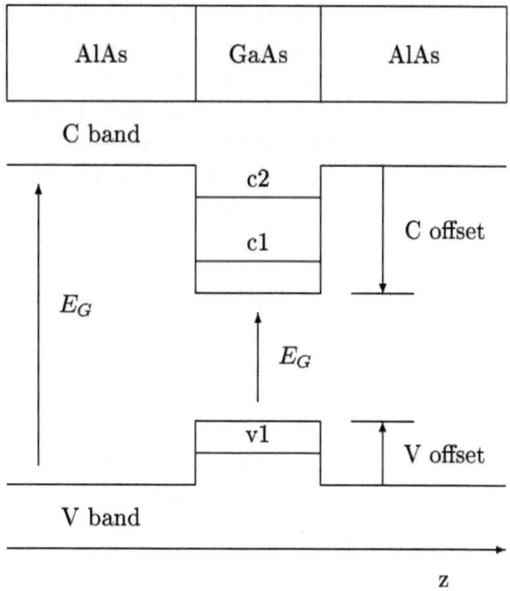

Fig. 6.6 The alternating semiconducting material AlAs and GaAs creates a heterostructure. C and V refer to conduction and valence bands, respectively. Since the band gaps are of different sizes, this mimics the one-dimensional potential well. The intermediate energy levels c1, c2, and v1 are in the quantum wells, the offset is the potential well depth. Reprinted with permission from [20]. Copyright 2010, American Institute of Physics

above the AlAs valence band (energy level v1 in Fig. 6.6). Similarly, energy levels arise between the conduction bands of GaAs and AlAs (energy level c1 and c2 in Fig. 6.6). As the number of quantum wells increases, these energy levels will split and spread into bands. Such bands occur as intermediate bands between the band gap of AlAs in the AlAl/GaAs heterostructure. Discrete energy levels occur in the z-direction, while there is no confinement in the quantum well in-plane direction. This confines charge carriers in one direction and produces the energy spectrum of the form $E_{\mathrm{d}} + \hbar^2(k_x^2 + k_y^2)/2m^*$, where E_{d} are discrete energies associated with confinement in the z-direction, $\hbar k_x$ and $\hbar k_y$ are the in-plane momentum components, and m^* is the charge carrier effective mass. In order to find the discrete energy levels within each band, we substitute the effective mass for the mass in Schrödinger's time-independent equation and solve for the z-direction subject to boundary conditions. In Cartesian coordinates, using the effective mass approximation, Schrödinger's equation is

$$\frac{-\hbar^2}{2m^*}\left(\frac{d^2}{dx^2} + \frac{d}{dy^2} + \frac{d^2}{dz^2} + V(z)\right)\Psi(x, y, z) = E\Psi(x, y, z) \tag{6.13}$$

where $V(z)$ is the conduction or valence band offset in the z-direction. In order to solve, appropriate continuity conditions are applied at the heterostructure boundary. A semiconductor heterostructure that confines charge carriers in two dimensions such that the energy spectrum is of the form $E_{\mathrm{d}_x} + E_{\mathrm{d}_y} + \hbar^2(k_z^2)/2m^*$, where $E_{\mathrm{d}_x} + E_{\mathrm{d}_y}$ are discrete energies associated with confinement in the traverse direction and $\hbar^2(k_z{}^2)/2m^*$ is the longitudinal momentum component, is called a quantum wire. The discrete energy levels are found by solving Eq. (6.13) with a potential $V(x, y)$ that represents the conduction or valence band offset. Semiconductor heterostructures that restrict the motion of charge carriers in all three spatial directions are referred to as quantum dots (QDs) [8]. Unlike the quantum well, the QD material is completely surrounded by a material with a larger band gap (barrier material) so the energy spectrum is discrete. If the number of QDs is increased and arranged in a periodic lattice, the energy levels will split and spread out into bands. The width of the band(s) depends on the spacing of the QDs within the lattice and wavefunction overlap.

Due to the energy levels that appear between the conduction and valence band of the barrier material, heterostructures are considered for the IBSC. All three types of heterostructures introduced above contain the necessary intermediate levels required for the IBSC. However, only one of the three types allows for the proper operation of the IBSC. The quantum well's intermediate levels provide a finite density of states lying between the conduction and intermediate bands because of the term in the energy spectrum $\hbar^2(k_x^2 + k_y^2)/2m^*$. Photogenerated charge carriers will lose energy through thermalization due to the high density of states available in the in-plane direction. The Fermi level will not split into the number of total bands. Rather, the Fermi level will split into two levels similar to the solar conversion device with a single energy gap. The lowest lying intermediate band will act as the conduction band and heterostructure will behave as a single energy gap conversion device. These ideas have been reviewed by Anderson and Luque et al. in [9, 10]. Similar to the quantum well, the quantum wire's intermediate levels provide a finite density of states lying between the conduction and intermediate bands because of the term $\hbar^2(k_z{}^2)/2m^*$ and photogenerated charge carriers will lose energy through thermalization due to the density of states available in the longitudinal direction. Again, the Fermi level will not split into the number of total bands but rather split into two levels. As opposed to the quantum well and wire, a QD heterostructure will provide zero density of states between the excited bands due to the discrete energy spectrum. With each band thermally isolated, the Fermi level will split into the number of bands and charge carrier concentration in each of the bands will be described by its own chemical potential as required for proper IBSC operation. This will reduce thermalization and is the main reason why QDs are considered for the physical realization of the IBSC.

The idea of the quantum dot intermediate band solar cell (QD-IBSC) was first proposed by Marti et al. [11], where they describe the structure of the design and offer us a proposed QD/barrier system that, when simplistically modeled, would produce one intermediate band in the conduction band offset at the ideal energetic location determined by the detailed balance method. The QD-IBSC is shown schematically in Fig. 6.7, with the quantum dot material located in-between the p–n junction. The p–n emitters prevent the intermediate band(s) from touching the deposited contacts on the external faces so no current is directly extracted from the intermediate bands in normal operation [12]. As described in Sect. 6.2, this is a necessary requirement for the proper operation of the IBSC.

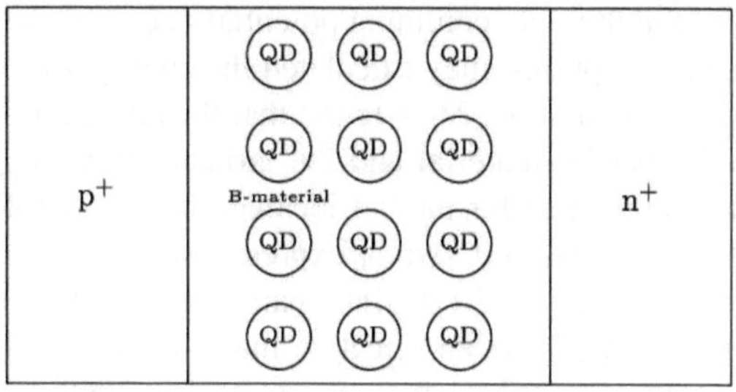

Fig. 6.7 A schematic of the proposed QD-IBSC with the barrier material surrounding the QDs. The QD material is arranged periodically within the barrier material and sandwiched in between the normal p–n junction. Reprinted with permission from [20]. Copyright 2010, American Institute of Physics

The QD-IBSC is a variation on the p–i–n junction theme used in certain types of solar cells today. The layer of semiconductor material between p and n, instead of being left undoped or *intrinsic*, contains a periodic array of QDs embedded within the intrinsic, or *barrier* material. It is important for the size and spacing of the quantum dots to be uniform. This helps establish well-placed intermediate band boundaries. If the dots are organized in a random array or are diverse in size, the energy levels will tend to be irregularly spaced throughout the offsets due to symmetry breaking [13]. The intermediate bands would not be created, rather single energy levels varied throughout the band gap aiding in recombination via thermalization.

One way to grow uniform coherent QD arrays is through the Stanski and Krastanov (S&K) method, which leads to self-organized growth [14–17]. A requirement to achieve this growth is a lattice constant mismatch, Δ_{lc}, between the quantum dot and barrier material. The formulation of the quantum dots are a result of the Δ_{lc}: as the quantum dot material is deposited on the surface of barrier material, the quantum dot material will compress to fit the smaller barrier lattice and 2D growth will continue until it becomes energetically favorable for quantum dots (islands) to spontaneously form. The layer thickness at which 2D–3D growth occurs is called the critical thickness and depends on the Δ_{lc} [18]. The QD characteristics such as size and shape are highly dependent on certain parameters such as growth temperature and growth interruptions. A discussion of the methods that are used to grow the self-assembled QDs is outside the scope of this chapter. There are various growth techniques, such as Metal Organic Chemical Vapor Deposition (MOCVP) and Molecular Beam Epitaxy (MBE), with each one very different from the next and constantly evolving. We note that QD growth technology is available and can produce highly uniform QD arrays. One main requirement for doing so is a lattice mismatch between the quantum dot and barrier material.

Marti et al. made the assumption that the geometry of the QD is spherical and is characterized by a QD radius, *a*, measured in Angstroms, Å. In addition, for simplicity, they assumed that the energy corresponding to the top of the valence band is the same both in the barrier and in the QD material. This means there is no valence band offset and the only confining potential occurs at the conduction band offset. Under these assumptions, they calculated the energy levels in the QD using the effective mass approximation. They found that the ternary alloy $Al_{0.40}Ga_{0.60}As$ could be used as the barrier material and the ternary alloy $In_{0.42}Ga_{0.58}As$ as the QD material. Band parameters for the barrier material are as follows: a band gap of 1. 95 eV, an effective mass of 0. 096m_0 expressed as the rest mass of an electron m_0, and a lattice constant of 5. 5654 Å. Band parameters for the QD material are as follows: a band gap of 0. 87 eV, an effective mass of 0. 045m_0 expressed as the rest mass of an electron m_0, and a lattice constant of 5. 7060 Å. The ground state energy level $E_{0,\,0}$ for a QD radius of $a = 39$ Å lies 1. 24 eV above the valence band, which is the optimal position for achieving 63. 2 % conversion efficiency as determined from the detailed balance argument. In addition, there is the necessary lattice mismatch between the QD and barrier material needed to manufacture the QD array.

Although the proposed QD-IBSC materials seemingly have the desired parameters identified for an IBSC under fully concentrated light, there exist unjustified assumptions that should be noted. The valence band offset in the present QD heterostructure is not negligible, rather the depth is 0. 37 eV. This confining potential will support energy levels, in fact it will support many energy levels because holes are generally heavy with a magnitude similar to m_0. This could potentially be a problem because as these energy levels spread out into bands, some will overlap or those next to the barrier's valence band will merge with that band. This result might reduce the efficiency through unwanted transitions or shrink the barrier energy gap due to bands merging with the barrier's valence band. A second assumption is that the conduction band offset will only support one energy level. However, the conduction band offset will not only support the ground state energy level but also support two additional energy levels $E_{0,\,1}$ and $E_{0,\,2}$. Additional energy levels will affect efficiency. Similar to the valence band multiple energy level discussion, when these energy levels spread out into bands they might overlap with each other or merge into the barrier's conduction band. Efficiency could be reduced. If, for example, one of the bands overlaps with the conduction band, the energy gap of the barrier material will decrease. This will degrade the cell's operating voltage and decrease the power output of the cell. If the additional bands are optimally placed, as seen in the previous section, they could be used to enhance efficiency. As such, it is important to calculate the placement of the bands.

6.4 Material Selection

Levy et al. also found potential QD-IBSC materials that contain an optimally placed intermediate band with a detailed balance conversion efficiency of 60 % assuming the geometry of the QD is spherical [19]. As part of their design criteria, they selected QD-IBSC materials that have a negligible valence band offset. However, solving the effective mass equation for locating the energy levels and realistic size of quantum dots, two intermediate bands will most likely emerge. This second band, if strategically placed, will increase the maximum efficiency of the solar cell from 63. 2 % (single intermediate band) to 72. 4 % (two intermediate bands) for an IBSC under full concentration. When searching for materials, this additional band should be taken advantage of to maximize the QD-IBSC capability. Using the assumptions in Sect. 6.3, the bound state energy levels $E_{n,\,l}$ obtained for typical quantum dot parameters are organized as follows: $E_{0,\,0} < E_{0,\,1} < E_{0,\,2} < E_{1,\,0} \cdots$. This is an important result for two reasons:

- The first intermediate energy level is found when $n = 0, l = 0$ and the second intermediate energy level is found when $n = 0, l = 1$.
- The quantum dot will support two intermediate energy levels when the second excited state, $n = 0, l = 2$, is not bound but first excited state, $n = 0, l = 1$, is bound.

This knowledge is useful when searching through numerous materials; only the values $n = 0$ and $l = 0$, 1, and 2 are used in calculations to find energy levels.

For an efficiency of an IBSC under full concentration of at least 70 %, the barrier energy gap must be greater than 2. 0 eV. This is the first design criterion and begins to narrow down possible barrier materials.

As mentioned, a valence band offset might form and the confining potential will support many energy levels. This result might reduce the efficiency through unwanted transitions or shrink the barrier energy gap due to bands merging with the barrier's valence band. Therefore, the materials determined have a negligible valence band offset in order to eliminate these possibilities. This will assure that no minibands are created outside those that are known and calculated in the conduction band offset.

As a final design criterion, it is desirable to produce solar cells as thin as possible. Thinner solar cells require less material and will reduce manufacturing costs. One way to reduce the thickness of the cell is to use direct band gap materials so the absorption coefficient is strong. Only direct band gap materials are used in this search.

Potential QD-IBSC materials are found that satisfy the design considerations and optimized two intermediate energy levels to achieve a theoretical efficiency greater than 70 % [20]. The two energy levels located in the conduction band offset are found by solving the effective mass equation in conjunction with band parameters (based on the review from Vurgaftman et al. [21]), all of which were calculated at room temperature, 300 K. The barrier and quantum dot materials, along with the energy transitions, lattice mismatch, quantum dot radius, and efficiency, are given in Table 6.1. This table lists possible QD-IBSC with efficiencies in excess of 70 %.

The transition energies will spread out into bands as the quantum dots are grown. As mentioned, it is important that the intermediate bands do not overlap with each other or with the conduction/valence bands. Qualitatively, the closer the quantum dots are spaced, the larger the width of the intermediate energy bands due to the overlap of their wavefunctions. Quantitatively, there should be some minimum distance between the quantum dots in order to prevent the bands from overlapping.

Using tight-binding theory and assuming the quantum dots are arranged in a simple cubic lattice, minimum distances are determined [20]. This distance is

Table 6.1 Barrier and quantum dot materials (QD) that produce an efficiency, η, greater than 70 % and have two intermediate bands

Barrier material	$AlAs_{0.17}Sb_{0.83}$	$AlP_{0.05}Sb_{0.95}$	$AlAs_{0.05}Sb_{0.95}$
QD material	$InP_{0.35}Sb_{0.65}$	$GaP_{0.37}Sb_{0.63}$	$GaAs_{0.72}Sb_{0.28}$
QD radius (Å)	35	34	33
E_1 (eV)	0. 73	1. 23	1. 31
E_2 (eV)	1. 31	0. 43	0. 43
E_3 (eV)	0. 26	0. 57	0. 55
η(%)	70. 0	72. 0	70. 0
Δ_{lc}(%)	3. 45	4. 30	5. 78

The energy transitions E_1, E_2, and E_3 refer to those in Fig. 6.2. Reprinted with permission from [20]. Copyright 2010, American Institute of Physics

Table 6.2 Maximum bandwidth associated with the two intermediate energy levels in a simple cubic lattice

Barrier	$AlAs_{0.17}Sb_{0.83}$	$AlP_{0.05}Sb_{0.95}$	$AlAs_{0.05}Sb_{0.95}$
QD	$InP_{0.35}Sb_{0.65}$	$GaP_{0.37}Sb_{0.63}$	$GaAs_{0.72}Sb_{0.28}$
QD radius (Å)	35	34	33
D (Å)	77	68	66
Δs (eV)	0. 033	0. 129	0. 133
Δp (eV)	0. 38	0. 287	0. 296

Δs is this bandwidth associated with the first energy level, Δp is the bandwidth associated with the second energy level, and D is the minimum distance as measured from the center of one quantum dot to center of next quantum dot needed to prevent overlapping. Reprinted with permission from [20]. Copyright 2010, American Institute of Physics

measured from the center of one quantum dot to the center of the next quantum dot. Table 6.2 displays the minimum distance, D, that is required to prevent an overlap of the minibands and the widths of minibands appearing in the QD, Δs and Δp, at the distance D. For the barrier/QD materials AlPSb/GaPSb and AlAsSb/GaAsSb, the minibands will not overlap even when the QDs touch each other in the lattice. In this case, the minimum distance is determined by the physical constraint of the QDs rather than the prevention of miniband overlap.

Although materials were found that produce the intermediate bands necessary to achieve an efficiency of greater than 70 %, these results are based on certain restrictive assumptions: the shape of the quantum dot is spherical; the quantum dot lattice is cubic; the Hamiltonian includes only the offset potential; and the QD-IBSC is under full concentration.

6.4.1 Assumptions

Evaluating the assumptions used to find materials for the QD-IBSC becomes of critical importance to the credibility of making the claim that certain material combinations are better QD-IBSC candidates than others. In order to leverage

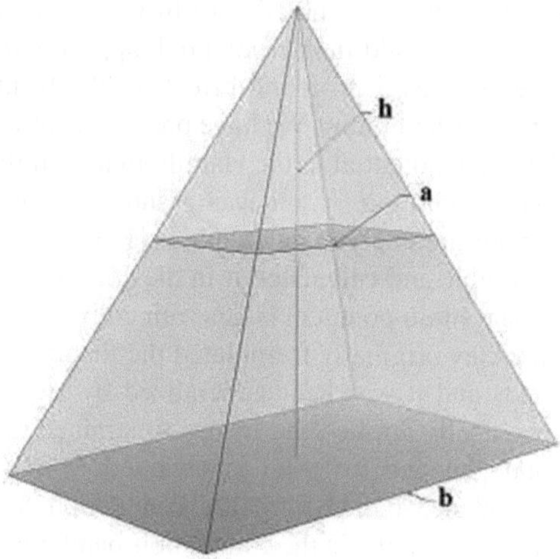

Fig. 6.8 The two types of geometry used in our analysis: pyramid and truncated pyramid. The pyramid is defined by the two dimensions *b* and *h* in the figure, while the truncated pyramid is defined by three dimensions *b*, *h*, and *a*

theoretical models to investigate the electronic structure of S&K grown QD's, assumptions based on experimental observation will help justify model results. In instances where experimental observations are either difficult to quantify or have not been performed, assumptions should be based on a known similar observation, numerical simulations, or some other "best estimate." In the previous section, we assumed the QD was spherically symmetric with a diameter that could be varied. The shape was chosen because of the relative ease in determining the energy levels and respective wavefunctions rather than experimental evidence suggesting a spherical nature to the QD. Reported shapes of QDs are of wide variety, including pyramidal, truncated pyramidal, lens, hemispherical, multifaceted domes, etc. [22–27]. However, to our knowledge, spherical S&K QDs have not been observed. Therefore, the assumption that QDs are spherical should be updated to include more realistic geometry. Nevertheless, there is merit to performing calculations on the spherical QD, as it helps us understand some of the more general features and could possibly be manufactured in the future with technological advances.

Based on the reported experimental observations that detail the geometry of the QD, we propose to limit the scope of this analysis to two QD geometries: a pyramid with square base and a truncated pyramid with square base. The scope of geometry is limited to the numerous calculations that must be performed and guided by a number of experimental observations that have been reported for the proposed geometry. The geometry is highlighted in Fig. 6.8. The pyramid with square base is defined by two dimensions, while the truncated pyramid is defined by three dimensions (see Fig. 6.8).

The growth of self-assembled QDs in the S&K mode is made possible by an initial strain caused by lattice mismatch between two materials. Take QDs of InAs on barrier material of GaAs, both of which are of the zinc blende structure and characterized by a single lattice constant a_{lc}. The lattice constants differ by about 7. 2 %, which causes a large stress to develop within the system. The stress is partly relieved by the formulation of InAs pyramidal dots. This strain enters the Hamiltonian as an additional potential appropriately called the *strain potential*. This potential has a profound effect on the band structure and must be included for serious calculations. We have previously assumed strain was negligible and omitted the strain potential in the Hamiltonian. Similar to the spherically symmetric geometry assumption, strain was assumed negligible because of the relative ease in performing analytic calculations. For our model, we do not assume that strain is negligible and calculated it in the context of continuum mechanics.

The strain potential is determined by deformation potential theory. Bardeen and Shockley originally formulated the theory in the context of non-degenerate energy bands and it was later generalized by Herring and Vogt [28, 29]. Pikus and Bir specifically applied the theory to strained materials [30]. The modification of the band gap due to strain is obtained from the strain tensor, $\varepsilon_{i\ j}$, by decoupling the valence bands from the conduction or spin-orbit bands. In the conduction band, the change in the conduction-band-edge energy due to strain can be described by one deformation potential constant, a_{c}, and the hydrostatic strain such that the strain potential is

$$V_s^c = -a_{\mathrm{c}}(\varepsilon_{xx} + \varepsilon_{yy} + \varepsilon_{zz}) \tag{6.14}$$

where the superscript refers to the respective band, in this case the conduction band. For our analysis, we are only concerned with the valence band maximum in both the barrier and QD materials, instead of locating bound states caused by the valence band offset. As noted in Sect. 6.4, valence band offsets will support bound states, in fact many bound states, and as a criterion for potential QD-IBSC materials, only material combinations with negligible valence band offsets were considered. However, when strain is introduced, the likelihood of this condition is greatly diminished because the valence band offset caused by a difference in energy gaps would need to negate the strain potential caused by the lattice mismatch. In addition, the valence band offset is uniform but due to the nonuniform strain for QD geometry considered, the strain potential will be nonuniform furthering the difficulty to satisfy this criterion. For the purposes of this analysis, we make the assumption that bound states formed in any valence band offset will merge together as they spread out into bands, effectively shrinking the band gap of the barrier material. The modified band gap of the barrier material will be determined by the following formula

$$E_{\mathrm{g}} = E_{\mathrm{g}}^{\mathrm{B}} - (V_{\mathrm{o}} + V_{\mathrm{s}}^{\mathrm{v}}) \tag{6.15}$$

where $E_{\mathrm{g}}^{\mathrm{B}}$ is the barrier's unmodified direct band gap at $k = 0$, V_{o} is the valence band offset caused by difference in energy gaps at $k = 0$, and $V_{\mathrm{s}}^{\mathrm{v}}$ is the strain potential occurring at the valence band at $k = 0$. In the valence band, the change in the valence-band-edge energy due to strain is determined by diagonalizing the 4×4 Luttinger Hamiltonian with strain interaction for the light and heavy hole bands (see [31–34]), which lift the degeneracy between the bands at $k = 0$

$$E_{\mathrm{hh}}(k=0) = -a_v\left(\varepsilon_{xx} + \varepsilon_{yy} + \varepsilon_{zz}\right) + B^{1/2} \tag{6.16}$$

$$E_{\mathrm{lh}}(k=0) = -a_v\left(\varepsilon_{xx} + \varepsilon_{yy} + \varepsilon_{zz}\right) - B^{1/2} \tag{6.17}$$

where

$$B = \frac{b^2}{2}\left[\left(\varepsilon_{xx} - \varepsilon_{yy}\right)^2 + \left(\varepsilon_{xx} - \varepsilon_{zz}\right)^2 + \left(\varepsilon_{yy} - \varepsilon_{zz}\right)^2\right] + d^2\left(\varepsilon_{xy}^2 + \varepsilon_{xz}^2 + \varepsilon_{yz}^2\right), \tag{6.18}$$

and a_v, b, and d are deformation potential constants. Values for the deformation potential constants are such that $a_v < 0$, $b < 0$, and $d < 0$ and for compressive strain, as in the QD, the valence band moves down in energy (and conduction band moves up in energy). Under these conditions, the heavy hole energy at $k = 0$ sits energetically higher than the light hole at $k = 0$ and will be used as the strain potential V_s^v in Eq. (6.15) to calculate the barrier's modified band gap.

6.4.2 Application of Finite Element Method to Schödinger's Equation

Analytic solutions to Schödinger's equation are few. Two of the more notable ones are the quantum oscillator potential and Coulomb potential, which Schrödinger solved in his first paper on quantum mechanics [35]. Other analytic solutions exist for simple potentials that include square or spherically symmetric potential wells. These solutions are useful in terms of our understanding of the atomic universe. However, analytic solutions are not available for practical applications, especially those that involve emerging technology.

The application of the finite element method normally takes place in two distinct steps. In the first step, the geometry of the problem is defined and tessellated into simplices or elements. In the second step, reformulate the dynamics into a variational problem and the geometry defined in the first step is used to transform the variational problem into a matrix problem. Once the problem is in a matrix representation, the solution is obtained through matrix operations that typically involve an inversion or diagonalization. This is the general idea.

Schödinger originally derived his equation from the action integral by placing a variation on function ψ such that any variation of it is stationary [35]. The result was the following

$$\delta\int_{\Omega}\left(\frac{\hbar^2}{2m}(\nabla\psi^*(\mathbf{x}))(\nabla\psi(\mathbf{x})) + \psi^*(\mathbf{x})\,V(\mathbf{x})\psi(\mathbf{x}) - \psi^*(\mathbf{x})\,E\,\psi(\mathbf{x})\right)d\Omega = 0 \tag{6.19}$$

where the Lagrange multiplier, E, interpreted as energy, is used to enforce the normalizing condition $\int\psi^*(\mathbf{x})\psi(\mathbf{x}) = 1$. The above equation, not normally expressed in its integral form but rather in its equivalent second order partial differential equation form, can be exploited using mesh elements. The action integral is discretized and the unknown function, $\psi(\mathbf{x})$ in our case, is solved for using a linear combination of elemental wave functions. These elemental functions are a linear combination of basis functions and chosen such that the unknown function has the value A_i at node i, i.e. $\psi(x_i) = A_i$. To be explicit, if the number of elements is $N_{\rm E}$ and each elemental wave function is ψ_α

$$\psi(\mathbf{x}) = \sum_{\alpha=1}^{N_{\rm E}} \psi_\alpha(\mathbf{x}). \tag{6.20}$$

By substituting Eq. (6.20) into the variational equation Eq. (6.19), each of the terms begins to simplify. The kinetic energy term is as follows:

$$\begin{aligned}\int_\Omega \frac{\hbar^2}{2m(\mathbf{x})}(\nabla\psi^*)(\nabla\psi)d\Omega &= \int_\Omega \frac{\hbar^2}{2m(\mathbf{x})}\sum_{\beta=1}^{N_{\rm E}} \nabla\psi_\beta(\mathbf{x})^* \sum_{\alpha=1}^{N_{\rm E}} \nabla\psi_\alpha(\mathbf{x})d\Omega \\ &\Rightarrow \sum_{\alpha=1}^{N_{\rm E}} \int_{\Omega_\alpha} \frac{\hbar^2}{2m(x)} \nabla\psi_\alpha^*(x)\nabla\psi_\alpha(x)d\Omega\end{aligned} \tag{6.21}$$

The elemental wave function $\psi_\alpha(\mathbf{x})$ is zero outside its element α and similarly $\psi_\beta(\mathbf{x})$ is zero outside its element β. Only when $\alpha = \beta$, the kinetic energy term is nonzero and this causes the double sum to be reduced to the single sum. The potential energy and energy eigenvalue terms are treated the same way:

$$\sum_{\alpha=1}^{N_{\rm E}} \int_{\Omega_\alpha} \psi_\alpha^*(\mathbf{x})\, V(\mathbf{x})\, \psi_\alpha(\mathbf{x})d\Omega \tag{6.22}$$

$$E\sum_{\alpha=1}^{N_{\rm E}} \int_{\Omega_\alpha} \psi_\alpha^*(\mathbf{x})\psi_\alpha(\mathbf{x})d\Omega \tag{6.23}$$

In principle, the three expressions above set up the finite element approximation. In order to see this, we introduce the elemental wave function as:

$$\psi_\alpha(\mathbf{x}) = \sum_{i=1}^{n} A_i^\alpha \phi_i^\alpha(\mathbf{x}) \tag{6.24}$$

where A_i^α are unknown coefficients that represent the amplitude of the wave function at a particular node,[2] $\phi_i^\alpha(\mathbf{x})$ are basis functions, and the summation's

[2] The coefficients might also represent the amplitude of the derivative of the wave function at a particular node.

upper limit is the number of basis functions per element. Inserting Eq. (6.23) into Eqs. 6.21 and 6.22 gives the following:

$$\sum_{\alpha=1}^{N_E}\sum_{i,j}^{n} A_i^{\alpha*}\underbrace{\left(\frac{\hbar^2}{2m^\alpha}\int_\Omega \left(\nabla\phi_i^\alpha(\mathbf{x})\right)^*\left(\nabla\phi_j^\alpha(\mathbf{x})\right)d\Omega\right)}_{KE_{ij}^\alpha} A_j^\alpha,$$

$$\sum_{\alpha=1}^{N_E}\sum_{i,j}^{n} A_i^{\alpha*}\underbrace{\left(\int_\Omega \phi_i^{\alpha*}(\mathbf{x})\,V(\mathbf{x})\,\phi_j^\alpha(\mathbf{x})d\Omega\right)}_{PE_{ij}^\alpha} A_j^\alpha,$$

$$E\cdot\sum_{\alpha=1}^{N_E}\sum_{i,j}^{n} A_i^{\alpha*}\underbrace{\left(\int_\Omega \phi_i^{\alpha*}(\mathbf{x})\phi_j^\alpha(\mathbf{x})d\Omega\right)}_{O_{ij}^\alpha} A_j^\alpha \tag{6.25}$$

The integrals are carried out and inserted into Eq. (6.19),

$$\sum_{\alpha=1}^{N_E}\sum_{i,j}^{n} A_i^{\alpha*}\left(KE_{ij}^\alpha + PE_{ij}^\alpha - E\cdot O_{ij}^\alpha\right)A_j^\alpha = 0, \tag{6.26}$$

resulting in a generalized eigenvalue equation with E representing the eigenvalues and $\psi_i{}^\alpha$ representing the eigenvectors. This equation represents a discretized version of action integral and we have to place a variation on the variable $\psi^{\alpha*}$ to exercise the principle of stationary action.

$$\frac{\delta J}{\delta A^{\alpha*}} = 0 \Rightarrow \sum_{\alpha=1}^{N_E}\sum_{i,j}^{n}\left(KE_{ij}^\alpha + PE_{ij}^\alpha - E\cdot O_{ij}^\alpha\right)A_j^\alpha \tag{6.27}$$

Once the boundary conditions are properly treated within the matrix, it is input into any standard generalized eigenvalue solver. Eigenvalues represent the energy and corresponding eigenvectors represent nodal wavefunction values.

6.4.3 Theoretical Model

The theoretical model used to investigate promising QD-IBSC materials begins by defining the geometry. We have discussed the two types of geometry that will be used in the calculations: a pyramid with square base and a truncated pyramid with square base. A three-dimensional mesh program written in MATLAB is used to tessellate the stated geometry into tetrahedral elements. Next, we calculate the strain for given QD and barrier materials using the finite element method applied to continuum mechanics subject to initial strains and boundary conditions. The initial strain is taken as the difference between the lattice constants of QD and the barrier with respect to the barrier such that the initial strain ε_0 is [36]

$$\varepsilon_0 = \frac{a_{\text{lc}}^{\text{QD}} - a_{\text{lc}}^{\text{B}}}{a_{\text{lc}}^{\text{B}}} \tag{6.28}$$

where $a_{\text{lc}}^{\text{QD}}$ and a_{lc}^{B} are the lattice constants of the QD and barrier material, respectively. The initial strain is measured with respect to the barrier material and is modeled such that it only occurs within the dot [37–39]. Thus, elements within the barrier material do not experience an initial strain. In addition, during self-assembled QD growth, the in-plane direction experiences this initial strain only as a dilation, meaning shear components are zero [40]. It is generally assumed that the in-plane direction initial strain is accompanied by a purely dilation out-of-plane initial strain proportional to ε_0 [41, 42]. The initial out-of-plain strain is found using Hooke's law in three dimensions

$$\begin{aligned} 0 &= C_{12}(\varepsilon_0 + \varepsilon_0) + C_{11}\varepsilon_{zz} \\ \varepsilon_{zz} &= -\frac{2C_{12}}{C_{11}}\varepsilon_0 \end{aligned} \tag{6.29}$$

where C_{11} and C_{12} are two of three independent elastic constants that are experimentally measured and presented in the literature, especially for the technologically important compound semiconductors. The resulting strain and deformation constants are then used to calculate the respective strain potential within the QD/barrier material system. The heavy hole strain potential, Eq. (6.16), will be used to calculate the shift in the valence band due to strain and the conduction band strain potential, Eq. (6.14), will be used to calculate the shift in the conduction band due to strain. The Hamiltonian consisting of a spatially varying effective mass and the conduction band offset potential plus the conduction band strain potential is used in the context of the finite element method to calculate the bound states appearing in the conduction band. As mentioned, the valence band offset is assumed to merge with the barrier's valence band, essentially shrinking the barrier's band gap. This assumption will be valid for "small" valence band offsets or $\approx$ 50 meV.

We have decoupled the valence band from the conduction band for the purpose of performing a large number of numerical calculations, thereby reducing the computational cost. It has been suggested that more complicated Hamiltonians, i.e. coupling the valence to the conduction band (8 $\times$8 Hamiltonian) or including additional physics (piezoelectric effect, where solutions are affected by less than 1 meV [23, 43]), give results that are no better than the input band parameters used [44].

Instead of looking for materials to produce the maximum efficiency, we choose slightly less efficient systems that are more robust under parameter perturbation to account for uncertainties. As in our previous work, band parameters are taken from the Vurgaftman et al., comprehensive review for the III–V semiconductors. A particular material system will then be compared to the theoretical models outlined in Sect. 6.2. Only those material systems that match the IBSC efficiency greater than 46 % for unconcentrated light and greater than 62 % for fully concentrated light are identified as potential candidates.

6.5 Results and Discussion

The calculations were made on six QDs, two for each type of geometry, each labeled structure A, B, C, D, E, and F and summarized in Table 6.3. The dimensions of structures C and D are selected based on having a similar in volume to the QD materials found in Table 6.1. The dimensions of structures A and B are chosen to have dimensions 3 / 4 that of structures C and D, while the dimensions of structures E and F are chosen to have dimensions 4 / 3 that of structures C and D. This will give us a good indication as to how potential materials for the QD-IBSC respond to changes in size and will help bound our analysis.

Band gaps in semiconductors are temperature-dependent quantities that are often empirically fitted to the Varshni function [45]

$$E_g(T) = E_g(0) - \frac{\alpha T^2}{T + \beta} \tag{6.30}$$

where α and β are adjustable Varshni parameters found in semiconductor band parameter tables. For the QD-IBSC, it is assumed that the device will be operating at ambient temperature and the band gaps respond accordingly. To get a sense of how the temperature affects the band gap, consider the binary semiconductor GaAs, which has a direct band gap of 1. 519 at 0 K. The Varshni parameters are taken as

Table 6.3 The geometric properties of the QD structures used in this analysis

Geometric property	A	B	C	D	E	F
Shape	Pyramid	Truncated pyramid	Pyramid	Truncated pyramid	Pyramid	Truncated pyramid
b (nm)	7. 5	6	10	8	13. 3	10. 6
h (nm)	4. 5	4. 5	6	6	8	8
a (nm)	–	3. 75	–	5	–	6. 66
Volume (nm^3)	84. 4	108. 8	200	258	471. 7	606. 2
r (Å)	27. 2	29. 6	36. 3	39. 5	48. 3	52. 5

The geometric properties b, h, and a refer to the dimensions in Fig. 6.8, while r refers to a sphere of radius r having the same volume as the respective structure. Additionally, we include the volume of each structure. Structures C and D have a similar volume to those QDs found in Table 6.1, while structures A, B, E, and F are based on the dimensions of structures C and D to help bound the analysis

Table 6.4 Potential QD-IBSC material systems that produce the desired efficiency, η, for structure A under fully concentrated light, $X = 1/F_{sun}$

Barrier/QD	E_1 (eV)	E_2 (eV)	E_3 (eV)	$\Delta_{lc}(\%)$	$\eta(\%)$	X
$Al_{0.68}In_{0.32}As/InAs_{0.82}N_{0.18}$	1. 22	0. 71	–	1. 31	≥ 62	$1/F_{sun}$
$Al_{0.66}In_{0.34}As/InAs_{0.82}N_{0.18}$	1. 21	0. 69	–	1. 17	≥ 62	$1/F_{sun}$
$Al_{0.64}In_{0.36}As/InAs_{0.84}N_{0.16}$	1. 16	0. 66	–	1. 40	≥ 62	$1/F_{sun}$
$Al_{0.64}In_{0.36}As/InAs_{0.82}N_{0.18}$	1. 20	0. 66	–	1. 03	≥ 62	$1/F_{sun}$
$Al_{0.62}In_{0.38}As/InAs_{0.84}N_{0.16}$	1. 15	0. 63	–	1. 26	≥ 62	$1/F_{sun}$
$Al_{0.76}In_{0.24}As/InP_{0.92}N_{0.08}$	1. 26	0. 75	0. 22	0. 73	≥ 70	$1/F_{sun}$
$Ga_{0.40}In_{0.60}P/InP_{0.90}N_{0.10}$	1. 12	0. 61	–	1. 38	≥ 62	$1/F_{sun}$
$GaAs_{0.50}P_{0.50}/InP_{0.90}N_{0.10}$	1. 26	0. 71	–	4. 12	≥ 62	$1/F_{sun}$
$GaAs_{0.48}P_{0.52}/InP_{0.90}N_{0.10}$	1. 26	0. 74	–	4. 20	≥ 62	$1/F_{sun}$

Material systems under unconcentrated light, $X = 1$, did not meet the desired efficiency. The energy transitions E_1, E_2, and E_3 refer to those in Fig. 6.2. The lattice mismatch between the QD and barrier material is designated as Δ_{lc}

Table 6.5 Potential QD-IBSC material systems that produce the desired efficiency, η, for structure B under unconcentrated, $X = 1$

Barrier/QD	E_1 (eV)	E_2 (eV)	E_3 (eV)	$\Delta_{lc}(\%)$	$\eta(\%)$	X
$Al_{0.50}In_{0.50}P/InAs_{0.56}N_{0.44}$	1. 00	1. 55	–	1. 49	≥ 46	1
$AlP_{0.70}Sb_{0.30}/InAs_{0.56}N_{0.44}$	1. 02	1. 57	–	1. 48	≥ 46	1

Material systems under fully concentrated light, $X = 1/F_{sun}$, did not meet the desired efficiency. The energy transitions E_1, E_2, and E_3 refer to those in Fig. 6.2. The lattice mismatch between the QD and barrier material is designated as Δ_{lc}

$\alpha = 0.5405\,\text{meV/K}$ and $\beta = 204$ K, and from Eq. (6.29) the band gap of GaAs is adjusted to 1. 42 at 300 K. For all ternary alloys, the band gap is assumed to fit the simple quadratic form [46]

$$E_g(A_{1-x}B_x) = (1-x)E_g(A) + xE_g(B) - x(1-x)C \tag{6.31}$$

where x is the molar concentration $(0 \leq x \leq 1)$ and C is called the bowing parameter. The bowing parameter accounts for the deviation from linear interpolation between the two binary semiconductors. In general, the bowing parameter for the III–V semiconductors is positive meaning the band gap is smaller than would be if linear interpolation was used. So to derive the band gap of a ternary alloy, we first correct the binary semiconductor for temperature and then calculate its band gap using Eq. (6.30).

Summarized in Tables 6.4–6.9 specific potential QD-IBSC material systems are identified for each QD structure that satisfies the stated design criteria. In each table, we display the barrier/QD material, the independent energy transition levels E_i, the lattice mismatch Δ_{lc}, the efficiency threshold the system meets η, and the concentration factor X. Due to the symmetry of the QD for both geometries, the first and second excited energy levels are doubly degenerate, so all the material systems found support either one or three energy levels. As a practical matter, this helps us

Table 6.6 Potential QD-IBSC material systems that produce the desired efficiency, η, for structure C under unconcentrated, $X = 1$

Barrier/QD	E_1 (eV)	E_2 (eV)	E_3 (eV)	Δ_{lc}(%)	η(%)	X
$Al_{0.50}In_{0.50}P/InAs_{0.56}N_{0.44}$	1.54	1.00	–	1.49	≥ 46	1
$AlAs_{0.80}Sb/InAs_{0.84}N_{0.16}$	1.05	1.63	–	2.25	≥ 46	1
$AlP_{0.70}Sb_{0.30}/InAs_{0.56}N_{0.44}$	1.01	1.57	–	1.48	≥ 46	1

Material systems under fully concentrated light, $X = 1/F_{sun}$, did not meet the desired efficiency. The energy transitions E_1, E_2, and E_3 refer to those in Fig. 6.2. The lattice mismatch between the QD and barrier material is designated as Δ_{lc}

Table 6.7 Potential QD-IBSC material systems that produce the desired efficiency, η, for structure D under unconcentrated, $X = 1$, and/or fully concentrated light, $X = 1/F_{sun}$

Barrier/QD	E_1 (eV)	E_2 (eV)	E_3 (eV)	Δ_{lc}(%)	η(%)	X
$AlP_{0.36}Sb_{0.64}/InAs_{0.82}N_{0.18}$	0.78	1.34	–	0.52	≥ 46	1
$AlP_{0.36}Sb_{0.64}/InAs_{0.82}N_{0.18}$	0.78	1.34	–	0.52	≥ 62	$1/F_{sun}$
$AlP_{0.34}Sb_{0.66}/InAs_{0.82}N_{0.18}$	0.80	1.32	–	0.75	≥ 62	$1/F_{sun}$
$AlP_{0.32}Sb_{0.68}/InAs_{0.84}N_{0.16}$	0.79	1.33	–	0.61	≥ 46	1
$AlP_{0.32}Sb_{0.68}/InAs_{0.84}N_{0.16}$	0.79	1.33	–	0.61	≥ 62	$1/F_{sun}$
$GaP_{0.90}N_{0.10}/InAs_{0.52}N_{0.48}$	0.93	1.49	–	3.46	≥ 46	1
$GaP_{0.88}Sb_{0.12}/InAs_{0.52}N_{0.48}$	0.88	1.46	–	3.83	≥ 46	1

The energy transitions E_1, E_2, and E_3 refer to those in Fig. 6.2. The lattice mismatch between the QD and barrier material is designated as Δ_{lc}

Table 6.8 Potential QD-IBSC material systems that produce the desired efficiency, η, for structure E under unconcentrated, $X = 1$, and/or fully concentrated light, $X = 1/F_{sun}$

Barrier/QD	E_1 (eV)	E_2 (eV)	E_3 (eV)	Δ_{lc}(%)	η(%)	X
$AlAs_{0.78}Sb_{0.22}/InAs_{0.84}N_{0.16}$	0.90	1.38	0.42	2.08	≥ 50	1
$AlAs_{0.78}Sb_{0.22}/InAs_{0.84}N_{0.16}$	0.90	1.38	0.42	2.08	≥ 70	$1/F_{sun}$
$AlP_{0.36}Sb_{0.64}/InAs_{0.82}N_{0.18}$	0.79	1.33	–	0.52	≥ 46	1
$AlP_{0.36}Sb_{0.64}/InAs_{0.82}N_{0.18}$	0.79	1.33	–	0.52	≥ 62	$1/F_{sun}$
$AlP_{0.34}Sb_{0.66}/InAs_{0.82}N_{0.18}$	0.80	1.32	–	0.75	≥ 62	$1/F_{sun}$
$GaP_{0.94}N_{0.06}/InAs_{0.56}N_{0.44}$	0.99	1.54	–	3.52	≥ 46	1
$GaP_{0.92}N_{0.08}/InAs_{0.56}N_{0.44}$	0.94	1.50	–	3.89	≥ 46	1
$GaP_{0.92}N_{0.08}/InAs_{0.54}N_{0.46}$	0.95	1.54	–	3.49	≥ 46	1
$GaP_{0.88}N_{0.12}/InAs_{0.54}N_{0.46}$	0.86	1.44	–	4.23	≥ 46	1

The energy transitions E_1, E_2, and E_3 refer to those in Fig. 6.2. The lattice mismatch between the QD and barrier material is designated as Δ_{lc}

speak qualitatively to what material systems might be potential QD-IBSCs (see Table 6.10 below and accompanying discussion).

In Table 6.4, we notice all the potential QD-IBSC material systems for structure A would be designed to operate for fully concentrated light. Although many of material systems supported either one or three energy levels, energy transitions E_1 or E_2 were too large for barrier materials suitable for the $X = 1$ QD-IBSC design. We also observed, in general, that material systems supporting three energy

Table 6.9 Potential QD-IBSC material systems that produce the desired efficiency, η, for structure F under unconcentrated, $X = 1$, and/or fully concentrated light, $X = 1/F_{sun}$

Barrier/QD	E_1 (eV)	E_2 (eV)	E_3 (eV)	$\Delta_{lc}(\%)$	$\eta(\%)$	X
$Al_{0.68}In_{0.32}As/InAs_{0.82}N_{0.18}$	0. 68	1. 24	-	1. 31	≥ 62	$1/F_{sun}$
$Al_{0.66}In_{0.34}As/InAs_{0.82}N_{0.18}$	0. 68	1. 20	-	1. 17	≥ 62	$1/F_{sun}$
$Al_{0.64}In_{0.36}As/InAs_{0.84}N_{0.16}$	0. 65	1. 16	-	1. 40	≥ 62	$1/F_{sun}$

The energy transitions E_1,E_2, andE_3 referto those in Fig. 6.2. The lattice mismatch between the QD and barrier material is designated as Δ_{lc}

Table 6.10 An exhaustive list of potential QD-IBSC material systems that meet the design criteria for various molar concentrations

Barrier material	QD material	Barrier material	QD material
AlGaAs	InAsN	AlAsSb	InAsN
AlGaAs	InPN	AlAsSb	InAsSb
AlInAs	InPN	AlAsSb	InPSb
AlInAs	InAsN	GaAsSb	InAsN
AlInAs	InPSb	GaAsSb	InPN
GaInP	InPN	AlAsP	InAsN
AlInP	InAsN	GaPSb	InPN
AlGaP	InAsN	AlPSb	InAsN
AlInSb	InASb	AlPSb	InAsSb
AlInSb	InPSb	AlPSb	InPSb
AlGaSb	InAsSb	GaPN	InAsN
AlAsSb	InPN	GaPN	InPN

transitions with band gaps ideal for $X = 1/F_{sun}$ achieved efficiencies between $60 - 65$ %, thereby precluding them from consideration due to our high efficiency criterion. In terms of the total number of material systems supporting one or three energy levels, structure A had the most relative to its two pyramid counterparts.

The potential QD-IBSC materials for structure B are displayed in Table 6.5. We observe the number of potential QD-IBSC materials meeting the efficiency criterion is two and designed for unconcentrated light. This is in contrast to structure A, which is of similar volume. A comparison with structure A indicates that the energy states are more tightly bound and in many cases, in material systems that support a single energy level for the geometry of structure A, the same material systems in structure B support three levels. As might be expected, the total number of material systems supporting one or three energy levels is similar to structure A and has the most relative to its two truncated pyramid counterparts.

The potential QD-IBSC materials for structure C are displayed in Table 6.6. There are three potential QD-IBSC materials that meet all the design criteria for unconcentrated light. We observe that the potential QD-IBSC material systems are similar to that of structure B. This is interesting because the two structures have different geometries, which suggests that the material systems are robust in terms of acting as a QD-IBSC device. The material system AlAsSb/InAsN that appears in

Table 6.6 would appear in Table 6.5 if the efficiency criterion was 45 %, further confirming the similarities between the two structures.

The potential QD-IBSC materials for structure D are displayed in Table 6.7. We notice that there are two potential QD-IBSC materials that could be designed for both a concentration factor of $X = 1$ and $X = 1/F_{\text{sun}}$, indicating the flexibility of the materials. This is the first occurrence of potential QD-IBSCs that could operate at the optimal efficiency levels of both concentration factors.

The potential QD-IBSC materials for structure E are displayed in Table 6.8. Similar to structure D, there are two potential QD-IBSC materials that could be designed for both a concentration factor of $X = 1$ and $X = 1/F_{\text{sun}}$. The AlPSb/InAsN combination is most flexible in the sense that its design is suitable for both concentration factors and both structures D and E. The material combination GaPN/InAsN also appears in structure D indicating geometric flexibility. We should note that the total number of material systems satisfying the design criteria was much less than its two geometric counterpart structures A and C.

The potential QD-IBSC materials for structure F are displayed in Table 6.9. We notice that the material combinations are a subset of the material combinations found in structure A. The difference being the energy level in structure F is more tightly bound such that the energy transitions E_1 and E_2 are now opposite of structure A. This larger structure has taken advantage of the flexibility of the IBSC to accommodate energy transition permutations. Like structure B and C, structure F did not have many material systems satisfying the design criteria. The reason is as the QD increases in size, it will support more energy levels thereby limiting the potential QD-IBSCs. Although not stated explicitly but implied, we limit the QD-IBSC to one or two energy levels[3] because they are less likely to merge with one another or the conduction band. In addition, more energy levels add complexity to the device and increase the likelihood it will not operate under the ideal IBSC conditions. The reasoning is similar to why we look for negligible valence band offsets.

In the tables above, we considered all the III–V ternary alloys and did not dismiss any material system that is considered developmental. The aspiration is to find a material breakthrough that will spur additional research on the QD-IBSC design rather than only search those material systems that could be manufactured today. The potential material systems found should focus the research of the QD-IBSC to those combinations and spur technological advances in QD growth. Nevertheless we briefly identify previous work with InAsN, which appears to be very important in designing a QD-IBSC device, to illustrate that the ternary has been grown as a QD. Bais et al. grew InAsN QDs embedded in GaAs and GaAsN [47], while Schumann et al. produced InAsN QDs with nitrogen concentrations $0 - 4.3$ % [48]. They found that as the amount of nitrogen increases, the QDs increase in size as well.

[3] The QD-IBSC could support more energy levels if one of the energy levels is degenerate, as seen in our analysis.

Although we found potential QD-IBSC materials satisfying the design criteria and in some cases exhibiting flexibility across the QD structures, material systems are QD size dependent. In addition, we put a strict efficiency design criterion that the material systems must meet in order to be considered. As a practical matter, material systems will most likely not be grown to the exact dimensions of the structures, and we might decide to select a material system that has a slightly less design efficiency than our efficiency criterion. In response, we show an exhaustive list of potential barrier/QD combinations in Table 6.10 that satisfied all but the efficiency design considerations and only support one or three energy levels. They were found during our search on structure A, which has the smallest volume of all the structures. If a material system supports more than three independent energy transitions in structure A, it was discarded because as the structure increases in volume, it will support more energy levels and thus not be considered for the QD-IBSC. We identify the QD/barrier combinations summarized in Table 6.10 that should be considered materials for the QD-IBSC.

6.6 Final Remarks

We justified using a less sophisticated Hamiltonian by only including physics that was of real significance due to the number of possible material permutations. However, now that we have identified potential QD-IBSC materials, we could further refine that pool of materials by employing a more sophisticated Hamiltonian, i.e. coupling the valence to the conduction band, including piezoelectric effect and Coulomb interaction, as computational efficiency becomes less important. In addition to a more sophisticated Hamiltonian, it would be appropriate to refine the geometry of a specific material system if there is previous work justifying to do so.

The aspiration of this work is to create a material breakthrough that will spur additional research on the QD-IBSC design attempting to mature the technology. The QD-IBSC is a novel device that will not only need the right material selection but must operate based on the IBSC operating principles. Therefore, additional research on how closely the identified material systems match the IBSC operating principles is needed. However, the basic ideas for designing the device are not much different than from those described in this chapter.

References

1. Shockley, W., Queisser, H.J.: Detailed balance limit of efficiency of p–n junction solar cells. J. Appl. Phys. **169**, 510–519 (1961)
2. Araujo, G.L., Marti, A.: Absolute limiting efficiencies for photovoltaic energy conversion. Sol. Energy Mater. Sol. Cell. **33**, 213–240 (1994)
3. Nelson, J.: The Physics of Solar Cells. Imperial College Press, London (2003)

4. Luque, A., Marti, A.: Increasing the Efficiency of Ideal Solar Cells by Photon Induced Transitions at Intermediate Levels. Phys. Rev. Lett. **78**, 5014–5017 (1997)
5. Green, M.A.: Third Generation Photovoltaics. Springer, New York (2006)
6. Ballentine, L.E.: Quantum Mechanics: A Modern Development. World Scientific, Singapore (1998)
7. Kittel, C.: Introduction to Solid State Physics. Wiley, Hoboken (2005)
8. Matagne, P., Leburton, J.P.: Quantum dots: artificial atoms and molecules. In: Bandyopadhyay S., Nalwa H.S. (eds.) Quantum Dots and Nanowires, p. 3. American Scientific, Stevenson Ranch (2003)
9. Anderson, N.G.: On quantum well solar cell efficiencies. Physica E **14**, 126–131 (2002)
10. Luque, A., Marti, A.: Thermodynamic consistency of sub-bandgap absorbing solar cell proposals. IEEE Trans. Electron Dev. **49**, 2118–2124 (2001)
11. Marti, A., Cuandra, L., Luque, A.: Quantum Dot Intermediate Band Solar Cell. Paper presented at the 28th IEEE photovoltaics specialist conference, Anchorage, Alaska, 15–22 September 2000
12. Luque, A., Marti, A.: A metallic intermediate band high efficiency solar cell. Prog. Photovoltaics Res. Appl. **9**, 73–86 (2001)
13. Gilmore, R.: Elementary Quantum Mechanics in One Dimension. John Hopkins, Baltimore (2004)
14. Tersoff, J., Teichert, C., Lagally, M.G.: Self-organization in growth of quantum dot superlattices. Phys. Rev. Lett. **76**, 1675–1678 (1996)
15. Leonard, D., Krishnamurthy, M., Reaves, C.M., Denbaars, S.P., Petroff, P.M.: Direct formation of quantum-sized dots from uniform coherent islands of InGaAs on GaAs surfaces. Appl. Phys. Lett. **63**, 3203–3205 (1993)
16. Notzel, R., Temmyo, J., Tamamura, T.: Self-organized growth of strained InGaAs quantum disks. Nature **369**, 131–133 (1994)
17. Nishi, K., Anan, T., Gomyo, A., Kohmoto, S., Sugou, S.: Spontaneous lateral alignment of $In_{0.25}Ga_{0.75}As$ self-assembled quantum dots on (311)B GaAs grown by gas source molecular beam epitaxy. Appl. Phys. Lett. **70**, 3579–3581 (1997)
18. Sears, K., Mokkapati, S., Tan, H.H., Jagadish, C.: In(Ga)As/GaAs quantum dots grown by MOCVD for opto-electronic device applications. In: Wang Z.M. (ed.) Self-Assembled Quantum Dots. Lecture Notes in Nanoscale Science and Technology, vol. 1, pp. 359–404. Springer, New York (2008)
19. Levy, M., Honsberg C., Marti, A., Luque, A.: Quantum dot intermediate band solar cell material systems with negligible valence band offsets. Paper presented at the 31st IEEE photovoltaics specialist conference, Orlando, Florida, 3–7 January 2005
20. Jenks, S., Gilmore, R.: Quantum dot solar cell: materials that produce two intermediate bands. J. Renew. Sustain. Energ. **2**, 013111–013120 (2010)
21. Vurgaftman, I., Meyer, J.R., Ram-Mohan, L.R.: Band parameters for III-V compound semiconductors and their alloys. J. Appl. Phys. **89**, 5816–5875 (2001)
22. Miller, M.S., Malm, J.O., Pistol, M.E., Jeppesen, S., Kowalski, B., Georgsson, K., Samuelson, L.: Stacking InAs islands and GaAs layers: Strongly modulated one-dimensional electronic systems. J. Appl. Phys. **80**, 3360–3365 (1996)
23. Grundmann, M., Stier, O., Bimberg, D.: InAs/GaAs pyramidal quantum dots: Strain distribution, optical phonons, and electronic structure. Phys. Rev. B **52**, 11969–11981 (1995)
24. Solomon, G.S., Trezza, J.A., Marshall, A.F., Harris, J.S.: Vertically aligned and electronically coupled growth induced InAs islands in GaAs. Phys. Rev. Lett. **76**, 952–955 (1996)
25. Xie, Q., Madhukar, A., Chen, P., Kobayashi, N.P.: Vertically self-organized InAs quantum box islands on GaAs (100). Phys. Rev. Lett. **75**, 2542–2545 (1995)
26. Moreno, M., Trampert, A., Jenichen, B., Daweritz, L., Ploog, K.H.: Correlation of structure and magnetism in GaAs with embedded Mn(Ga)As magnetic nanoclusters. J. Appl. Phys. **92**, 4672–4678 (2002)

27. Lenz, A., Timm, R., Eisele, H., Hennig, C., Becker, S.K., Sellin, R.L., Pohl, U.W., Bimberg, D., Dahne, M.: Reversed truncated cone composition distribution of $In_{0.8}Ga_{0.2}As$ quantum dots overgrown by an $In_{0.1}Ga_{0.9}As$ layer in a GAs matrix. J. Appl. Phys. **81**, 5150–5152 (2002)
28. Bardeen, J., Shockley, W.: Deformation potentials and mobilities in non-polar crystals. Phys. Rev. **80**, 72–80 (1950)
29. Herring, C., Vogt, E.: Transport and deformation-potential theory for many-valley semiconductors with anisotropic scattering. Phys. Rev. **101**, 944–961 (1956)
30. Bir, G.L., Pikus, G.E.: Symmetry and Strain Induced Effects in Semiconductors. Wiley, New York (1974)
31. Luttinger, J.M., Kohn, W.: Motion of electrons and holes in perturbed periodic fields. Phys. Rev. **97**, 869–883 (1954)
32. Pikus, G.E., Bir, G.L.: Effect of deformation on the hole energy spectrum of germanium and silicon. Sov. Phys. Solid State **1**, 1502–1517 (1960)
33. Chuang, S.L.: Physics of Optoelectronic Devices. Wiley, New York (1995)
34. Bahder, T.B.: Analytic dispersion relations near the Gamma point in strained zinc-blende crystals. Phys. Rev. B **45**, 1629–1637 (1992)
35. Schödinger, E.: Quantization as an eigenvalue problem (part I). Ann. Phys. **79**, 1–12 (1926)
36. Freund, L.: The mechanics of electronic materials. Int. J. Solid. Struct. **37**, 185–196 (2000)
37. Chang, S.: The Physics of Photonic Devices. Wiley, Hoboken (2009)
38. Eshelby, J.D.: Distortion of a crystal by point imperfections. J. Appl. Phys. **25**, 255–262 (1954)
39. Eshelby, J.D.: The continuum theory of lattice defects. Solid State Phys. **3**, 79–144 (1956)
40. Kuo, M., Lin, T., Hong, K., Liao, B., Lee, H., Yu, C.: Two step strain analysis of self assembled InAs/GaAs quantum dots. Semicond. Sci. Tech. **21**, 626–632 (2006)
41. Boxberg, F., Tulkki, J.: Theory of the electronic structure and carrier dynamics of strain-induced (Ga,In)As quantum dots. Rep. Progr. Phys. **70**, 1425–1471 (2007)
42. Andreev, A., Downes, J., Faux, D., O'Reilly, E.: Strain distributions in quantum dots of arbitrary shape. J. Appl. Phys. **86**, 297–305 (1999)
43. Holm, M., Pistol, M., Pryor, C.: Calculations of the electronic structure of strained InAs quantum dots in InP. J. Appl. Phys. **92**, 932–936 (2002)
44. Pryor, C.: Eight-band calculations of strained InAs/GaAs quantum dots compared with one-, four-, and six-band approximations. Phys. Rev. B **57**, 7190–7195 (1998)
45. Varshni, Y.P.: Temperature dependence of the energy gap in semiconductors. Physica **34**, 149–154 (1967)
46. Bosio, C., Staehli, J.L., Guzzi, M., Burri, G., Logan, R.A.: Direct-energy-gap dependence on Al concentration in AlGaAs. Phys. Rev. B **38**, 3263–3268 (1988)
47. Bais, G., Cristofoli, A., Jabeen, F., Piccin M., Carlino, E., Rubini, S., Martelli, F., Franciosi, A.: InAsN/GaAsN quantum-dot and InGaNAs/GaAs quantum-well emitters: a comparison. Appl. Phys. Lett. **86**, 233107–233110 (2005)
48. Schumann, O., Geelhaar, L., Riechert, H., Cerva, H., Abstreiter, G.: Morphology and optical properties of InAsN quantum dots. J. Appl. Phys. **96**, 2832–2840 (2004)

Chapter 7
AlGaInAs Quantum Dots for Intermediate Band Formation in Solar Cell Devices

Stefan Kremling, Christian Schneider, Sven Höfling, Martin Kamp, and Alfred Forchel

Abstract In this chapter, we focus on the integration of composition tailored AlGaInAs quantum dots (QDs) into AlGaAs p–i–n diode structures for solar cell (SC) applications. This type of QDs can absorb a wide range of the solar spectrum from the red visible to the near infrared spectral range. Moreover, the size, density, and eigenenergies of these quaternary QDs can be easily adjusted via varying their material composition, e.g. the aluminum content. We anticipate that the suitable choice of material composition, dot shape, and inter-dot barrier thickness can compensate the built-in potential of the p-i-n junction and electronic coupling between QD layers will be established to generate an electronically isolated intermediate band (IB). In order to probe the operation principle of an IBSC utilizing AlGaInAs QDs, two photon absorption measurements were performed. We show that two sub-bandgap photons produce photocurrent while the first photon pumps electrons from the valence band (VB) to the IB and a second photon lifts the electrons from the IB to the conduction band (CB). Based on our result we propose a route towards intermediate band solar cells (IBSCs) with these quaternary QDs.

7.1 Introduction

Solar radiation is an energy source for mankind with almost unlimited resources. Conversion of solar energy has the potential to provide a significant part of the world's energy production. Since the beginning of research in solar energy conversion using semiconductor solar cells (SCs), increasing the efficiency has been one of the most important challenges in this field to reduce the costs for the generation of electrical power (price/watt) [1, 2].

S. Kremling • C. Schneider • S. Höfling • M. Kamp • A. Forchel (✉)
Technische Physik, Physikalisches Institut and Wilhelm Conrad Röntgen-Research Center for Complex Material Systems, Universität Würzburg, Am Hubland, 97074 Würzburg, Germany
e-mail: forchel@physik.uni-wuerzburg.de

J. Wu and Z.M. Wang (eds.), *Quantum Dot Solar Cells*, Lecture Notes in Nanoscale Science and Technology 15, DOI 10.1007/978-1-4614-8148-5_7,

In general, the conversion efficiency of SCs is related to the number of generated electron–hole pairs caused by incident photons and their probability of recombination before the charge carriers were separated into external contacts [3]. A large variety of power loss mechanisms in SC devices are limiting the transformation of solar energy into useful electrical energy. The most important losses are:

- The inability to absorb photons with energy less than the bandgap; the so-called red losses.
- Thermalization effects of carriers generated by photons far exceeding the band gap energy; the so-called blue losses.

In 1960, Shockley and Queisser presented detailed balance limit calculations which derive the theoretical maximum conversation efficiency in SCs. For a single gap device considering earth and sun temperatures of 300 and 6,000 K, respectively, the maximum efficiency achieve either 31 or 41 % for isotropic illumination, depending on the concentration factor [4].

During the last few years, researchers have been proposing and developing novel concepts for SCs with the aim to overcome the Shockley–Queisser limit. Such devices are called to belong to the third-generation photovoltaics [5, 6]. One of the most common approaches is the use of multijunction SC, consisting of several single junction SCs with different bandgap energies connected via tunnel junctions [7]. Detailed balance limit calculations of devices with infinite number of sub-cells yield maximum efficiencies of 68 % for unconcentrated illumination and 87 % for maximum concentration, respectively [8]. Currently, another widely discussed topic for increasing efficiency in SC is the improvement of light management for incident photons. Recent advances in the control of light at micrometer and nanometer length scale promise solar cells with efficiencies up to 70 % [9].

However, another concept which has been attracted a lot of attention in recent years was introduced by Luque and Martí in 1997, the so-called intermediate band solar cell (IBSC) [10]. In addition to the absorption of photons with energies above the host materials bandgap, this approach allows the absorption of below-bandgap photons via an intermediate band (IB) located between valence band (VB) and conduction band (CB). Hence, IBSC generates photocurrent through the absorption of two sub-bandgap photons while keeping the output voltage on the high level of the matrix material. It is essential that the IB is electronically isolated from CB and VB to achieve the high output voltage. This assumes that the IB has its own quasi-Fermi level (QFL). A schematic band diagram of an IBSC is depicted in Fig. 7.1, showing the respective bandgap of the matrix E_G, of the two intermediate band transitions E_H, E_L, as well as the VB, CB, and IB QFLs. As indicated in Fig. 7.1, photons with energies less than the bandgap energy E_G can be absorbed via a two-step process, whereby electrons are successively pumped from VB to IB (1) and further from IB to CB (2). Conclusively two photons are necessary to produce one high energy electron to preserve the output voltages on high level.

Detailed balance limit calculations yield IBSC efficiency values up to 63 % under concentrated illumination within a single intermediate band [10]. This clearly exceeds the Shockley–Queisser limit of 41 % for single gap devices operating

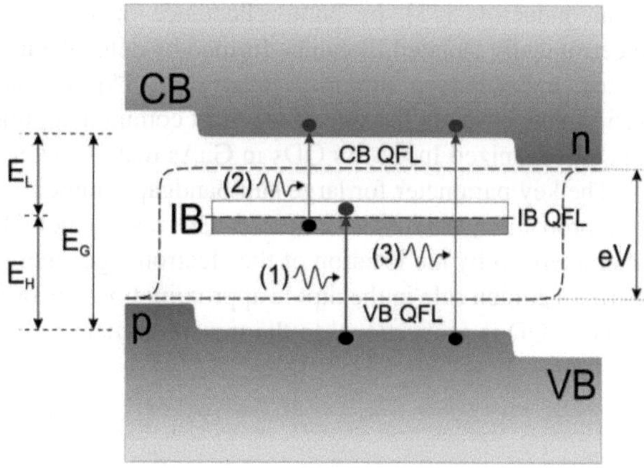

Fig. 7.1 Schematic band diagram of an IBSC with the band gap E_G and the sub band gaps E_H and E_L, respectively. Via a two photon absorption process VB–IB (1) and IB–CB (2) photons with energies less than the bandgap energy can produce electrons with high output voltage

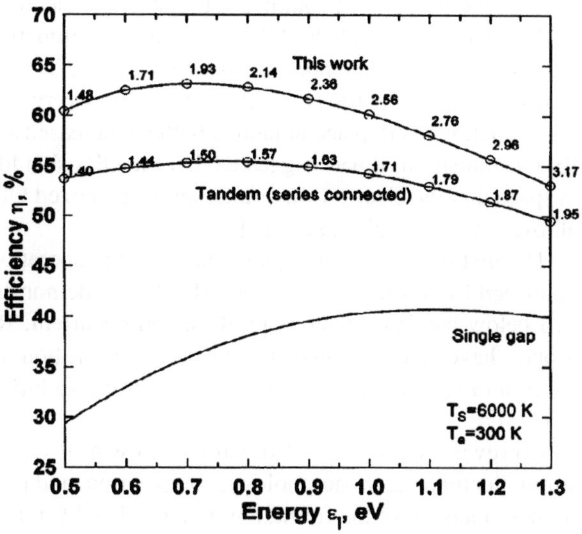

Fig. 7.2 Efficiency limit for an IBSC as a function of the low bandgap energy E_L. The efficiency for a single gap solar cell is also presented (reproduced from Luque et al. [10])

under the same condition [4]. Furthermore, detailed balance limit calculations of the efficiency for solar cell devices within multiple intermediate bands approach maximum efficiencies of nearly 80 % [11]. For achieving maximum efficiency, calculations assume an ideal bandgap of the matrix material for an IBSC of $E_G = 1.93$ eV, which is split in the two sub-bandgaps of $E_H = 1.22$ eV between VB–IB and $E_L = 0.71$ eV for the IB–CB [10]. The correlation between (sub-) bandgap energies and the efficiency for the according IBSC and a single gap solar cell is presented in Fig. 7.2.

Different approaches have been proposed and realized for practical implementation of an IB within a single gap device. On the one hand, some synthesized bulk materials exhibit multiple band formation within the host bandgap. This was first demonstrated for GaNAsP quaternary alloys [12] or diluted II–VI oxide semiconductors [13], by photoreflectance measurements. On the other side, an electronically isolated IB can be formed by delocalization of the energy of confined states, e.g. of deep impurity levels [14, 15] or stacked quantum dots (QDs) [16]. In the case of QDs one of the most common approaches is the implementation of self-organized In(Ga)As QDs in GaAs matrix material.

The key parameter for large sub-bandgap photocurrent is a high electron transition probability from VB–IB and IB–CB, respectively. In the case of QDs, the IB is characterized by the location of the electron eigenenergies in the dot. According to Fermi's golden rule in the dipole approximation, the probability of absorption of an isolated QD is proportional to the transition matrix element given by [17]

$$M_{\mathrm{i,j}} = \langle \psi_{\mathrm{i}} | -e\vec{r} \cdot \vec{E} | \psi_{\mathrm{j}} \rangle,$$

where ψ_{i} and ψ_{j} are the involved wavefunctions, $\vec{r}$ is the position, and $\vec{E}$ the related electric field. In confined QD states, the electron wavefunction is localized to the space region of the QD while the wavefunction in the CB continuum is delocalized, i.e. is extended all over the space. This implies that the absorption probability from the transition IB–CB becomes very weak. In order to enhance the wavefunction overlap accompanied with an increased absorption probability for the IB–CB transition, it was suggested that the QDs should be periodically distributed in space that the electron wavefunctions of localized QD states became delocalized all over the other QD states [17].

The first experimental demonstration of the operation principle of an IBSC was published by Martí et al. in 2006 [18]. It was demonstrated that the absorption of two below-bandgap photons produces photocurrent. Recently, also several other groups have demonstrated an enhanced absorption in the near infrared (NIR) spectral range with an increase of photocurrent in In(Ga)As/GaAs QD SC devices [19–22].

However, due to the self-organized growth of QDs, these nanostructures have strong fluctuations in morphology, composition, and position within the device. All of these facts make the formation of delocalized bands with energetically degenerate QD states very challenging. Recently, Reid et al. showed for realistic QDs that the formation of a subband with delocalized states strongly depends on the QD uniformity and spacer thickness. For example, for QDs separated by 10 nm spacer layers delocalization occurs for energy distributions $\sigma/\sigma_0 > 10^{-4}$, where σ (σ_0) is the FWHM of the single (ensemble) QD [23]. Even if the uniformity of the integrated QDs is perfect, the built-in potential in the intrinsic region of p–i–n diode SC unavoidably lifts the approximate degeneracy between the QD electron eigenenergies. Hence, the wave functions in adjacent layers would be electronically decoupled, resulting in strong carrier localization.

In order to facilitate subband formation even in the presence of built-in potential, our work is focused on the integration of composition tailored quaternary AlGaInAs QDs with various aluminum (Al) content [24]. These QDs allow one to engineer the size and eigenenergies by varying the Al content of adjacent QD layers.

Consequently, a wide spectralConse range from the red visible to the NIR can be covered by QD luminescence. Conversely, this promises a broad absorption range of the solar spectrum entirely due to the QDs. In addition, AlGaInAs QDs achieve high surface densities up to 10^{11} cm^{-2} which is crucial for enhanced photon absorption. We anticipate that the suitable choice of QD composition, shape, and barrier thickness can be employed for electronic coupling to generate a subband in the built-in potential region of the device.

7.2 Growth, Morphology, and Optical Properties of AlGaInAs QDs

AlGaInAs is a material system which allows one to engineer the emission wavelength in a wide spectral range by varying the composition via the In and/or Al content. Consequently, QDs formed by AlGaInAs have been utilized as gain material for high performance QD laser applications [25, 26].

A comprehensive study of the influence of Al and In content on the growth, morphology, and optical properties of quaternary AlGaInAs QDs was given by Schlereth et al. [24]. Various samples with varying In and Al content of $Al_xGa_{1-x-y}In_yAs$ QDs were grown using solid source molecular beam epitaxy (MBE) on GaAs. Morphological characteristics were investigated by scanning electron microscopy (SEM). In Fig. 7.2 a number of SEM images of uncapped $Al_xGa_{0.4-x}In_{0.6}As$ QD samples with varying Al content x between 0 (a) and 0.19 (e) are shown. Generally, two principal characteristics change with varying Al content: first, the QD surface density gradually increases with increasing Al content from 3.9×10^{10} cm^{-2} for 0 % Al up to 1.1×10^{11} cm^{-2} for 19 % Al content. The increase of QD density for higher Al concentrations within the QDs is attributed to the stronger bonds of Al to the substrate leading to a lower surface migration ability [27]. Second, for the same range of Al variation the geometrical size of the QDs decreases from 23 nm in diameter down to 16 nm. In contrast, moderate variations of the In content around 60 % for constant Al contents showed no significant change in density and size distribution. This indicates that density and size of AlGaInAs QDs are mainly determined by the Al concentration.

Low temperature PL measurements showed a strong dependence of the wavelength with varying Al content. In general, with increasing Al content the wavelength decreases for an arbitrary In content. Figure 7.3 shows an interpolated contour plot of the PL wavelength of $Al_xGa_{1-x-y}In_yAs$ QD samples with different Al and In contents. For example, the QD luminescence with an In content of 43 % covers a broad spectral range between 660 nm for 34 % Al content and 880 nm for 0 % Al content. Additionally, some data points in Fig. 7.3 are labeled with their corresponding QD densities. This demonstrates that not only the wavelength can be tailored for a chosen In concentration but also the QD density. The lattice mismatch between AlAs and GaAs of ~0.14 % is about 50 times smaller than that of InAs and

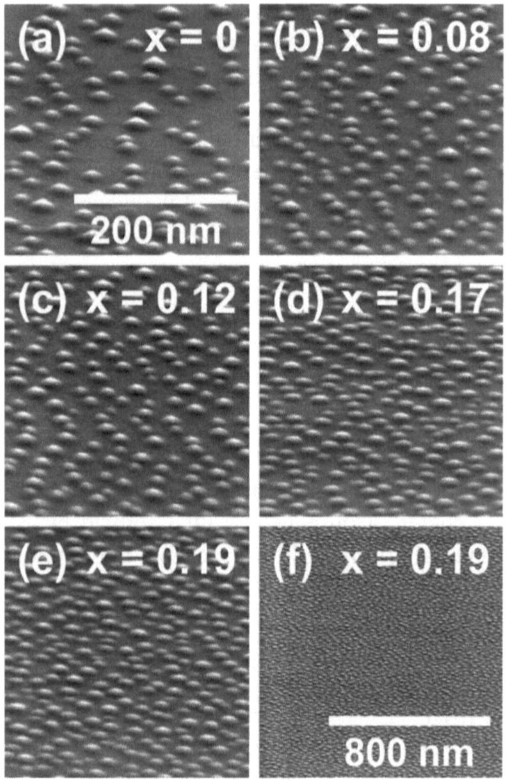

Fig. 7.3 Scanning electron microscopy images of uncapped $Al_xGa_{0.4-x}In_{0.6}As$ QD samples with nominal AL content x of (**a**) 0, (**b**) 0.08, (**c**) 0.12, (**d**) 0.17, (**e**) 0.19, (**f**) shows the same sample as (**e**) with lower magnification to point out the homogeneous distribution of the QD size over the whole sample (reproduced from Schlereth et al. [24])

GaAs with ~7.16 %. The contribution of Al to the strain can be neglected and Al can be used as a degree of freedom in QD engineering. Therefore QDs can be tailored in a wide range of surface densities and emission wavelength by varying the Al content. For solar cell applications it is important to note that the incorporation of Al to the QDs does not affect the strain significantly. It is well known that the accumulation of strain in QDSC results in a degradation of the open circuit voltage accompanied by a reduction of efficiency. For the compensation of strain-related open circuit voltage degradation in InAs/GaAs QDSC, strain balancing layers were implemented between the QD layers to reduce the strain and improve the performance of QDSC [19, 28].

7.3 AlGaInAs QDs in Solar Cells

As mentioned above, the most common approach for the realization of IBSC using QDs for the IB formation is by employing InAs QDs in GaAs matrix material. In contrast, here we focus on the integration of composition tailored AlGaInAs QDs in

Table 7.1 Overview of the investigated QD solar cell samples

Sample	Intrinsic region	Doping	QD arrangement	Spacer (nm)
REF	AlGaAs bulk	–	–	–
SC-1	5 QD layers	–	Normal	20
SC-2	5 QD layers	δ-Doping	Normal	20
SC-3	25 QD layers	δ-Doping	Normal	20
SC-4	5 QD layers	δ-Doping	Inverted	20
SC-5	5 QD layers	δ-Doping	Normal	10

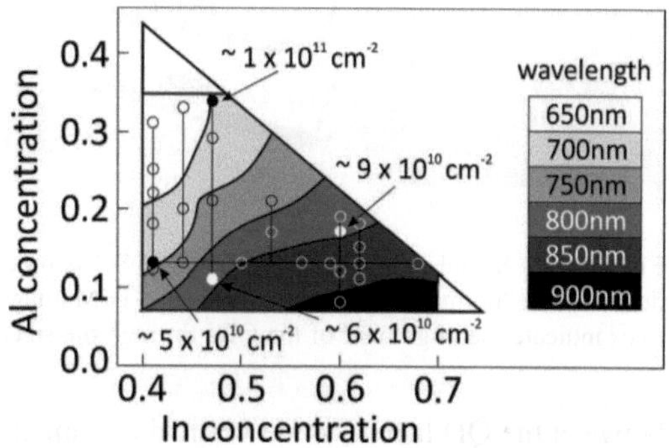

Fig. 7.4 Contour plot of PL wavelength as a function of In and Al concentration. Some data points (*filled circles*) are labeled with their corresponding QD surface densities (reproduced from Schlereth et al. [24])

AlGaAs matrix material with varying Al content to exploit the beneficial properties discussed above.

We have fabricated and investigated six different SC samples with varying intrinsic regions and active media. For all samples, the layer sequence starts with a highly n-doped GaAs buffer layer followed by a silicon (n-type) dopeda silicon (n-type) doped $Al_{0.33}Ga_{0.67}As$ layer with a thickness of 300 nm grown on a silicon doped (100) GaAs substrate. Afterwards the intrinsic region of the device consisting of undoped $Al_{0.33}Ga_{0.67}As$ with a total thickness of 300 nm was deposited. The QD layers were grown within the center of the intrinsic region separated by $Al_{0.33}Ga_{0.67}As$ spacer layers. The growth sequence ends with a 250 nm carbon-doped (p-type) $Al_{0.33}Ga_{0.67}As$ layer, a 50 nm p-type $Al_{0.70}Ga_{0.30}As$ layer as etch stop and a highly carbon-doped GaAs contact layer. An overview of all samples is given in Table 7.1.

Sample REF serves as an AlGaAs bulk reference cell without QDs to investigate the influence of the AlGaAs matrix material. In sample SC-1 the active region is designed as follows: four layers of AlGaInAs QDs with 43 % In content were integrated in the middle of the intrinsic region, yet the aluminum amount was varied from 30 % over 20 % and 10 to 0 % from the bottom to the top QD layer. On top, a layer of InAs QDs was deposited, sandwiched between two 5 nm GaAs layers. In this sample the QD layers are separated by 20 nm thick spacer layers. A cross-sectional scanning transmission electron microscopy (STEM) image of the QD region shown in Fig. 7.4b indicates that even despite the 20 nm barrier thickness

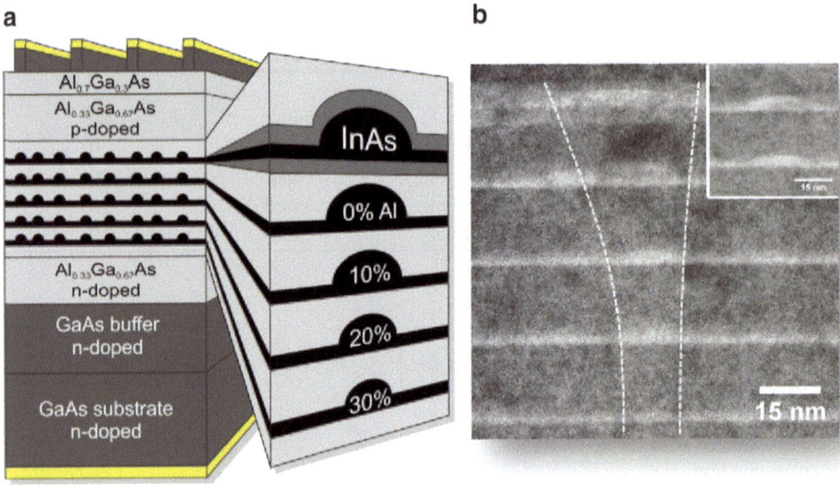

Fig. 7.5 (**a**) Sketch of the layer sequence of SC samples and magnification of the QD region of devices with "normal" QD arrangement. (**b**) STEM image of the QD region of SC-1. *White dashed lines* indicate the alignment of the QDs through the stack

between the QD layers, some vertical alignment is still present due to strain fields. Sample SC-2 is identical to sample SC-1, but silicon sheet layers as δ-doping were deposited in between the QD layers with a density similar to the QD density of 5×10^{10} 1/cm^2. These δ-doping layers are important to partially fill the intermediate band with additionally free charge carriers [16]. In sample SC-3 we have integrated five layers of each QD species in order to study the prospects of efficiency enhancement with increasing number of QD layers. In contrast, sample SC-4 has an inverted QD arrangement compared to sample SC-1. This sample gives us the opportunity to experimentally investigate the effect of built-in potential compensation with tailored QDs on the device performance. Sample SC-5 is a replica of sample SC-2 but with reduced spacer layer thickness of only 10 nm within the QD layers (see schematics in Fig. 7.5).

A key parameter for an enhancement of QD absorption in SC devices certainly is the surface density of the QDs. High densities are beneficial with respect to a high QD absorption by an increased electronic density of states. First we have fabricated reference samples to check the QD growth process and determine the QD surface densities. For example, the QD density of an uncapped reference sample with a single $Al_{0.20}Ga_{0.37}In_{0.43}As$ QD layer was determined to be 5×10^{10} cm^{-2} favourably comparing to 1×10^{10} cm^{-2} for standard InAs/GaAs QDs due to the low migration length of aluminum. Atomic force microscopy (AFM) images of these two control samples are shown in Fig. 7.6. It's worth noting, that AlGaInAs QDs have also been exploited to serve as high-density nucleation centers to increase the sheet density of strain coupled In(Ga)As QDs [29]. Furthermore, the QDs bandgap and therefore the absorption band can be adjusted via the aluminum content of the QDs, while maintaining a high QD density.

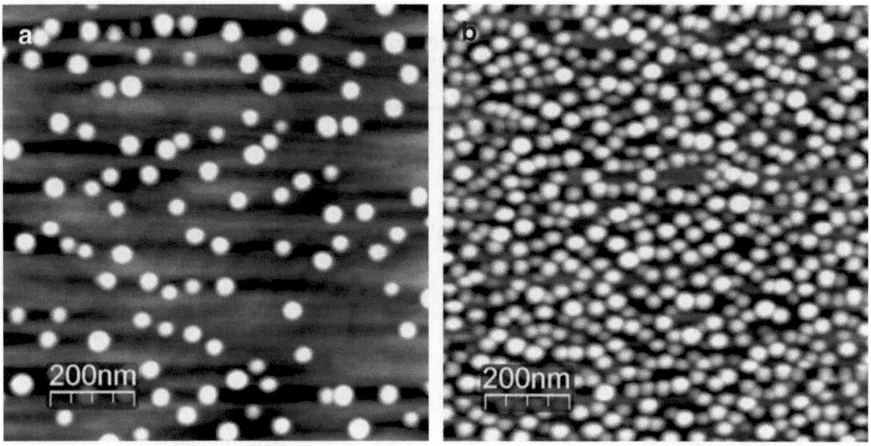

Fig. 7.6 AFM images of uncapped reference samples with (**a**) InAs QDs and (**b**) $Al_{0.20}Ga_{0.37}In_{0.43}As$ QDs

7.4 Experimental Results

7.4.1 Luminescence Measurements

Basic optical properties of the QDs in the devices were investigated by means of photoluminescence (PL) at sample temperatures of 30 K to prevent thermal carrier escape from the shallow nanostructures. A 532 nm laser was used for photo-excitation. The PL signal was dispersed by a grating monochromator and detected by an Si-CCD camera.

An ensemble PL spectrum of device SC-1 is shown in Fig. 7.7. One can identify various luminescence peaks attributed to the spectrally detuned QDs and different barrier materials, as indicated in the spectrum. The first peak labeled as QD1 at a wavelength of 700 nm can be assigned to Al_xGaInAs QDs with 30 % Al content. Systematic PL investigations of planar samples reveal similar PL energies for this material composition, compared to the contour PL plot in Fig. 7.4 above. We furthermore assign peak QD2 at a wavelength of 750 nm to AlxGaInAs QDs with an Al content of 20 %, peak QD3 at 800 nm to 10 % Al and peak QD4 at 890 nm to the GaInAs QDs. The long wavelength PL peak at 1,000 nm originates from InAs QDs sandwiched between two GaAs layers and is labeled as peak QD5. Additionally, the luminescence of the AlGaAs barrier material at a wavelength of 620 nm as well as the GaAs barrier material at 830 nm were detected and labeled in Fig. 7.6. This broad spectral range of QD luminescence between 650 and 1,050 nm promises a broad QD absorption range of the solar cell device in the same spectral range. This is much larger than in devices containing only one kind of QDs, e.g. InAs/GaAs.

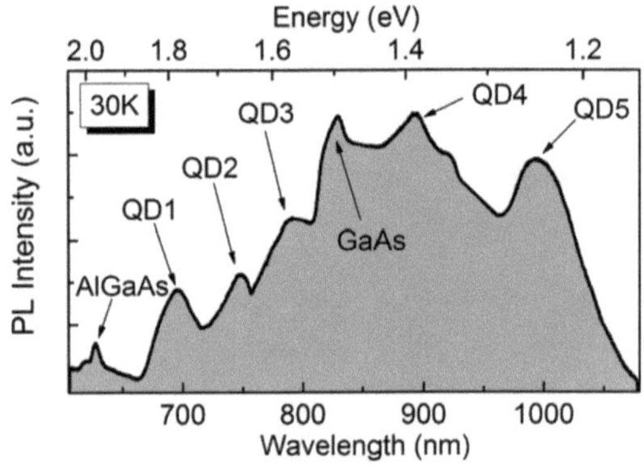

Fig. 7.7 Ensemble PL spectrum of device SC-1 shows a broad luminescence range between 650 and 1,050 nm. In addition to the two matrix materials GaAs and AlGaAs, five QD luminescence peaks are clearly identified

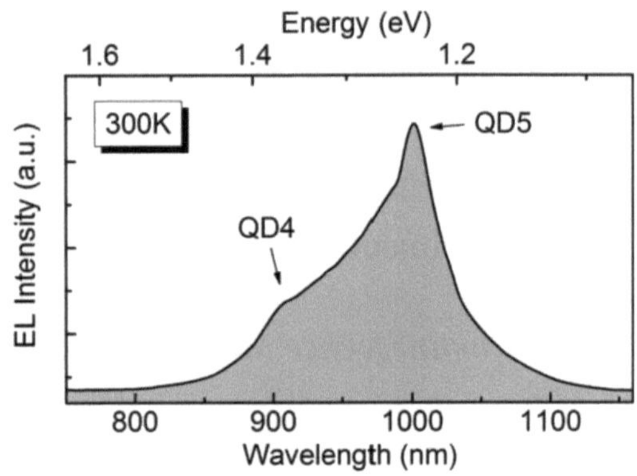

Fig. 7.8 EL spectrum of device SC-1, recorded at room temperature. Only QD4 and QD5 exhibit EL at 300 K

In addition to PL we have also measured the electroluminescence (EL) of the device at room temperature. The corresponding EL spectrum is shown in Fig. 7.8. EL was only detected between 850 and 1,050 nm, corresponding to the luminescence of QD4 and QD5. No EL was observed below 850 nm, indicating that the thermal carrier escape is the dominating process in the small QDs.

7.4.2 External Quantum Efficiency

A key parameter for solar cell characterization is the external quantum efficiency (EQE), which is crucial for understanding current generation and recombination mechanisms in SC devices. In general, the EQE is defined as the number of incident photons that are converted to electrons for the whole spectral range when the device operates under short circuit condition. In our case, the EQE was determined using an Xe-lamp dispersed by a grating monochromator and measuring the short circuit current $I_{SC}(\lambda)$ directly with a picoamperemeter. Then the EQE can be calculated via the equation:

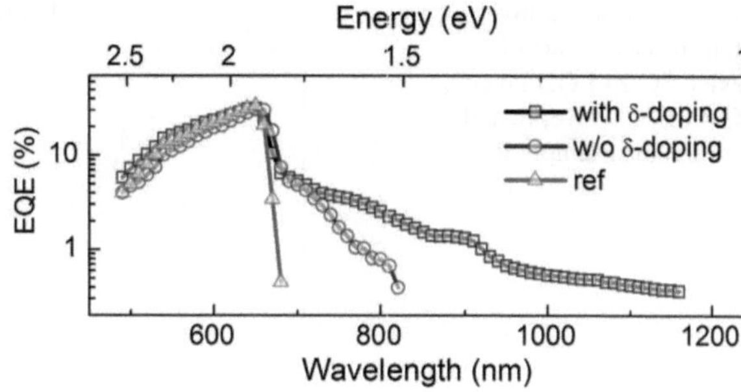

Fig. 7.9 Comparative EQE measurements of device SC-1 (w/o δ-doping), SC-2 (w δ-doping) and the reference sample REF (w/o QDs)

$$\mathrm{EQE}(\lambda) = \frac{I_{\mathrm{SC}}(\lambda)}{P(\lambda) \cdot \lambda} \cdot \frac{hc}{e},$$

where h is the Planck constant, c the speed of light in vacuum, and e the elementary charge. The illumination power $P(\lambda)$ was measured with a powermeter and two different detector heads for the visible and infrared spectral range.

In the following part we present and discuss comparative EQE measurements of the varying QDSC devices. Figure 7.9 shows room temperature EQE measurements for three different samples SC-1, SC-2, and REF. As expected, the EQE of sample REF without QDs is peaked around the wavelength of 650 nm, which corresponds to the bandgap energy of the $Al_{0.33}Ga_{0.67}As$ matrix material. For incident photons with energies below the bandgap the EQE abruptly drops below the detection resolution. At higher energies, i.e. far above the material bandgap, the EQE decreases again due to hot carrier losses in the device.

In device SC-1 the EQE is extended by the absorption of the QDs into the NIR range up to 820 nm. Remarkably, no photocurrent could be detected above this wavelength for this device. This is indicating a very inefficient carrier transfer from the electron states of the long wavelength emitting QDs into the conduction band. In contrast, the EQE of sample SC-2 with additional δ-doping between the QD layers features a significantly enhanced EQE in the spectral range above 800 nm. Furthermore, the absorption range is extended up to a wavelength of 1,150 nm. We anticipate that this enhancement was facilitated by introducing additional electrons via silicon δ-doping layers in between the QD layers. Those additional charges in the QDs can enhance the photo-absorption especially from the deeper QD states into the matrix conduction band [5], making prior population of these states by photo-absorption from the valence band not inherently necessary.

Next, we investigated the influence of the number of QDs on the EQE of the solar cell devices. The results of comparative EQE measurements for the corresponding devices are depicted in Fig. 7.10. Here, one has to distinguish between two regions: for photons with energies below the AlGaAs barrier an even further enhancement of the EQE was observed in sample SC-3 with a fivefold stack of each QD layer, overall 25 QD layers. This enhancement amounts to about a factor of two throughout most of the QD absorption band compared to sample SC-2.

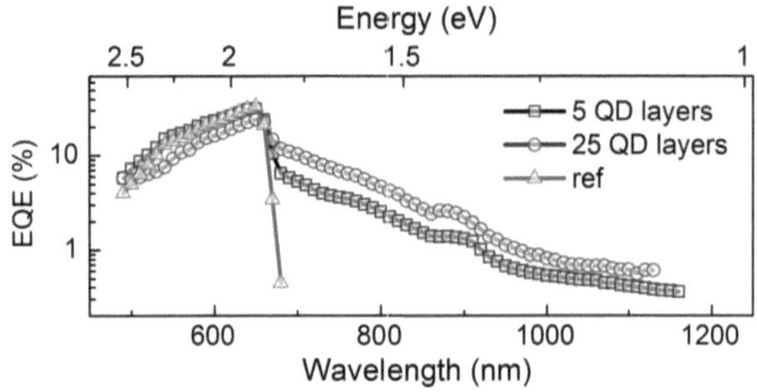

Fig. 7.10 Comparative EQE measurements of device SC-2 (5 QD layers), SC-3 (25 QD layers), and the reference sample REF (w/o QDs)

On the other side, for photons with energies above the bandgap of the p-AlGaAs emitter, the EQE is reduced due to less photocurrent generated in the p-emitter region. This can likely be explained by a decrease of the carrier lifetime caused by strain-induced dislocations from the QD region into the emitter region [30]. Consequently, the losses in the p-emitter region exceed the gain in the QD region and the overall efficiency indeed decreases. Previous works on QDSC also reported a correlation of EQE with the number of QD layers in the device. For example, Sugaya et al. reported a degradation of efficiency from 12.2 % for 10 layers of InGaAs QD down to 7.7 % for 50 QD layers [21]. This means that advanced growth techniques are necessary to reduce dislocations and take advantage of enhanced QD absorption due to more QD layers.

The design of our solar cell devices is based on different considerations. As mentioned above, one is the compensation of the gradient in the conduction band based on the built-in potential in p–i–n devices to facilitate extended wavefunctions between adjacent QD layers. In order to check the influence of the built-in compensation we have fabricated SC-4 with an inverted QD arrangement. In this device, InAs/GaAs QDs with the largest bandgap were grown nearest to the p-contact and the QDs with the smallest bandgap were grown closest to the n-contact. Band structure calculations in the next section point out that the energy of the electronic levels in the QDs are strongly separated through the QD stack.

Figure 7.11 depicts a comparison of the EQE from device SC-4 with a reversed arrangement of QD layers and SC-2 with the "normal" arrangement. The device with the inverted QD layer arrangement shows a drastically reduced EQE in the NIR range compared to all other devices, and almost a complete absence of EQE above 800 nm. This unambiguously demonstrates the importance of choosing a suitable arrangement of the QD layers in the device. Conduction band calculations in the next part will emphasize this in detail.

Finally, we probe the influence of the spacer layer thickness between the QD layers. For example, Shoji et al. have shown by time-resolved PL techniques that for InGaAs QDs grown on GaAs (311) the coupling between adjacent QD layers strongly depends on the spacer thickness [31]. In Fig. 7.12 we have compared the EQE for sample SC-2 with 20 nm spacer between the QD layers with sample SC-5 with only 10 nm spacer layers. Surprisingly, the EQE of device SC-5 with 10 nm

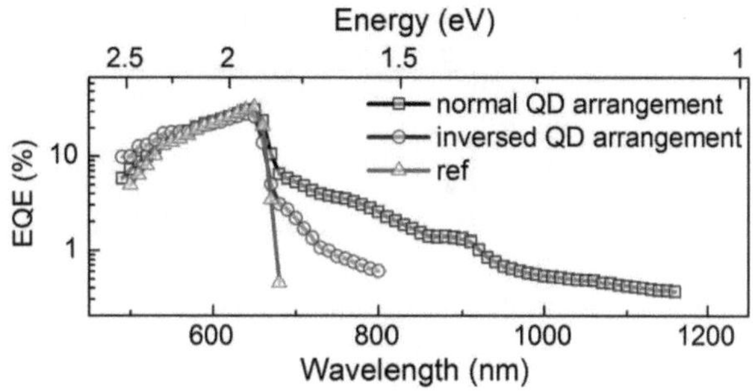

Fig. 7.11 Comparative EQE measurements of device SC-2 (normal QD arrangement), SC-4 (inverted QD arrangement) and the reference sample REF (w/o QDs)

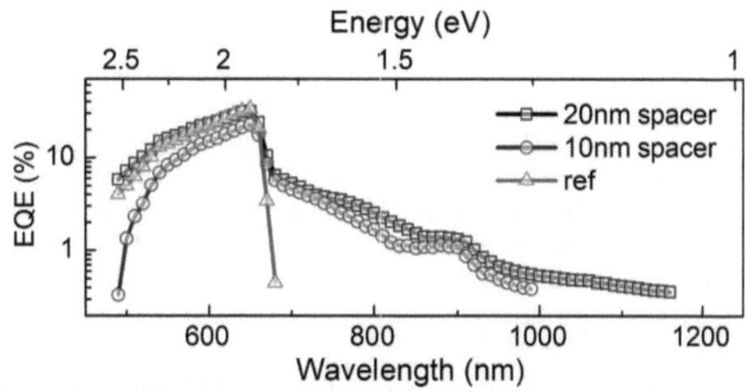

Fig. 7.12 Comparative EQE measurements of device SC-2 (20 nm spacer layers), SC-5 (10 nm spacer layers), and the reference sample REF (w/o QDs)

spacer layers decreases over the entire spectral range. In fact, one would expect an enhanced electronic coupling between the QDs for smaller spacer layers and therefore an enhanced absorption due to extended wavefunctions and increased transition probabilities. But on the other side, smaller spacer layers results in an increase of strain within the device. Once again, this results in more dislocations in the emitter region. This results in a reduced EQE above the host materials bandgap due to a reduction of the photocurrent generated by the emitter region [28, 30].

7.4.3 *Two Photon Absorption*

The measurements presented above demonstrate the absorption of photons with energies lower than the host materials bandgap. During recent years, also several groups have reported an enhanced absorption and photocurrent generation in the NIR spectral region when implementing QDs within SC devices [19–22]. Indeed, sub-bandgap photons were absorbed in the QD region but the excited electrons often thermally escape to the CB continuum because all of these measurements were performed at room temperature. However, all these measurements did not finally confirm the IB operating principle yet, where thermodynamic laws predict the absorption of two sub-bandgap photons for the generation of high energy electrons, which should enable notably improved overall efficiencies.

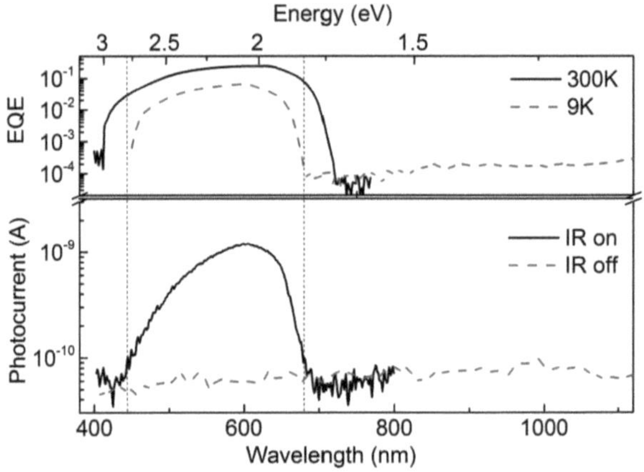

Fig. 7.13 External quantum efficiency (EQE) of device REF for 300 K (*solid curve*) and for 9 K (*dashed curve*). The *lower graph* shows the generated photocurrent produced by an additionally pulsed IR light source. If the IR source is on, an IR photocurrent was measured in the range between 440 and 680 nm (*black solid curve*). Without IR illumination, no additional photocurrent was detected (*dashed gray curve*)

In order to verify basics of the IBSC operating principle in our samples, we designed an experiment similar to Martí et al. presented in 2006 [18]. For the first time, they demonstrated the absorption of two photons using an additional chopped infrared (IR) light source and measured the IR generated current with a lock-in amplifier. In our case we have used a frequency modulated 1.3 μm laser diode operating at 477 Hz to synchronize a lock-in amplifier. The energy of the laser (~0.95 eV) is much smaller than the bandgap energy of AlGaAs barrier (~1.85 eV) and the ground state energy of the long wavelength InAs/GaAs QDs (~1.2 eV). This arrangement allows the measurement of the photocurrent produced by the IR source in the presence of the overall photocurrent produced by the primary light source on short circuit condition. All the measurements were performed at low temperatures ($T = 9$ K) to suppress thermal carrier escape from the QDs to the CB continuum.

In the upper part of Fig. 7.13 the EQE of device REF is presented for room temperature (300 K, black solid curve) and for low temperature (9 K, dashed gray curve). The maximum absorption wavelength decreases with decreasing temperature due to the temperature dependence of the bandgap energy [32]. For low temperature no EQE was measured for the region above 670 nm, which corresponds to the bandgap of the AlGaAs barrier material. In the lower part of Fig. 7.13 the photocurrent generated by the additional IR light source is shown. Surprisingly, we have measured a photocurrent produced by the IR light source when the sample is irradiated by the Xe-lamp with a wavelength between 440 and 680 nm (black solid curve), and no current if the IR source is blocked (dashed gray curve). The origin of this current is possibly based on localized impurities and/or doping centers near the conduction band edge. Photons from the primary light source excite electrons from

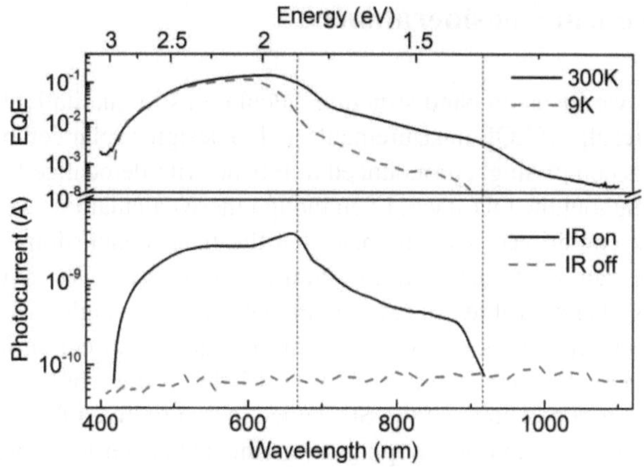

Fig. 7.14 External quantum efficiency (EQE) of device SC-2 for 300 K (*solid curve*) and for 9 K (*dashed curve*). The *lower graph* shows the photocurrent generated of an additionally pulsed infrared (IR) light source. If the IR source is on, a photocurrent produced by this IR source was measured in the range between 420 and 920 nm (*black solid curve*). Without IR illumination, no additional photocurrent was detected (*dashed gray curve*)

deeper VB states to the localized states. Now, low energy photons of the IR light source can excite these carriers to the CB continuum and generate an additional IR photocurrent. Note that for photons with smaller energies than the AlGaAs bandgap from the primary light source no additional IR photocurrent was measured.

Now let us focus on the samples containing QD to investigate the operation principle of IBSC. The experimental measurement of two sub-bandgap photon absorption process for device SC-2 is depicted in Fig. 7.14. Remember, this device was designed as an IBSC with AlGaInAs QDs separated by 20 nm spacers for IB band formation. In analogy with the measurements on sample REF, the EQE for room temperature (black solid curve) and for low temperature (9 K, gray dashed curve) is shown in the upper part of Fig. 7.14. In the range of QD absorption, i.e. for photons with energies below the AlGaAs bandgap, the EQE decreases with decreasing temperature due to the suppression of thermalization effects of confined carriers in the CB continuum. The lower part of Fig. 7.14 presents the photocurrent generated by the modulated IR light source (black solid curve). Here, an enhanced IR photocurrent for the entire spectral range was observed. Furthermore, the range is extended between 680 and 920 nm compared to the REF sample in Fig. 7.13, indicated by the vertical dotted lines. In this spectral range, corresponding to primary photon energies below the AlGaAs bandgap of 1.85 eV, photocurrent can only be caused by the absorption of two sub-bandgap photons via VB–IB by the primary light source and IB–CB by the IR light source. If the IR source is switched off, no additional current was observed (dashed gray curve). This measurement clearly demonstrates the operation principle of the IBSC by absorption of two sub-bandgap photons in our QD solar cell.

7.5 Theoretical Considerations

In this section we focus on band structure calculations to qualitatively explain the experimental results of EQE measurements and to design a solar cell based on these AlGaInAs QDs comprising a pronounced miniband with delocalized electron wave functions throughout the QD stack. Even though these calculations were carried out assuming dark conditions, we anticipate that the main conclusions drawn by the considerations can be transferred to illuminated devises. We have calculated the conduction band profile together with the squared single particle electron ground state wavefunctions of the QDs along the growth direction in a vertically aligned QD stack. The upper graph of Fig. 7.15 shows the whole intrinsic region between n- and p-doped areas of the SC device including five QD layers. Between the QD layers, the position of δ-dopings are indicated as vertical dotted lines. The simulations were performed using nextnano® software package and the following assumptions:

- All of the QDs are located on top of a 0.5 nm thick wetting layer of the same material composition as the QDs.
- The shape of each QD was simulated as semi-ellipsoid with 20 nm in diameter and a constant material composition.
- The heights of the QDs vary from 2 nm for the three different $Al_xGa_{1-x}InAs$ QDs, 2.8 nm for InGaAs QDs and 5 nm for InAs/GaAs QDs, respectively.

These values are not arbitrarily chosen, but based on the structural data in previous investigations of reference structures [24], the data of STEM image (Fig. 7.5b), and a comparison between PL measurements and simulations. As expected, the p–i–n structure generates an electric field strongly tilting the AlGaAs band edge in the intrinsic region.

Now let us focus on the QD region within the devices. The bottom graphs of Fig. 7.15 shows a magnification of the QD region. For SC-2, the QDs tailored bandgap and hence groundstate eigenenergies energetically partially compensate the tilt of the conduction band, resulting in almost equi-energetic electron energy levels. The squared single particle wavefunctions are almost on the same energy level. Yet, due to rather large distance of 20 nm between the individual QD layers, electronic coupling of the electron wave function is rather weak and only observable under perfect energetic alignment. No delocalized subband formation could be observed in this calculation. Note that for simulation with reduced spacer layers of only 10 nm also no delocalization for the electron wavefunctions were observed, which is not shown here. In contrast, the band configuration of device SC-4 with reversed QD arrangement shows a totally different behavior. Here, the energies of the electron levels in the QD layers are energetically strongly separated. Additionally, the InAs/GaAs QDs closest to the n-doped AlGaAs with an approximately doubled electron confinement potential compared to the AlGaInAs QDs literally generate an electron sink explaining the very low EQE of this device.

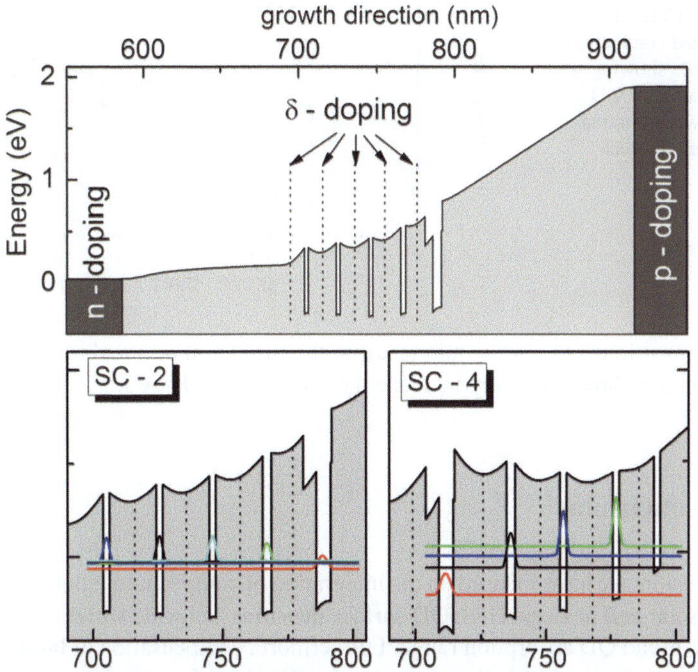

Fig. 7.15 *Top*: Simulated conduction band profile with n- and p-doped regions indicated with *darker contrast*. The QDs are located in the center of the intrinsic region. δ-dopings are indicated as *dotted lines*. *Bottom*: Magnification of the QD region from device SC-2 (normal QD arrangement) and device SC-4 (inversed QD arrangement) and squared wave functions of the respective QDs. The QD eigenenergies in device SC-2 are nearly equi-energetic whereas the eigenenergies in SC-4 were strongly separated

A design based on SC-2 could basically be employed to fabricate a QD solar cell comprising a truly delocalized intermediate band. Therefore, we carried out several simulations varying the thickness of the spacer layer between the QDs and slightly modified QD heights until we observed full wavefunction hybridization within the stack. The finally result of the simulation is shown in Fig. 7.16, simulated with the following assumptions:

- The device features $Al_{0.33}Ga_{0.67}As$ barriers of 5, 6, 6, and 6 nm between the AlGaInAs QDs, respectively.
- The long wavelength absorbing InAs QDs are still being surrounded by only 1 and 3 nm of GaAs.
- The QD heights were chosen to be 1.6, 2.2, 2.5, 2.5, and 2.0 nm throughout the QD stack which are reasonable values for buried QDs.

In such an idealized device, the electron wave function is delocalized. We anticipate that the experimental realization of such a QD scheme, possibly based on advanced growth techniques providing a high degree of control over the QD

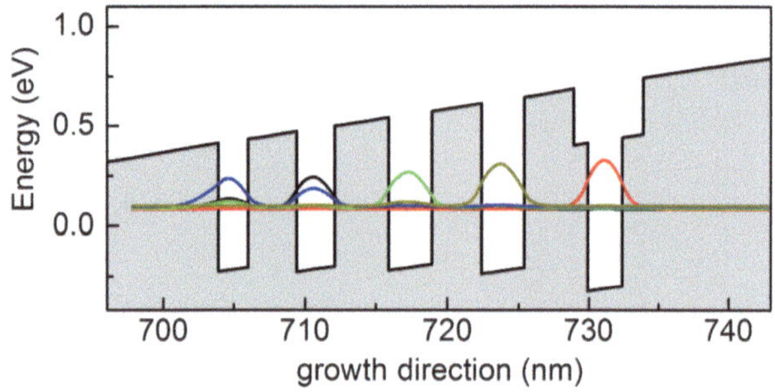

Fig. 7.16 Calculation of an idealized conduction band profile showing some delocalization of QD electron wavefunction within the QD stack

energy transitions [33, 34], would result in the formation of a real subband. Such a QD solar cell should strongly increase the efficiency of IBSCs.

7.6 Conclusion

In conclusion, we have focused on the integration of composition tailored AlGaInAs QDs in solar cell devices. High QD surface densities and wide variation of QD size exhibit a broad QD absorption range. Furthermore, compensation of built-in potential facilitate electronic coupling between the QD states for an enhanced absorption. Comparative EQE measurements of different QD solar cell devices comprising spectrally detuned AlGaInAs/AlGaAs and InAs/GaAs QD layers were presented and discussed. We have experimentally investigated the influence of δ-doping, QD arrangement, varying number of QD layers and influence of spacer layer thickness on the device performance. In addition, the absorption of two sub-bandgap photons was measured and clearly demonstrates the operation principle of an IBSC including AlGaInAs QDs for IB formation. Theoretical band and energy level calculations could qualitatively explain the behavior of the devices and furthermore give perspective for the exploitation of similar scheme in QD IBSCs of the next generation. The subband formation would be based on a compensation of the intrinsic potential in the p–i–n solar cell device by adjustable QDs.

Acknowledgment This work was supported by the state of Bavaria. The authors would like to thank N. Tarakina, T. Braun, M. Adams, M. Lermer, T. Steinl, A. Härtl, S. Reitzenstein, and L. Worschech for sample preparation, measurements, and fruitful discussions.

References

1. Lewis, N.S.: Toward cost-effective solar energy use. Science **315**(5813), 798–801 (2007). doi:10.1126/science.1137014
2. Green, M.A.: Third generation photovoltaics: ultra-high conversion efficiency at low cost. Prog. Photovolt. Res. Appl. **9**(2), 123–135 (2001). doi:10.1002/pip.360

3. Chapin, D.M., Fuller, C.S., Pearson, G.L.: A new silicon p–n junction photocell for converting solar radiation into electrical power. J. Appl. Phys. **25**(5), 676–677 (1954)
4. Shockley, W., Queisser, H.J.: Detailed balance limit of efficiency of p–n junction solar cells. J. Appl. Phys. **32**(3), 510–519 (1961)
5. Conibeer, G.: Third-generation photovoltaics. Mater. Today **10**(11), 42–50 (2007). doi:10.1016/s1369-7021(07)70278-x
6. Brown, G.F., Wu, J.: Third generation photovoltaics. Laser Photon. Rev. **3**(4), 394–405 (2009). doi:10.1002/lpor.200810039
7. Gee, J.M.: A comparison of different module configurations for multi-band-gap solar cells. Solar Cells **24**(1–2), 147–155 (1988). doi:10.1016/0379-6787(88)90044-0
8. De Vos, A., Pauwels, H.: On the thermodynamic limit of photovoltaic energy conversion. Appl. Phys. Mater. Sci. Process. **25**(2), 119–125 (1981). doi:10.1007/bf00901283
9. Polman, A., Atwater, H.A.: Photonic design principles for ultrahigh-efficiency photovoltaics. Nat. Mater. **11**(3), 174–177 (2012)
10. Luque, A., Martí, A.: Increasing the efficiency of ideal solar cells by photon induced transitions at intermediate levels. Phys. Rev. Lett. **78**(26), 5014–5017 (1997)
11. Nozawa, T., Arakawa, Y.: Detailed balance limit of the efficiency of multilevel intermediate band solar cells. Appl. Phys. Lett. **98**(17), 171108-1–171108-3 (2011)
12. Yu, K.M., Walukiewicz, W., Ager III, J.W., Bour, D., Farshchi, R., Dubon, O.D., Li, S.X., Sharp, I.D., Haller, E.E.: Multiband GaNAsP quaternary alloys. Appl. Phys. Lett. **88**(9), 092110–092113 (2006)
13. Yu, K.M., Walukiewicz, W., Wu, J., Shan, W., Beeman, J.W., Scarpulla, M.A., Dubon, O.D., Becla, P.: Diluted II–VI oxide semiconductors with multiple band gaps. Phys. Rev. Lett. **91**(24), 246403 (2003)
14. Antolin, E., Marti, A., Olea, J., Pastor, D., Gonzalez-Diaz, G., Martil, I., Luque, A.: Lifetime recovery in ultrahighly titanium-doped silicon for the implementation of an intermediate band material. Appl. Phys. Lett. **94**(4), 042115-1–042115-3 (2009)
15. Wahnón, P., Tablero, C.: Ab initio electronic structure calculations for metallic intermediate band formation in photovoltaic materials. Phys. Rev. B **65**(16), 165115 (2002)
16. Martí, A., López, N., Antolín, E., Cánovas, E., Stanley, C., Farmer, C., Cuadra, L., Luque, A.: Novel semiconductor solar cell structures: the quantum dot intermediate band solar cell. Thin Solid Films **511–512**, 638–644 (2006). doi:10.1016/j.tsf.2005.12.122
17. Marti, A., Luque, A.: Next Generation Photovoltaics – High Efficiency Through Full Spectrum Utilization. Optics and Optoelectronics. Institute of Physics Publishing, Bristol (2004)
18. Martí, A., Antolín, E., Stanley, C.R., Farmer, C.D., López, N., Díaz, P., Cánovas, E., Linares, P.G., Luque, A.: Production of photocurrent due to intermediate-to-conduction-band transitions: a demonstration of a key operating principle of the intermediate-band solar cell. Phys. Rev. Lett. **97**(24), 247701 (2006)
19. Bailey, C.G., Forbes, D.V., Raffaelle, R.P., Hubbard, S.M.: Near 1 V open circuit voltage InAs/GaAs quantum dot solar cells. Appl. Phys. Lett. **98**(16), 163105-1–163105-3 (2011)
20. Guimard, D., Morihara, R., Bordel, D., Tanabe, K., Wakayama, Y., Nishioka, M., Arakawa, Y.: Fabrication of InAs/GaAs quantum dot solar cells with enhanced photocurrent and without degradation of open circuit voltage. Appl. Phys. Lett. **96**(20), 203507-1–203507-3 (2010)
21. Sugaya, T., Furue, S., Komaki, H., Amano, T., Mori, M., Komori, K., Niki, S., Numakami, O., Okano, Y.: Highly stacked and well-aligned $In_{0.4}Ga_{0.6}As$ quantum dot solar cells with $In_{0.2}Ga_{0.8}As$ cap layer. Appl. Phys. Lett. **97**(18), 183104-1–183104-3 (2010)
22. Laghumavarapu, R.B., El-Emawy, M., Nuntawong, N., Moscho, A., Lester, L.F., Huffaker, D. L.: Improved device performance of InAs/GaAs quantum dot solar cells with GaP strain compensation layers. Appl. Phys. Lett. **91**(24), 243115-1–243115-3 (2007)
23. Reid, W.M., Driscoll, T., Doty, M.F.: Forming delocalized intermediate states with realistic quantum dots. J. Appl. Phys. **111**(5), 056102-1–056102-3 (2012)
24. Schlereth, T.W., Schneider, C., Höfling, S., Forchel, A.: Tailoring of morphology and emission wavelength of AlGaInAs quantum dots. Nanotechnology **19**(4), 045601 (2008)

25. Schlereth, T.W., Gerhard, S., Kaiser, W., Hofling, S., Forchel, A.: High-performance short-wavelength (~760 nm) AlGaInAs quantum-dot lasers. IEEE Photon. Technol. Lett. **19**(18), 1380–1382 (2007). doi:10.1109/lpt.2007.902925
26. Schlereth, T.W., Schneider, C., Kaiser, W., Hofling, S., Forchel, A.: Low threshold, high gain AlGaInAs quantum dot lasers. Appl. Phys. Lett. **90**(22), 221113-1–221113-3 (2007)
27. Baklenov, O., Huffaker, D.L., Anselm, A., Deppe, D.G., Streetman, B.G.: Influence of Al content on formation of InAlGaAs quantum dots grown by molecular beam epitaxy. J. Appl. Phys. **82**(12), 6362–6364 (1997)
28. Hubbard, S.M., Cress, C.D., Bailey, C.G., Raffaelle, R.P., Bailey, S.G., Wilt, D.M.: Effect of strain compensation on quantum dot enhanced GaAs solar cells. Appl. Phys. Lett. **92**(12), 123512–123513 (2008)
29. Kovsh, A.R., Zhukov, A.E., Egorov, A.Yu., Ustinov, V.M., Shernyakov, Yu.M., Maksimov, M.V., Tsatsul'nikov, A.F., Volovik, B.V., Lunev, A.V., Ledentsov, N.N., Kop'ev, P.S., Alferov, Zh.I.: Effect of the quantum-dot surface density in the active region on injection-laser characteristics Semiconductors **32**, 997–1000 (1998)
30. Marti, A., Lopez, N., Antolin, E., Canovas, E., Luque, A., Stanley, C.R., Farmer, C.D., Diaz, P.: Emitter degradation in quantum dot intermediate band solar cells. Appl. Phys. Lett. **90**(23), 233510–233513 (2007)
31. Shoji, Y., Narahara, K., Tanaka, H., Kita, T., Akimoto, K., Okada, Y.: Effect of spacer layer thickness on multi-stacked InGaAs quantum dots grown on GaAs (311)B substrate for application to intermediate band solar cells. J. Appl. Phys. **111**(7), 074305-1–074305-4 (2012)
32. Varshni, Y.P.: Temperature dependence of the energy gap in semiconductors. Physica **34**(1), 149–154 (1967). doi:10.1016/0031-8914(67)90062-6
33. Huggenberger, A., Heckelmann, S., Schneider, C., Hofling, S., Reitzenstein, S., Worschech, L., Kamp, M., Forchel, A.: Narrow spectral linewidth from single site-controlled In(Ga)As quantum dots with high uniformity. Appl. Phys. Lett. **98**(13), 131104-1–131104-3 (2011)
34. Mohan, A., Gallo, P., Felici, M., Dwir, B., Rudra, A., Faist, J., Kapon, E.: Record-low inhomogeneous broadening of site-controlled quantum dots for nanophotonics. Small **6**(12), 1268–1272 (2010). doi:10.1002/smll.201000341

Chapter 8
Requisites for Highly Efficient Hot-Carrier Solar Cells

Yasuhiko Takeda

Abstract We have constructed new models based on detailed balance of particle and energy fluxes to clarify the operating principle of hot-carrier solar cells (HC-SCs) and find the requisites for high conversion efficiency. Energy dissipation due to thermalization of photogenerated carriers can be significantly reduced, even though the thermalization time is not sufficiently long. Instead, the energy dissipation related to entropy generation associated with hot-carrier extraction is remarkable. The thermalization time must be several nanoseconds to exceed the Shockley–Queisser limit under the 1 sun solar irradiation and over 10 ns to compete against triple-junction solar cells at 1,000 sun. The other requisites unique to hot-carrier extraction are a short carrier equilibration time being around one-thousandth of the thermalization time, and an energy-selection width of energy-selective contacts (ESCs) for mono-energetic carrier extraction, which is needed to match the quasi-Fermi levels in the hot absorber and in the cold electrodes, being narrower than 0.1 eV. It seems extremely challenging to fulfill all the requisites, although investigations for material development as well as new concepts for post conventional HC-SCs are underway.

8.1 Introduction

In a conventional solar cell, a photon whose energy is higher than the bandgap of the light-absorbing material used in the cell is absorbed to generate a carrier with an energy equal to the photon energy. However, carrier energies in excess of the bandgap of the absorber immediately dissipate by emitting phonons whose temperature equals the room temperature, namely, thermalization of carriers occurs within several picoseconds in most cases. Therefore, the excess carrier energies cannot be

Y. Takeda (✉)
Toyota Central Research and Development Laboratories, Inc., 41-1 Yokomichi, Nagakute, Aichi 480-1192, Japan
e-mail: takeda@mosk.tytlabs.co.jp

J. Wu and Z.M. Wang (eds.), *Quantum Dot Solar Cells*, Lecture Notes in Nanoscale Science and Technology 15, DOI 10.1007/978-1-4614-8148-5_8,

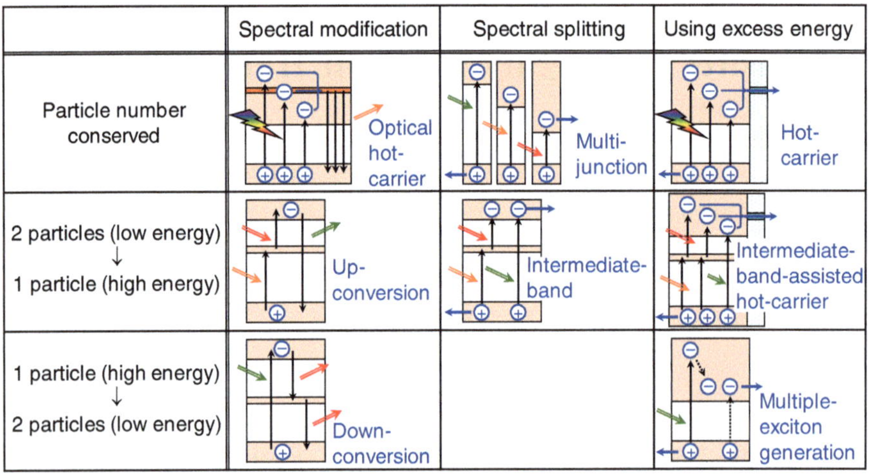

Fig. 8.1 Concepts of solar cells to exceed the Shockley–Queisser (SQ) limit

converted into electricity. This thermalization is one of the most significant energy dissipation processes in photovoltaic conversion [1]. The limiting conversion efficiency of single-junction solar cells (1J-SCs), i.e., the Shockley–Queisser (SQ) limit [2], is determined assuming a single-bandgap absorber, quantum efficiency being unity (one generated carrier per one absorbed photon), and extraction of completely thermalized carriers.

Figure 8.1 summarizes previously proposed concepts to exceed the SQ limit. The most straightforward method based on existing technologies is a multi-junction solar cell using plural absorbers with different bandgaps for spectral splitting, to reduce the excess carrier energies. In particular, triple-junction solar cells (3J-SCs) consisting of III–V compound semiconductors have successfully realized conversion efficiencies over 40 % under concentrated solar irradiation [3, 4], which are significantly higher than the best record of 1J-SCs using GaAs (29 %) [3]. New materials used in 4-junction solar cells have been intensively investigated for further higher efficiency toward 50 % [5, 6]. However, the complicated multilayered structures required for 3J-SCs (and for 4-junction cells) raise the costs and therefore 3J-SCs are currently used only for solar-concentrator systems, which employ Fresnel lenses and concave mirrors to reduce the cell area relative to the module area, and for space applications.

Third-generation solar cells attempt to achieve very high conversion efficiency over 50 % with only a small increase in areal costs compared with those of conventional 1J-SCs. Among them, the idea to achieve the spectral splitting using a single absorber has been implemented with an intermediate-band solar cell (IB-SC) that uses an absorber with an intermediate band (IB) in its fundamental bandgap [7] [8, and references therein]. In an IB-SC, photons of low energies excite electrons from the valence band (VB) to the IB and then from the IB to the conduction band (CB), i.e., two low-energy particles generate a single high-energy particle, like spectral up-conversion [9]. However, IB-SCs involve various issues. Especially, "photon selectivity" that the spectral absorption ranges for the three interband transitions have no overlap with each other is needed to minimize the excess carrier energies and to achieve high conversion efficiency close to those of 3J-SCs [10, 11] but seems difficult to realize.

On the other hand, hot-carrier solar cells (HC-SCs) and solar cells using multiple-exciton generation (MEG) convert the excess carrier energies into electricity. In MEG solar cells, a single high-energy photon can generate two or more carriers [12, 13], which is similar to the mechanism in spectral down-converters [14] because a single high-energy particle is converted to two (or more) low-energy particles. However, previously demonstrated quantum yields for QDs dispersed in solvents [15, and references therein] are not sufficiently high to boost the conversion efficiency [16]. Recently, significantly improved MEG quantum yields have been realized for Si QDs densely dispersed in SiO_2 matrices, in which QDs are strongly coupled to each other, although the origin of the improvement has not yet been elucidated [17, 18].

In contrast, carriers of high energies are directly extracted from HC-SCs [19, 20], which is the topic of this chapter. In the next section, I summarize the points unique to hot-carrier extraction that cause energy dissipation in photovoltaic conversion. Then, in Sects. 8.3 and 8.4, I construct new models based on detailed balance of particle and energy fluxes to involve the unique points, to clarify the operating principle of HC-SCs and to find the requisites for high conversion efficiency. Current status of development of materials used in HC-SCs, including semiconductor quantum dots and quantum wells, and issues to fulfill the requisites are reviewed in Sect. 8.5. New concepts, including optical HC-SCs and intermediate-band-assisted HC-SCs illustrated in Fig. 8.1, to solve some parts of these issues are overviewed in Sect. 8.6. Finally, in Sect. 8.7, I summarize important features and present concluding remarks.

8.2 Frequently Asked Questions on HC-SCs

First, let us consider time evolution of energy distribution of carriers in a light-absorbing material, when the material is solar-irradiated for just a moment. Immediately after the pulsed solar irradiation, the energy distribution of photogenerated carriers reflects the solar spectrum, as schematically depicted in Fig. 8.2. After the generation, due to carrier–carrier elastic scattering, the distribution evolves towards the Fermi–Dirac distribution described by a quasi-Fermi level splitting $\Delta\mu$ and a carrier temperature T_c equivalent to the average of the absorbed photon energies, i.e., around 6,000 K. This "equilibration" process occurs within the equilibration time τ_{eq}, with no energy dissipation from the carrier subsystem. However, due to carrier–phonon scattering, in other words, energy

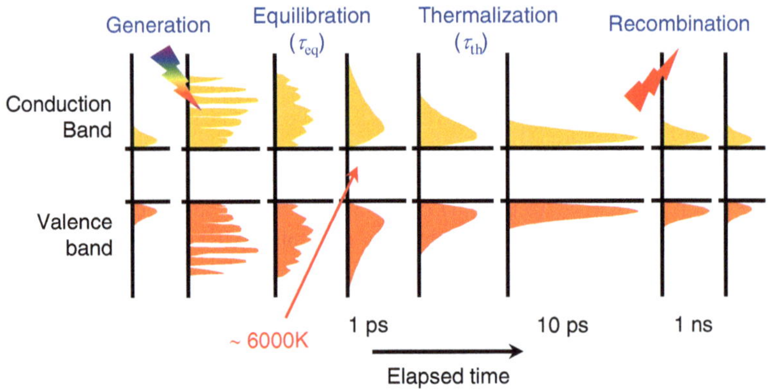

Fig. 8.2 Time evolution of carrier energy distribution after pulsed solar irradiation. τ_{eq}: carrier equilibration time, τ_{th}: carrier thermalization time

dissipation from the carrier subsystem to the phonon subsystem, T_c lowers, eventually to the room temperature T_{RT}. This is "thermalization" process characterized by the thermalization time τ_{th}. Finally, the carriers recombine to emit photons and disappear.

In most cases, the carrier equilibration time τ_{eq} is sub-picoseconds and the thermalization time τ_{th} is several picoseconds. Therefore, in usual solar cells, both τ_{eq} and τ_{th} are extremely shorter than the carrier retention time τ_{re} [average time from generation to extraction, see Eq. (8.13) and Fig. 8.8] and hence carriers are extracted after they are completely thermalized. Consequently, the carrier energies in excess of the bandgap dissipate, as described in the Sect. 8.1. Thus, τ_{eq} and τ_{th} are the points unique to hot-carrier extraction to be considered for high conversion efficiency.

The basic configuration of an HC-SC schematically illustrated in Fig. 8.3 has been established by Ross and Nozik [19], and Würfel [20]. Hot carriers generated in the absorber are extracted through the energy-selective contacts (ESCs) to the metal electrodes on the both sides. The mono-energetic extraction is required to match the quasi-Fermi level splitting $\Delta\mu$ in the absorber at T_c and that in the electrodes at T_{RT}. Therefore, the energy-selection width of the ESCs w_{esc} is the third unique point for hot-carrier extraction.

There are frequently asked questions (FAQs) on HC-SCs:

- FAQ 1. Cannot carriers be extracted from the energy levels outside the narrow energy-selection ranges of the ESCs?
- FAQ 2. Should the ESCs be located at the average carrier energy in the absorber?
- FAQ 3. Is the carrier temperature T_c around 6,000 K, if the carrier thermalization is negligible?
- FAQ 4. In reality, the carrier thermalization cannot be completely suppressed. Therefore, should carriers be extracted as rapidly as possible to reduce the energy dissipation due to the thermalization?

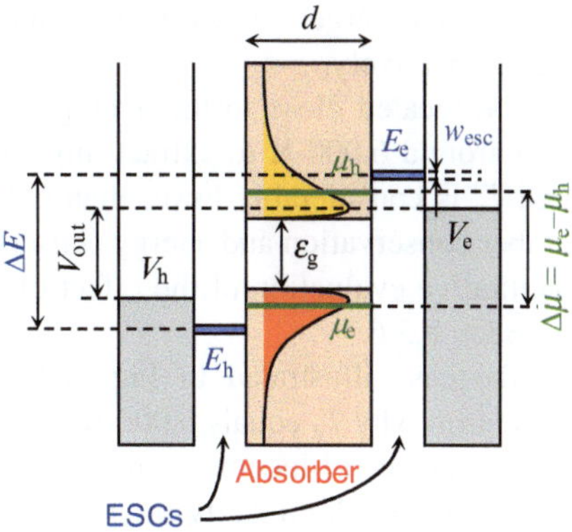

Fig. 8.3 Schematic energy diagram of a hot-carrier solar cell (HC-SC). ε_g: bangap, d: thickness, μ_e, μ_h: quasi-Fermi levels for electrons and holes, respectively, E_e, E_h: ESC energy levels, V_e, V_h: Fermi levels in the electrodes, w_{esc}: ESC energy-selection width. Reproduced with permission from [21]. © 2009, The American Institute of Physics

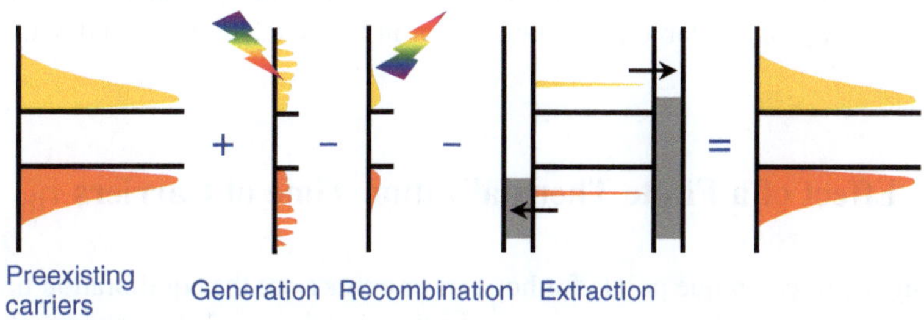

Fig. 8.4 Balance of particle and energy fluxes in a steady-state operation. Energy dissipation due to thermalization is excluded for simplification

All the answers to the FAQs 1–4 are No. The energy distribution of carriers in an HC-SC in a steady-state operation is determined by the balance of preexisting carriers, newly generated carriers, disappearing carriers due to radiative recombination, and extracted carriers, as depicted in Fig. 8.4. The resultant distribution should be significantly different from that shown in Fig. 8.2 under pulsed irradiation and the open-circuit condition. The followings are the summarized answers to the FAQs. Detailed answers will be drawn from the new models constructed in Sects. 8.3 and 8.4.

- A 1. The energy-selection width of the ESCs w_{esc} must be as narrow as possible, or entropy generation associated with energy transport from the hot absorber to the cold electrodes becomes significant, resulting in a low conversion efficiency. Just after the extraction of carriers through the ESCs, "dips" are created in the energy distribution of carriers. However, when the equilibration immediately occurs, i.e., τ_{eq} is sufficiently short, other carriers fill the dips and thus all the carriers can be extracted. The target values of w_{esc} and τ_{eq} will be derived in Sects. 8.4.2 and 8.4.3, respectively.

- A 2. The ESCs must be located close to the average of the absorbed photon energies equivalent to around 6,000 K to extract most of the absorbed photon energies, even though T_c is considerably lower than 6,000 K. This is required from the particle number conservation and energy conservation in the photovoltaic conversion. Quantitative evaluation of the effect of the ESC energy levels will be carried out in Sect. 8.3.6.
- A 3. In contrast to the case illustrated in Fig. 8.2 under the open-circuit condition, there is no reason why T_c equals 6,000 K in a steady-state operation. The requisite is only the balance of incoming and outgoing particle and energy fluxes. The T_c value depends predominantly on the ESC energy levels and the output voltage, which will be shown in Sect. 8.3.6.
- A 4. Rapid extraction of carriers results in a low carrier density, which in turn decreases the quasi-Fermi level splitting $\Delta\mu$ and the output voltage. It holds true for usual solar cells. Therefore, carriers must not be extracted too rapidly. Of course, carriers must be extracted rapidly, i.e., the carrier retention time τ_{re} must be much shorter than τ_{th}. The target values of τ_{th} and the relationship between τ_{th} and suitable τ_{re} will be drawn in Sects. 8.3.5, 8.3.7, and 8.4.4.

8.3 Effect of a Finite Thermalization Time of Carriers τ_{th}

Among the three unique points for hot-carrier extraction: the equilibration time τ_{eq} and thermalization time τ_{th} of carriers in the absorber, and the energy-selection width of the ESCs w_{esc}, let us first concentrate on the effect of τ_{th}. For this purpose, a negligibly short τ_{eq} and an extremely narrow w_{esc} are assumed. The effects of all the three points will be quantitatively evaluated in Sect. 8.4.

8.3.1 Detailed Balance of Particle and Energy Fluxes

To derive the limiting conversion efficiency, the SQ model and detailed-balance models for usual solar cells assume conservation of particle fluxes. In addition, energy fluxes must be considered for HC-SCs. Ross and Nozik established the particle conservation (PC) model in 1982, in which energy conservation along with particle number conservation like the SQ model is involved [19]. Nonradiative recombination processes including Auger recombination (AR), as well as impact ionization (II), are neglected. Another model, referred to as the impact-Auger (IA) model, has been constructed by Würfel in 1997, assuming that extremely frequent II and AR breaks the particle number conservation [20]. However, both the two models deal with the ideal case excluding the thermalization of photogenerated carriers.

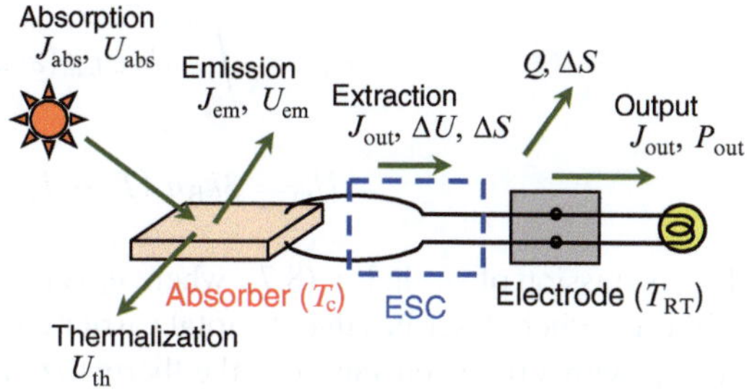

Fig. 8.5 The particle (J), energy (U, Q), and entropy (ΔS) fluxes involved in the present model. Reprinted with permission from [22]. © 2011, Elsevier B.V.

We construct a new model to introduce the energy flux of the thermalization, employing the PC model as a base of the new model [21, 22]. The particle, energy, and entropy fluxes involved in the present model are depicted in Fig. 8.5. Validity of the PC model and IA model will be determined in Sect. 8.3.8 [23].

The output current density J_{out} is represented by the difference between the absorbed and emitted photon fluxes, as is the case for the SQ formulation,

$$\begin{aligned} J_{out} &= q \int_{\varepsilon_g}^{\infty} d\varepsilon \, (j_{sun}(\varepsilon) - j_{em}(\varepsilon; \Delta\mu, T_c)) \\ &\equiv J_{abs} - J_{em}, \end{aligned} \tag{8.1}$$

where $j_{sun}(\varepsilon)$ is the solar spectral photon flux, ε_g the bandgap of the absorber, and q the elementary charge. The assumption of a negligibly short τ_{eq} ensures that the energy distribution of carriers in the absorber is characterized by quasi-Fermi levels in the conduction band (CB) and valence band (VB), μ_e and μ_h, respectively, similar to conventional solar cells, and a common carrier temperature T_c that is higher than the room temperature T_{RT}. This in turn allows us to describe the emitted spectral photon flux $j_{em}(\varepsilon;\Delta\mu,T_c)$ using the generalized Planck's law [24],

$$j_{em}(\varepsilon; \Delta\mu, T_c) = \frac{2\pi}{h^3 c^2} \frac{\varepsilon^2}{\exp[(\varepsilon - \Delta\mu)/(k_B T_c)] - 1}, \tag{8.2}$$

$$\Delta\mu = \mu_e - \mu_h, \tag{8.3}$$

where h denotes the Plank constant, c the velocity of light in vacuum, k_B the Boltzmann constant.

The balance of the energy fluxes describing the absorption, emission, and thermalization, U_{abs}, U_{em} and U_{th}, respectively, determines the energy flux extracted from the absorber ΔU,

$$\Delta U = U_{abs} - U_{em} - U_{th}, \tag{8.4}$$

$$U_{abs} = \int_{\varepsilon_g}^{\infty} d\varepsilon \, \varepsilon \, j_{sun}(\varepsilon), \tag{8.5}$$

$$U_{\mathrm{em}} = \int_{\varepsilon_{\mathrm{g}}}^{\infty} d\varepsilon\, \varepsilon\, j_{\mathrm{em}}(\varepsilon; \Delta\mu, T_{\mathrm{c}}), \tag{8.6}$$

$$U_{\mathrm{th}} = 3k_{\mathrm{B}} n_{\mathrm{c}} (T_{\mathrm{c}} - T_{\mathrm{RT}}) d/\tau_{\mathrm{th}}. \tag{8.7}$$

The expression of U_{th} in Eq. (8.7), where n_{c} is the carrier density and d the thickness of the absorber, describes that the total carrier energy at T_{c} evolves toward the value at T_{RT}, with a time constant, i.e., the thermalization time τ_{th}. Here, τ_{th} is dealt with as a given constant parameter, because it depends on n_{c} and T_{c} only slightly, as long as T_{c} is sufficiently higher than the Debye temperature and n_{c} is not large enough to cause saturation effects [25, 26].

A parabolic two-band model for the VB and CB in the absorber simplifies the relationship between n_{c} and μ_{e},

$$n_{\mathrm{c}} = 8\sqrt{2}\,\pi\, m_{\mathrm{e}}^{3/2}/h^3 \int_{\varepsilon_{\mathrm{CBM}}}^{\infty} d\varepsilon \sqrt{\varepsilon - \varepsilon_{\mathrm{CBM}}}\, \frac{1}{\exp[(\varepsilon - \mu_{\mathrm{e}})/(k_{\mathrm{B}} T_{\mathrm{c}})] + 1}, \tag{8.8}$$

Where m_{e} is the effective mass of electrons in the CB, $\varepsilon_{\mathrm{CBM}}$ the conduction-band minimum in the absorber. The similar expression for n_{c} and μ_{h} is derived using the effective mass of holes m_{h} and the valence-band maximum $\varepsilon_{\mathrm{VBM}}$.

The energy extracted from the absorber by one carrier equals the ratio of ΔU to J_{out}/q, which is exactly the difference between the ESC energy levels for electrons and holes ΔE,

$$\Delta E = \Delta U/(J_{\mathrm{out}}/q). \tag{8.9}$$

When carriers are extracted from the hot absorber at T_{c} to the cold electrodes, in which the carrier temperature equals T_{RT}, entropy is generated, although the extremely narrow energy-selection width of the ESC w_{esc} minimizes the amount of the entropy generation ΔS [19, 27]. This relates to the other energy dissipation process: thermodynamically derived unavoidable heat flux to the ambient surroundings Q. Therefore, the output energy flux P_{out} is lower than ΔU by Q,

$$\begin{aligned} Q &= T_{\mathrm{RT}}\,\Delta S \\ &= (J_{\mathrm{out}}/q)(\Delta E - \Delta\mu)\, T_{\mathrm{RT}}/T_{\mathrm{c}} \\ &= (\Delta U - (J_{\mathrm{out}}/q)\Delta\mu) T_{\mathrm{RT}}/T_{\mathrm{c}}, \end{aligned} \tag{8.10}$$

$$\begin{aligned} P_{\mathrm{out}} &= \Delta U - Q \\ &= (J_{\mathrm{out}}/q)(\Delta E(1 - T_{\mathrm{RT}}/T_{\mathrm{c}}) + \Delta\mu T_{\mathrm{RT}}/T_{\mathrm{c}}) \\ &\equiv J_{\mathrm{out}}\, V_{\mathrm{out}}. \end{aligned} \tag{8.11}$$

The conversion efficiency η is the ratio of P_{out} to the incident energy flux,

$$\eta = P_{\text{out}} \Big/ \int_0^\infty d\varepsilon\, \varepsilon\, j_{\text{sun}}(\varepsilon)$$
$$\equiv P_{\text{out}}/P_{\text{in}}. \tag{8.12}$$

We introduce the carrier retention time τ_{re} which equals the average time from generation to extraction, for comparison with the thermalization time τ_{th},

$$\tau_{\text{re}} = n_{\text{c}}\, d/(J_{\text{abs}}/q)\,, \tag{8.13}$$

where $n_{\text{c}}\, d$ equals the carrier number per unit area whereas J_{abs}/q represents the carrier generation rate per unit area.

For numerical evaluations in the following, $m_{\text{e}} = 0.04\, m_0$ and $m_{\text{h}} = 0.4\, m_0$ (m_0 denotes the electron rest mass), which are close to the values of $In_{0.53}Ga_{0.47}As$ and GaSb [28], were used, because a small m_{e}, i.e., a small density of states (DOS) in the CB is required for high conversion efficiency [21]. Thin absorbers ($d = 500$ nm) were employed so that n_{c} was as large as possible (see Sect. 8.3.7) [21]. AM1.5G solar spectrum was used up to the 10 sun solar irradiation intensity, and AM1.5D for higher intensity considering the use of solar concentrators [29, 30] [31, and references therein].

8.3.2 Qualitative Answer to FAQ 2: ESC Energy Level ΔE

Equation (8.9) derived from the particle and energy conservation gives the straightforward answer to the FAQ 2. To achieve a high conversion efficiency, the emission from the cell must be insignificant, meaning ΔU and J_{out} are close to U_{abs} and J_{abs}, respectively, and consequently $\Delta E \cong U_{\text{abs}}/(J_{\text{abs}}/q)$ must stand. The optimal ΔE values will be derived in Sect. 8.3.6.

8.3.3 Qualitative Answer to FAQ 3: Carrier Temperature T_{c}

When an HC-SC is operating at the open-circuit voltage and if the thermalization is negligible, T_{c} must be around 6,000 K and the emission spectrum must be equivalent to the absorbed solar spectrum, because the emission is the only energy dissipation process from the cell. However, we are interested in the conditions around the maximal power points. Equations (8.8)–(8.11) show that T_{c} depends on V_{out} and $\Delta\mu$, whereas $\Delta\mu$ is a function on T_{c}. Therefore, T_{c} depends on ΔE and V_{out} in a very complicated manner. Quantitative evaluation will be carried out in Sect. 8.3.6.

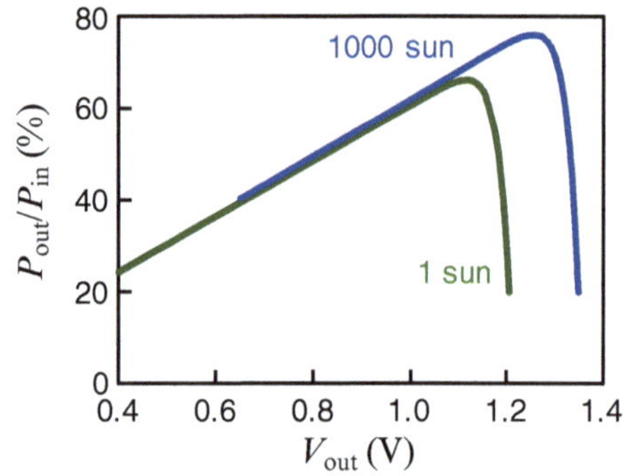

Fig. 8.6 Conversion efficiency P_{out}/P_{in} as a function of the output voltage V_{out}. $\varepsilon_g = 0.7$ eV, $\Delta E = 1.58$ eV (1 sun), 1.54 eV (1,000 sun), no thermalization

8.3.4 Qualitative Answer to FAQ 4: Carrier Retention Time τ_{re}

In the present subsection, the energy dissipation due to the thermalization U_{th} is excluded for straightforward understanding of HC-SC operation.

Figure 8.6 shows the $P_{out}/P_{in} - V_{out}$ relationships of HC-SCs at $\varepsilon_g = 0.7$ eV, where ΔE is set to be exactly the average of the absorbed photon energies $U_{abs}/(J_{abs}/q)$. At the maximal power points, the conversion efficiencies are as high as 66 % (1 sun) and 76 % (1,000 sun), respectively. In the low V_{out} ranges, P_{out}/P_{in} increases with increasing V_{out}, and both the results at 1 sun and 1,000 sun are almost identical to each other. However, P_{out}/P_{in} exhibits the maximum at a higher V_{out} at 1,000 sun and starts to decrease from there. Therefore, the maximal value at 1,000 sun is higher than that at 1 sun.

A high V_{out} and a large $\Delta\mu$ originate from a large n_c. Therefore, P_{out}/P_{in} is replotted as a function of n_c in Fig. 8.7a. From Eq. (8.8), a linear dependence of $\Delta\mu$ on $\log[n_c]$ is derived using the Maxwell–Boltzmann distribution function instead of the Fermi–Dirac distribution function, whereas P_{out} is linearly dependent on $\Delta\mu$ as shown in Eq. (8.11). Therefore, P_{out}/P_{in} exhibits a linear dependence on $\log[n_c]$ in the small n_c range. However, a too large n_c causes a significant effect of the energy dissipation due to the emission U_{em}, lowering in P_{out}/P_{in}. These features are quite similar to those for conventional 1J-SCs represented by the SQ formulation shown in Fig. 8.8.

In Eq. (8.8), the use of a bulk material is assumed. To achieve suppressed thermalization, low-dimensional materials such as quantum dots and multiple quantum wells (MQWs) are attractive candidates (see Sect. 8.5.1). Even using the DOS of 0-, 1-, or 2-dimensional structures, the linear dependence of $\Delta\mu$ on $\log[n_c]$ is also derived from the relationship between $\Delta\mu$ and n_c similar to Eq. (8.8).

However, no thermalization ($U_{th} = 0$) is supposed here. Therefore, a question arises; where does U_{abs} dissipate at a small n_c? Fig. 8.7b show the effects of the two energy dissipation processes: the emission U_{em} and $Q = T_{RT}\Delta S$ related to the entropy generation associated with hot-carrier extraction normalized by P_{in}, depending on n_c. As described above, U_{em}/P_{in} is negligibly small in the small n_c range, and therefore, $Q = T_{RT}\Delta S$ is the dominant energy dissipation process.

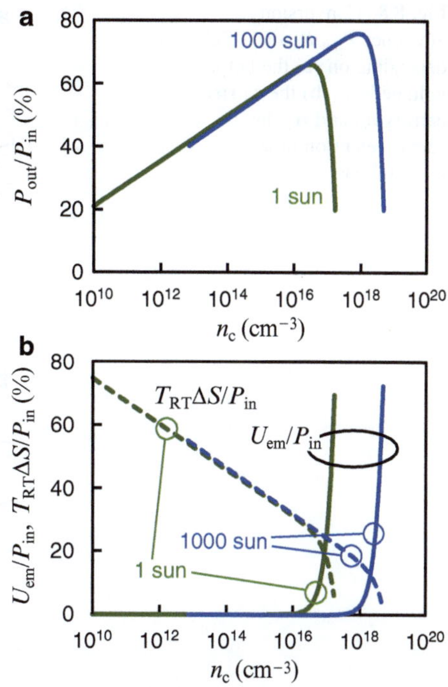

Fig. 8.7 (**a**) Conversion efficiency P_{out}/P_{in}, (**b**) energy dissipation due to the emission U_{em} and related to the entropy generation associated with hot-carrier extraction $Q = T_{RT}\Delta S$ normalized by P_{in}, depending on the carrier density n_c. $\varepsilon_g = 0.7$ eV, $\Delta E = 1.58$ eV (1 sun), 1.54 eV (1,000 sun), no thermalization

Equation (8.10) shows that Q exhibits liner dependence on $\Delta\mu$, and consequently Q is a decreasing function of n_c linearly dependent on $\log[n_c]$. In addition, Q is proportional to J_{abs}, as long as U_{em} is negligible. Thus, the $Q/P_{in} = T_{RT}\Delta S/P_{in}$ values at 1 sun and 1,000 sun are identical to each other at a given n_c, apart from a slight difference originating from the solar spectra: AM1.5G and AM1.5D.

In contrast, it is derived from Eqs. (8.2), (8.6) and (8.8) that U_{em} is approximately proportional to n_c and, of course, has no explicit relation to P_{in}. Therefore, U_{em}/P_{in} starts to increase rapidly at a larger n_c for a larger P_{in}: 10^{16} cm^{-3} at 1 sun and 10^{18} cm^{-3} at 1,000 sun. This is the reason why the conversion efficiency P_{out}/P_{in} at the maximal power point is higher for more intense irradiation. The impact of irradiation intensity or solar concentration ratio has been more explicitly elucidated in terms of entropy generation due to eténdue expansion [32, 33].

The carrier retention time τ_{re} has been determined by Eq. (8.13), which will be compared with τ_{th} in Sect. 8.3.5. The dependence of P_{out}/P_{in}, U_{em}/P_{in}, and $T_{RT}\Delta S/P_{in}$ on τ_{re} is shown in Fig. 8.9. The maximal value of P_{out}/P_{in} is determined from the trade-off relationship that $T_{RT}\Delta S/P_{in}$ decreases whereas U_{em}/P_{in} increases with increasing τ_{re}. Comparing at a given τ_{re}, $T_{RT}\Delta S/P_{in}$ is lower and consequently P_{out}/P_{in} is higher at 1,000 sun than those at 1 sun because of a larger n_c, as long as U_{em} is negligible. The maximum of P_{out}/P_{in} appears at around $\tau_{re} = 10$ μs at 1 sun and $\tau_{re} = 100$ ns at 1,000 sun, where U_{em}/P_{in} starts to decrease. These τ_{re} values correspond to $n_c = 10^{16}$ cm^{-3} and 10^{18} cm^{-3}, respectively.

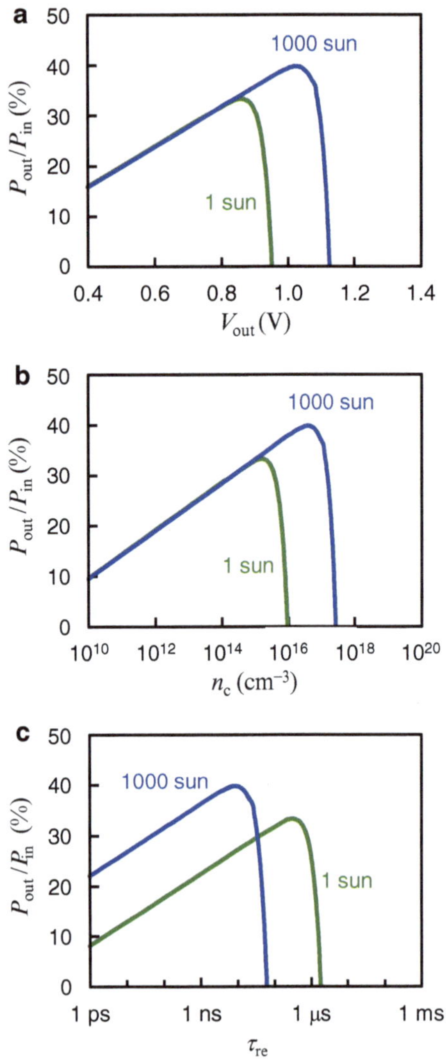

Fig. 8.8 Conversion efficiency P_{out}/P_{in} of 1J-SCs depending on (**a**) the output voltage V_{out}, (**b**) the carrier density n_c, and (**c**) the carrier retention time τ_{re}. $\varepsilon_g = 1.2$ eV

In reality, it is difficult to completely suppress the thermalization, i.e., τ_{th} should be finite. When τ_{re} is close to and longer than τ_{th}, the energy dissipation due to the thermalization U_{th} is remarkable and consequently the output power P_{out} is lowered than the value at $\tau_{th} = \infty$ ($U_{th} = 0$). Therefore, τ_{re} must be much shorter than τ_{th}. In other words, carriers must be extracted rapidly, which agrees with our common sense.

On the other hand, if hot carriers are extracted too rapidly, i.e., τ_{re} is too short, P_{out} is again lowered, because of significant $T_{RT}\Delta S$. This stands for a conventional

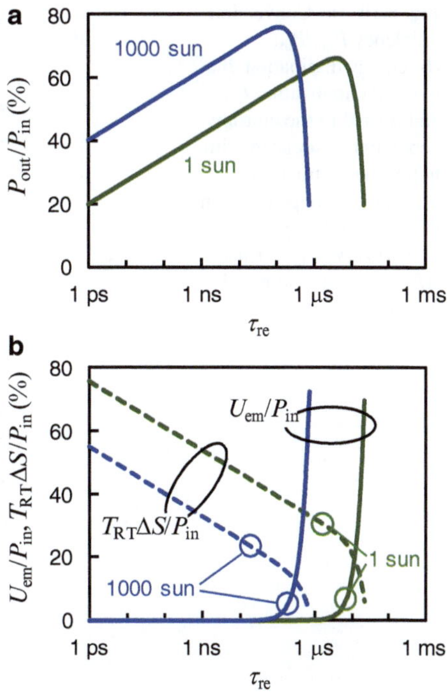

Fig. 8.9 (**a**) Conversion efficiency P_{out}/P_{in}, (**b**) energy dissipation due to the emission U_{em} and related to the entropy generation associated with hot-carrier extraction $Q = T_{RT}\Delta S$ normalized by P_{in}, depending on the carrier retention time τ_{re}. $\varepsilon_g = 0.7$ eV, $\Delta E = 1.58$ eV (1 sun), 1.54 eV (1,000 sun), no thermalization

solar cell and corresponds to a low V_{out}, which is found from a comparison of Figs. 8.6, 8.7, 8.8, and 8.9.

Thus, the conclusion of the present subsection is that hot carriers must be extracted rapidly, but not too rapidly. This is exactly the answer to the FAQ 4.

8.3.5 Quantitative Answer to FAQ 4: Carrier Retention Time τ_{re} vs. Thermalization Time τ_{th}

The next question arises; how long is the optimal τ_{re} value for high conversion efficiency?

Because previously demonstrated values of τ_{th} are several hundred picoseconds at most, which will be reviewed in Sect. 8.5.1, we use here $\tau_{th} = 1$ ns with expectation. The results at $\varepsilon_g = 0.7$ eV including the energy dissipation due to the thermalization U_{th} with $\tau_{th} = 1$ ns are shown in Fig. 8.10. Note that the ΔE value was optimized for each case. The maximal P_{out}/P_{in} values appear at $n_c = 10^{12}$ cm^{-3} for the 1 sun solar irradiation and at $n_c = 10^{15}$ cm^{-3} for 1,000 sun, respectively.

The values of $Q = T_{RT}\Delta S$ for 1 sun and 1,000 sun are very close to each other in the small n_c range, as is the case of no thermalization shown in Fig. 8.7b. The values

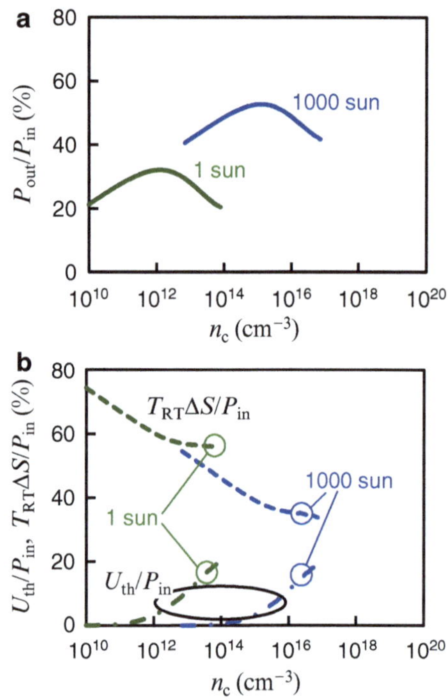

Fig. 8.10 (**a**) Conversion efficiency P_{out}/P_{in}, (**b**) energy dissipation due to the thermalization U_{th} and related to the entropy generation associated with hot-carrier extraction $Q = T_{RT}\Delta S$, depending on the carrier density n_c. $\varepsilon_g = 0.7$ eV, $\tau_{th} = 1$ ns. ΔE value was optimized for each case

of U_{em} have been confirmed to be negligibly small. In the present model, U_{th} is proportional to n_c as described in Eq. (8.7), and of course not related explicitly to P_{in}. Therefore, U_{th}/P_{in} starts to increase at a three-order larger n_c for 1,000 sun than that for 1 sun. This is exactly the reason why P_{out}/P_{in} is higher at 1,000 sun.

Note that the optimal n_c values are extremely smaller than those for no thermalization shown in Fig. 8.7. Consequently, the maximal P_{out}/P_{in} values are remarkably lowered compared with those for no thermalization: 32 % at 1 sun ($\varepsilon_g = 0.7$ eV is not the optimal for 1 sun, see Fig. 8.13) and 53 % at 1,000 sun.

The optimal τ_{re} values are around 100 ps that are 1/10 of τ_{th} for both 1 sun and 1,000 sun, where U_{th} starts to increase, as shown in Fig. 8.11. This quantitative answer to the FAQ 4 is a quite natural consequence drawn from the relationship among τ_{re}, τ_{th} and U_{th} derived from Eqs. (8.7) and (8.13),

$$U_{th} = 3\,k_B\,(J_{abs}/q)\,(T_c - T_{RT})\,\tau_{re}/\tau_{th}. \tag{8.14}$$

Note again that the factors that determine the maximal points in the $P_{out}/P_{in} - n_c$ and $P_{out}/P_{in} - \tau_{re}$ relationships are different between the two cases: $\tau_{th} = \infty$ ($U_{th} = 0$) and $\tau_{th} = 1$ ns. As long as U_{em} and U_{th} are negligible, P_{out} is linearly dependent on log[n_c]. For the former shown in Figs. 8.7 and 8.9, when the effect of U_{em} becomes significant with increasing n_c and τ_{re}, P_{out} starts to decrease. This stands for conventional solar cells represented by the SQ formulation, as shown in Fig. 8.8.

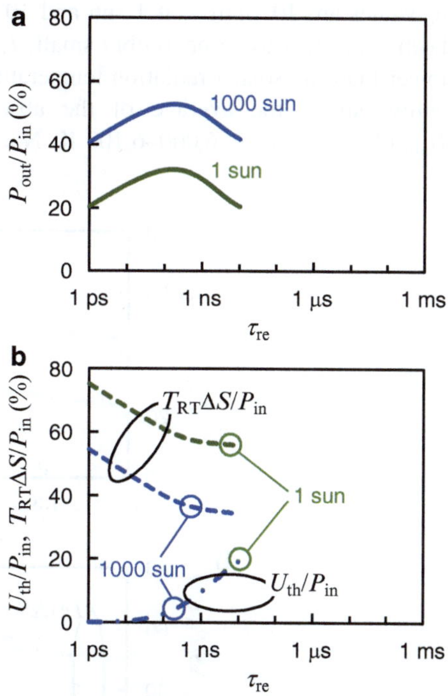

Fig. 8.11 (**a**) Conversion efficiency P_{out}/P_{in}, (**b**) energy dissipation due to the thermalization U_{th} and related to the entropy generation associated with hot-carrier extraction $Q = T_{RT}\Delta S$, depending on the carrier retention time τ_{re}. $\varepsilon_g = 0.7$ eV, $\tau_{th} = 1$ ns. ΔE value was optimized for each case

In contrast, U_{th} starts to increase prior to U_{em} being remarkable at $\tau_{th} = 1$ ns, and therefore the maximal point is determined from the trade-off relationship between U_{th} and $T_{RT}\Delta S$. A too long τ_{re} results in remarkable U_{th} and a low P_{out}/P_{in}. In short, carriers must be extracted rapidly; otherwise, U_{th} is detrimental. In addition, carriers must not be extracted too rapidly for both the two cases so that $T_{RT}\Delta S$ is insignificant.

8.3.6 *Quantitative Answers to FAQs 2 and 3: ESC Energy Level* ΔE *and Carrier Temperature* T_c

In Sect. 8.3.5, the ΔE value was optimized at a given n_c and τ_{re}. Here, changing the perspective, let us deal with ΔE as a given constant. This seems a natural way, because ΔE is determined from the materials and structures employed in the cells, whereas n_c and τ_{re} are affected by the operating conditions.

Figure 8.12a, b shows the dependence of P_{out}/P_{in}, U_{th}/P_{in} and $Q/P_{in} = T_{RT}\Delta S/P_{in}$ on ΔE. The values of T_c, n_c, and τ_{re} optimized for each ΔE are plotted in Fig. 8.12c–e. The two energy dissipation processes: U_{th} and $T_{RT}\Delta S$ again exhibit a trade-off relationship with changing ΔE, and as a result, P_{out}/P_{in} reaches the maximum when ΔE is slightly lower than $U_{abs}/(J_{abs}/q)$ by 0.06 eV. At this point, the

n_c values are 10^{12} cm^{-3} at 1 sun and 10^{15} cm^{-3} at 1,000 sun. Although U_{th} has been confirmed to be negligibly small, T_c is around 1,700 K that is considerably lower than the solar irradiation temperature around 6,000 K and the temperature equivalent to the average of the absorbed photon energies represented by $U_{abs}/(J_{abs}/q)/(3k_B) = 6{,}000$–$6{,}100$ K. Note that the difference between T_c around

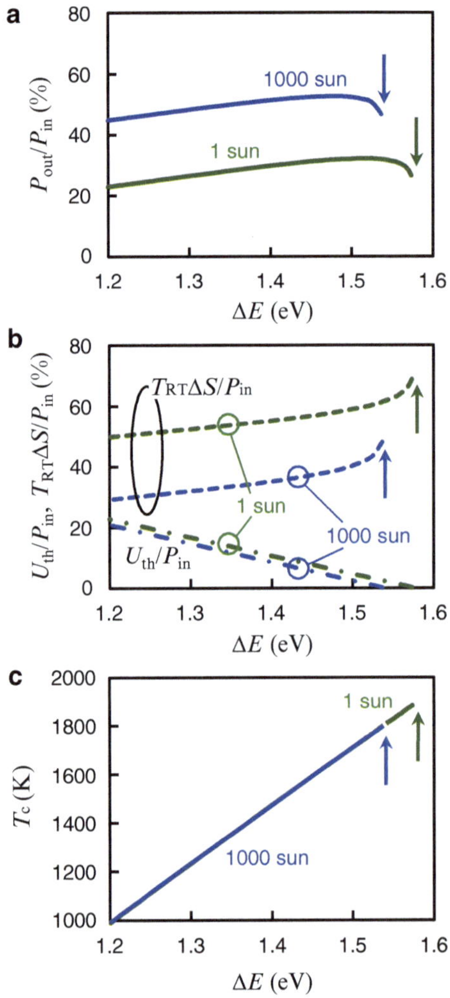

Fig. 8.12 (**a**) Conversion efficiency P_{out}/P_{in}, (**b**) energy dissipation due to the thermalization U_{th} and related to the entropy generation associated with hot-carrier extraction $Q = T_{RT}\Delta S$, depending on the ESC energy level ΔE. The optimized (**c**) carrier temperature T_c, (**d**) carrier density n_c, (**e**) carrier retention time τ_{re}, and (**f**) quasi-Fermi level splitting $\Delta\mu$. *Arrows* indicate the $U_{abs}/(J_{abs}/q)$ values: 1.58 eV for 1 sun (AM1.5G) and 1.54 eV for 1,000 sun (AM1.5D). $\varepsilon_g = 0.7$ eV, $\tau_{th} = 1$ ns

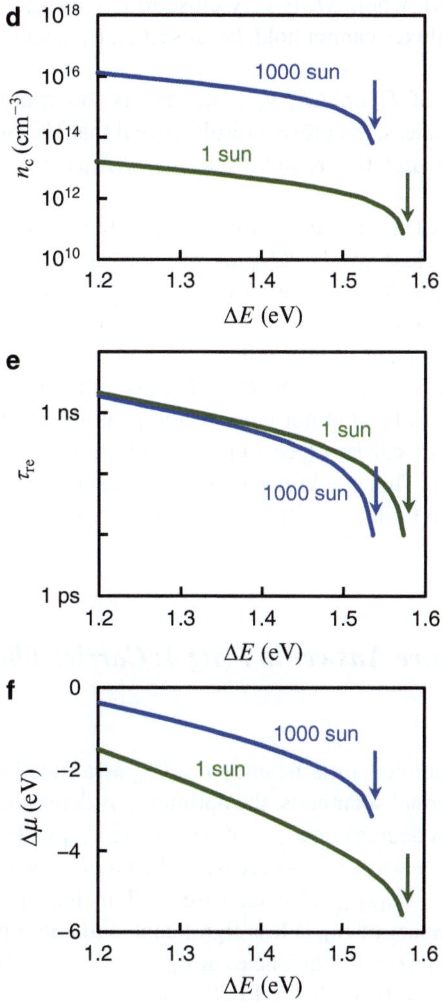

Fig. 8.12 (continued)

1,700 K and $U_{abs}/(J_{abs}/q)/(3k_B)$ around 6,000 K is not due to the thermalization of carriers but the results of the balance between the preexisting, generated, and extracted carrier energies.

Even if ΔE is considerably smaller than $U_{abs}/(J_{abs}/q)$, a steady-state operation is possible. However, energy conservation requires that the difference between ΔE and $U_{abs}/(J_{abs}/q)$ equals U_{th}, leading to a large n_c, a long τ_{re} and a low P_{out}/P_{in}. The resultant T_c is even lower.

On the other hand, when ΔE is very close to $U_{abs}/(J_{abs}/q)$, conservation of the particle and energy fluxes cannot hold, because J_{em}, U_{em}, and U_{th} are very small but not zero.

The dependence of T_c on ΔE, V_{out}, τ_{re}, etc. is too complicated to be simply interpolated. Nonetheless, the present results reveal that HC-SCs can operate at a T_c considerably lower than 6,000 K and $U_{abs}/(J_{abs}/q)/(3k_B)$ even though U_{th} is negligibly small.

The conclusions of the present subsection and the answers to the FAQs 2 and 3 that ΔE must be close to $U_{abs}/(J_{abs}/q)$ whereas the optimal T_c is considerably lower than $U_{abs}/(J_{abs}/q)/(3k_B)$ raise an issue. At $\varepsilon_g = 0.7$ eV, $U_{abs}/(J_{abs}/q)$ equals 1.58 eV for AM 1.5G and 1.54 eV for AM 1.5D, and ΔE must be close to these values. On the other hand, the average carrier energy at the optimal T_c around 1,700 K represented by $3k_BT_c$ is 0.85 eV, which is remarkably lower than ΔE. This means that the ESC is located not close to the peak energy but on the tail of the energy distribution of carriers (see Fig. 8.3), where the carrier number per unit energy is quite small. This would make the equilibration of carriers a more serious issue. The effect of a finite τ_{eq} will be evaluated in Sect. 8.4.3.

8.3.7 Quantitative Answer to FAQ 4: Carrier Thermalization Time τ_{th}

Figure 8.13 shows conversion efficiency P_{out}/P_{in} as a function of ε_g at various τ_{th} values. For conventional solar cells, the optimal ε_g is determined from the trade-off relationship between decreasing J_{out} and increasing V_{out} with increasing ε_g. This is interpreted in another way; U_{abs} decreases whereas U_{th}, which is represented by $U_{abs} - (J_{abs}/q)(\varepsilon_g - 3k_BT_{RT})$, also decreases with increasing ε_g. When hot carriers are extracted, the impact of U_{th} is less significant, and consequently the optimal ε_g shifts to lower. The maximal efficiencies at $\tau_{th} = 1$ ns are 35 % at $\varepsilon_g = 1.12$ eV for 1 sun and 53 % at $\varepsilon_g = 0.70$ eV for 1,000 sun.

The dependence of P_{out}/P_{in} on τ_{th}, as well as the impact of the energy dissipation processes: the thermalization U_{th} and $Q = T_{RT}\Delta S$ related to the entropy generation associated with hot-carrier extraction at the optimal ε_g, is shown in Fig. 8.14. Here, these two values are normalized by U_{abs}, because the optimal ε_g and hence U_{abs} are different depending on τ_{th}. The effect of U_{th} is significantly reduced to be only 4 % throughout the τ_{th} range considered here, while it accounts for as high as 26 % for a 1J-SC with $\varepsilon_g = 1.34$ eV, and 35 % with $\varepsilon_g = 1.12$ eV. Instead, the impact of $T_{RT}\Delta S$ is as high as around 50 % at $\tau_{th} = 0.1$–10 ns for 1 sun. Although it decreases with increasing τ_{th} and the solar irradiation intensity, $T_{RT}\Delta S$ is still detrimental considerably even at $\tau_{th} = 100$ ns and for 1,000 sun.

Under the 1 sun solar irradiation, $\tau_{th} = 1$ ns at least is required to achieve conversion efficiency higher than that of 1J-SCs. As shown in Fig. 8.11a, the maximal values of P_{out}/P_{in} appear at around $\tau_{re} = \tau_{th}/10$, not depending on the

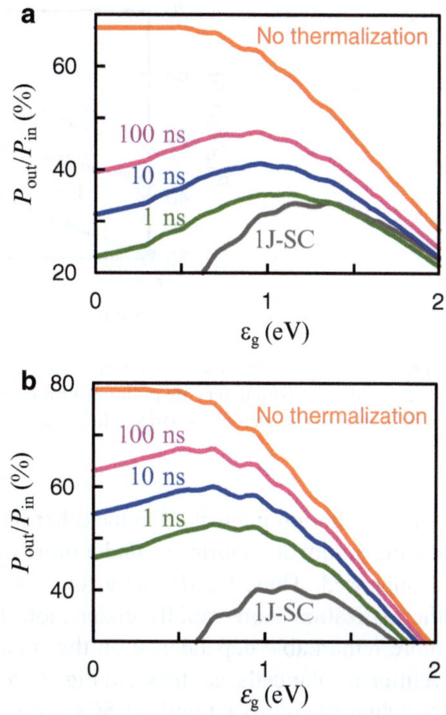

Fig. 8.13 Conversion efficiency P_{out}/P_{in} depending on the bandgap ε_g and carrier thermalization time τ_{th} (1, 10, 100 ns). ΔE and other parameters are optimized for each case. Results for 1J-SCs are also shown for comparison. (**a**) 1 sun, (**b**) 1,000 sun

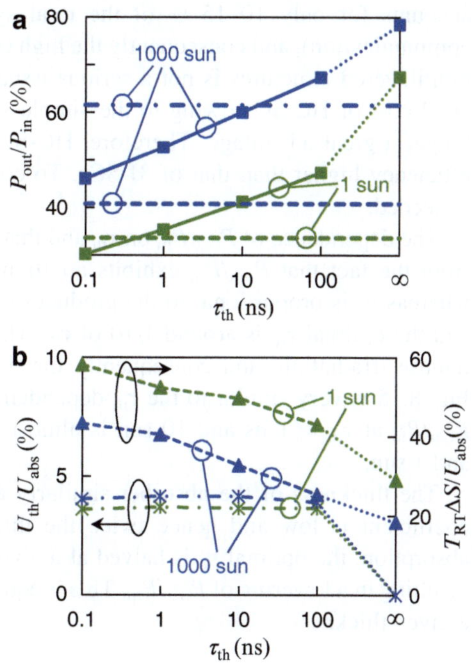

Fig. 8.14 (**a**) Conversion efficiency P_{out}/P_{in} depending on the thermalization time τ_{th}. ε_g, ΔE and other parameters are optimized for each case. *Broken lines* and a *dashed–dotted line* are the results for 1J- and 3J-SCs, respectively. (**b**) Energy dissipation due to the thermalization U_{th} and related to the entropy generation associated with hot-carrier extraction $Q = T_{RT}\Delta S$ normalized by the absorbed energy flux U_{abs}

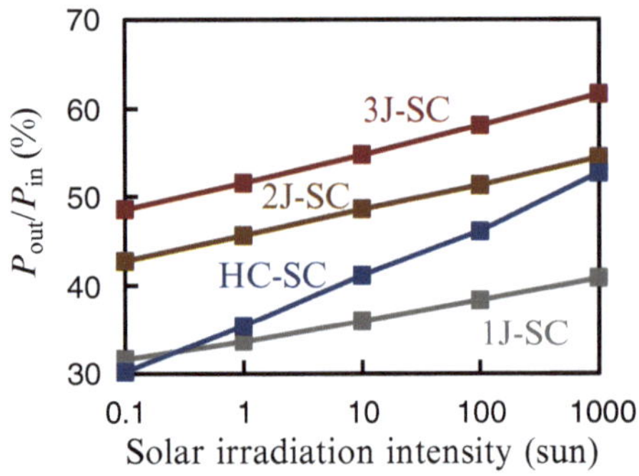

Fig. 8.15 Dependence of conversion efficiency P_{out}/P_{in} on the solar irradiation intensity. $\tau_{th} = 1$ ns. ε_g, ΔE and other parameters are optimized for each case. Results for 1J-, 2J- (double-junction), and 3J-SCs are also shown for comparison. Reproduced with permission from [22]. © 2011, Elsevier B.V.

solar irradiation intensity. On the other hand, Fig. 8.9a shows that P_{out}/P_{in} reaches the maximum at a shorter τ_{re} under more intense irradiation when no thermalization is supposed. Thus, P_{out}/P_{in} as a function of τ_{th} reaches close to the value for no thermalization more rapidly under more intense irradiation. This in turn leads to more remarkable dependence on the solar irradiation intensity than those of conventional solar cells, as shown in Fig. 8.15, and the result that a remarkable gain can be achieved compared with 1J-SCs even at $\tau_{th} = 0.1$ ns for 1,000 sun.

However, when solar cells are coupled with solar concentrators, the cell cost accounts for only 10–15 % of the total system cost (Araki, K, 2007, private communication), and consequently the high cost of 3J-SCs due to their complicated multilayered structures is not a serious issue. In other words, the potentially low areal cost of HC-SCs owing to the simple structure compared with 3J-SCs is no longer a great advantage. Therefore, HC-SCs are required to achieve conversion efficiency higher than that of 3J-SCs. To compete against 3J-SCs, τ_{th} over 10 ns is needed.

The dependence of P_{out}/P_{in} on τ_{th} and that on the solar irradiation intensity arise from the fact that P_{out}/P_{in} exhibits approximately linear dependence on log[n_c], whereas n_c is proportional to the product of τ_{re} and the solar irradiation intensity, and the optimal τ_{re} is around 1/10 of τ_{th}. Therefore, a long τ_{th} is equivalent to an intense irradiation, and consequently the dependence on the intensity shown in Fig. 8.15 is very similar to the τ_{th}-dependence shown in Fig. 8.14a. For example, P_{out}/P_{in} at $\tau_{th} = 1$ ns and 10 sun is almost the same as the value at $\tau_{th} = 10$ ns and 1 sun.

The thickness of the absorber similarly affects P_{out}/P_{in}. When the absorption coefficient is low and hence twice the thickness is needed for sufficient light absorption, the optimal n_c is halved at a given τ_{th} and a solar irradiation intensity, resulting in a lowering of P_{out}/P_{in}. This is equivalent to halving τ_{th} or the intensity at a given thickness.

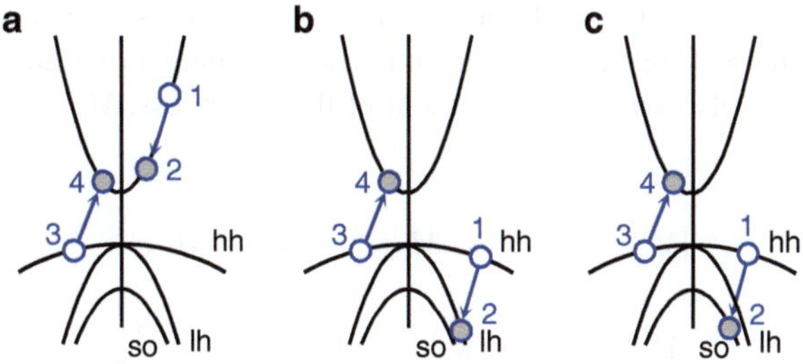

Fig. 8.16 Three major processes of impact ionization in the absorber: (**a**) CCCH, (**b**) CHHL, (**c**) CHHS. Note that all the *circles* represent electrons (not holes) in the valence bands and conduction bands. Reprinted with permission from [23]. © 2009, Elsevier B.V.

Note that the detrimental effects of τ_{eq} and w_{eq}, which will be evaluated in Sect. 8.4, are excluded in the results shown in Figs. 8.13, 8.14, and 8.15. Therefore, the conclusion of the present subsection is that τ_{th} of around 1 ns at 1 sun and over 10 ns are, at least, required to compete against existing solar cells.

8.3.8 Validity of the Particle Conservation Model and the Impact-Auger Model

As stated in Sect. 8.3.1, two contrasting detailed-balance models have been established: the PC model and the IA model. We have employed the PC model as a base of the improvement to involve the thermalization of carriers. The PC model and present model lead to negative values of the quasi-Fermi level splitting in the absorber $\Delta\mu$ in steady-state operations at moderate T_c and n_c, which would be against our common sense. The $\Delta\mu$ values corresponding to Fig. 8.12a–e are shown in Fig. 8.12f. In general, impact ionization (II) and Auger recombination (AR) are frequent when T_c is high and n_c is large. To evaluate the effect of II and AR, τ_{re} is again employed to be compared with II and AR lifetimes, τ_{II} and τ_{AR}, respectively.

We calculate τ_{II} and τ_{AR} as functions of T_c and n_c for $In_{0.53}Ga_{0.47}As$ ($\varepsilon_g = 0.73$ eV), considering the three major processes depicted in Fig. 8.16 [23]. The transition probability of the II-CCCH process, R_{II}(CCCH), is expressed as follows [34],

$$R_{II}(\text{CCCH}) = R_0(\text{CCCH}) \int d\mathbf{k}_1\, d\mathbf{k}_2\, d\mathbf{k}_3\, d\mathbf{k}_4\, |M|^2 \times f_e(\mathbf{k}_1)\,(1 - f_e(\mathbf{k}_2))\,(1 - f_h(\mathbf{k}_3))\,(1 - f_e(\mathbf{k}_4)) \times \delta(\mathbf{k}_1 - \mathbf{k}_2 + \mathbf{k}_3 - \mathbf{k}_4)\,\delta(\varepsilon_1 - \varepsilon_2 + \varepsilon_3 - \varepsilon_4), \quad (8.15)$$

where f_e and f_h are the Fermi–Dirac distribution functions of electrons in the CB and holes in the VB, respectively, $\mathbf{k}$ and ε denote momentum and energy of the carriers, respectively. The matrix element of the II process, M, is approximated as follows [35],

$$|M|^2 = |M_C|^2 + |M_{EX}|^2 + |M_C - M_{EX}|^2, \tag{8.16}$$

$$|M_C|^2 = |\mathbf{k}_1 - \mathbf{k}_2|^2 \Big/ \left(|\mathbf{k}_1 - \mathbf{k}_2|^2 + \lambda^2 \right), \tag{8.17}$$

$$|M_{EX}|^2 = |\mathbf{k}_1 - \mathbf{k}_4|^2 \Big/ \left(|\mathbf{k}_1 - \mathbf{k}_4|^2 + \lambda^2 \right), \tag{8.18}$$

$$\lambda^2 = n_c\, q^2 / \varepsilon_0\, k_B\, T_c, \tag{8.19}$$

where M_C and M_{EX} denote the matrix elements of the Coulomb interaction and exchange interaction, respectively, λ the screen length derived using the Thomas–Fermi approximation, ε_0 the dielectric constant in vacuum. The expressions of the CHHL and CHHS processes are similarly derived.

The II lifetime τ_{II} is inversely proportional to R_{II},

$$\begin{aligned} \tau_{II} &= n_c / R_{II} \\ &= n_c\, (R_{II}(\text{CCCH}) + R_{II}(\text{CHHL}) + R_{II}(\text{CHHS}))^{-1}. \end{aligned} \tag{8.20}$$

AR is the inverse process of II; therefore, the AR lifetime τ_{AR} is similarly expressed. The proportionality constant R_0(CCCH), etc. were determined so that the calculated results agree with experimental values [36–38].

The calculated results shown in Fig. 8.17 manifest that τ_{II} is determined by T_c with being scarcely dependent on n_c, whereas n_c is the predominant factor of τ_{AR}. The ratios of τ_{II}/τ_{re} and τ_{AR}/τ_{re} are sufficiently larger than unity in the region of $T_c < 1{,}800$ K and $n_c < 10^{18}$ cm^{-1} when τ_{re} equals 100 ps assuming $\tau_{th} = 1$ ns. This means that the effects of II and AR are negligibly small and hence the PC model can be applied under the present conditions with finite τ_{th} values. This is exactly the reason why HC-SCs are categorized in the row "particle number conserved" in Fig. 8.1, although a slight correction for the PC model would be required depending on operating conditions [39–41].

Würfel et al. pointed out that T_c calculated using the PC model depends on ΔE too sensitively, which seems physically meaningless, under the ideal condition of full concentration and no thermalization [42]. They concluded that II and AR are frequent and hence the particle conservation breaks, when T_c is close to or higher than ε_g/k_B. However, it has been revealed that the IA model, which assumes extremely frequent II and AR resulting in $\Delta\mu = 0$, does not stand either, under the ideal condition [23].

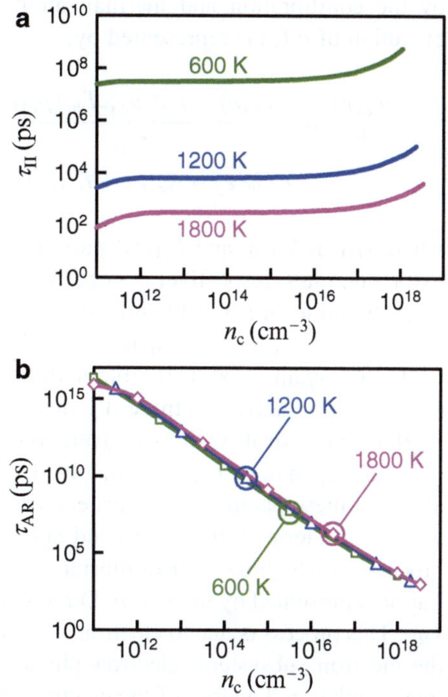

Fig. 8.17 Dependence of (**a**) impact ionization lifetime τ_{II}, (**b**) Auger recombination lifetime τ_{II} on the carrier density n_c and carrier temperature T_c (600, 1,200, 1,800 K). Reprinted with permission from [23]. © 2009, Elsevier B.V.

8.4 Requisites for High Conversion Efficiency Competitive to Existing Solar Cells

8.4.1 A Rate Equation Model

For quantitative evaluation of the three points unique to hot-carrier extraction: the equilibration time τ_{eq} and thermalization time τ_{th} in the absorber, and the energy-selection widths of the ESCs w_{esc}, we formulate a rate equation to describe the energy distribution of electrons in the CB per unit area $n_e(\varepsilon)$ [43]. Holes in the VB are assumed to be separated from the electron subsystem so that holes scarcely interact with electrons (see Sect. 8.5.1.2) and to be immediately thermalized obeying the Fermi–Dirac statistics at T_{RT} for simplification, because most of the absorbed photon energies are stored in the electron subsystem and consequently thermalization of holes has a slight effect on the conversion efficiency when m_h is sufficiently larger than m_e [21].

In a steady-state operation of an HC-SC, electrons in the CB are constantly generated by photoabsorption. Some of these electrons radiatively recombine with holes and the remaining are extracted to the electrode. In addition, $n_e(\varepsilon)$ is affected by the equilibration and thermalization. Considering these processes, the time evolution of $n_e(\varepsilon)$ is represented by,

$$\frac{\partial n_e(\varepsilon)}{\partial t} = -\frac{n_e(\varepsilon) - f(\varepsilon;\mu_{eq},T_{eq})\,\rho_e(\varepsilon)}{\tau_{eq}} - \frac{n_e(\varepsilon) - f(\varepsilon;\mu_{th},T_{RT})\,\rho_e(\varepsilon)}{\tau_{th}} + G_e(\varepsilon) - R_e(\varepsilon) - E_e(\varepsilon;V_e), \quad (8.21)$$

where $G_e(\varepsilon)$, $R_e(\varepsilon)$, and $E_e(\varepsilon;V_e)$ are the generation, radiative recombination, and extraction rates, respectively, $f(\varepsilon;\mu,T)$ the Fermi–Dirac distribution function with a quasi-Fermi level μ and a temperature T, and $\rho_e(\varepsilon)$ the DOS per unit area in the CB (not per unit volume). Impact ionization and Auger recombination have been confirmed again to be negligible under the conditions employed here. The Fermi level in the negative electrode, V_e, directly affects $E(\varepsilon;V_e)$ [see Eq. (8.27)].

Here, τ_{th} is dealt with as a given constant, as stated in Sect. 8.3.1. The dependence of τ_{eq} on n_c or T_{eq} is not involved, either, for a straightforward interpretation, although these parameters intricately affect τ_{eq} [44].

The first term in the right-hand side of Eq. (8.21) describes the equilibration process, in which the nonequilibrium $n_e(\varepsilon)$ evolves toward the equilibrated distribution represented by the Fermi–Dirac statistics with μ_{eq} and T_{eq} that is higher than T_{RT}. This process is due to electron–electron elastic scattering (equilibration within the electron subsystem; electron–phonon scattering is excluded). Therefore, the total number and energy of the electrons are conserved,

$$\int_{\varepsilon_{CBM}}^{\infty} d\varepsilon f(\varepsilon;\mu_{eq},T_{eq})\,\rho_e(\varepsilon) = \int_{\varepsilon_{CBM}}^{\infty} d\varepsilon\; n_e(\varepsilon), \quad (8.22)$$

$$\int_{\varepsilon_{CBM}}^{\infty} d\varepsilon\, \varepsilon f(\varepsilon;\mu_{eq},T_{eq})\,\rho_e(\varepsilon) = \int_{\varepsilon_{CBM}}^{\infty} d\varepsilon\; \varepsilon\, n_e(\varepsilon), \quad (8.23)$$

where ε_{CBM} is the conduction-band minimum in the absorber. These relationships determine the values of μ_{eq} and T_{eq}.

In contrast, the thermalization process described by the second term originates from inelastic scattering of electrons with phonons at T_{RT} (in other words, equilibration between electrons and phonons). Therefore, $n_e(\varepsilon)$ evolves toward the distribution at T_{RT} and the total electron energy dissipates in this process, although conservation of the total electron number holds,

$$\int_{\varepsilon_{CBM}}^{\infty} d\varepsilon f(\varepsilon;\mu_{th},T_{RT})\,\rho_e(\varepsilon) = \int_{\varepsilon_{CBM}}^{\infty} d\varepsilon\; n_e(\varepsilon). \quad (8.24)$$

A parabolic two-band model is again used to describe $\rho_e(\varepsilon)$, $G_e(\varepsilon)$, and $R_e(\varepsilon)$. An electron with an energy ε is generated by absorbing a photon with an energy $\hbar\omega = \varepsilon_g + (1 + m_e/m_h)\,\varepsilon$. Therefore, $G_e(\varepsilon)$ is expressed as,

$$G(\varepsilon) = j_{sun}(\varepsilon_g + (1 + m_e/m_h)\,\varepsilon)\,. \quad (8.25)$$

The generalized Planck's law [24] can be modified to determine $R_e(\varepsilon)$ using the occupation probability in the CB $n_e(\varepsilon)/\rho_e(\varepsilon)$, instead of the Fermi–Dirac distribution function,

$$R_e(\varepsilon) = \frac{2\pi}{h^3 c^2} \frac{\left(\varepsilon_g + (1 + m_e/m_h)\,\varepsilon\right)^2}{(\rho_e(\varepsilon)/n_e(\varepsilon) - 1)\,\exp\left[\left(\varepsilon_g + m_e/m_h\,\varepsilon + \mu_h\right)/(k_B\,T_{RT})\right] - 1}. \tag{8.26}$$

When a resonant tunneling diode (RTD) consisting of semiconductor quantum dots (QDs) imbedded in a barrier material is used as the ESC, electrons of a specific energy in the absorber can be extracted via the quantum-confined states in the QDs. However, in reality, size nonuniformity of the QDs causes a finite value of w_{esc}. Assuming that the distribution of the quantum-confined levels is represented by a Gauss function centered at E_{esc}, $E_e(\varepsilon;V_e)$ is derived using the standard expression of a tunneling current [45],

$$E_e(\varepsilon; V_e) = A\,\frac{2}{\pi\,w_{esc}}\,\exp\left[-\left(\frac{\varepsilon - E_{esc}}{w_{esc}/2}\right)^2\right]\varepsilon\,(n_e(\varepsilon)/\rho_e(\varepsilon) - f(\varepsilon; V_e, T_{RT})), \tag{8.27}$$

where $f(\varepsilon;V_e,T_{RT})$ is the Fermi–Dirac distribution function in the negative electrode. The proportionality constant A is dependent on the detailed structure of the ESC, such as the areal density of the QDs, and the thickness and potential height of the barriers, as well as the QD and barrier materials [46, 47].

The total extraction rate equals the difference between the total generation and recombination rates,

$$\int_{\varepsilon_{CBM}}^{\infty} d\varepsilon\,E_e(\varepsilon; V_e) = \int_{\varepsilon_{CBM}}^{\infty} d\varepsilon\,G_e(\varepsilon) - \int_{\varepsilon_{CBM}}^{\infty} d\varepsilon\,R_e(\varepsilon). \tag{8.28}$$

Therefore, it is an additional requisite for high conversion efficiency to realize an A value that satisfies Eq. (8.28). The total energy-extraction rate is affected by the energy dissipation due to the thermalization,

$$\begin{aligned}\int_{\varepsilon_{CBM}}^{\infty} d\varepsilon\,\varepsilon\,E_e(\varepsilon; V_e) &= \int_{\varepsilon_{CBM}}^{\infty} d\varepsilon\,\varepsilon\,G_e(\varepsilon) - \int_{\varepsilon_{CBM}}^{\infty} d\varepsilon\,\varepsilon\,R_e(\varepsilon)\\ &\quad - \int_{\varepsilon_{CBM}}^{\infty} d\varepsilon\,\varepsilon\,\frac{n_e(\varepsilon) - f(\varepsilon; \mu_{th}, T_{RT})\,\rho_e(\varepsilon)}{\tau_{th}}.\end{aligned} \tag{8.29}$$

By solving the simultaneous equations of $\partial n_e(\varepsilon)/\partial t = 0$ and Eqs. (8.28) and (8.29), $n_e(\varepsilon)$ and other values in a steady-state operation are determined. The Fermi level in the positive electrode V_h equals μ_h, because the hole temperature is assumed

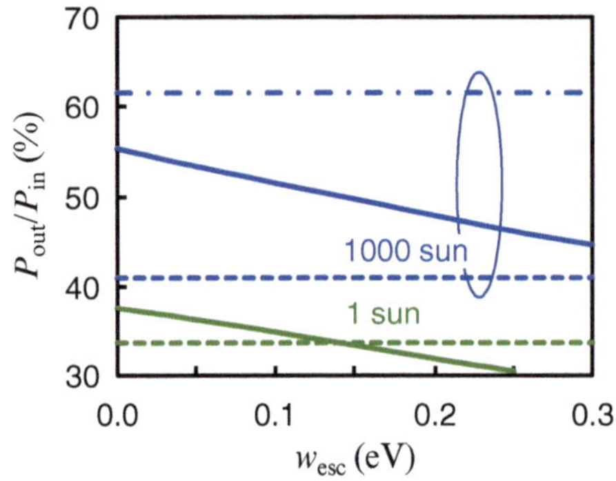

Fig. 8.18 Conversion efficiency P_{out}/P_{in} depending on the energy-selection with of the ESCs w_{esc}. τ_{eq} (ps) $= 10\ w_{esc}$ (eV), $\tau_{th} = 1$ ns. ε_g, E_{esc} and other parameters are optimized for each case. *Broken lines* and a *dashed–dotted line* are the results for 1J- and 3J-SCs, respectively. Reprinted with permission from [43]. © 2010, The Japan Society of Applied Physics

to be T_{RT}. Thus, the conversion efficiency η is the product of $V_e - V_h$ and the total extraction rate divided by the total incident power,

$$\eta = (V_e - V_h) \int_{\varepsilon_{CBM}}^{\infty} d\varepsilon\, E_e(\varepsilon; V_e) \bigg/ \int_0^{\infty} d\varepsilon'\, \varepsilon'\, j_{sun}(\varepsilon')$$
$$\equiv P_{out}/P_{in}. \tag{8.30}$$

In the following numerical procedures, $m_e = 0.04\,m_0$, $m_h = 0.4\,m_0$, $d = 500$ nm were again used. The AM1.5G solar spectrum was used up to the 10 sun solar irradiation intensity, and AM1.5D for higher intensity.

8.4.2 Quantitative Answer to FAQ 1: Energy-Selection Width of ESCs w_{esc}

In Sect. 8.3, the energy-selection width of the ESCs w_{esc} was assumed to be extremely narrow, which ensures isoentropic extraction of hot carriers [27]. Figure 8.18 shows the effect of increasing w_{esc}, where τ_{eq} is set to be sufficiently short depending on the w_{esc} value so that it scarcely affects P_{out}/P_{in} (see Sect. 8.4.3). When w_{esc} has a finite value, additional entropy generation lowers P_{out}/P_{in}. Another interpretation is that the influence of the reverse flow of electrons from the electrode to the absorber represented by a negative value of $n_e(\varepsilon)/\rho_e(\varepsilon) - f(\varepsilon; V_e, T_{RT})$ in Eq. (8.27) at a lower ε is more significant when w_{esc} is wider [48]. If the target of the conversion efficiency is relatively 5 % lower than the value at $w_{esc} = 0$, a w_{esc} around 0.1 eV is required. This is the quantitative answer to the FAQ 1 on w_{esc}.

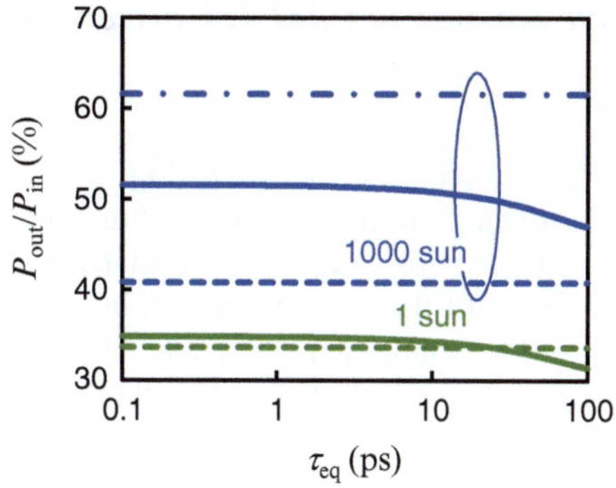

Fig. 8.19 Conversion efficiency P_{out}/P_{in} depending on the carrier equilibration time τ_{eq}. $\tau_{th} = 1$ ns, $w_{esc} = 0.1$ eV. ε_g, E_{esc} and other parameters are optimized for each case. *Broken lines* and a *dashed–dotted line* are the results for 1J- and 3J-SCs, respectively. Reprinted with permission from [43]. © 2010, The Japan Society of Applied Physics

8.4.3 Quantitative Answer to FAQ 1: Carrier Equilibration Time τ_{eq}

Figure 8.19 shows the dependence of the conversion efficiency P_{out}/P_{in} on the carrier equilibration time τ_{eq} at $\tau_{th} = 1$ ns, where w_{esc} was set to be 0.1 eV according to the conclusion in Sect. 8.4.2. The profiles of $n_e(\varepsilon)$ is illustrated in Fig. 8.20. As long as electrons are sufficiently supplied into the empty states located around the ESC energy level E_{esc}, τ_{eq} does not have a significant effect on P_{out}/P_{in}. When τ_{eq} equals 1 ps, although a slight dip appears in $n_e(\varepsilon)$ at E_{esc}, it scarcely affects P_{out}/P_{in}. In the energy range lower than E_{esc}, $n_e(\varepsilon)$ almost overlaps with the equilibrated distribution represented by the product of $f(\varepsilon;\mu_{eq},T_{eq})$ and $\rho_e(\varepsilon)$. On the other hand, $n_e(\varepsilon)$ is remarkably different from the equilibrated distribution in the higher energy range, because the carrier generation rate $G_e(\varepsilon)$ corresponding to the solar irradiation temperature around 6,000 K is extremely different from the Fermi–Dirac distribution function at T_{eq} being around 1,600–1,700 K. However, the difference scarcely affects P_{out}/P_{in}. Note again that the optimal T_c remarkably lower than the solar irradiation temperature is not due to the thermalization of carriers but the results of the balance between the preexisting, generated, and extracted carrier energies.

In contrast, at $\tau_{eq} = 10$ ps, $n_e(\varepsilon)$ exhibits a considerable dip at E_{esc}, i.e., a low occupation probability $n_e(\varepsilon)/\rho_e(\varepsilon)$, with small bumps on the both sides of the dip. To prevent the reverse flow of electrons from the electrode to the absorber, the occupation probability $f(E_{esc};V_e,T_{RT})$ in the electrode must also be low, leading to lowering in V_e and consequently P_{out}/P_{in}. The target value of τ_{eq} to eliminate the dip formation is in the order of 1/1,000 of τ_{th} (e.g., several picoseconds at $\tau_{th} = 1$ ns), although the precise value should depend on w_{esc} and E_{esc}. This is the quantitative answer to the FAQ 1 on τ_{eq}.

8.4.4 Quantitative Answer to FAQ 4: Carrier Thermalization Time τ_{th}

The dependence of P_{out}/P_{in} on τ_{th} is shown in Fig. 8.21. The difference from Fig. 8.14a in Sect. 8.3.7 mainly originates from the finite w_{esc} value being 0.1 eV, because the τ_{eq} was set to be 1 ps = τ_{th}/1,000 so that it scarcely affect P_{out}/P_{in}. Figure 8.22 shows the effect of the solar irradiation intensity on P_{out}/P_{in} at τ_{th} = 1ns, with those for 1J-, 2J- (double-junction), and 3J-SCs for comparison. The results are similar to those shown in Fig. 8.15 but the P_{out}/P_{in} values are lowered due to the finite w_{esc}.

A τ_{th} of several nanoseconds is required to compete against 1J-SCs when HC-SCs are used under the 1 sun solar irradiation. The efficiency cannot reach the value of 3J-SCs at 1,000 sun even if τ_{th} = 10 ns is assumed, although it can exceed the SQ limit at τ_{th} = 0.1 ps. Note again that the target value of the

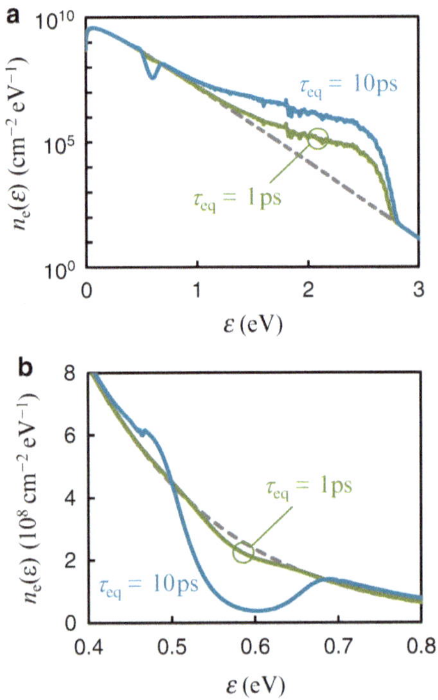

Fig. 8.20 Carrier energy distribution $n_e(\varepsilon)$ depending on the carrier equilibration time τ_{eq}. τ_{th} = 1 ns, w_{esc} = 0.1 eV. ε_g, E_{esc} and other parameters are optimized for each case. (**a**), (**b**) 1 sun, ε_g = 1.12 eV, E_{esc} = 0.58 eV above the CBM, T_{eq} = 1,590 K. (**c**), (**d**) 1,000 sun, ε_g = 0.93 eV, E_{esc} = 0.61 eV above the CBM, T_{eq} = 1,700 K. *Broken lines* show the equilibrated distribution $f(\varepsilon;\mu_{eq},T_{eq})\,\rho_e(\varepsilon)$

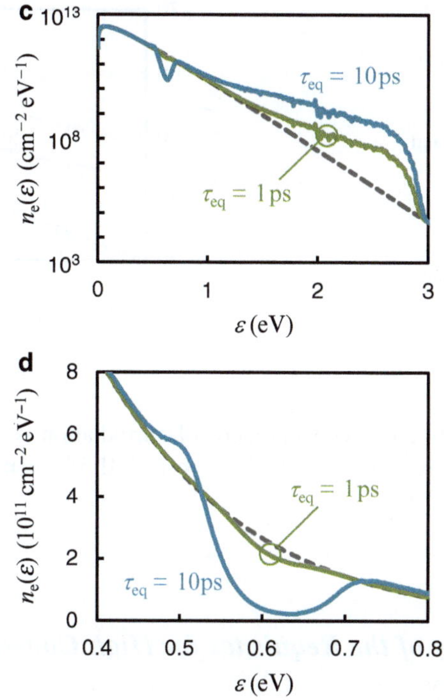

Fig. 8.20 (continued)

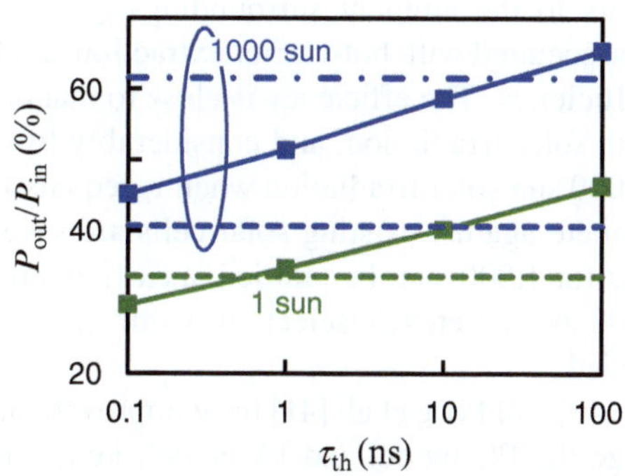

Fig. 8.21 Conversion efficiency P_{out}/P_{in} depending on the carrier thermalization time τ_{th}. $\tau_{eq} = \tau_{th}/1{,}000$, $w_{esc} = 0.1$ eV. ε_g, E_{esc} and other parameters are optimized for each case. *Broken lines* and a *dashed–dotted line* are the results for 1J- and 3J-SCs, respectively

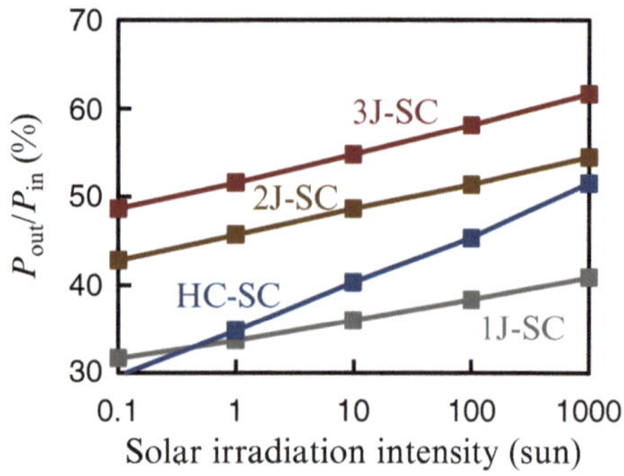

Fig. 8.22 Dependence of conversion efficiency P_{out}/P_{in} on the solar irradiation intensity. $\tau_{eq} = 1$ ps, $\tau_{th} = 1$ ns, $w_{esc} = 0.1$ eV. ε_g, E_{esc} and other parameters are optimized for each case. Results for 1J-, 2J- (double-junction) and 3J-SCs are also shown for comparison

conversion efficiency under concentrated solar irradiation is not the SQ limit, but the efficiency of 3J-SCs, because the high cost of 3J-SCs is not a serious issue, as discussed in Sect. 8.3.7.

8.4.5 Summary of the Requisites for High Conversion Efficiency

We have evaluated the effects of the three points unique to hot-carrier extraction. The energy dissipation due to the thermalization of photogenerated carriers U_{th} can be significantly reduced, even though the thermalization time τ_{th} is not sufficiently long. Instead, another energy dissipation process, thermodynamically unavoidable heat flux to the ambient surroundings Q, which is related to the entropy generation associated with hot-carrier extraction ΔS, has a serious impact on the conversion efficiency. The efficiency is close to that of 1J-SCs, i.e., the SQ limit under the 1 sun solar irradiation, and considerably lower than the value of 3J-SCs under the 1,000 sun solar irradiation when τ_{th} equals 1 ns. Thus, the target values of τ_{th} to compete against existing solar cells are several nanoseconds for 1 sun and over 10 ns for 1,000 sun. In addition, a carrier equilibration time τ_{eq} in the order of $\tau_{th}/1{,}000$ and an energy-selection width w_{esc} of the ESCs narrower than 0.1 eV are needed.

Aliberti et al. [39, 40] and Feng et al. [41] have improved the present model used in Sect. 8.3 to bridge the PC model and IA model, i.e., to involve the effect of impact ionization and Auger recombination quantitatively, assuming immediate equilibration, and applied to InN absorbers. Le Bris et al. have proposed another model of the energy dissipation due to the thermalization to involve the dependence of the thermalization rate on the carrier temperature T_c [49, 50]. These analyses have drawn similar conclusions.

8.5 Quantum Dots and Other Candidate Materials for HC-SCs

I will give an overview of current status of development of materials used in HC-SCs and summarize the issues to realize the target values of the carrier equilibration time τ_{eq} and thermalization time τ_{th} in the absorbers, and the energy-selection width of the ESCs w_{esc} for high conversion efficiency.

8.5.1 Materials for Absorbers

8.5.1.1 Carrier Equilibration in Bulk Materials

First, let us consider τ_{eq}. Transient optical spectroscopy has been carried out to evaluate τ_{eq} in GaAs films. It has been found that τ_{eq} is a decreasing function of n_c; proportional to $n_c^{-1/2}$ in the n_c range larger than 10^{14} cm^{-3}, being shorter than 1 ps at $n_c > 10^{15}$ cm^{-3} [51–53]. The $n_c^{-1/2}$ dependence has also been theoretically derived [44]. In contrast, τ_{eq} scarcely depends on n_c being several tens picoseconds at $n_c \sim 10^{12}$–10^{14} cm^{-3} [52].

The target value of τ_{th} is over 10 ns corresponding to $n_c \sim 10^{16}$ cm^{-3}, when HC-SCs are used under the 1,000 sun solar irradiation. For high conversion efficiency, $\tau_{eq} \sim \tau_{th}/1{,}000 = 10$ ps is required. The measured values of $\tau_{eq} \sim 1$ ps at $n_c \sim 10^{15}$ cm^{-3} suggest that the requisite of τ_{eq} can be fulfilled at 1,000 sun, if τ_{th} over 10 ns is achieved.

On the other hand, n_c is in the order of 10^{12} cm^{-3} under the 1 sun solar irradiation, when τ_{th} is several nanoseconds. Considering the previous experimental results described above, it would be difficult to realize the target value of $\tau_{eq} = \tau_{th}/1{,}000 \sim 1$ ps at 1 sun.

Kirk and Fischetti employed a Monte Carlo method to deal with carrier–carrier scattering and carrier–phonon scattering in GaAs absorbers and highlighted the serious impact of insufficient carrier equilibration when τ_{th} is not sufficiently long and hence n_c is moderate [54].

8.5.1.2 Quantum Dot Materials with Phonon-Bottleneck Effects

Among the three points unique to hot-carrier extraction, the most essential one seems a long τ_{th}. The photogenerated carriers emit optical phonons, followed by acoustic phonon generation. Namely, the carrier energies dissipate to optical phonons, and then to acoustic phonons. In usual bulk materials, τ_{th} is several picoseconds, which is far from the target values being in the order of 1–10 ns. To realize such a long τ_{th}, at least, one of the sequential energy dissipation processes must be suppressed.

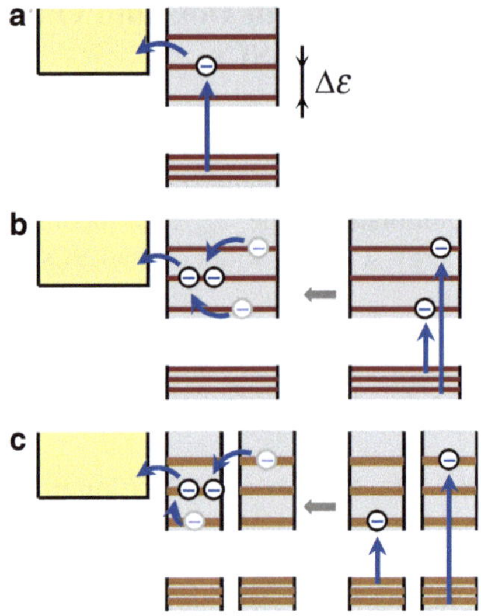

Fig. 8.23 (**a**) Hot-carrier extraction from a single QD. Carrier equilibration processes in (**b**) a single QD and (**c**) electronically coupled QDs, followed by hot-carrier extraction

Semiconductor quantum dots (QDs) attract much attention for the absorber materials, because the first energy dissipation, i.e., from photogenerated carriers to optical phonons can be suppressed due to the phonon-bottleneck effect [55, 56]. When the separation of neighboring energy levels in a QD ($\Delta\varepsilon$ in Fig. 8.23a) is wider than the optical phonon energy, a carrier located at a high-energy level cannot relax to the next lower level by emitting a single optical phonon. In particular, long lifetimes of carriers at high-energy levels in the order of 100 ps have been demonstrated for InAs and $In_xGa_{1-x}As$ QDs deposited on GaAs layers using the Stranski–Krastanov thin-film growth mode [57–59].

In most semiconductors, holes have larger effective masses than electrons, and consequently $\Delta\varepsilon$ for holes would not be sufficiently wide. Therefore, electron energies transfer to holes via electron–hole elastic scattering and dissipate immediately [60]. Spatial separation between electrons and holes can suppress the energy dissipation. Urayama et al. compared the results for resonant excitation (an electron–hole pair is generated and confined in a single QD) and non-resonant excitation (an electron and a hole generated in the matrix are captured in different QDs), and observed τ_{th} being 600 ps for the latter that is extremely longer than that for the former [61].

In a QD material with a type II band alignment, electrons are confined in the QDs whereas holes are itinerant in the matrix, i.e., they are spatially separated from each other. When thin potential walls, through which holes generated in the QDs can escape into the matrix by tunneling, are inserted between the QDs and matrix, electrons and holes are farther separated, leading to less interaction between them [62, 63]. Pandy et al. designed CdSe/ZnS/ZnSe/CdSe concentric QDs with organic ligands on the QD surfaces to trap the holes [64]. They realized a τ_{th} as long as 1.7 ns at the room temperature.

However, there are other studies reporting short τ_{th} values close to those in bulk materials due to strong carrier–phonon coupling referred to as polaron formation [65, 66] and other mechanisms [67, and references therein]. The requisite for the presence of the phonon-bottleneck effect and a resultant long τ_{th} has not yet been clarified.

The step next to a long τ_{th} is extraction of hot carriers. Tisdale et al. demonstrated extraction of hot electrons generated in PbSe QDs into TiO_2 electrodes on which the QDs are deposited [68]. Although the QD size was designed so that the lowest energy levels in the QDs were located blow the conduction-band minimum of the TiO_2 electrodes as shown in Fig. 8.23a and electrons remained in the QDs if hot carriers were immediately thermalized, i.e., relaxed to the lowest energy levels, electrons were successfully extracted.

However, for all the photogenerated carriers to be extracted, the equilibration is essential, as quantified in Sect. 8.4.3. For the equilibration, existence of plural carriers in a single QD (Fig. 8.23b), or strong electronic coupling between neighboring QDs (Fig. 8.23c) is required. Considering absorption coefficients of usual materials and solar irradiation intensity practically up to 1,000 sun [30, 31], the former seems not feasible. Or, it is not clear for the latter whether a short τ_{eq} close to the values in bulk materials is possible. Note that the τ_{eq} values even in bulk materials would not be sufficiently long at 1 sun. In addition, inter-QD coupling forms minibands in the QD array system, which decreases the energy separation between the neighboring minibands [69] and accelerates the thermalization of carriers [70].

Other concerns on QDs as the absorber materials are carrier transport properties. Inter-QD coupling promotes the carrier transport, as well as enhances the carrier equilibration, but can lead to rapid carrier thermalization as stated above. Thus, it has not yet been elucidated whether all the requisites for the absorbers can be fulfilled compatibly with each other using QD materials.

8.5.1.3 Bulk Materials with Wide Phononic Bandgaps

Another way to realize a long τ_{th} is to suppress the second energy dissipation from optical phonons to acoustic phonons, namely, to achieve hot optical phonons, so that the phonon energy can transfer back to carriers [71]. The fastest one among the decay mechanisms of optical phonons is illustrated in Fig. 8.24 referred to as the Klemens decay, in which a longitudinal optical (LO) phonon located at the Brillouin zone center generates two longitudinal acoustic (LA) phonons with half the energy and opposite momenta [72]. Therefore, when the phononic bandgap between the optical modes and acoustic modes is wide enough so that no LA phonon states exist at the half of the LO phonon energy, the Klemens decay is inhibited [73, 74]. The next most likely mechanism is the Ridley decay also depicted in Fig. 8.24, in which an LO phonon splits into a transverse optical (TO) phonon and an LA phonon [75]. This also could be suppressed by the wide phononic bandgap [76]. Materials with highly symmetric crystal structures and resultant narrow bandwidths of optical modes are more suitable to suppress the Ridley decay [73, 74].

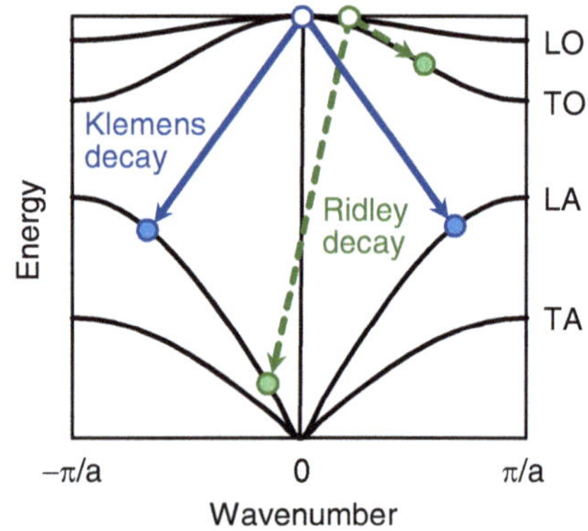

Fig. 8.24 Schematic phononic dispersion, and the Klemens and Ridley decay mechanisms

InN has a sufficiently wide phononic bandgap [77]. Chen et al. demonstrated τ_{th} values as long as 300–400 ps at the room temperature [78]. However, other reported values are considerably scattered ranging from sub-100 fs to around 100 ps [79–81], which would be due to the difference of quality (defects, impurities, etc.) between the samples.

The wide phononic bandgaps are formed for compound materials consisting of heavy and light elements, like InN. Therefore, other nitride semiconductors are also the candidates. Clady et al. compared InP and GaAs that have similar electronic band structures but different phononic properties [82]. Although InP has a phononic band gap not sufficiently wide to completely inhibit the Klemens decay, it exhibited a longer τ_{th} than GaAs in which the atomic masses of the two constituent elements are very close to each other.

In addition, the strength of electron–phonon interaction characterized by the Fröhlich constants is also a criterion for the material choice [71, 74]. Covalent materials such as SnC and GeC [83], and pseudo-covalent ones like BBi and BiP [74] could realize longer τ_{th} values than polar materials.

8.5.1.4 Phononic Engineered Materials

Similar to the zone-folding effect on electronic states in MQWs, phononic dispersion can also be modified using nanostructures consisting of two materials with different acoustic impedance, resulting in the appearance of new bandgaps at the Brillouin zone edge of the nanostructures. The idea of "phononic engineering" to design a nanostructure so that the new phononic bandgap is located at the half of the optical phonon energy has been proposed by Conibeer et al. to inhibit the Klemens decay [73]. Interface layers much softer or stiffer than the two constituent materials can further widen the phononic bandgap.

However, periodicity as small as 1–2 nm is needed for a sufficiently wide new phononic bandgap [84]. Therefore, fabrication of suitable phononic-engineered materials of practically large sizes seems extremely challenging. Using mesoporous materials with other materials introduced in the mesopores would be a way to fabricate.

8.5.1.5 Phonon Confinement Effects in MQWs

Long τ_{th} values were also demonstrated in MQWs of GaAs/AlGaAs by transient optical spectroscopy [25]. Carrier temperatures higher than the room temperature under intense excitation were also realized in GaInAsSb/AlGaAsSb MQWs and InGaAs/GaAsP QWs [50, 85]. Rodière et al. fabricated GaInAsSb/GaSb and GaInAsP/InP MQW-based devices and demonstrated to control the carrier temperature by applied voltage (see Sect. 8.3.6) [86].

However, the origin of the suppressed thermalization in MQWs has not yet been clarified. A possible reason is the phonon confinement effect due to acoustic impedance mismatch, i.e., few overlap between the phonon modes of the well and barrier materials, as well as confinement of hot carriers [83].

8.5.2 Materials for ESCs

8.5.2.1 RTDs Using Quantum Dots

RTDs using QDs are the first choice for the ESCs. König et al. demonstrated hot-carrier extraction from an Si absorber through an RTD consisting of Si QDs imbedded in an SiO_2 matrix [87]. They observed a significant increase in the tunneling current under Xe-lamp illumination compared with that in the dark, reflecting the hot-carrier distribution in the absorber.

A discrete energy level in a QD can realize an extremely narrow energy-selection width w_{esc}. In particular, a sufficiently narrow peak in a current (I)–voltage (V) relationship has been realized using an RTD consisting of InAs QDs fabricated using the Stranski–Krastanov thin-film growth mode with electrodes of several square micrometers in size [88]. However, in reality, size nonuniformity of the QDs causes widespread energy levels and consequently widens w_{esc}. Therefore, it still remains challenging to realize a narrow w_{esc} using practically large sized devices.

In addition, a low areal density of the QDs would limit the conductance of the ESCs. On the other hand, a dense QD array with strong inter-QD coupling form minibands and widens w_{esc} [69]. The value of the areal density needed for the suitable conductance has not yet been elucidated [46, 47].

8.5.2.2 RTDs Using Quantum Wells

Another candidate for the ESCs is an RTD using quantum wells (QWs) [89]. Yagi et al. demonstrated hot-carrier extraction from a GaAs absorber through an RTD consisting of a GaAs/AlGaAs QW [90]. They observed shifts of the thresholds and peaks in the I–V relationships depending on the photon energy of monochromatic illumination, i.e., the average energy of hot carriers generated in the absorber.

Well-established fabrication technologies of III–V compound semiconductor thin layers enable structural uniformity. In addition, a tunneling current density as large as 50 A/cm^2, which is close to J_{out} under the 1,000 sun irradiation at $\varepsilon_g = 1$ eV, has been demonstrated under the resonance condition for a GaAs/AlGaAs QW-RTD [91].

However, carriers whose energies are higher than a certain threshold are extracted through a QW-RTD due to the step-like DOS of a 2-dimensional structure, which is contrasting to a QD-RTD. This energy selectivity is equivalent to a wide w_{esc} for a QD-RTD. It is practically an important subject which are more promising, QD-RTDs with considerable QD size nonuniformity or QW-RTDs with uniform thickness [92].

8.5.2.3 An Open Question: Conductance of ESCs

As previously stated, carriers must be rapidly extracted but not too rapidly. From a perspective of tunneling transport, the carrier extraction rate is proportional to the conductance of the ESCs and the A value in Eq. (8.27). The results in Sects. 8.3 and 8.4 are obtained based on the existence of the suitable conductance values.

The conductance of QD- and QW-RTDs remarkably depends on the band offset of materials, QD size or QW thickness, barrier thickness, etc. [46, 47]. It is an open question whether the suitable value of the conductance is possible compatibly with the ESC energy level ΔE and energy-selection width w_{esc} required for high conversion efficiency.

8.6 Post Conventional HC-SCs

As discussed above, it seems extremely challenging to fulfill all the requisites for high conversion efficiency with conventional HC-SC structures. Recently new concepts to solve some parts of the issues have been proposed.

8.6.1 Hot-Carrier Absorbers with Energy-Selective Carrier Transport

To solve the issues on the ESCs, König et al. have proposed a new concept of hot-carrier absorbers using MQWs [93, 94]. The schematic structure is illustrated in Fig. 8.25. The first miniband in the MQWs determines the absorption edge, whereas the second one provides energy-selective electron transport. This absorber eliminates the needs of the ESCs; only carrier selective contacts (CSCs) as in a conventional solar cell are needed. The CSCs transfer one of the carrier types but block the other, which is a task realized by highly n- and p-doped contact regions in

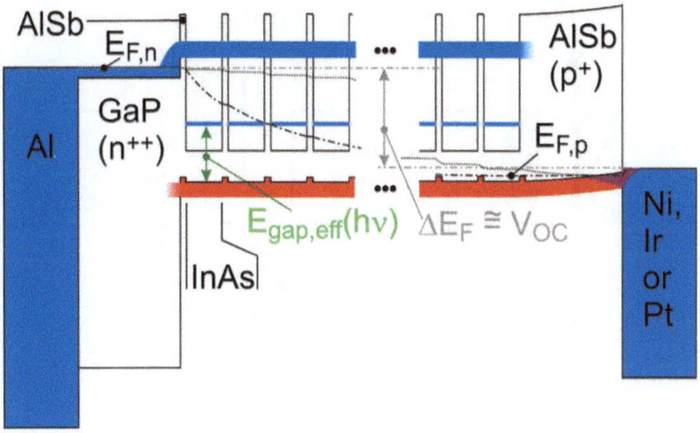

Fig. 8.25 Schematic energy diagram of a hot-carrier solar cell using an MQW absorber. Reprinted with permission from [93]. © 2012, The American Institute of Physics

conventional solar cells. Requisites for the carrier equilibration and light absorption can be fulfilled due to continuous DOS in the lateral direction in the MQWs. The confinement effect on optical phonons is also feasible.

König et al. designed a concrete material combination of InAs/AlSb MQWs with an n^{++}-GaP layer and a p^{+}-AlSb layer for electron- and hole-selective contacts, respectively, considering technological feasibility as well as the suitable energy level alignment. Most of the absorbed photon energies are stored in the electron subsystem due to $m_e \ll m_h$ in InAs [28], so that only hot-electron characteristics should be considered. The low Fröhlich constant of InAs [95] and mismatch of optical phonon bands between InAs and AlAs [28] are suitable features to suppress the carrier thermalization. This structure can be epitaxially grown, because of a small lattice mismatch 1.3 %. They carried out Monte Carlo simulation for electron transport and evaluated robustness against structural fluctuation, suggesting a practical implementation of the new concept.

8.6.2 *Optical HC-SCs*

Another new concept to solve the issues on the ESCs is optical extraction of hot-carrier energies proposed by Farrell et al. [96]. In general, radiative recombination lifetime τ_{rad} ranges from several hundred picoseconds to several nanoseconds, which is significantly longer than τ_{th} (see Fig. 8.2). However, recent developments of plasmonic and optical antenna technologies can greatly accelerate the radiative recombination at resonant energies [97] [98, and references therein]. If the resonant energy, i.e., selection energy is suitably designed and the relationship $\tau_{eq} \ll \tau_{rad} \ll \tau_{th}$ is achieved, hot-carrier energies can be converted into monochromatic emission with slight energy dissipation. The optical ESC coupled with a hot-carrier absorber

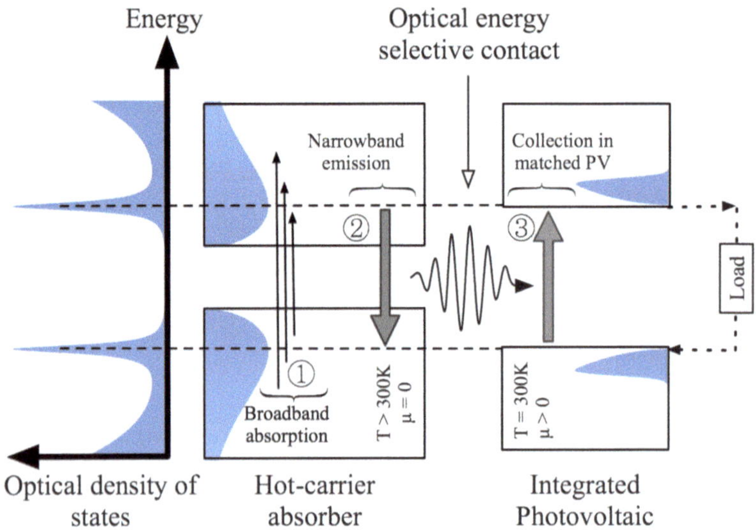

Fig. 8.26 Operating principle of an optical hot-carrier solar cell. Reprinted with permission from [96]. © 2011, The American Institute of Physics

can be used with conventional solar cells, like up- and down-converters, as depicted in Fig. 8.26. Another advantage of this new concept is that it eliminates the need to transport hot carriers through the device, instead hot-carrier energies are partitioned and transported in the enhanced photon DOS in the system.

Farrell et al. carried out detailed-balance calculations and revealed that high conversion efficiency close to that of conventional HC-SCs can be achieved with around 1,000 times-accelerated radiative recombination rate, i.e., ${\tau_{\mathrm{rad}}}^{-1}$, or in other words, 1,000 times-enhanced photon DOS. They employed the impact-Auger model in which particle number is not conserved, as described in Sect. 8.3.8. However, in principle the concept could conserve the particle number depending on the material properties of the absorber and also the choice of the optical ESC energy level. Therefore, optical HC-SCs are categorized in the row "particle number conserved" in Fig. 8.1. The goal of combining a hot-carrier absorber with an optical ESC is to convert the solar spectrum to monochromatic emission with slight energy dissipation. In this respect, this concept is a form of "spectral modification" but without the need for the multiphoton processes which are essential with conventional up- and down-converters.

8.6.3 Intermediate-Band-Assisted HC-SCs

The most serious issue to realize high conversion efficiency by hot-carrier extraction is to achieve a long τ_{th} in the absorber. When τ_{th} is not sufficiently long, the entropy generation ΔS associated with hot-carrier extraction is remarkable, which

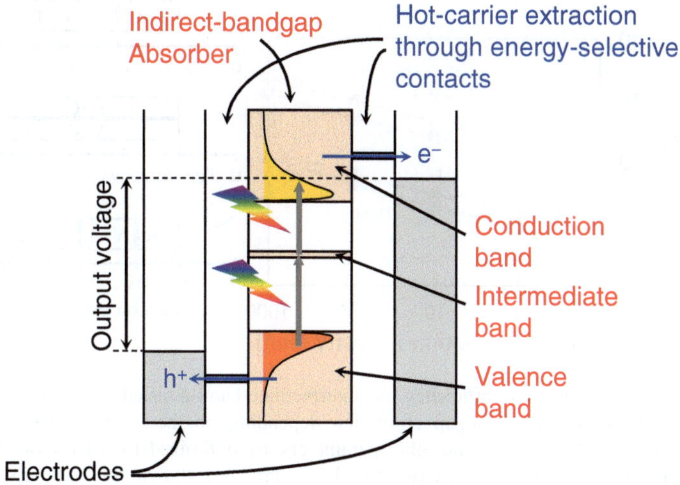

Fig. 8.27 Schematic energy diagram of an intermediate-band-assisted hot-carrier solar cell (IB-HC-SC) and operating principle. Reprinted with permission from [99]. © 2012, John Wiley & Sons, Ltd.

is the dominant energy dissipation process, whereas the energy dissipation due to the thermalization U_{th} is significantly reduced by choosing a suitable ESC energy level ΔE, as shown in Figs. 8.11b, 8.12b, and 8.14b.

To reduce ΔS, it is found from Eq. (8.10) that a decrease in the output current density J_{out} (note that the quasi-Fermi level splitting in the absorber $\Delta\mu$ is negative) and/or an increase in $\Delta\mu$ are needed. We have proposed that the use of an absorber that has an intermediate band (IB) in its fundamental bandgap instead of a single-gap absorber, namely, an intermediate-band-associated HC-SCs (IB-HC-SCs) shown in Fig. 8.27 is promising for this purpose [22, 92].

In an IB absorber, carriers are generated by two-step excitation via the IB and consequently J_{out} is remarkably smaller than that of a single-gap absorber with the same absorption edge (the fundamental bandgap for a single-gap absorber, the smaller one between the CB–IB gap and IB–VB gap for an IB absorber) [7, 8]. In addition, a wide fundamental bandgap of an IB absorber enlarges $\Delta\mu$ at the same n_{c} and T_{c}. The use of an indirect-bandgap IB absorber, in which the optical transition from the VB to the CB is substantially forbidden in the appreciable spectral range and hence all the carriers are generated by two-step excitation via the IB, further decreases J_{out}, resulting in even higher conversion efficiency. Although the carrier energies in excess of the bandgap are larger than those for a direct-bandgap IB absorber, the excess carrier energies are efficiently converted to electricity by hot-carrier extraction.

Another advantage of this new concept is that it eliminates the need of "photon selectivity" that the spectral absorption ranges for the different interband transitions have no overlap with each other [99]. For conventional IB-SCs, when photon selectivity is being broken, carrier energies in excess of the bandgap are increased and

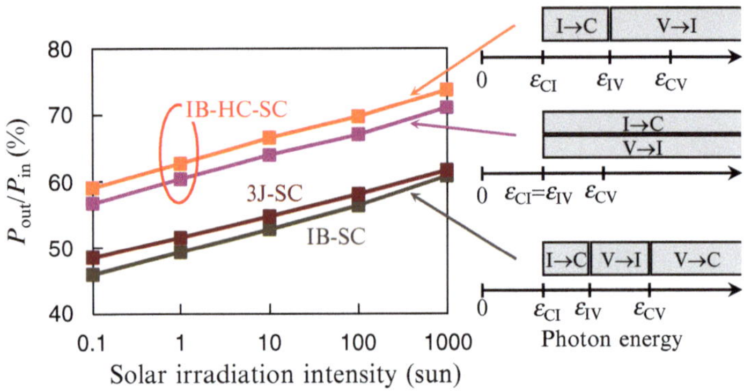

Fig. 8.28 (*Left*) Conversion efficiency of intermediate-band-assisted hot-carrier solar cells (IB-HC-SCs) using indirect-bandgap absorbers, depending on the solar irradiation intensity. $\tau_{th} = 1$ ns. The bandgap values and other parameters are optimized for each case. Results for IB-SCs with perfect photon selectivity and 3J-SCs are also shown for comparison. (*Right*) Absorption ranges for each interband transition. I→C: IB to CB (bandgap ε_{CI}), V→I: VB to IB (bandgap ε_{IV}), V→C: VB to CB (bandgap ε_{CV}). Reproduced with permission from [99]. © 2012, John Wiley & Sons, Ltd.

consequently conversion efficiency is lowered, as stated in the Sect. 8.1 [10, 11]. By extracting hot carriers, the increased excess carrier energy is not an issue. Thus, the advantage and disadvantage of HC-SCs and IB-SCs are complementary.

We have carried out detailed-balance calculations on IB-HC-SCs, using a model with a sufficiently short τ_{eq} and an extremely narrow w_{esc} as is the case in Sect. 8.3. Even though photon selectivity is completely broken, the conversion efficiency assuming $\tau_{th} = 1$ ns is higher than that of conventional IB-SCs with perfect photon selectivity by around 10 % throughout the practical intensity range of 0.1–1,000 sun, as shown in Fig. 8.28. These are contrasting to the results for conventional HC-SCs of slight gain compared with 1J-SCs at 1 sun.

8.7 Concluding Remarks

The concept of HC-SCs has been previously proposed and has attracted increasing attention to achieve high conversion efficiency, as well as a potentially low areal cost owing to the simple structures compared with triple-junction solar cells (3J-SCs). Two detailed-balance models to evaluate limiting conversion efficiency have been established: the particle conservation (PC) model by Ross and Nozik in 1982, and the impact-Auger model by Würfel in 1997. However, both the two models deal with the ideal case excluding thermalization of photogenerated carriers in the absorbers used in HC-SCs. Another concern is the fact that these two models are contrasting; the PC model excludes impact ionization (II) and Auger recombination (AR) whereas the IA model assumes extremely frequent II and AR leading to no quasi-Fermi level splitting in the absorber, while the coverage where each model can be applied has not yet been clarified.

We have constructed new models to evaluate the effect of partial thermalization of photogenerated carriers and revealed that the PC model including the carrier thermalization stands under the practical conditions of finite carrier thermalization times and the solar irradiation intensity up to 1,000 sun.

The impact of a finite thermalization time τ_{th} is serious. The energy dissipation due to the thermalization can be significantly reduced even though τ_{th} is not sufficiently long. Instead, the energy dissipation due to thermodynamically derived unavoidable heat flux to the ambient surroundings, which is related to the entropy generation associated with hot-carrier extraction, is remarkable. The requisite of τ_{th} is several nanoseconds to compete against single-junction solar cells under the 1 sun solar irradiation. A τ_{th} over 10 ns is needed at 1,000 sun for which 3J-SCs are the competitors.

We have further improved the new model to involve the effects of the carrier equilibration time τ_{eq} in the absorber and the energy-selection width w_{esc} of the ESCs. The target values are τ_{eq} in the order of $\tau_{th}/1,000$ and w_{esc} narrower than 0.1 eV.

It seems extremely challenging to realize the target value of each parameter. Furthermore, it should be a more serious issue to fulfill all the requisites, including light absorption and carrier transport properties that are not considered in the present models, compatibly with each other.

Investigations to address the issues are underway. Much progress to understand the energy dissipation mechanisms has been made in recent years, thanks to the widespread of ultrashort-pulsed lasers for transient optical spectroscopy. However, the target values are far from being reached. New concepts of post conventional HC-SCs have been proposed to solve some parts of the issues, although the proof-of-concepts has not yet been demonstrated.

Acknowledgments I wish to thank Dr. T. Motohiro, Dr. K. Higuchi, Dr. T. Ito, Dr. T. Ikuno, Dr. S. Ogawa, Dr. K. Yamanaka, K. Nishikawa (Toyota Central R&D Labs., Inc., Japan), T. Nagashima, K. Okumura, D. Sato, J. Ota (Toyota Motor Corp., Japan), Professor G. Conibeer, Dr. D. König, Dr. P. Aliberti (Univ. New South Wales, Australia), Dr. N. J. Ekins-Daukes, Dr. D. J. Farrell (Imperial College London, UK), Professor Y. Okada, Dr. N. Miyashita (Univ. Tokyo, Japan), Dr. S. Yagi (Saitama Univ., Japan), and Dr. R. Oshima (National Institute of Advanced Industrial Science and Technology, Japan) for many valuable discussions.

References

1. Green, M.A.: Third Generation Photovoltaics. Springer, Berlin (2003)
2. Shockley, W., Queisser, H.J.: Detailed balance limit of efficiency of p–n junction solar cells. J. Appl. Phys. **32**, 510–519 (1961)
3. Green, M.A., Emery, K., Hishikawa, Y., Warta, W., Dunlop, E.D.: Solar cell efficiency tables (version 42). Prog. Photovolt. Res. Appl. **21**, 827–837 (2013)

4. Yoshida, A., Agui, T., Nakaido, K., Murasawa, K., Juso, H., Sasaki, K., Takamoto, T.: Development of InGaP/GaAs/InGaAs inverted triple junction solar cells for concentrator application. In: Technical Digest of the 21st Photovoltaic Science and Engineering Conference, 4B-4O-01 (2011), Fukuoka, Japan, Nov. 28–Dec.2, 2011
5. Yamaguchi, M., Suzuki, H., Oshita, Y., Kojima, N., Takamoto, T.: Recent R&D topics on concentrator multi-junction solar cells and materials under innovative solar cells's project. In: Proceedings of the 35th IEEE Photovoltaic Specialists Conference, pp. 1237–1242 (2010), Honolulu, HI, USA, June 20–25, 2010
6. King, R.R., Bhusari, D., Larrabee, D., Liu, X.-Q., Rehder, E., Edmondson, K., Cotal, H., Jones, R.K., Ermer, J.H., Fetzer, C.M., Law, D.C., Karam, N.H.: Solar cell generations over 40% efficiency. Prog. Photovolt. Res. Appl. **20**, 801–815 (2012)
7. Luque, A., Martí, A.: Increasing the efficiency of ideal solar cells by photon induced transitions at intermediate levels. Phys. Rev. Lett. **78**, 5014–5017 (1997)
8. Luque, A., Martí, A.: The intermediate band solar cell: progress toward the realization of an attractive concept. Adv. Mater. **22**, 160–174 (2010)
9. Trupke, T., Green, M.A.: Improving solar cell efficiencies by up-conversion of sub-band-gap light. J. Appl. Phys. **92**, 4117–4122 (2002)
10. Cuadra, L., Martí, A., Luque, A.: Influence of the overlap between the absorption coefficients on the efficiency of the intermediate band solar cell. IEEE Trans. Electron Devices **ED-51**, 1002–1007 (2004)
11. Navruz, T.S., Saritas, M.: Efficiency variation of the intermediate band solar cell due to the overlap between absorption coefficients. Solar Energy Mater. Solar Cells **92**, 273–282 (2008)
12. Kim, S.J., Kim, W.J., Cartwright, A.N., Prasad, P.N.: Carrier multiplication in a PbSe nanocrystal and P3HT/PCBM tandem cell. Appl. Phys. Lett. **92**, 191107 (2008)
13. Sukhovatkin, V., Hinds, S., Brzozowski, L., Sargent, E.H.: Colloidal quantum-dot photodetectors exploiting multiexciton generation. Science **324**, 1542–1544 (2009)
14. Trupke, T., Green, M.A., Würfel, P.: Improving solar cell efficiencies by up-conversion of sub-band-gap light. J. Appl. Phys. **92**, 1668–1674 (2002)
15. Beard, M.C.: Multiple exciton generation in semiconductor quantum dots. J. Phys. Chem. **2**, 1282–1288 (2011)
16. Takeda, Y., Motohiro, T.: Requisites to realize high conversion efficiency of solar cells utilizing carrier multiplication. Solar Energy Mater. Solar Cells **94**, 1399–1405 (2010)
17. Timmerman, D., Valenta, J., Dohnalová, K., de Boer, W.D.A.M., Gregorkiewicz, T.: Step-like enhancement of luminescence quantum yield of silicon nanocrystals. Nat. Nanotechnol. **6**, 710–713 (2011)
18. Trinh, M.T., Limpens, R., de Boer, W.D.A.M., Schins, J.M., Siebbeles, L.D.A., Gregorkiewicz, T.: Direct generation of multiple excitons in adjacent silicon nanocrystals revealed by induced absorption. Nat. Photonics **6**, 316–321 (2012)
19. Ross, R.T., Nozik, A.J.: Efficiency of hot-carrier solar energy converters. J. Appl. Phys. **53**, 3813–3818 (1982)
20. Würfel, P.: Solar energy conversion with hot electrons from impact ionization. Solar Energy Mater. Solar Cells **46**, 43–52 (1997)
21. Takeda, Y., Ito, T., Motohiro, T., König, D., Shrestha, S., Conibeer, G.: Hot carrier solar cells operating under practical conditions. J. Appl. Phys. **105**, 074905 (2009)
22. Takeda, Y., Motohiro, T.: Highly efficient solar cells using hot carriers generated by two-step photo-excitation. Solar Energy Mater. Solar Cells **95**, 2638–2644 (2011)
23. Takeda, Y., Ito, T., Suzuki, R., Motohiro, T., Shrestha, S., Conibeer, G.: Impact ionization and Auger recombination at high carrier temperatures. Solar Energy Mater. Solar Cells **93**, 797–802 (2009)
24. Würfel, P.: The chemical potential of radiation. J. Phys. C Solid State Phys. **15**, 3967–3985 (1982)

25. Rosenwaks, Y., Hanna, M.C., Levi, D.H., Szmyd, D.M., Ahrenkiel, R.K., Nozik, A.J.: Hot-carrier cooling in GaAs: quantum wells versus bulk. Phys. Rev. B **48**, 14675–14678 (1993)
26. Luque, A., Martí, A.: Electron–phonon energy transfer in hot-carrier solar cells. Solar Energy Mater. Solar Cells **94**, 287–296 (2010)
27. Humphrey, T.E., Newbury, R., Taylor, R.P., Linke, H.: Reversible quantum Brownian heat engines for electrons. Phy. Rev. Lett. **89**, 116801 (2002)
28. Madelung, O.: Semiconductors: Data Handbook. Springer, Berlin (2003)
29. Web site for NREL'S AM1.5 Standard Dataset. http://rredc.nrel.gov/solar/spectra/am1.5/. Accessed 25 Oct 2012
30. Araki, K., Kondo, M., Uozumi, H., Ekins-Daukes, N.J., Egami, T., Hiramatsu, M., Miyazaki, Y., Yamaguchi, M.: Packaging III–V tandem solar cells for practical terrestrial applications achievable to 27% of module efficiency by conventional machine assemble technology. Solar Energy Mater. Solar Cells **90**, 3320–3326 (2006)
31. Luque, A.: Will we exceed 50% efficiency in photovoltaics? J. Appl. Phys. **110**, 031301 (2011)
32. Markvart, T.: Solar cell as a heat engine: energy-entropy analysis of photovoltaic conversion. Physica Status Solidi A **12**, 2752–2756 (2008)
33. Hirst, L.C., Ekins-Daukes, N.J.: Fundamental losses in solar cells. Prog. Photovolt. Res. Appl. **19**, 286–293 (2011)
34. Dutta, N.K., Nelson, R.J.: The case for Auger recombination in $In_{1-x}Ga_xAs_yP_{1-y}$. J. Appl. Phys. **53**, 74–92 (1982)
35. Chiu, L.C., Chen, P.C., Yariv, A.: Interband Auger recombination in InGaAsP. IEEE J. Quantum Electron. **QE-18**, 938–941 (1982)
36. Hausser, S., Fuchs, G., Hangleiter, A., Streubel, K., Tsang, W.T.: Auger recombination in bulk and quantum well InGaAs. Appl. Phys. Lett. **56**, 913–915 (1990)
37. Rees, R., Blood, P., Vanhommerig, M.J.H., Davies, G.J., Skevington, P.J.: The temperature dependence of threshold current of chemical beam epitaxy grown InGaAs-InP lasers. J. Appl. Phys. **78**, 1804–1807 (1995)
38. Ahrenkiel, R.K., Ellingson, R., Johnston, S., Wanlass, M.: Recombination lifetime of $In_{0.53}Ga_{0.47}As$ as a function of doping density. Appl. Phys. Lett. **72**, 3470–3472 (1998)
39. Aliberti, P., Feng, Y., Takeda, Y., Shrestha, S.K., Green, M.A., Conibeer, G.: Investigation of theoretical efficiency limit of hot carriers solar cells with a bulk indium nitride absorber. J. Appl. Phys. **108**, 094507 (2010)
40. Aliberti, P., Feng, Y., Shrestha, S.K., Green, M.A., Conibeer, G., Tu, L.W., Tseng, P.H., Clady, R.: Effects of non-ideal energy selective contacts and experimental carrier cooling rate on the performance of an indium nitride based hot carrier solar cell. Appl. Phys. Lett. **99**, 223507 (2011)
41. Feng, Y., Aliberti, P., Veettil, B.P., Patterson, R., Shrestha, S., Green, M.A., Conibeer, G.: Non-ideal energy selective contacts and their effect on the performance of a hot carrier solar cell with an indium nitride absorber. Appl. Phys. Lett. **100**, 053502 (2012)
42. Würfel, P., Brown, A.S., Humphrey, T.E., Green, M.A.: Particle conservation in the hot-carrier Solar Cell. Prog. Photovolt. Res. Appl. **13**, 277–285 (2005)
43. Takeda, Y., Motohiro, T., König, D., Aliberti, P., Feng, Y., Shrestha, S., Conibeer, G.: Practical factors lowering conversion efficiency of hot carrier solar cells. Appl. Phys. Exp. **3**, 104301 (2010)
44. Snoke, D.W.: Density dependence of electron scattering at low density. Phys. Rev. B **50**, 11583–11591 (1994)
45. Davis, J.H.: The Physics of Low-Dimensional Semiconductors. Cambridge University Press, Cambridge (1998). Chapter 5
46. O'Dwyer, M.F., Humphrey, T.E., Lewis, R.A., Zhang, C.: Electronic and thermal transport in hot carrier solar cells with low-dimensional contacts. Nanoelectron. J. **39**, 656–659 (2008)
47. Veetill, B.P., Patterson, T., König, D., Conibeer, G., Green, M.A.: Optimized resonant tunneling structures with high conductivity and selectivity. Europhys. Lett. **96**, 57006 (2011)

48. Conibeer, G., Ekins-Daukes, N., Guillemoles, J.-F., König, D., Cho, E.-C., Jiang, C.-W., Shrestha, S., Green, M.: Progress on hot carrier cells. Solar Energy Mater. Solar Cells **93**, 713–719 (2009)
49. Le Bris, A., Guillemoles, J.-F.: Hot carrier solar cells: achievable efficiency accounting for heat losses in the absorber and through contacts. Appl. Phys. Lett. **97**, 113506 (2010)
50. Le Bris, A., Lombez, L., Laribi, S., Boissier, G., Christold, P., Guillemoles, J.-F.: Thermalisation rate study of GaSb-based heterostructures by continuous wave photoluminescence and their potential as hot carrier solar cell absorbers. Energy Environ. Sci. **5**, 6225–6232 (2012)
51. Elsaesser, T., Shah, J., Rota, L., Lugli, P.: Initial thermalization of photoexcited carriers by femtosecond luminescence spectroscopy. Phys. Rev. Lett. **13**, 1757–1760 (1991)
52. Snoke, D.W., Rühle, W.W., Lu, Y.-C., Bauser, E.: Evolution of a nonthermal electron energy distribution in GaAs. Phys. Rev. B **45**, 10979–10989 (1992)
53. Leitenstorfer, A., Fürst, C., Laubereau, A., Kaiser, W.: Femtosecond carrier dynamics in GaAs far from equilibrium. Phys. Rev. Lett. **76**, 1545–1548 (1996)
54. Kirk, A.P., Fischetti, M.V.: Fundamental limitations of hot-carrier solar cells. Phys. Rev. B **86**, 165206 (2012)
55. Bockelmann, B., Bastard, G.: Phonon scattering and energy relaxation in two-, one-, and zero-dimensional electron gasses. Phys. Rev. B **42**, 8947–8951 (1990)
56. Benisty, H., Sotomayor-Torres, C.M., Weisbuch, C.: Intrinsic mechanism for the poor luminescence properties of quantum-box systems. Phys. Rev. B **44**, 10945–10948 (1990)
57. Heitz, R., Born, H., Guffarth, G., Stier, O., Schliwa, A., Hoffmann, A., Bimberg, D.: Existence of a phonon bottleneck for excitons in quantum dots. Phys. Rev. B **64**, 241305 (2001)
58. Kitamura, T., Ohtsubo, R., Murayama, M., Kuroda, T., Yamaguchi, K., Tackeuchi, K.: Direct observation of phonon relaxation bottleneck in InAs quantum dots of high uniformity. Physica Status Solidi C **0**, 1165–1168 (2003)
59. Sanguinetti, S., Guzzi, M., Grilli, E., Gurioli, M., Seravalli, L., Frigeri, P., Franchi, S., Capizzi, M., Mazzuccato, S., Polimeni, A.: Effective phonon bottleneck in the carrier thermalization of InAs/GaAs quantum dots. Phys. Rev. B **80**(78), 085313 (2008)
60. Hendry, E., Koeberg, M., Wang, F., Zhang, H., Mello Donegá, C., Vanmaekelbergh, D., Bonn, M.: Direct observation of electron-to-hole energy transfer in CdSe quantum dots. Phys. Rev. Lett. **96**, 057408 (2006)
61. Urayama, J., Norris, T.B., Singh, J., Bhattacharya, P.: Observation of phonon bottleneck in quantum dot electronic relaxation. Phys. Rev. Lett. **86**, 4930–4933 (2001)
62. Nishikawa, K., Takeda, Y., Motohiro, T., Sato, D., Ota, J., Miyashita, N., Okada, Y.: Extremely long carrier lifetime over 200 ns in GaAs wall-inserted type II InAs quantum dots. Appl. Phys. Lett. **100**, 113105 (2012)
63. Sato, D., Ota, J., Nishikawa, K., Takeda, Y., Motohiro, T., Miyashita, N., Okada, Y.: Extremely long carrier lifetime at intermediate states in wall-inserted type II quantum dot absorbers. J. Appl. Phys. **112**, 094305 (2012)
64. Pandy, A., Guyot-Sionnest, P.: Slow electron cooling in colloidal quantum dots. Science **322**, 929–932 (2008)
65. Li, X.-Q., Nakayama, H., Arakawa, Y.: Phonon bottleneck in quantum dots: role of lifetime of the confined optical phonons. Phys. Rev. B **59**, 5069–5073 (1999)
66. Bahir, G., Finkman, E., Fossard, F., Julien, F.H., Brault, J., Gendry, M., Schacham, S.E.: Intraband polaron dynamics of excited carriers in InAs/$In_xAl_{1-x}As$ quantum dots. Phys. Rev. B **71**, 075327 (2005)
67. Kambhampati, P.: Hot exciton relaxation dynamics in semiconductor quantum dots: radiationless transitions on the nanoscale. J. Phys. Chem. C **115**, 22089–22109 (2011)
68. Tisdale, W.A., Williams, K.J., Timp, B.A., Norris, D.J., Aydil, E.S., Zhu, X.-Y.: Hot-electron transfer from semiconductor nanocrystals. Science **328**, 1543–1547 (2010)
69. Petrosyan, L.S., Kirakosyan, A.S., Shahbazyan, T.V.: Extraordinary electron transmission through a periodic array of quantum dots. Phys. Rev. B **107**, 196802 (2011)

70. Gao, Y., Talgorn, E., Aerts, M., Trinh, M.T., Schins, J.M., Houtepen, A.J., Siebbeles, L.D.A.: Enhanced hot-carrier cooling and ultrafast spectral diffusion in strongly coupled PbSe quantum-dot solids. Nano Lett. **11**, 5471–5476 (2011)
71. Tea, E., Hamzeh, H., Aniel, F.: Hot carriers relaxation in highly excited polar semiconductors: hot phonons versus phonon–plasmon coupling. J. Appl. Phys. **110**, 113108 (2011)
72. Klemens, P.G.: Anharmonic decay of optical phonons. Phys. Rev. **148**, 845–848 (1966)
73. Conibeer, G.J., König, D., Green, M.A., Guillemoles, J.-F.: Slowing of carrier cooling in hot carrier solar cells. Thin Solid Films **516**, 6948–6953 (2008)
74. König, D., Casalenuovo, K., Takeda, Y., Conibeer, G., Guillemoles, J.-F., Patterson, R., Huang, L.M., Green, M.A.: Hot carrier solar cells: principles, materials and design. Physica E **42**, 2862–2866 (2010)
75. Ridley, B.K.: The LO phonon lifetime in GaN. J. Phys. Condens. Matter **8**, L511 (1996)
76. Pomeroy, J.W., Kuball, M., Lu, H., Schaff, W.J., Wang, X., Yoshikawa, A.: Phonon lifetimes and phonon decay in InN. Appl. Phys. Lett. **86**, 223501 (2005)
77. Davydov, V., Emtsev, V., Goncharuk, I., Smirnov, A., Petrikov, V., Mamutin, V., Vekshin, V., Ivanov, S., Smirnov, M.B., Inushima, T.: Experimental and theoretical studies of phonons in hexagonal InN. Appl. Phys. Lett. **75**, 3297–3299 (1999)
78. Chen, F., Cartwright, A.N., Lu, H., Schaff, W.J.: Time-resolved spectroscopy of recombination and relaxation dynamics in InN. Appl. Phys. Lett. **83**, 4984–4986 (2003)
79. Sun, S.-Z., Wen, Y.-C., Guol, S.-H., Lee, H.-M., Gwo, S., Sun, C.-K.: Observation of femtosecond carrier thermalization time in indium nitride. J. Appl. Phys. **103**, 123513 (2008)
80. Fukunaga, K., Hashimoto, M., Kunugita, H., Kamimura, J., Kikuchi, A., Kishino, K., Ema, K.: Energy- and density-dependent dynamics of photoexcited carriers in InN films. Appl. Phys. Lett. **95**, 232114 (2009)
81. Su, Y.-E., Wen, Y.-C., Lee, H.-M., Gwo, S., Sun, C.-K.: Observation of sub-100 femtosecond electron cooling time in InN. Appl. Phys. Lett. **96**, 052108 (2010)
82. Clady, R., Tayebjee, M.J.Y., Aliberti, P., König, D., Ekins-Daukes, N.J., Conibeer, G.J., Schmidt, T.W., Green, M.A.: Interplay between the hot phonon effect and intervalley scattering on the cooling rate of hot carriers in GaAs and InP. Prog. Photovolt. Res. Appl. **20**, 82–92 (2012)
83. Conibeer, G., Shrestha, S., Huang, S., Patterson, R., Aliberti, P., Xia, H., Feng, Y., Gupta, N., Smyth, S., Liao, Y., Kamikawa, Y., Green, M.A.: Hot carrier solar cell absorbers: Superstructure materials and mechanisms for slowed carrier cooling. In: Proceedings of the 27th European Photovoltaic Solar Energy Conference, pp 81–84 (2012), Frankfurt, Germany, Sept. 24–28, 2012
84. Patterson, R., Kirkengen, M., Puthen-Veettil, B., König, D., Green, M.A., Conibeer, G.: Phonon lifetimes in model quantum dot superlattice systems with applications to the hot carrier solar cell. Solar Energy Mater. Solar Cells **94**, 1931–1935 (2010)
85. Hirst, L., Führer, M., Farrell, D.J., Le Bris, A., Guillemoles, J.-F., Tayebjee, M.J.Y., Clady, R., Schmidt, T.W., Wang, Y., Sugiyama, M., Ekins-Daukes, N.J.: Hot carrier dynamics in InGaAs/GaAsP quantum well solar cells. In: Proceedings of the 37th IEEE Photovoltaic Specialists Conference, pp. 3302–3306 (2011), Seattle, WA, USA, June 19–24, 2011
86. Rodière, J., Levard, H., Delamarre, A., Le Bris, A., Laribi, S., Lombez, L., Guillemoles, J.-F., Perez, J.-P., Christol, P., Folliot, H., Tavernier, K., Durand, O., Colin, C., Collin, S., Pelouard, J.-P.: Hot carrier solar cells: from simulation to devices. In: Proceedings of the 27th European Photovoltaic Solar Energy Conference, pp. 89–90 (2012), Frankfurt, Germany, Sept. 24–28, 2012
87. König, D., Hiller, D., Zacharias, S., Michard, M., Flynn, C.: Static hot carrier populations as a function of optical excitation energy detected through energy selective contacts by optically assisted IV. Prog. Photovolt. Res. Appl. (2013). doi:10.1002/pip.2367
88. Bryllert, T., Borgstrom, M., Sass, T., Gustafson, B., Landin, L., Wernersson, L.-E., Seifert, W., Samuelson, L.: Designed emitter states in resonant tunneling through quantum dots. Appl. Phys. Lett. **80**, 2681–2683 (2002)

89. O'Dwyer, M.F., Lewis, R.A., Zhang, C., Humphrey, T.E.: Electronic efficiency in nanostructured thermionic and thermoelectric devices. Phys. Rev. B **72**, 205330 (2005)
90. Yagi, S., Oshima, R., Okada, Y.: Evaluation of selective energy contacts for hot carrier solar cells based on III–V semiconductors. In: Proceedings of the 34th IEEE Photovoltaic Specialists Conference, pp. 530–533 (2009)
91. Deshpande, M.R., Sleight, J.W., Reed, M.A., Wheeler, R.G., Matyi, R.J.: Spin splitting of single 0D impurity states in semiconductor heterostructure quantum wells. Phys. Rev. Lett. **76**, 1328–1331 (1996)
92. Takeda, Y., Motohiro, T.: Hot-carrier extraction from intermediate-band absorbers through quantum-well energy-selective contacts. Jpn. J. Appl. Phys. **51**, 10ND03 (2012)
93. König, D., Takeda, Y., Puthen-Veettil, B.: Technology-compatible hot carrier solar cell with energy selective hot carrier absorber and carrier-selective contacts. Appl. Phys. Lett. **101**, 153901 (2012)
94. König, D., Takeda, Y., Puthen-Veettil, B., Conibeer, G.: Lattice-matched hot carrier solar cell with energy selectivity integrated into hot carrier absorber. Jpn. J. Appl. Phys. **51**, 10ND02 (2012)
95. Nash, K.J., Skolnick, M.S., Bass, S.J.: Electron–phonon interactions in indium gallium arsenide. Semicond. Sci. Technol. **2**, 329–336 (1987)
96. Farrell, D.J., Takeda, Y., Nishikawa, N., Nagashima, T., Motohiro, T., Ekins-Daukes, N.J.: A hot-carrier solar cell with optical energy selective contacts. Appl. Phys. Lett. **99**, 111102 (2011)
97. Sau, T.K., Rogach, A.L., Jäckel, F., Klar, T.A., Feldmann, J.: Properties and applications of colloidal nonspherical noble metal nanoparticles. Adv. Mater. **22**, 1805–1825 (2010)
98. Biagioni, P., Huang, J.-S., Hecht, B.: Nanoantennas for visible and infrared radiation. Rep. Prog. Phys. **75**, 024402 (2012)
99. Takeda, Y., Motohiro, T.: Intermediate-band-assisted hot-carrier solar cells using indirect-bandgap absorbers. Prog. Photovolt. Res. Appl. **21**, 1308–1318 (2013)

Chapter 9
Increasing Efficiency with Multiple Exciton Generation

N. McElroy, M. Cadirci, A. Al-Otaify, R. Page, and D.J. Binks

Abstract In a conventional solar cell the energy of an absorbed photon in excess of the band gap is lost as heat as the photo-generated charges cool rapidly to the band edge; for a silicon-based cell under standard insolation this loss is equivalent to 47 % of incident power. However, in colloidal quantum dots an alternative process can become significant—multiple exciton generation (MEG). Here, immediately after the absorption of a photon photo-generated carriers can instead use this excess energy to produce additional electron–hole pairs, which can contribute to the photocurrent and thereby increase the efficiency of a solar cell. In this chapter, this phenomenon will be introduced and its potential for increasing the efficiency of quantum dot solar cells discussed. MEG was first investigated spectroscopically and the challenging techniques used will be described. Progress since the first demonstration of MEG in 2004 will then be reviewed, including the controversy over quantum yields and its resolution. The recent demonstration of devices benefiting from MEG will be detailed. Finally, several possible future directions for research will be discussed.

9.1 Introduction

The solar spectrum reaching the surface of the earth is essentially a black-body spectrum modified somewhat by Rayleigh scattering and the absorption spectra of various gases, principally atmospheric ozone, water vapour and carbon dioxide. This spectral breadth, extending from the near ultraviolet (UV) to the mid-infrared, is a key factor in determining the efficiency of solar cells. Most of the photons from this broad spectrum absorbed by a semiconductor will have an energy, $h\nu$,

N. McElroy • M. Cadirci • A. Al-Otaify • R. Page • D.J. Binks (✉)
School of Physics and Astronomy & Photon Science Institute, University of Manchester, Manchester M13 9PL, UK
e-mail: David.Binks@manchester.ac.uk

J. Wu and Z.M. Wang (eds.), *Quantum Dot Solar Cells*, Lecture Notes in Nanoscale Science and Technology 15, DOI 10.1007/978-1-4614-8148-5_9,

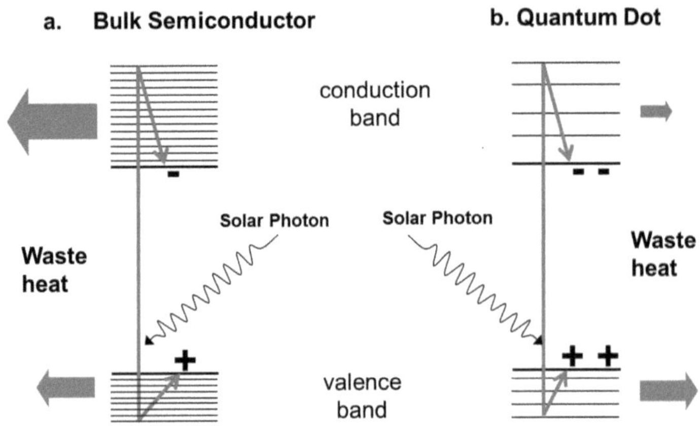

Fig. 9.1 (**a**) Absorption of a solar photon in a bulk semiconductor initially produces a "hot" electron–hole pair that rapidly cools to the band edge so that the absorbed photon energy in excess of the band gap is lost as waste heat. (**b**) In a quantum dot, however, the process of MEG can become competitive with cooling and uses some or all of the excess photon energy to produce additional excitons that contribute to the photocurrent and thereby improve the performance of a solar cell

significantly greater than the band gap, E_g, and hence produce initially "hot" carriers i.e. electrons and holes not yet thermally relaxed to the conduction band minimum (CBM) or valance band maximum (VBM), respectively. In conventional semiconductors, the energy of the absorbed photon in excess of the band gap, $h\nu - E_g$, is very rapidly (within ~1 ps) lost to heat as these hot carriers thermalize to the CBM or VBM, largely by the emission of phonons [1]. This process is the largest source of energy loss in a conventional solar cell; for example, in the case of a silicon-based cell, 47 % of the incident solar energy is lost in this way [1]. Clearly, obviating or ameliorating this mechanism is a route to a significant increase in solar cell efficiency.

The process of impact ionization is an alternative relaxation mechanism for hot carriers in which some or all of the excess photon energy is instead used to generate additional electron–hole pairs by an inverse Auger interaction. These additional carriers can contribute to the photocurrent and thereby improve the efficiency of a solar cell. Unfortunately, impact ionization only achieves significant efficiency in bulk semiconductors for high energy photons, of which there are too few in the solar spectrum for this effect to have an appreciable effect on solar cell performance [2–4]. However, the confinement of the hot carriers within a colloidal quantum dot (QD) increases the efficiency of the impact ionization process so that it can compete effectively with carrier cooling at much lower photon energies. This enables impact ionization, which has become known as multiple exciton generation (MEG) or carrier multiplication (CM) when observed in QDs, to have a significant potential impact on the efficiency of solar cells based on quantum dots (Fig. 9.1).

In this chapter, we will describe MEG in QDs and the potential benefit it could bring to solar cell performance. We detail the development of this topic over the last decade, beginning with its initial identification and characterization by ultrafast spectroscopic techniques, and the associated challenges and controversies that emerged. We will then describe the more recent demonstrations of MEG in devices, including a quantum dot-based solar cell. Finally, we will discuss the current status of the subject and possible future directions.

9.2 Potential Benefit to Solar Cell Efficiency

What is the potential benefit of MEG to the efficiency of a solar cell? This has been quantitatively answered [5–7] by using the "Detailed Balance Model" of solar cell operation first developed by Shockley and Quiesser in their seminal work on solar efficiency in 1961 [8]. This model considers an ideal cell in which only three intrinsic loss mechanisms are significant; these are:

1. The thermalization of hot carriers.
2. Solar photons with energy less than E_g not absorbed.
3. The radiative recombination of carriers before extraction to the external circuit.

As discussed above, hot carrier thermalization is the most important of these loss mechanisms; in comparison, the second and third of these processes are responsible for the loss of 1.6 and 19 % of the incident solar energy in a conventional silicon cell [1]. The effects of the onset of absorption above E_g, the thermalization of carriers and MEG can all be incorporated into the model via a quantum yield function that varies with both $h\nu$ and E_g, $\mathrm{QY}(h\nu, E_g)$. For an ideal conventional solar cell exhibiting no MEG, QY is a unit step function at E_g i.e. no photo-generated charges are produced for values of $h\nu$ below E_g and only one electron–hole pair is produced per absorbed photon for $h\nu \geq E_g$ (implying that any photon energy in excess of E_g is lost as heat). In contrast, for an ideal cell in which MEG is the dominant mechanism by which hot carriers relax when energetically possible, $\mathrm{QY}(h\nu, E_g)$ would have the same functional form for values of $h\nu$ below the threshold value for MEG, $h\nu_2$, but then increase stepwise to 200 % at this threshold, which corresponds to when the energy of the absorbed photon in excess of E_g becomes sufficient to produce a second exciton i.e. the second exciton creation energy, ε_2. Further step increases in QY of 100 % each occur when the excess photon energy rises sufficiently to allow the creation of an additional exciton, with the nth exciton requiring an energy of ε_n more than the $(n-1)$th one to be created, where n is an integer. This can be expressed as:

$$\mathrm{QY}(h\nu,\ E_g) = \sum_{n=1}^{\infty} \theta(h\nu, h\nu_n) \tag{9.1}$$

$$h\nu_n = \sum_{i=1}^{n} \varepsilon_n$$

where θ is the Heaviside unit step function, and $\varepsilon_1 = E_g$.

The excess photon energy is divided between photo-generated electron and hole according to the inverse of their effective masses [1]. Since the MEG process is an inverse Auger interaction involving one or other of the hot carriers, it is more efficient for one of the two initially photo-generated to gain most of the excess photon energy. This corresponds to the situation when there is a large ratio of effective masses, as is the case, for example, of InAs QDs for which the hole effective mass exceeds that of the electron by a factor of 19 resulting in the electron gaining 95 % of the excess energy immediately after the photon has been absorbed [1]. A large effective mass ratio minimizes the photon energy threshold for MEG which has been shown to be crucial if the solar spectrum is to be effectively exploited [7]. Consequently, the analysis presented here will assume that a material with a large effective mass ratio has been selected and used to produce the QDs. In this case, the creation energy for the second exciton (i.e. the first additional one produced by the MEG process) is given by $\varepsilon_2 = E_g + \Delta_{xx}$, where Δ_{xx} is the biexciton interaction energy which may be positive or negative depending on whether the net interaction between the two excitons is repulsive or attractive, respectively. An attractive interaction is beneficial to the performance of a solar cell utilizing MEG because it reduces the threshold value of photon energy and thereby allows more efficient exploitation of the solar spectrum. The prospects for engineering QDs for large negative values of Δ_{xx}, and thereby improve the efficiency of a solar cell, are discussed in Sect. 9.5.1 below. The values of $\varepsilon_{n>2}$ depend both on the interaction energies of higher-order multi-excitons and on the electronic structure of the QDs used. Many QD types have CBM that are twofold degenerate [9] (the notable exceptions to this are lead chalcogenide QDs, such as PbX (X = S, Se, Te), which are eightfold degenerate [10]) and so the creation of a third exciton may involve the promotion of an electron from the valance band to the CBM-1 level, which requires more energy than excitation to the CBM. Thus, $\varepsilon_{n>2}$ will be of order E_g or greater. However, it has been shown [5] that most of the benefit of MEG for solar cell performance is gained from the creation of the second exciton, with each additional exciton beyond this being of diminishing importance, and so for simplicity $\varepsilon_{n>2} = \varepsilon_2$ will be assumed in the following analysis.

The enhancement of QY by MEG increases the photo-generated current density, j_{pg}, in a solar cell which is described by [11]:

$$j_{pg} = e \int_0^{\infty} \mathrm{QY}(h\nu, h\nu_n)\phi(h\nu)d(h\nu) \tag{9.2}$$

where $\phi(h\nu)$ is the solar photon flux density and e is the charge on the electron. In the Detailed Balance Model, j_{pg} is offset by a recombination current density, j_r, associated with the black-body radiation from the cell and described by [11]:

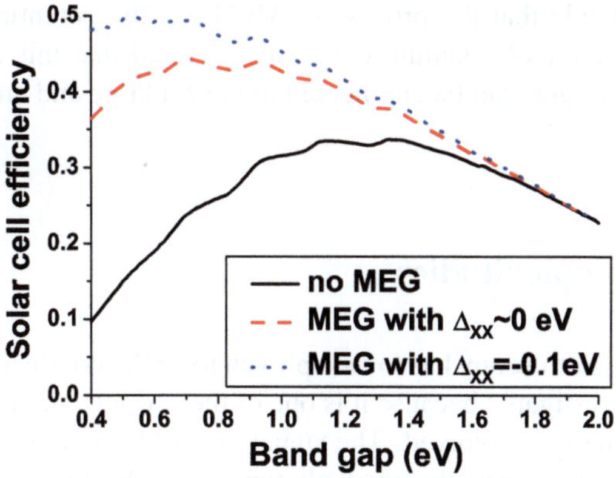

Fig. 9.2 Maximum theoretical solar cell efficiency for a range of band gaps, E_g, calculated using the Detailed Balance Model for a conventional solar cell (*solid line*), and cell in which MEG is efficient but for which the biexciton interaction energy, Δ_{XX}, is negligible (*dashed line*), and a cell which combines efficient MEG and a large, attractive biexciton energy of $\Delta_{XX} = -0.1$ eV (*dotted line*)

$$j_r = \frac{2\pi e}{c^2 h^3} \int_0^\infty \mathrm{QY}[\exp(\beta(h\nu - \mathrm{QY}eV)) - 1]^{-1}(h\nu)^2 d(h\nu) \tag{9.3}$$

where h is Planck's constant, c is the speed of light, and $\beta^{-1} = kT$ with k being Boltzmann's constant and T being the temperature (set to 300 K here); V is the operating voltage of the cell and is determined by finding the value which optimizes the efficiency of the cell, $\eta = (j_{pg} - j_r)V/I$, where I is the intensity of the incident sunlight.

Figure 9.2 compares the maximum solar cell efficiency under standard AM1.5 insolation, as calculated from Eqs. (9.1)–(9.3), obtainable for a range of band gaps for a conventional solar cell (i.e. no MEG and thus $n = 1$ only in Eq. (9.1) to one in which MEG is the dominant hot carrier cooling mechanism). Two cases are considered for the MEG device: in the first, the biexciton interaction is negligible, i.e. $\Delta_{XX} \sim 0$, and hence the exciton creation energy is constant and equal to the band gap, $\varepsilon_n = E_g$ for all n; in the second case, Δ_{xx} is set to -0.1 eV, the magnitude (but not sign) of which corresponds to the largest biexciton interaction energy yet reported for a QD [12], and thus $\varepsilon_1 = E_g$ and $\varepsilon_{n>1} = E_g + \Delta_{xx}$. The efficiency curve for the conventional cell peaks at 33 % for a band gap of $E_g \sim 1.3$ eV, reproducing the original result of Shockley and Queisser [8]. For the case of the device exhibiting efficient MEG but with negligible interaction energy, the peak efficiency increases to 44 % at $E_g \sim 0.7$ eV; however, the efficiency of this cell only changes gradually with band gap and remains greater than 40 % between $E_g \sim 0.5$ and 1.2 eV. For the device which combines both efficient MEG and a large, attractive biexciton interaction, the maximum efficiency rises to 50 % and exceeds 40 % from at least $E_g \sim 0.4$ to ~1.2 eV.

Thus, we conclude that the process of MEG has the potential to significantly increase the efficiency of quantum dot solar cells and that this efficiency can be further improved if QDs can be engineered to have a large and attractive biexciton interaction energy.

9.3 Spectroscopic Studies

Ultrafast optical spectroscopy has to date proven to be the dominant tool with which MEG has been investigated because it is one of the few techniques able to meet the experimental challenges involved. The signature of MEG is the presence of more than one electron–hole per absorbed photon per QD. However, multi-excitons undergo rapid Auger recombination on a typical timescale of 10–100s of picoseconds [13] (cf. the recombination lifetime of single excitons is on the timescale of 10s ns [14]) and thus techniques with a time-resolution significantly better than this are required in order to be able to measure the multi-exciton population before recombination. Ultrafast lasers are commercially available that produce pulses with a duration of 100 fs or less, thus offering the potential of a time-resolution that is more than adequate. The major experimental challenge in studying MEG in fact lies in ensuring the ultrafast spectroscopic techniques employed have sufficient sensitivity. Multiple excitons per QD can be created without MEG if the pump fluence is sufficiently high such that a single QD absorbs more than one photon per pump pulse. To avoid creating multi-excitons in this way, MEG studies thus involve exciting the sample with low pump fluences such that the probability of absorbing more than one photon per QD is much less than unity. However, these low fluences result in low signal levels and necessitate the use of sensitive detection techniques.

9.3.1 Ultrafast Transient Absorption Spectroscopy

An ultrafast transient absorption (TA) experiment is a pump probe technique where an intense pump pulse excites the sample to create hot electrons. A subsequent probe pulse monitors the dynamics of the charges photo-generated by the pump pulse as a function of the delay time between arrivals of the pump and probe pulses. The probe beam is kept sufficiently weak so that it does not perturb the system. Typically, both beams emerge from the same source, enabling synchronization, and spatially overlap at the sample position. This method is widely used to detect MEG in QDs because the laser pulse duration, which can be 100 fs or less, determines the experimental time-resolution and is thus much less than the multi-exciton recombination time, as required.

Ultrafast TA experiments can be categorized as one of the three types based on the wavelength of the probe beam. The first of them, and the most common, is

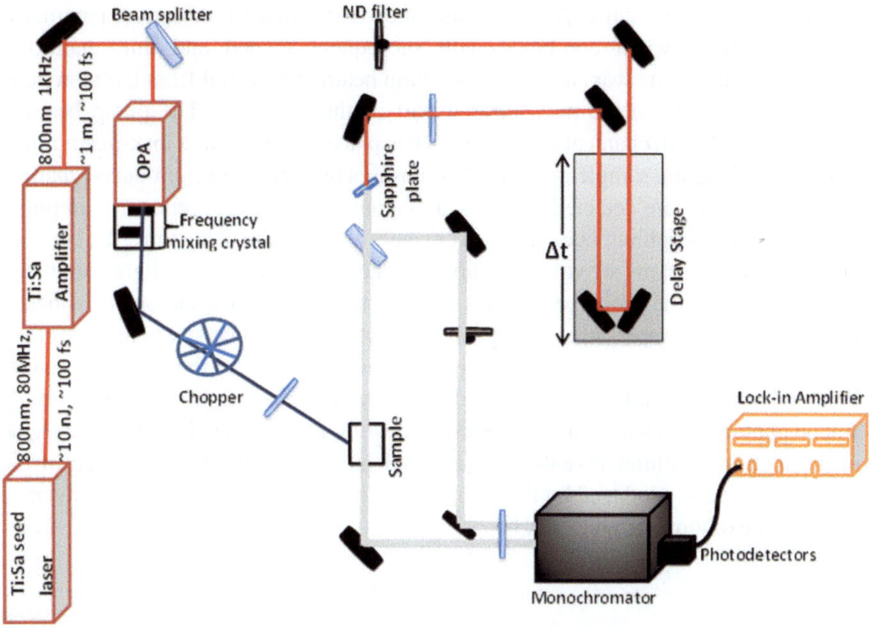

Fig. 9.3 Experimental set-up for a typical ultrafast transient absorption spectrometer

visible TA in which the energy of probe beam is tuned to the lowest 1S absorption feature i.e. an interband absorption corresponding to the bandgap of the QD. This method measures the average number of electrons residing in the $1S_e$ state, i.e. the CBM, and was the technique used to observe MEG in QDs for the first time [15]. The second type is infrared TA in which the probe beam monitors the exciton dynamics between the CBM and a higher state in the conduction band i.e. an intraband transition. In this system, because the probe photons are at an energy less than the band edge, the probe beam is only absorbed by the QDs following perturbation of the system by the pump beam and monitors only optically allowed transitions, enabling the detection of clear MEG signals in QDs [16]. The third type of TA utilizes THz radiation as the probe beam and monitors changes in average conductivity due to the generation of electron–hole pairs. Due to the low electromagnetic field of the THz probe (~1 kV/cm), it is an effective way to investigate the internal fine structure of holes and compare MEG in QDs and in bulk semiconductors as phonon modes in bulk semiconductor and internal energy levels of holes in QDs lie in the THz region [17].

An example ultrafast visible TA experiment is shown in Fig. 9.3. Low energy (~10 nJ), ultrafast (~100 fs) seed pulses are produced by a mode-locked Ti:Sapphire oscillator operating at central wavelength of 800 nm with 80 MHz repetition rate. A Ti: Sapphire regenerative amplifier is employed to increase the energy of the seed pulses to 1 mJ whilst simultaneously decreasing the repetition rate down to about 1 kHz.

The majority of the output from the amplifier is directed to an optical parametric amplifier which, when combined with subsequent second and third harmonic generation, allows the wavelength of the pump beam to be tuned from the ultraviolet through the visible to the near-infrared part of the spectrum. The pump beam is modulated with a mechanical chopper which is phase-locked with a lock-in amplifier, prior to reaching the sample position. This ensures that only pump-produced changes in the probe beam are detected. The remaining output from the regenerative amplifier is routed onto a sapphire plate via a delay stage which sets the delay time, Δt, between pump and probe beams arrivals. The sapphire plate generates a white light continuum which is used as the probe beam. This beam is split into two arms; one of which is focused on to the sample before being routed to the monochromator, whereas the other, the reference beam, bypasses the sample and goes directly to the monochromator. Finally, both the sample and reference beams are sent to a pair of detectors (usually silicon photodiodes) which are connected to the lock-in amplifier. The observed signal in the lock-in amplifier reveals the fractional increase in the transmittance of the sample, or *bleach*, produced by the pump beam, $\Delta T/T$, which is related to the average number of excitons per QD, $\langle N \rangle$, by

$$\frac{\Delta T(t)}{T} = A\frac{\langle N \rangle}{g} \tag{9.4}$$

where A is the absorbance of the sample and g is the degeneracy of the CBM [18]. The QY is equal to the ratio, R, of the maximum initial fractional transmittance change, $(\Delta T/T)_{\mathrm{MX}}$, (corresponding to a point after cooling and MEG and before multi-exciton recombination) to its value, $(\Delta T/T)_{\mathrm{X}}$, after multi-exciton decay but before significant recombination of single excitons (see Fig. 9.4), *in the limit of vanishing excitation* i.e. QY $\rightarrow R$ as the pump pulse fluence tends to zero [1]. This can be found by measuring R for a range of pump fluences and extrapolating to zero by fitting to the expression [1]:

$$R = \mathrm{QY}\sigma J(1 - \exp(-\sigma J))^{-1} \tag{9.5}$$

where σ is the absorption cross-section of the QD at the pump wavelength, and J is the pump pulse fluence averaged over the probed volume of the sample (in units of photons per cm^2 per pulse). A procedure devised by Ruhman and co-workers [19] can also be used in which R is instead found at non-vanishing fluences for pump photon energies above and below the threshold for MEG. In this method, the samples are diluted to compensate for the rise of σ with increasing photon energy, thus ensuring that the absorption depth is constant for each pump wavelength. The pump fluence is also scaled to ensure that σJ is also constant at each pump wavelength. Any enhancement of R at the higher photon energy can thus be attributed to MEG rather than the increased probability that a QD absorbs more than one photon. A variation of this method has recently been devised which measures R at a single concentration of sample [20]. In this case, σJ is kept constant

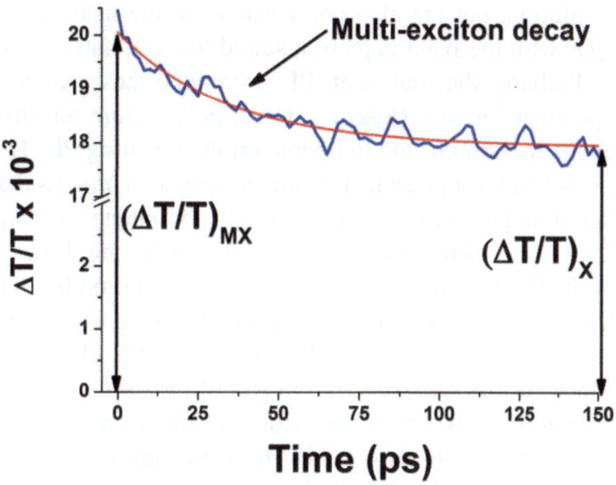

Fig. 9.4 A typical transmittance transient obtained from an ultrafast transient absorption experiment showing the decay of multi-excitons to single exictons in QDs. The MEG quantum yield can be determined from the ratio of the fractional transmittance change before $[(\Delta T/T)_{MX}]$ and after $[(\Delta T/T)_{X}]$ the decay

with pump wavelength by varying the pump fluence such that $\Delta T/T$ after multi-exciton decay is complete is the same for each wavelength.

9.3.2 Ultrafast Transient Photoluminescence Spectroscopy

Ultrafast transient photoluminescence (PL) spectroscopy has also been used to detect and measure MEG in QDs [21–24]. In a similar way to ultrafast TA, MEG results in an increased initial PL amplitude relative to the value after multi-exciton recombination is complete, although the measurement of QY is more complicated because the radiative recombination constants for single and multi-excitons are different [24]. Several different types of transient PL experiment have also been used to detect MEG, including time-correlated single photon counting (TCSPC) [21], streak cameras [23] and upconversion spectroscopy [22, 24]. TCSPC involves excitation of the sample by an ultrafast laser with a high repetition rate (typically a few MHz). The resultant PL is guided through a monochromator onto a multichannel plate (MCP). For each pulse, the time delay between the excitation pulse and the arrival of a PL photon at the MCP is recorded. This process is repeated over a large number of pulses, allowing a distribution of delays to be accumulated which corresponds to the PL decay transient. This technique benefits from the excellent sensitivity of MCPs, enabling very low excitation rates to be used and thus ensuring the probability of producing multi-excitons by a single QD absorbing more than one per pulse is negligible. However, the time-resolution of TCSPC is typically a few 10s of picoseconds and hence insufficient to well resolve the decay of multi-excitons in some materials. More, MCPs are only sensitive to photons with energies of ~1.5 eV or greater, which corresponds to emission from a QD with a band gap that is too large for the efficient harvesting of the solar spectrum, as discussed in Sect. 9.2. Streak cameras offer a similar sensitivity to MCPs and better time-resolution, but are also not sensitive to the near-infrared emission expected from QDs with the band gaps best suited to exploitation of the solar spectrum.

Perhaps the transient PL technique best suited to MEG measurement is upconversion spectroscopy. Here, pulses from an ultrafast laser are used both to excite the sample and to upconvert the resulting PL. This is achieved by mixing the PL with a laser pulse in a nonlinear crystal, a process that can only occur whilst both are simultaneously passing through the crystal. Changing the arrival time of the laser pulse with a delay stage (similar to the one shown in Fig. 9.3) allows the decay of the PL to be monitored with a time-resolution limited only by the duration of the laser pulse. Moreover, the upconversion process shifts the PL from the near-infrared region corresponding to QDs with ideal band gaps to the visible, and hence within the spectral range of sensitive photomultiplier tubes (the time-resolution of which is not critical in this experiment). Consequently, this method has been used to detect the MEG efficiency of some QDs emitting in the near-infrared [22, 24] although the nonlinear upconversion process has proven to result in a signal-to-noise ratio somewhat less than a TCSPC experiment [25].

9.3.3 Experimental Results

The history of impact ionization goes back to late 1950s when it was first observed in bulk semiconductors [26]. However, the overall efficiency was low in these materials due to both a high threshold energy ($5 \times E_g$ for bulk PbS [27]) and very fast electron–phonon scattering compared with the impact ionization rate. For example, the enhancement of a solar cell efficiency due to impact ionization in Si/Ge alloys was reported to be as low as 0.5 % [28]. However, in 2001, it was proposed that impact ionization in colloidal QDs might have the potential to increase the efficiency of photovoltaic cells significantly [29]. Shortly afterwards, efficient MEG in QDs was experimentally observed for the first time using an ultrafast transient absorption technique [15]. In that report, the QY for colloidal PbSe QDs was found to reach as much as 218 %, indicating that more than two electron–hole pairs were being created per absorbed photon. Soon after that, these pioneer results were confirmed by a study conducted by Ellingson et al. where similar results in colloidal PbSe and PbS QDs were obtained [30]. These developments encouraged other researchers to focus on the MEG concept in semiconductor colloidal QDs and led to spectroscopic studies of various QDs and reports of high MEG yields (up to seven exciton per absorbed photon) in a number of different QDs including: PbTe [31], CdSe [32], InAs [33] and Si [34].

However, later reports argued that MEG in colloidal QDs was less efficient than originally reported or does not occur at all [19, 23, 35]. For example, unlike the previously reported results, the MEG threshold energy for CdTe QDs was found to be more than the $3E_g$, much more than that of the previously reported value, and no evidence of MEG in CdSe QDs was found [23]. A period of controversy followed which was resolved by a series of careful studies [24, 36, 37] on PbSe QDs of the effect of sample flowing or stirring on the apparent efficiency of MEG. After excitation one or other of the photo-generated charges can become trapped on the surface of a QD for a time longer than the period between pump pulses. Thus, when a photon is absorbed by the QD during a subsequent pump pulse a trion is formed between the newly photo-generated electron–hole pair and the geminate partner of the surface-trapped charge, which has remained within the QD volume. Like a biexciton but unlike a single exciton, a trion can undergo rapid Auger recombination and so typically has a sub-nanosecond characteristic lifetime. Moreover, this process requires the absorption of only one photon per pulse by the QD concerned and so can occur at the low pump fluences typically used in MEG experiments. This process can therefore be easily mistaken for MEG in spectroscopic studies and likely led to the inconsistency between early MEG reports. However, these studies also showed that this process could be effectively suppressed by sufficient stirring or flowing of the sample, which sweeps the QDs with surface-trapped charges out of the excitation volume in between pump pulses. Other QD types for which sample stirring or flowing has been used to ensure that trion formation is not significant include: InP [38], PbS [18] and type II CdTe/CdSe [39].

Subsequent work on CdSe [40] and InAs [20] QDs also showed that the process of surface-trapping of electrons itself, which depopulates the CBM and thus reduces the PL or value of $\Delta T/T$, can also occur on a sub-nanosecond timescale and hence can be mistaken for MEG even when the trapped electron recombines before the next pump pulse. Fortunately, these studies also showed that the presence of a significant number of surface-traps produces a characteristic broad photo-induced absorption (PIA) feature in the transient transmittance spectrum at wavelengths longer than the 1S bleach. Thus, the greatest confidence can be had in MEG results in which the absence of such a PIA feature has been confirmed experimentally.

9.4 Devices

For the production of additional excitons by MEG to result in an enhanced photocurrent in a device, the electrons and/or holes must be extracted from the QD before significant multi-exciton recombination which, as discussed above, occurs on a timescale of 10s or 100s of picoseconds. Whether this would be possible or not, was a key question in determining if the efficiency of a solar cell could be increased by MEG in practise. However, measurements in 2010 [41] showed that electron injection from PbSe QDs into a TiO_2 surface was less than 50 fs i.e. an order of magnitude smaller than lifetime of a biexciton in PbSe QDs [36]. This immediately suggested that device architectures involving QDs adsorbed onto TiO_2 would be well suited to the exploitation of MEG, with an obvious candidate design being a Grätzel-type cell in which the QDs act as the sensitizer in place of the ruthenium dye.

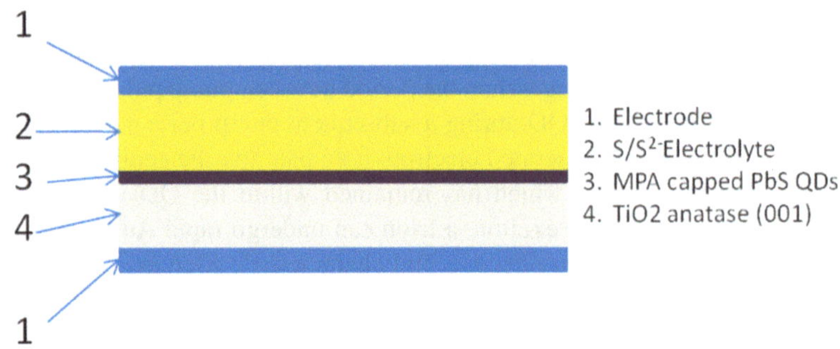

Fig. 9.5 Schematic of the device used to demonstrate a photocurrent enhanced by MEG [42] consisting of: (*1*) two electrodes; (*2*) an aqueous S/S^{2-} electrolyte; (*3*) MPA-capped PbS QD monolayer and (*4*) an atomically flat (0 0 1) anatase TiO_2 substrate (*MPA* mercaptoproprionic acid)

The first clear demonstration [42] of a photocurrent enhanced by MEG was indeed in just such a Grätzel-type device. The device consisted of a monolayer of mercaptoproprionic (MPA) coated PbS QDs of different band gaps adsorbed on to an atomically flat single crystal of anatase (0 0 1) TiO_2 crystal, and was completed by a sulphide-based electrolyte and a top electrode—see Fig. 9.5. The experimental study reported by Sambur et al. [42] involved illuminating this device with different wavelengths of monochromatic light whilst monitoring the absorbed photon to current efficiency (APCE). For photon energies of $2.9E_g$ and $3.1E_g$ the APCE exceeded unity demonstrating both that MEG was occurring and that the additional charges produced by MEG could be efficiently extracted into an external circuit, and was an important result for these reasons. However, the power conversion efficiency (PCE) for this device was very low due to the low fractional absorption of light by the QD monolayer on the atomically flat surface, and hence it was not a practical solar cell device design.

The demonstration of solar cell performance enhanced by MEG in a realistic device design came at the end of 2011 by Semonin et al. [43]. In this case, the cell was of a depleted heterojunction (DHJ) design. This approach replaces the electrolyte with further layers of QDs and hence obviates the problems of leakage of liquid electrolytes and corrosion of the QDs by the electrolyte. The particular architecture used in this study is shown in Fig. 9.6. Here a layer of ZnO is deposited onto indium tin oxide coated glass to form the transparent front electrode. On to this were deposited successive layers of PbSe QDs; the QDs closest to the ZnO acted as the light absorbing species (and hence the location of the MEG process) whilst the subsequent layers acts as the hole transporting medium; the device is completed with a gold top electrode. Under AM1.5 simulated sunlight this device had an overall PCE of 4.5 %, similar to the best efficiencies for QD-based devices at the time. MEG was demonstrated by illuminating the device with a range of photon energies and monitoring the APCE. Above $3E_g$ the APCE rose to 114 % (corresponding to an internal quantum efficiency of 130 %) demonstrating that functional photovoltaic devices could benefit from a photocurrent enhanced by MEG.

Fig. 9.6 Schematic of the PbSe QD sensitized bulk heterojunction device used to demonstrate MEG in a solar cell [43] consisting of: (*1*) A thermally deposited Au electrode; (*2*) 50–250 nm layer of PbSe QDs deposited in a layer by layer method with hydrazine and 1,2-ethandithiol; (*3*) 40–50 nm of ZnO deposited on (*4*); and (*5*) ITO coated glass

9.5 Future Directions

9.5.1 Nanoheterojunctions

One of the fundamental challenges of quantum dot-based solar cells (QDSCs) is to extract the electron from the photoexcited QD before exciton recombination occurs. Many initial QDSCs were based on simple QDs, such as CdS [44], CdSe [45] and PbS [46], and their efficiencies were significantly lower than organic solar cells. Important factors which limited their efficiency were the large exciton recombination rates and the low electron-injection efficiencies. Recent developments in material synthesis have allowed the synthesis of complex nanoheterojunctions composed of multiple component materials. These nanoheterojunctions allow control of the optical and physical properties of QDs by varying the thickness and nature of the junction materials. Careful consideration of the optical properties of the component materials and their orientations has enabled wave-function engineering to be developed, which has led to significant improvements in both electron extraction rates as well as greatly reducing exciton recombination rates ultimately leading to increased photovoltaic efficiency [47].

Nanoheterojunctions are created by synthesizing core/shell nanoparticles. These nanoheterojunctions can take three different distinct forms depending on the materials used to form the core/shell systems, each having their own advantages (see Fig. 9.7). The simplest is a type-I system where the electron and hole are confined in the same volume within the quantum dot whether this be in the core for type-I systems (e.g. CdSe/CdS), or in the shell, for inverse type-I systems (e.g. CdS/CdSe). A type-II system is one where the electron and hole reside in spatially different regions of the QD with one carrier being localized in the core and the other in the shell (e.g. CdSe/CdTe). The final system is a quasi-type-II system where one charge carrier is localized in one region of the QD whereas the other charge carrier is delocalized throughout the QD (e.g. CdSe/ZnSe).

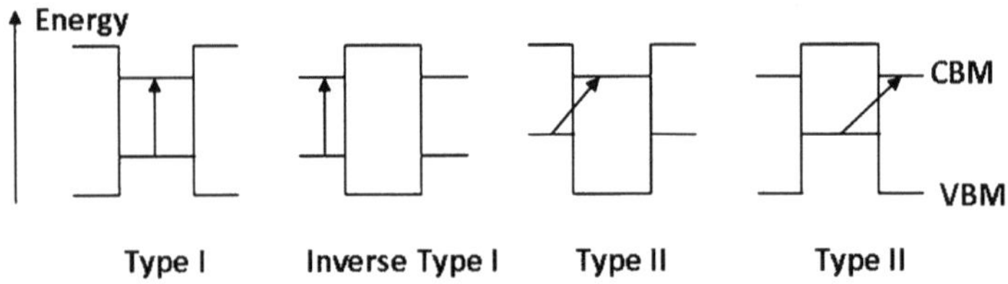

Fig. 9.7 Schematic showing the electronic band alignments for type-I, inverse type-I and type-II quantum dots, with the *arrows* indicating the lowest energy transitions. *CBM* conduction band minimum, *VBM* valance band maximum

The versatility of these nanoheterojunctions enables an increase in electron extraction rates by allowing control over the spatial positioning of the electron. In the simple QDSCs initially tested the electron is confined deep within the QD and extraction of this electron to the electrode proved to be a slow process compared to recombination. However, in nanoheterojunctions the electron can be forced to reside in the shell of the quantum dot and hence close to the surface to enable facile and rapid extraction. The simplest way to ensure the electron is in the shell, is to synthesize an inverse type-I system. This has been carried out using a CdS/CdSe quantum dot, whose electron injection rate increased fourfold when compared with its CdSe/CdS analogue [48]. The efficiency of this system was further increased by over coating the CdS/CdSe, after deposition onto the TiO_2, with a final CdS shell to form a quantum well structure. This increased efficiency was attributed to a reduced electron recombination rate with the redox couple in the electrolyte.

Although increased electron extraction rates have been realized with inverse type-I systems, this QD design is not well suited to the reduction of exciton recombination rates because carriers are confined to the same volume of the QD. To greatly reduce exciton recombination rates a more complex type-II quantum dot has to be synthesized. In these type-II systems (e.g. CdTe/CdSe) the staggered alignment of the component band gaps leads to charge separation and the localization of the electron in the shell and the hole in the core (or vice versa). This separation results in a greatly reduced overlap of the electron and hole wave-functions and hence a large decrease in exciton recombination rates [49]. A disadvantage of these type-II systems is the increased rate of non-radiative recombination caused by trap sites at the interface of the core/shell. However, further developments have been investigated involving the creation of graded interfaces by using alloys of the two components at the interface. This has been shown to "smooth" the potential difference at the interface and relax any lattice defects caused by a change in lattice constants of the parent materials. The formation of these graded structures has led to greatly reduced Auger recombination rates [50].

A further advantage of a type II QD architecture is that it allows a smaller E_g to be achieved than a type I QD made from either of the core or shell materials. As discussed in Sect. 9.2 and illustrated in Fig. 9.2, obtaining an effective band gap in the near infrared is key to the efficient exploitation of the solar spectrum. As shown

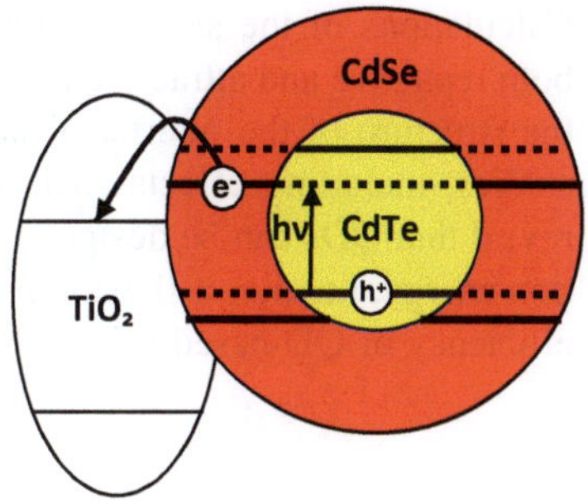

Fig. 9.8 Schematic showing a type-II quantum dot which undergoes charge separation under irradiation resulting in a long lived electron which can be extracted to a TiO_2 surface

in Fig. 9.8, unlike simple type-I quantum dots, where the lowest energy transition is confined to a single material, in a type-II system it is a core–shell transition or vice versa, due to the staggered nature of the component materials band gaps. Consequently this results in a narrower band gap than is obtainable by any of the component materials alone, allowing the E_g to be tuned to longer wavelengths.

Type II QD structures are also of particular benefit to MEG for two reasons. Firstly, the quantum yield of the MEG process results from the competition between the rate of MEG and the rate of electron cooling, which can be greatly decreased by using these type-II systems. In simple type-I quantum dots, the excess energy possessed by hot electrons is quickly transferred to the holes which have a much larger density of states resulting in rapid cooling. However, with the charge separation that occurs in type-II systems, there is a greatly reduced overlap of the wave-functions of the electron and the hole which greatly decreases the rate of cooling via this pathway. It has been shown that the $1P_e$ to $1S_e$ cooling rate for type-II systems can be decreased by three orders of magnitude, with cooling times exceeding 1 ns having been published [51]. This dramatic reduction in the rate of cooling could allow MEG processes to dominate and hence a step change in solar cell efficiency could be realized.

The second benefit of a type II structure to MEG in particular is that the separation of charges it produces increases the magnitude of Δ_{xx} which, as described in Sect. 9.2, reduces the exciton creation energy and thereby improves the potential improvement of photovoltaic efficiency produced by MEG (see Fig. 9.2). Upon the formation of a biexciton in type-II systems a strong localized electric field is produced across the interface due to the imbalance of positive and negative charge, resulting in an enhanced biexciton interaction energy. Experiments to date have found that this interaction to be repulsive for type-II QDs, i.e. $\Delta_{xx} > 0$ [12]. This is beneficial for lasing applications [12] but detrimental for photovoltaic devices because it would increase the threshold for MEG (see Sect. 9.2). These experiments are consistent with a simple perturbative model [52] that considers the Coulombic interactions between the four charges that comprise a biexciton as well as the interaction of each charge with the dielectric interface between the core and the shell. However, recent reports have suggested that this model does not include all the important physical effects [53] and does not fully describe the potential for engineering Δ_{xx} by control of the nanoheterojunction.

Calculations of the same ZnSe/CdS species but with different geometries led to both repulsive and attractive [54, 55] biexciton interaction energies, indicating that the structure of the interfacial layer may be important. More detailed theoretical studies, including effects such as band bending and electron correlation, could reveal that QDs can be designed to exhibit a large and attractive biexciton interaction energy, i.e. $\Delta_{xx} < 0$, which would enhance the potential benefit of MEG to the efficiency of QD-based solar cells.

9.5.2 *Semi-Metals*

The band gap of QDs is broadly given by the sum of the bulk band gap, E_g^{bulk}, and the confinement energy, E_C, i.e. $E_g = E_g^{bulk} + E_C$. However, in bulk semi-metal materials, such as HgTe and α-Sn, the VBM and CBM are overlapped by a very small value such that $E_g^{bulk} \sim 0$ and hence the band gap of semi-metal QDs is determined almost entirely by quantum confinement. This enables E_g for these QD to be widely size-tuned throughout the infrared, allowing it to be optimized for efficient exploitation of the solar spectrum [56]. There are also reasons to expect that MEG will be particularly efficient in semi-metal QDs, making them even more well suited for QDSCs.

9.5.2.1 High Density of States at the Band Edge

The rate of impact ionization and thus MEG quantum yield increases with the density of states in the region of the band edge. A recent theoretical study [57] showed that density of states increases much more rapidly from the band edge the more the band gap of the QD was greater than E_g^{bulk}. Example calculations were reported for the impact ionization lifetime (i.e. the reciprocal of the rate) for PbS QDs, which dropped more quickly from its band edge value as E_g increased with reducing QD diameter—see Fig. 9.9

This analysis suggests that for improved MEG efficiency QDs with a minimized value of E_g^{bulk} should be used i.e. semi-metals. The authors of [57] selected α-Sn as an example semi-metal material and calculated the MEG quantum yield as a function of photon energy for a range of QD sizes—see Fig. 9.10. It was found that the lifetime of impact ionization is smaller than 1 ps (the assumed thermalization lifetime) for a wide range of the excited carrier energies (2–4 eV). Furthermore, even at lower energies (i.e. less than 2 eV) it remains less than 1 ps for QDs with 3.6 nm size or greater. The quantum yield also shows sharp peaks not observed in similar calculated spectra for PbSe, PbS, Si and InAs QDs that are attributed to the high efficiency of the optical transitions between electron and hole states and the high rates of impact ionization. Moreover, decreasing the band gap of the α-Sn QDs

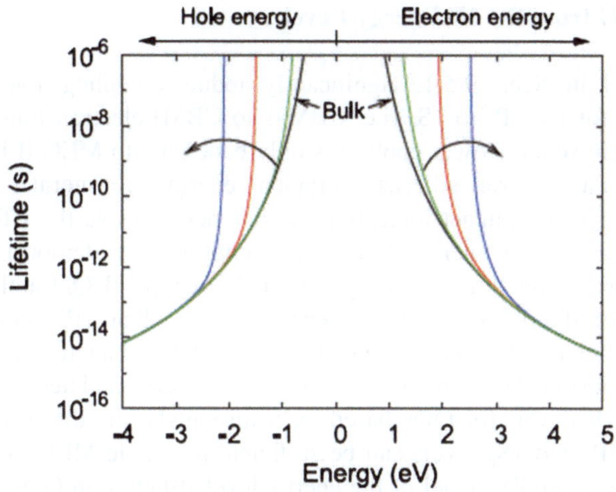

Fig. 9.9 Impact ionization lifetime against the energy of the photoexcited carriers for PbS. *Arrows* indicate the decrease in size of the QDs and consequent increase in the band gap. Reprinted with permission from [57]. Copyright 2011 American Chemical Society

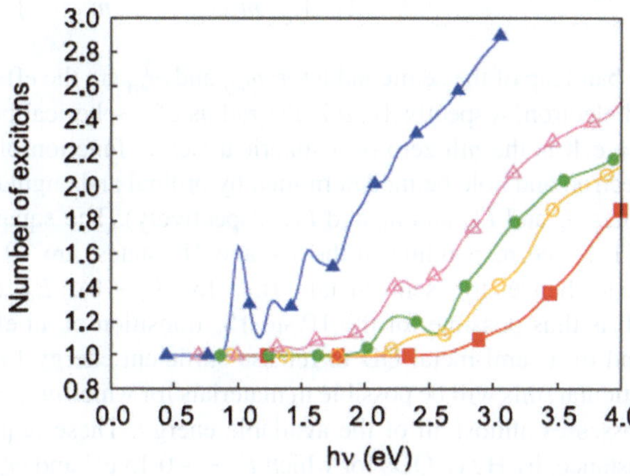

Fig. 9.10 Quantum yield (number of excitons created per absorbed photon) for different QDs sizes corresponding to $E_g = 0.44$ eV (*filled triangles*), $E_g = 0.69$ eV (*open triangles*), $E_g = 0.86$ eV (*filled circles*), $E_g = 0.94$ eV (*open circles*), and $E_g = 1.24$ eV (*filled squares*). Reprinted with permission from [57]. Copyright 2011 American Chemical Society

led to a decreasing MEG threshold (without the saturation that appeared above 1.8 eV in similar calculations for PbSe and PbS QDs) scaling as $h\nu_2 \simeq 2.15E_g$, which is very close to the ideal threshold ($2E_g$) and emphasizes the importance and the benefits of such zero band gap materials to achieve small MEG threshold. As a consequence of the low MEG threshold in α-Sn QDs, the energy efficiency stays above 38 % while it reaches only 25 % for PbSe and PbS. QDs [57].

9.5.2.2 MEG from the 1P Energy Level

As mentioned in Sect. 9.5.1, significantly reduced cooling rates have been demonstrated for the $1P_e$ to $1S_e$ (i.e. CBM-1 to CBM) electron transition in type II QDs [51]. However, slower cooling is only beneficial to MEG if it comes in to effect whilst the electron still has sufficient energy to generate an additional electron–hole pair by impact ionization i.e. its energy above the CBM is greater than the exciton creation energy. To date, it has not been yet demonstrated whether the reduced wave-function overlap produced by a type II QD architecture also produces reduced cooling at higher energy levels than $1P_e$. For most QDs, the energy difference between an electron in the $1P_e$ conduction band level and one in the $1S_e$ level is less than the exciton creation energy and hence not enough to allow MEG. However, for QDs based on semi-metals, the difference in energy between the $1P_e$ and $1S_e$ levels can be sufficient to enable MEG, as can be seen by considering a simple model of the energy level structure in QDs. For a spherically symmetric QD, the energy of an electron–hole pair, E_{ehp}, is approximately given by [9]:

$$E_{\mathrm{ehp}}(n_{\mathrm{h}}L_{\mathrm{h}}n_{\mathrm{e}}L_{\mathrm{e}}) = E_{\mathrm{g}} + \frac{\hbar^2}{2a^2}\left\{\frac{\varphi^2(n_{\mathrm{h}},L_{\mathrm{h}})}{m_{\mathrm{eff}}^{v}} + \frac{\varphi^2(n_{\mathrm{e}},L_{\mathrm{e}})}{m_{\mathrm{eff}}^{c}}\right\} \tag{9.6}$$

where E_{g} is the bandgap of the semiconductor; m_{eff}^{v} and m_{eff}^{c} are the effective masses of the hole and electron, respectively; a is the radius of a spherical potential well; and $\varphi(n_i, L_i) i = \mathrm{e,h}$ is the nth zero of a spherical Bessel function of order L (the states of the electron and hole being determined by ordinal and angular momentum quantum numbers n_{e} and L_{e}, and n_{h} and L_{h}, respectively). The squared values of $\varphi(1,0)$ and $\varphi(1,1)$, corresponding to the $1S_i$ and $1P_i$ states, are ~9.9 and ~20.3, respectively, and hence for semi-metals (i.e. for $E_{\mathrm{g}} \sim 0$), $E_{\mathrm{ehp}}(1P_h 1P_e) > 2$ $E_{\mathrm{ehp}}(1S_h 1S_e)$. It is thus possible for the $1P_e$-to-$1S_e$ transition of an electron in the conduction band of a semi-metal QD to release sufficient energy for MEG to be enabled. In particular, this will be possible in materials for which $m_{\mathrm{eff}}^{c} \ll m_{\mathrm{eff}}^{v}$ so that the electron possesses almost all of the available energy. These requirements are satisfied, for instance, by HgTe QDs for which $E_{\mathrm{g}} \sim -0.15$ eV and $m_{\mathrm{eff}}^{c} = 0.017m_0$ and $m_{\mathrm{eff}}^{v} = 0.5m_0$, where m_0 is the electron mass [58].

9.6 Conclusions

The process of MEG has the potential to significantly increase the efficiency of QD-based solar cells. It has been experimentally demonstrated that the additional charges produced by this process can be extracted very rapidly from a QD, well within the lifetime of the multi-exciton. Moreover, there has been a recent demonstration of a solar cell with an efficiency enhanced by MEG. However, the enhancement of overall solar cell performance demonstrated so far has been marginal. The potential benefit of MEG has not yet been fully realized because in the simple QDs studied so far the quantum yield of MEG is significantly reduced by competition with other hot carrier cooling processes, as demonstrated by the latest spectroscopic studies (in which the results are free from distortion by experimental artefacts that are now understood) [11]. However, as proposed in Sect. 9.5, there are a number of approaches that could be used to improve the quantum yield of MEG. These rely on the scope for engineering the photo-physical properties of QDs by control of their composition and structure, as well as their size. Once a high quantum yield QD design has been developed and tested spectroscopically then the successful device work already reported suggests that these improved QDs could be rapidly incorporated into a solar cell which would have an efficiency significantly enhanced by MEG.

References

1. Beard, M.C., Ellingson, R.J.: Multiple exciton generation in semiconductor nanocrystals: toward efficient solar energy conversion. Laser Photon. Rev. **2**(5), 377–399 (2008)
2. Shah, J.: Hot electrons and phonons under high intensity photoexcitation of semiconductors. Solid State Electron. **21**(1), 43–50 (1978)
3. Christensen, O.: Quantum efficiency of the internal photoelectric effect in silicon and germanium. J. Appl. Phys. **47**(2), 689–695 (1976)
4. Geist, J., Gardner, J.L., Wilkinson, F.J.: Surface-field-induced feature in the quantum yield of silicon near 3.5 eV. Phys. Rev. B **42**(2), 1262–1267 (1990)
5. Hanna, M.C., Nozik, A.J.: Solar conversion efficiency of photovoltaic and photoelectrolysis cells with carrier multiplication absorbers. J. Appl. Phys. **100**(7), 074510 (2006)
6. Klimov, V.I.: Detailed-balance power conversion limits of nanocrystal-quantum-dot solar cells in the presence of carrier multiplication. Appl. Phys. Lett. **89**(12), 123118 (2006)
7. Takeda, Y., Motohiro, T.: Requisites to realize high conversion efficiency of solar cells utilizing carrier multiplication. Solar Energy Mater. Solar Cells **94**(8), 1399–1405 (2010)
8. Shockley, W., Queisser, H.J.: Detailed balance limit of efficiency of p–n junction solar cells. J. Appl. Phys. **32**(3), 510–519 (1961)
9. Klimov, V.I. (ed.): Semiconductor and Metal Nanocrystals, p. 484. Marcel Dekker, New York (2004)
10. Kang, I., Wise, F.W.: Electronic structure and optical properties of PbS and PbSe quantum dots. J. Opt. Soc. Am. B **14**(7), 1632–1646 (1997)
11. Binks, D.J.: Multiple exciton generation in nanocrystal quantum dots – controversy, current status and future prospects. Phys. Chem. Chem. Phys. **13**(28), 12693–12704 (2011)
12. Klimov, V., et al.: Single-exciton optical gain in semiconductor nanocrystals. Nature **447**(7143), 441–446 (2007)
13. Klimov, V.I., Mikhailovsky, A.A., McBranch, D.W., Leatherdale, C.A., Bawendi, M.G.: Quantization of multiparticle Auger rates in semiconductor quantum dots. Science **287**, 1011 (2000)
14. Crooker, S.A., et al.: Multiple temperature regimes of radiative decay in CdSe nanocrystal quantum dots: intrinsic limits to the dark-exciton lifetime. Appl. Phys. Lett. **82**(17), 2793–2795 (2003)
15. Schaller, R.D., Klimov, V.I.: High efficiency carrier multiplication in PbSe nanocrystals: implications for solar energy conversion. Phys. Rev. Lett. **92**(18), 186601 (2004)

16. Ji, M.B., et al.: Efficient multiple exciton generation observed in colloidal PbSe quantum dots with temporally and spectrally resolved intraband excitation. Nano Lett. **9**(3), 1217–1222 (2009)
17. Pijpers, J.J.H., et al.: Assessment of carrier-multiplication efficiency in bulk PbSe and PbS. Nat. Phys. **5**(11), 811–814 (2009)
18. Hardman, S.J.O., et al.: Electronic and surface properties of PbS nanoparticles exhibiting efficient multiple exciton generation. Phys. Chem. Chem. Phys. **13**(45), 20275–20283 (2011)
19. Ben-Lulu, M., et al.: On the absence of detectable carrier multiplication in a transient absorption study of InAs/CdSe/ZnSe core/shell1/shell2 quantum dots. Nano Lett. **8**(4), 1207–1211 (2008)
20. Cadirci, M., et al.: Ultrafast exciton dynamics in InAs/ZnSe nanocrystal quantum dots. Phys. Chem. Chem. Phys. **14**(43), 15166–15172 (2012)
21. Schaller, R.D., et al.: High-efficiency carrier multiplication and ultrafast charge separation in semiconductor nanocrystals studied via time-resolved photoluminescence. J. Phys. Chem. B **110**(50), 25332–25338 (2006)
22. Nair, G., et al.: Carrier multiplication yields in PbS and PbSe nanocrystals measured by transient photoluminescence. Phys. Rev. B **78**(12), 125325 (2008)
23. Nair, G., Bawendi, M.G.: Carrier multiplication yields of CdSe and CdTe nanocrystals by transient photoluminescence spectroscopy. Phys. Rev. B **76**(8), 081304 (2007)
24. McGuire, J.A., et al.: Apparent versus true carrier multiplication yields in semiconductor nanocrystals. Nano Lett. **10**(6), 2049–2057 (2010)
25. Hadfield, R.H.: Single-photon detectors for optical quantum information applications. Nat. Photonics **3**(12), 696–705 (2009)
26. Koc, S.: The quantum efficiency of the photo-electric effect in germanium for the 0.3–2 μ wavelength region. Czechoslovak J. Phys. **7**(1), 91 (1957)
27. Smith, A., Dutton, D.: Behavior of lead sulfide photocells in the ultraviolet. J. Opt. Soc. Am. **48**(12), 1007–1009 (1958)
28. Wolf, M., et al.: Solar cell efficiency and carrier multiplication in Si[sub 1 − x]Ge[sub x] alloys. J. Appl. Phys. **83**(8), 4213–4221 (1998)
29. Nozik, A.J.: Spectroscopy and hot electron relaxation dynamics in semiconductor quantum wells and quantum dots. Annu. Rev. Phys. Chem. **52**(1), 193–231 (2001)
30. Ellingson, R.J., et al.: Highly efficient multiple exciton generation in colloidal PbSe and PbS quantum dots. Nano Lett. **5**(5), 865–871 (2005)
31. Murphy, J.E., et al.: PbTe colloidal nanocrystals: synthesis, characterization, and multiple exciton generation. J. Am. Chem. Soc. **128**(10), 3241–3247 (2006)
32. Schaller, R.D., Petruska, M.A., Klimov, V.I.: Effect of electronic structure on carrier multiplication efficiency: comparative study of PbSe and CdSe nanocrystals. Appl. Phys. Lett. **87**(25), 253102–253103 (2005)
33. Schaller, R.D., Pietryga, J.M., Klimov, V.I.: Carrier multiplication in InAs nanocrystal quantum dots with an onset defined by the energy conservation limit. Nano Lett. **7**(11), 3469–3476 (2007)
34. Beard, M.C., et al.: Multiple exciton generation in colloidal silicon nanocrystals. Nano Lett. **7**(8), 2506–2512 (2007)
35. Pijpers, J.J.H., et al.: Carrier multiplication and its reduction by photodoping in colloidal InAs quantum dots. J. Phys. Chem. C **112**(12), 4783–4784 (2008)
36. Midgett, A.G., et al.: Flowing versus static conditions for measuring multiple exciton generation in PbSe quantum dots. J. Phys. Chem. C **114**(41), 17486–17500 (2010)
37. McGuire, J.A., et al.: New aspects of carrier multiplication in semiconductor nanocrystals. Acc. Chem. Res. **41**(12), 1810–1819 (2008)
38. Stubbs, S.K., et al.: Efficient carrier multiplication in InP nanoparticles. Phys. Rev. B **81**(8), 081303 (2010)
39. Gachet, D., et al.: An upper bound to carrier multiplication efficiency in type II colloidal quantum dots. Nano Lett. **10**(1), 164–170 (2009)

40. Tyagi, P., Kambhampati, P.: False multiple exciton recombination and multiple exciton generation signals in semiconductor quantum dots arise from surface charge trapping. J. Chem. Phys. **134**(9), 10 (2011)
41. Tisdale, W.A., et al.: Hot-electron transfer from semiconductor nanocrystals. Science **328**(5985), 1543–1547 (2010)
42. Sambur, J.B., Novet, T., Parkinson, B.A.: Multiple exciton collection in a sensitized photovoltaic system. Science **330**(6000), 63–66 (2010)
43. Semonin, O.E., et al.: Peak external photocurrent quantum efficiency exceeding 100% via MEG in a quantum dot solar cell. Science **334**(6062), 1530–1533 (2011)
44. Peter, L.M., et al.: Photosensitization of nanocrystalline TiO2 by self-assembled layers of CdS quantum dots. Chem. Commun. (10), 1030–1031 (2002)
45. Kongkanand, A., et al.: Quantum dot solar cells. Tuning photoresponse through size and shape control of CdSe–TiO2 architecture. J. Am. Chem. Soc. **130**(12), 4007–4015 (2008)
46. Plass, R., et al.: Quantum dot sensitization of organic–inorganic hybrid solar cells. J. Phys. Chem. B **106**(31), 7578–7580 (2002)
47. Zhu, H., Lian, T.: Wavefunction engineering in quantum confined semiconductor nanoheterostructures for efficient charge separation and solar energy conversion. Energy Environ. Sci. **5**(11), 9406–9418 (2012)
48. Ning, Z.J., et al.: Wave-function engineering of CdSe/CdS core/shell quantum dots for enhanced electron transfer to a TiO_2 substrate. J. Phys. Chem. C **114**(35), 15184–15189 (2010)
49. Wang, C.H., et al.: Recombination dynamics in CdTe/CdSe type-II quantum dots. Nanotechnology **19**(11), 15702.1–15702.6 (2008)
50. Garcia-Santamaria, F., et al.: Breakdown of volume scaling in Auger recombination in CdSe/CdS heteronanocrystals: the role of the core–shell interface. Nano Lett. **11**(2), 687–693 (2011)
51. Pandey, A., Guyot-Sionnest, P.: Slow electron cooling in colloidal quantum dots. Science **322**(5903), 929–932 (2008)
52. Piryatinski, A., et al.: Effect of quantum and dielectric confinement on the exciton–exciton interaction energy in type II core/shell semiconductor nanocrystals. Nano Lett. **7**(1), 108–115 (2006)
53. Deutsch, Z., et al.: Energetics and dynamics of exciton–exciton interactions in compound colloidal semiconductor quantum dots. Phys. Chem. Chem. Phys. **13**(8), 3210–3219 (2011)
54. Sitt, A., et al.: Multiexciton engineering in seeded core/shell nanorods: transfer from type-I to quasi-type-II regimes. Nano Lett. **9**(10), 3470–3476 (2009)
55. Avidan, A., Deutsch, Z., Oron, D.: Interactions of bound excitons in doped core/shell quantum dot heterostructures. Phys. Rev. B **82**(16), 6 (2010)
56. Im, S.H., et al.: Efficient HgTe colloidal quantum dot-sensitized near-infrared photovoltaic cells. Nanoscale **4**(5), 1581–1584 (2012)
57. Allan, G., Delerue, C.: Optimization of carrier multiplication for more efficient solar cells: the case of Sn quantum dots. ACS Nano **5**(9), 7318–7323 (2011)
58. Berger, L.I.: Properties of semiconductors. In: Lide, D.R. (ed.) Handbook of Physics and Chemistry, p. 90. CRC, Boca Raton (2009)

Chapter 10
Graphene Quantum Dot-Based Organic Solar Cells

Vinay Gupta, Tanvi Upreti, and Suresh Chand

Abstract Current research in organic photovoltaic (OPV) is largely focused on the development of low cost OPV materials such as semiconductor quantum dots (QDs). Graphene quantum dots (GQDs) are a fascinating class of QDs having size below 10 nm. They have emerged as an alternative to semiconductor QDs in photovoltaics due to their size-dependent photoluminescence (PL) and tunable band gap properties. They are expected to be a versatile candidate due to their low cost, non-toxicity, and biocompatibility. Recently, it has been shown that they are promising for efficient light harvesting in solar cells. Keeping this in view, we present a comprehensive review of the progress made so far for the application of GQDs in organic solar cells.

10.1 Introduction

Graphene is a two-dimensional sheet of sp^2-hybridized carbon. Its extended honeycomb network can be stacked to form a 3D graphite, 1D nanotube, and 0D fullerenes. Discovered by Novoselov and his group in 2004 [1], it is considered to be one of the most promising materials for the future nanotechnology owing to its low cost coupled with the superior electronic, thermal, and mechanical properties as well as an enhanced chemical stability. The most exciting characteristics are ambipolar field effect [1], the quantum Hall effect [2–7], and extremely high carrier mobility [8–11] that can find several applications including in optoelectronic devices. The use of graphene sheets (GSs) in the solar cell devices is possible due to the efficient charge transfer in graphene. However, the zero band gap in graphene imposes limitations for optoelectronic applications, since a finite band gap is

V. Gupta (✉) • T. Upreti • S. Chand
Physics of Energy Harvesting Division, Organic and Hybrid Solar Cell Group, CSIR-National Physical Laboratory, New Delhi 110012, India
e-mail: guptavinay@nplindia.org; drvinaygupta@netscape.net

J. Wu and Z.M. Wang (eds.), *Quantum Dot Solar Cells*, Lecture Notes in Nanoscale Science and Technology 15, DOI 10.1007/978-1-4614-8148-5_10,

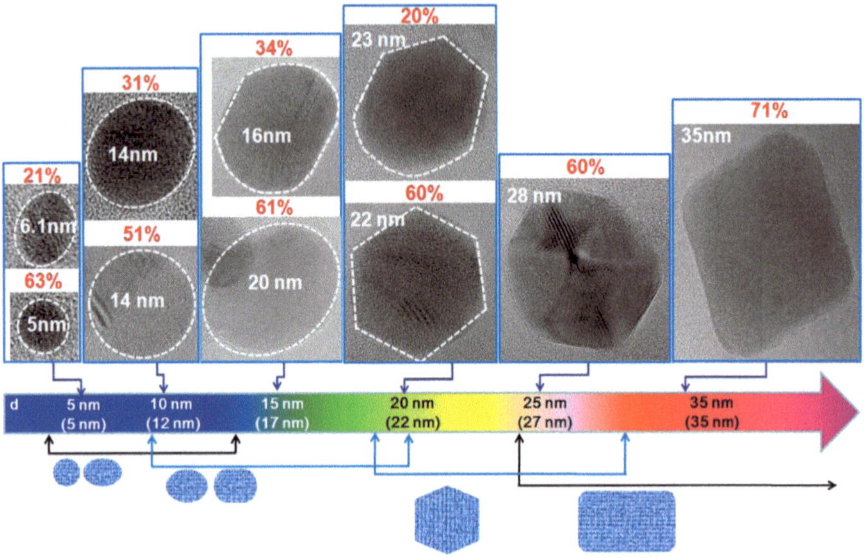

Fig. 10.1 HRTEM images of GQDs for their major shapes and corresponding populations (p) with increasing average size of GQDs. Here, the *dotted line* indicates the region of a GQD and p is defined as the ratio of number of GQDs with a major shape at each average size. Average sizes (d_a) of GQDs estimated from the HRTEM images at each d are indicated in the parentheses at the bottom of this figure. The connected *arrows* indicate the range of the average size in which GQDs with particular major shapes are found (reprinted with permission from [23]. Copyright (2012) American Chemical Society)

favorable to achieve sharp fluorescent peaks that are key in such applications. A monolayer of bulk graphene (~3.4 Å thick) remarkably absorbs 2.3 % of incident light [11, 12] and has an extinction coefficient as large as 10^5 M^{-1} cm^{-1} similar to the small aromatic compounds [11, 12]. The size-dependent bandgap of graphene [13, 14] and their large optical absorptivity [11] are some of the key factors for them being used in the light harvesting in photovoltaic devices. Also in bulk graphene having a zero bandgap, the absorption edge of the graphene can be tuned from zero to few eV by varying their sizes [13, 14], enabling it to cover the whole spectrum of sunlight (Fig. 10.1).

A band-gap in graphene can be modified/altered by lithography, nanoimprinting, or chemical modifications to generate graphene nanoribbon (GNR), graphene nanomesh (GNM), and graphene quantum dots (GQDs) [13–60]. These nanosized graphene structures show tunable band gap characteristics depending on the size, shape, and edge or surface-functionalization. Compared to 1D and 2D graphene nanostructures, zero-dimensional GQDs possesses stronger quantum confinement and edge-effect, resulting in different kinds of photoluminescence (PL) characteristics that are of significant importance for light harvesting. GQDs were first proposed and synthesized by Yan et al. [13, 20] by bottom-up approach, following which many unique quantum characteristics of GQDs were discovered.

The first top-down approach was described by Pan et al. [19]. The applications of GQDs in solar cells were shown by Gupta et al. [21] and by Li et al. [22] and since then the research on GQDs has intensified significantly and various synthetic routes have been proposed recently which can be classified into two main categories: top-down and bottom-up methods [23]. GQDs have the real potential of replacing toxic and expensive heavy-metals such as semiconductor nanocrystals with their low cost, optical tenability, and solution processability in solar cells which is a new key challenging area and shall be covered in this chapter.

10.2 Solar Cell Devices

Yan et al. [13] reported a novel solubilization strategy for solution-processable GQDs (Fig. 10.2) and showed that GQDs can be used as sensitizers in solar cells due to their high optical absorptivity and nearly optimized absorption in the visible and near-IR region. As shown in Fig. 10.3, they calculated that the energy levels of the highest occupied and the lowest unoccupied molecular orbitals (HOMO and LUMO) of GQDs are, respectively, at 5.3 and 3.8 eV below the vacuum level. These values, relative to the band levels of TiO_2 and reduction potential of I_3^-/I^- (Fig. 10.3b), make it possible to use quantum dot 1 as a sensitizer in dye-sensitized solar cells made up of nanocrystalline TiO_2 particles. Upon photoexcitation, in principle, GQD is capable of getting oxidized by injecting an electron to TiO_2 and then getting reduced by taking an electron from I^-. Figure 10.3c shows a current–voltage characteristics of such a device under a simulated AM 1.5 Global

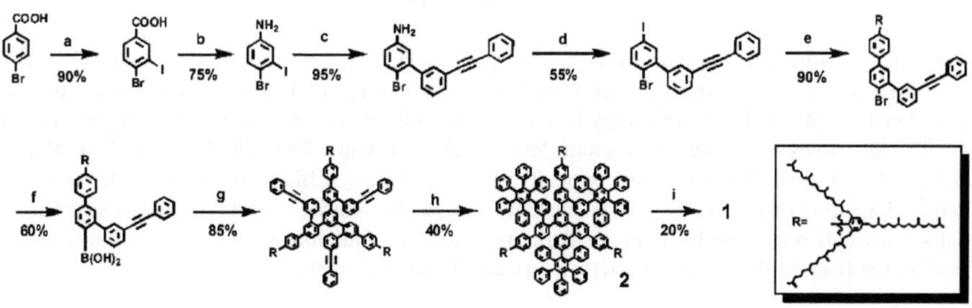

Fig. 10.2 Synthesis of graphene quantum dot 1. Steps are as follows: (**a**) $NaIO_4$, I_2, concentrated H_2SO_4, room temperature. (**b**) Heated with diphenylphosphoryl azide in triethylamine and *tert*-butanol at 80 °C, followed by treatment with CF_3COOH in dichloromethane at room temperature. (**c**) Suzuki condition with 3-(phenylethynyl) phenylboronic acid, $Pd(PPh_3)_4$, K_2CO_3 in water, ethanol, and toluene mixture, 60 °C. (**d**) Iodine and *tert*-butyl nitrite in benzene, 5 °C to room temperature. (**e**) Suzuki condition with substituted phenyl boronic acid, $Pd(PPh_3)_4$, K_2CO_3 in water, ethanol, and toluene mixture, 80 °C. (**f**) Treatment with butyllithium in tetrahydrofuran (THF) at −78 °C, then with triisopropyl borate at −78 °C, followed by treatment with acidic water at room temperature. (**g**) Suzuki condition with 1,3,5-triiodobenzene, $Pd(PPh_3)_4$, K_2CO_3 in water and toluene mixture, 80 °C. (**h**) Tetraphenylcyclopentadienone in diphenylether, 260 °C. (**i**) $FeCl_3$ in nitromethane and dichloromethane mixture, room temperature (reprinted with permission from [13]. Copyright 2010 American Chemical Society)

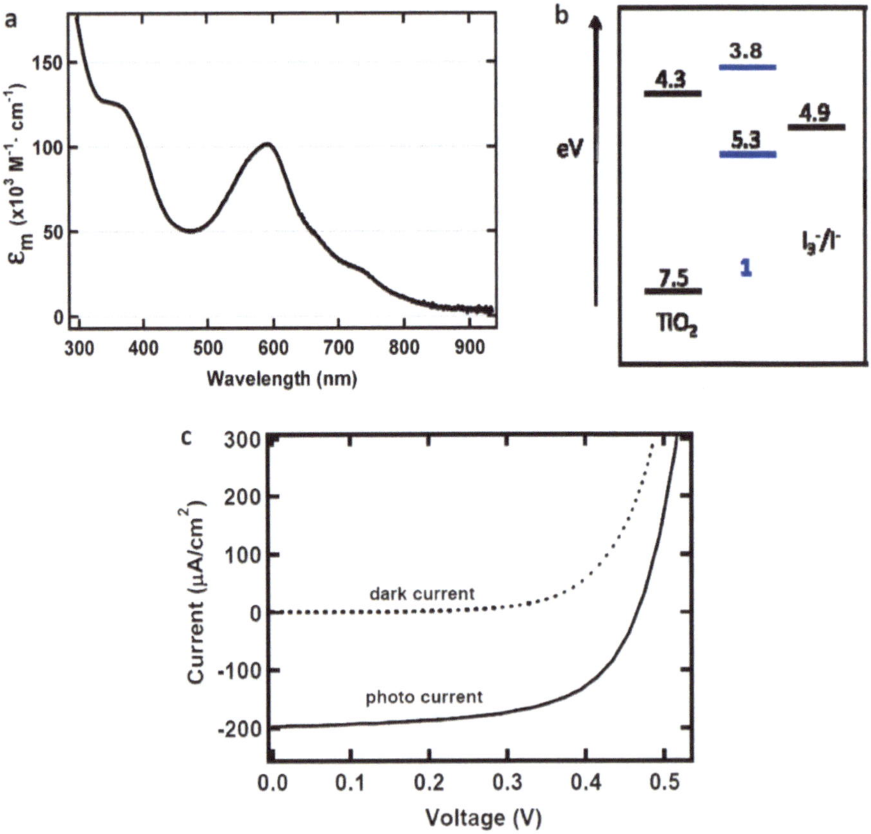

Fig. 10.3 Absorption spectrum of **1** and its application as a sensitizer in nanocrystalline solar cells. (**a**) Absorption spectrum (molar extinction coefficient ε_m) of **1** in dichloromethane. (**b**) The calculated HOMO and LUMO energy levels of **1**, band levels of TiO_2, and reduction potential of I_3^-/I^-. All values are below the vacuum level in electron volts. The calculated bandgap of **1** is 0.1 eV larger than that measured optically. It nevertheless suffices to illustrate the possible sensitizing effect of **1** on TiO_2. (**c**) The current–voltage characteristics of a typical nanocrystalline TiO_2 solar cell sensitized by **1**, in the dark and under illumination, respectively (reprinted with permission from [13]. Copyright 2010 American Chemical Society)

light and in the dark, respectively. A short-circuit current density of 200 μA cm^{-2} and an open-circuit voltage of 0.48 V were produced with a fill factor of 0.58. These GQDs can find significant applications in photoelectrochemical cells, especially for near-infrared light absorption.

Li et al. [22] fabricated the GQD-based bulk heterojunction (BHJ) polymer solar cell and demonstrated its potential. They applied colloidal GQDs with green luminescence as an electron-acceptor material in the conjugated polymer poly (3-hexylthiophene) (P3HT)-based thin film solar cells. A power conversion efficiency of 1.28 % was achieved without device optimization. Figure 10.4a shows the

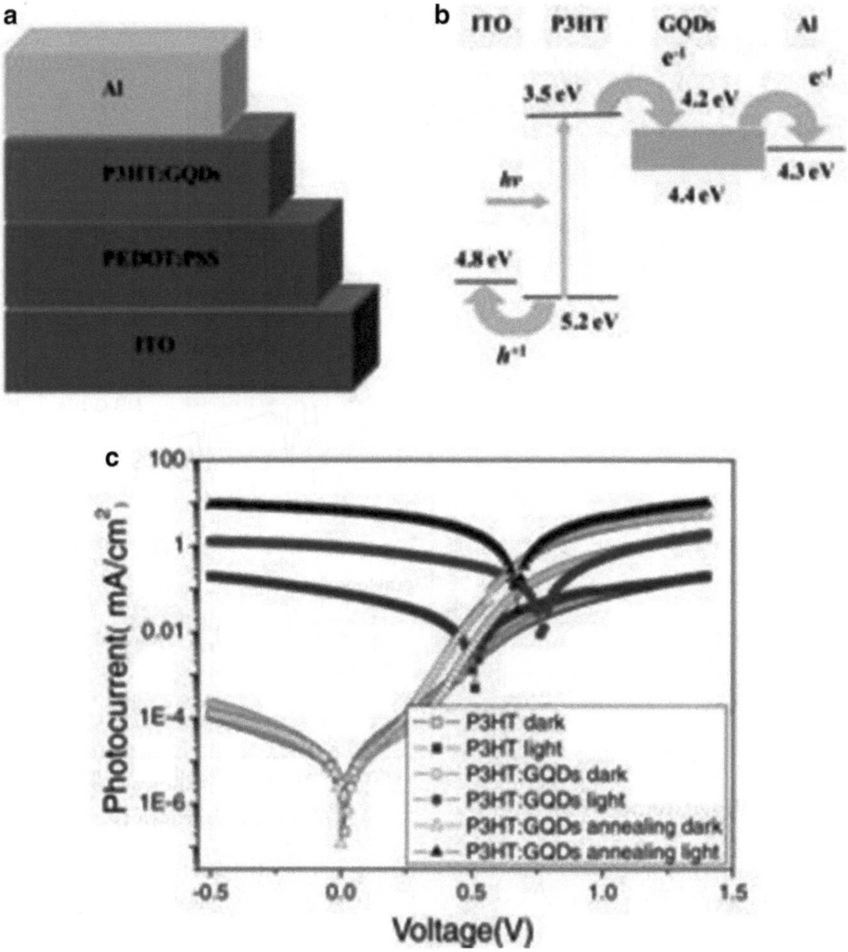

Fig. 10.4 Schematic (**a**) and energy band (**b**) diagrams of the ITO/PEDOT:PSS/P3HT:GQDs/Al device. (**c**) *J–V* characteristic curves for the ITO/PEDOT:PSS/P3HT/Al, ITO/PEDOT:PSS/P3HT: GQDs/Al and ITO/PEDOT:PSS/P3HT: GQDs/Al devices after annealing at 140 °C for 10 min, single log scale (reprinted with permission from [22]. Copyright (2011) John Wiley and Sons)

typical polymer photovoltaic cell configuration ITO|PEDOT:PSS|P3HT:GQDs|Al that was fabricated (ITO, PEDOT, PSS, and P3HT corresponds to indium tin oxide, poly (3, 4-ethylenedioxythiophene), poly (styrenesulfonate), and poly (3-hexylthiophene), respectively). The optimized GQDs (10 wt.%) condition provided an effective interface for the charge separation and a pathway for electron transport. The energy level diagram for the GQD-based photovoltaic cells is shown in Fig. 10.4b. The current density–voltage (*J–V*) characteristics of a typical undoped P3HT device with those of one blended with GQDs (before and after annealing) in the dark and under illumination are compared in Fig. 10.4c.

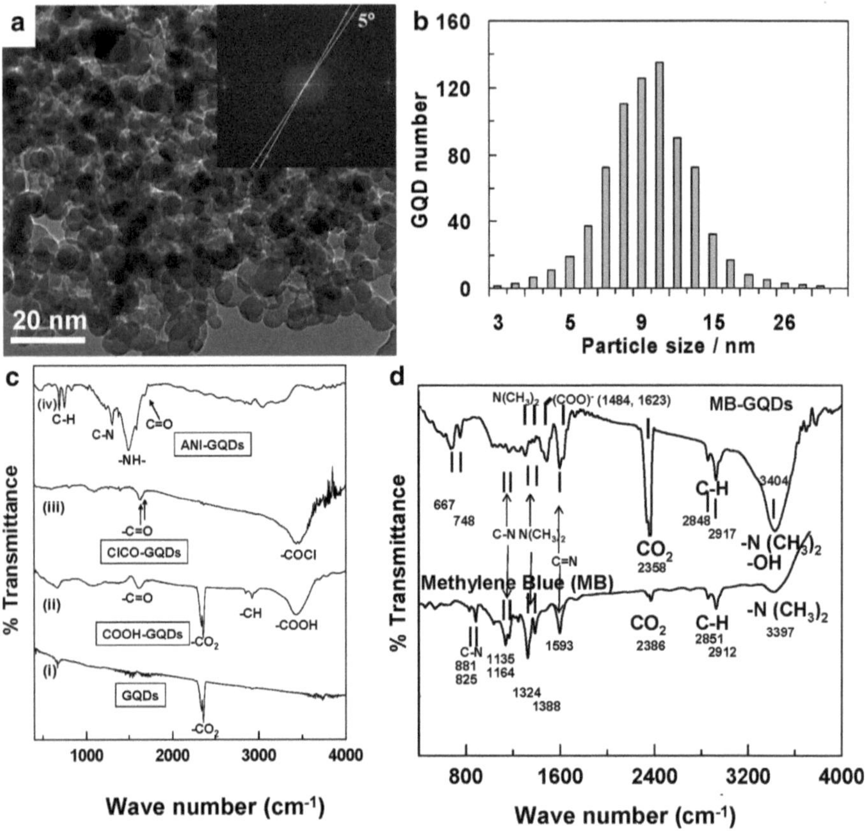

Fig. 10.5 (**a**) TEM image, (**b**) diameter distribution, and (**c**, **d**) FTIR spectra of the GQDs (reprinted with permission from [21]. Copyright (2011) American Chemical Society)

Gupta et al. [21] showed that solar cells made up of GQDs blended with a regioregular P3HT polymer gave a significant improvement in the efficiency over those containing graphene nanosheets (GS) blended conjugated polymers. The GQDs were synthesized from graphene nanosheets by the hydrothermal approach [19] as shown in Fig. 10.5, which were functionalized with aniline (ANI) or methylene blue (MB) (ANI-GQDs and MB-GQDs). The UV–vis absorption spectra and the optical images of GQDs, ANI-GQDs, and MB-GQDs (in ethanol) are shown in Fig. 10.6a. The spectrum of GQDs shows a typical absorption peak at 292 and 325 nm. The two electronic transitions at 325 nm (3.81 eV) and 292 nm (4.24 eV) observed in the photoluminescence (PL) spectrum (Fig. 10.6b) of GQDs can be regarded as the transition from σ and π orbitals, viz. highest occupied molecular orbital (HOMO) to lowest unoccupied molecular orbital (LUMO). The PL spectrum of ANI-GQDs (Fig. 10.6b) is extended from 400 to 700 nm while that for MB-GQDs (475–700 nm) is narrow and extended from 475 to 700 nm (centered at 525 nm), giving green luminescence [21]. The PL emission spectra

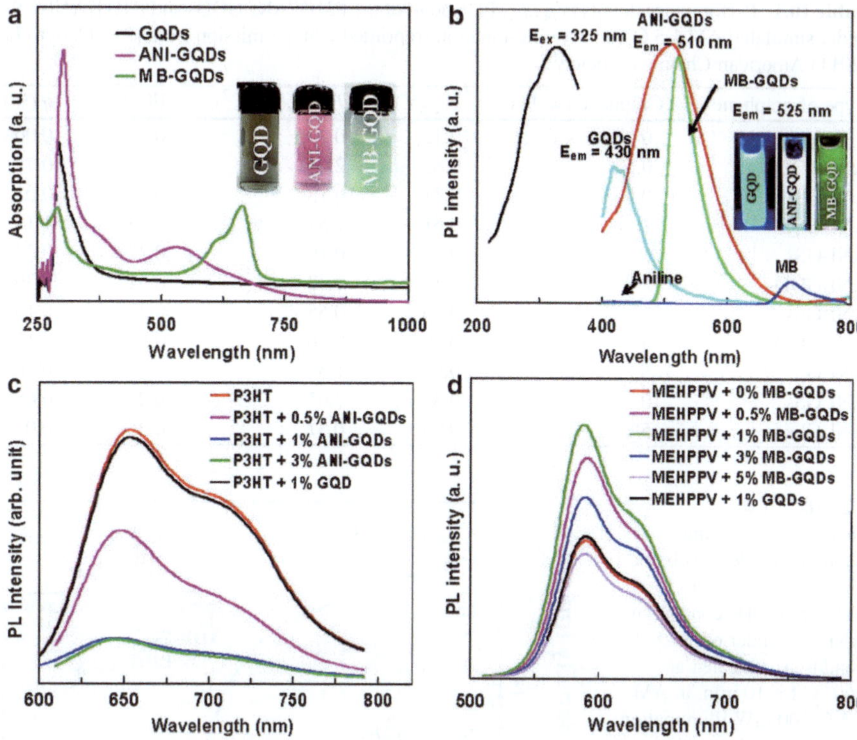

Fig. 10.6 (**a**) UV–vis absorption, (**b**) PL spectra of GQDs, ANI-GQDS, MB, Aniline, and MB-GQDs, (**c**) PL spectra of P3HT/ANI-GQDs, and (**d**) PL spectra of MEH-PPV/MB-GQDs (reprinted with permission from [21]. Copyright (2011) American Chemical Society)

(E_{ex} = 510 nm) of the films of poly(3-hexylthiophene-2,5-diyl) (P3HT) blended with ANI-GQDs in different wt.% ratios (0, 0.5, 1, 3, 5 wt.%) are shown in Fig. 10.6c [21]. The PL shows a quenching behavior with saturation at 1 wt.% ANI-GQDs. The PL emission spectra (E_{ex} = 450 nm) of the films of MB-GQDs blended with MEH-PPV in different wt.% ratios (0, 0.5, 1, 3, 5 wt. %) are shown in Fig. 10.6d. The solar cell characteristics were measured under 100 mW AM 1.5 G illumination. The obtained power conversion efficiency (η), open-circuit voltage (V_{oc}), short-circuit current density (J_{sc}), and fill factor (FF) are summarized in Table 10.1. It can be seen that the FF of the P3HT/ANI-GQDs heterojunction device is much higher (0.53) as compared to ~0.33 for GSs. Current density versus voltage (J–V) curves of the ANI-GQD and ANI-GS (optimized) PV devices are plotted in Fig. 10.7. The corresponding band diagram is given in the inset of Fig. 10.7. Maximum values of $\eta = 1.14$, $V_{oc} = 0.61$ V, $J_{sc} = 3.51$ mA cm^{-2}, and FF = 0.53 were obtained for 1 wt.% ANI-GQD in P3HT. In comparison, maximum values of $\eta = 0.65$, $V_{oc} = 0.88$ V, $J_{sc} = 2.65$ mA cm^{-2}, and a low FF = 0.28 were obtained for 10 wt.% ANI-GSs in P3HT. The corresponding band diagram is shown in the inset of Fig. 10.7. The position of the LUMO (or work function) of GQDs between those of P3HT and Al suggests their suitability for OPV applications.

Table 10.1 Performance details (V_{oc}, J_{sc}, FF, and η) of the P3HT/ANI-GQDs and P3HT/ANI-GSs under simulated AM 1.5 G 100 mW illumination (reprinted with permission from [21]. Copyright (2011) American Chemical Society)

Type of graphene	Graphene (wt.%)	V_{oc} (V)	J_{sc} (mA cm^{-2})	FF (%)	η (%)
–	0	0.43	0.037	0.21	0.003
ANI-GQDs	0.5	0.62	2.65	0.47	0.77
ANI-GSs	0.5	0.71	0.15	0.19	0.02
ANI-GQDs	1	0.61	3.51	0.53	1.14
ANI-GSs	1	0.72	0.19	0.22	0.03
ANI-GQDs	3	0.58	1.32	0.51	0.39
ANI-GSs	3	0.86	0.55	0.27	0.13
ANI-GQDs	5	0.59	0.36	0.52	0.12
ANI-GSs	5	0.94	1.5	0.33	0.46
ANI-GSs	10	0.88	2.65	0.28	0.65
ANI-GSs	15	0.95	0.31	0.25	0.07

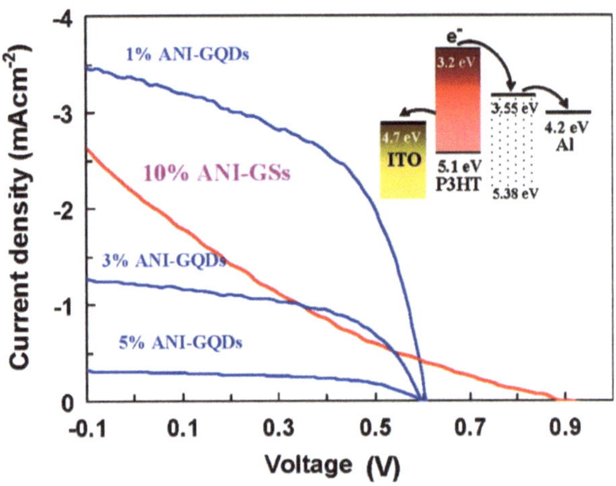

Fig. 10.7 *J–V* characteristics of the photovoltaic devices based on ANI-GQDs with different GQDs content and ANI-GS (under optimized condition) annealed at 160 °C for 10 min, in AM 1.5 G 100 mW illumination (reprinted with permission from [21]. Copyright (2011) American Chemical Society)

Dutta et al. [25] synthesized GQDs by a direct chemical method and in combination with ZnO nanowires. Figure 10.8a shows the microstructure and size distribution of the synthesized GQDs as observed by the TEM. Their sizes vary within 4–15 nm with an average diameter of ~8.5 nm. The Raman spectra of the GQDs in Fig. 10.8b show the D band at around 1,350 cm^{-1} as a breathing mode of k-point phonons of A_{1g} symmetry, which is assigned to the local defects and disorder, especially at the edges of graphene; and the G band at around 1,593 cm^{-1}, which is usually assigned to the E_{2g} phonon of C sp^2 atoms. The AFM image of the GQDs is shown in Fig. 10.8c. The topographic height distribution obtained from the AFM studies varies between 1 and 4 nm, as shown in Fig. 10.8d, suggesting that among these GQDs, more than 50 % are single layer, 30 % are bilayer, and 11 % are trilayer graphene.

Figure 10.9 shows the excitation-dependent PL behavior of the synthesized GQDs. The maximum of the emission band shifts toward the higher wavelength side as the excitation wavelength is increased from 280 to 420 nm while the

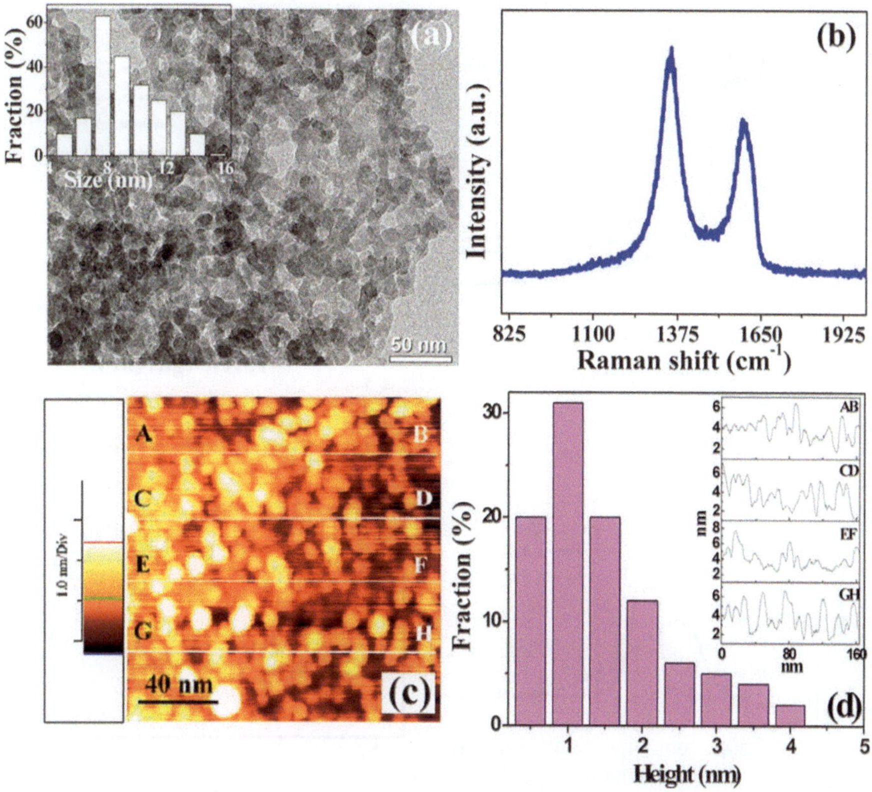

Fig. 10.8 (**a**) TEM image of GQDs. The inset shows the corresponding size distribution. (**b**) Raman spectrum of GQDs. (**c**) AFM image of the GQDs deposited on cleaned glass substrate. (**d**) Height distribution of the GQDs. (*Inset*) Height profile along the lines AB, CD, EF, and GH in *panel c* (reprinted with permission from [25]. Copyright (2012) American Chemical Society)

intensity decreases rapidly. According to Dutta et al. [25], the evolution of the luminescence behavior with a change in the excitation energy is due to the fact that the band gap in these QDs depends on the size and shapes of the sp^2 domains that result from the size distribution of the GQDs. When these GQDs are combined with ZnO NWs, the emission from these QDs is quenched fully in the visible region (Fig. 10.9c). The emission due to ZnO component in the composite material has been reduced as compared to the emission of bare ZnO because of the blocking effect by the black-colored GQDs layer on the NWs. From the PLE spectrum, the position of the LUMO levels is determined as 1.8–2.8 eV. The HOMO and LUMO levels of GQDs as well as the valence (7.6 eV) and conduction bands (4.35 eV) of ZnO w.r.t. vacuum level are shown schematically in Fig. 10.9d. The band alignment indicates that the electron transfer from LUMO of GQD to LUMO of ZnO NW & AZO is quite feasible. The maximum of the emission band shifts toward higher wavelength side as the excitation wavelength is varied from 280 to 420 nm and the intensity of the emission band decreases rapidly.

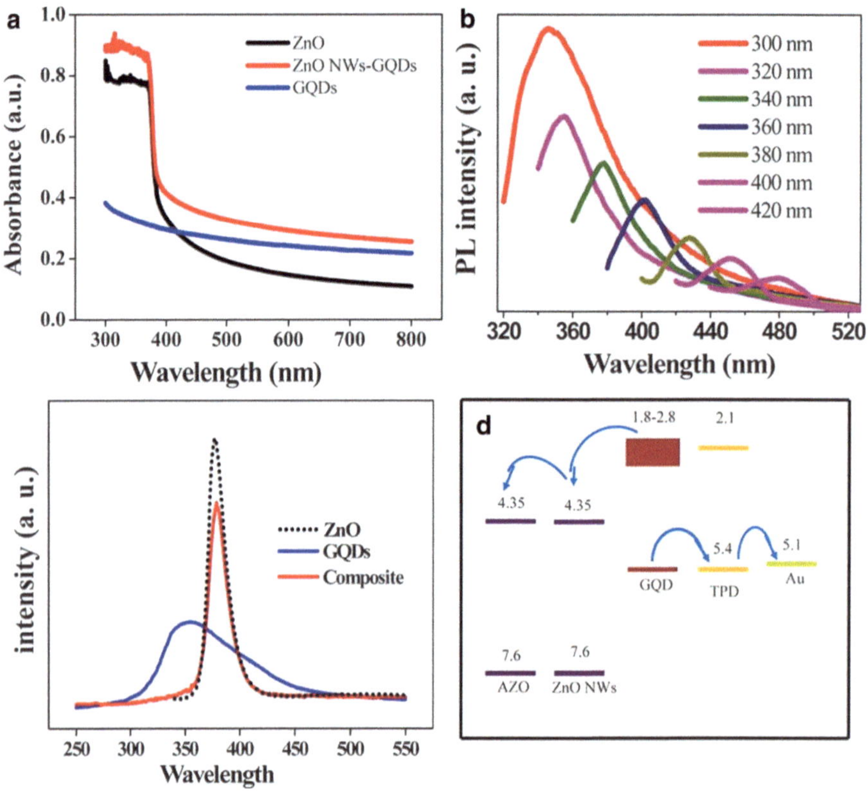

Fig. 10.9 Absorbance spectra of GQDs, ZnO NWs, and ZnO NWs–GQDs composites. (**b**) Excitation-dependent ZnO NWs, GQDs, and ZnO NWs–GQDs combined structure excited at 325 nm. (**d**) Schematic energy band diagram of the device with energy levels in eV relative to vacuum (reprinted with permission from [25]. Copyright (2012) American Chemical Society)

Figure 10.10 shows the charge transfer process at the interface in photovoltaic cells based on GQDs–ZnO NWs bulk heterojunction and the device structure. Figure 10.10a also shows the IQE of the devices made up of GQDs–ZnO NWs and ZnO NWs. The IQE reaches a value of 87 % at the absorption maximum. The I-V measurements (Fig. 10.10b) show the photocurrent spectra of the device made by the composite (ZnO NWs–GQDs) and the control sample (ZnO NWs only). While there is no photovoltaic effect shown by the control sample, a clear photovoltaic effect is observed from the *I–V* curve of the device made with the composite. This indicates and confirms that there is an electron injection into the NWs from the excited GQDs. The control device shows a short-circuit current density, $J_{sc} = 5.9\ \mu A\ cm^{-2}$ and $V_{oc} = 0.1$ V, whereas the composite device shows $J_{sc} = 0.45$ mA cm^{-2} with open-circuit voltage, $V_{oc} = 0.8$ V and FF = 0.5, leading to an estimated power conversion efficiency of 0.2 %. So, there is a 75 times increase in the J_{sc} value, which is also at par with the difference in the IQE values between ZnO and ZnO NWs–GQDs composites in the visible region (Fig. 10.10a).

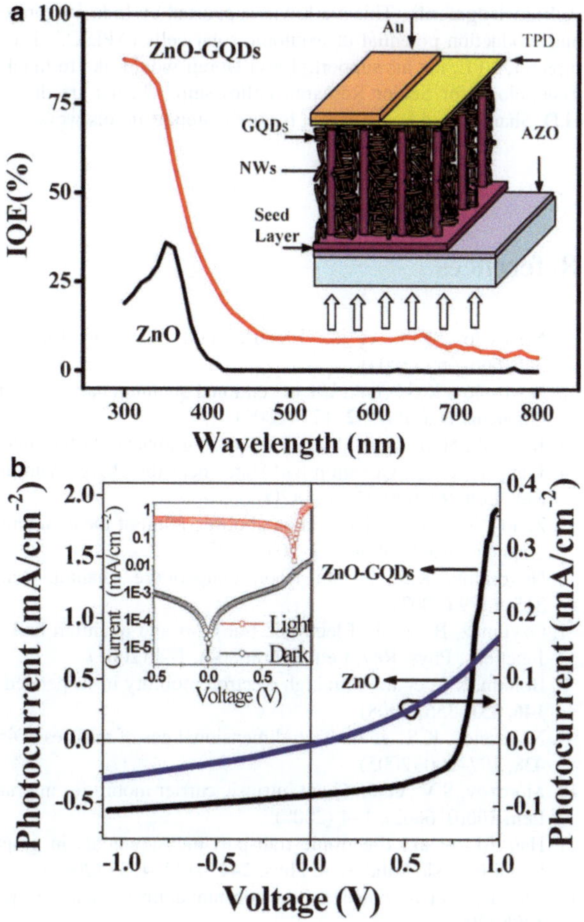

Fig. 10.10 (**a**) IQE of the samples. The inset shows the schematic model of the device. (**b**) Solar cell *I–V* characteristics under white light illumination. The inset shows the logarithmic plot of dark and photo current of the GQDs–ZnO cell (reprinted with permission from [25]. Copyright (2012) American Chemical Society)

The power conversion efficiency is, however, low due to the inefficient hole collection by TPD from GQDs [25]. Because the gaps in between the NWs were filled with the GQDs, and the TPD layer exists only at the top, the hole collection efficiency thus may not be good, leading to a poor cell performance.

10.3 Conclusions

Recently, GQDs have generated a great interest because of their size-dependent PL, low cost, chemical inertness, biocompatibility, and low toxicity. A lot of efforts have been put to synthesize them in large scales and with controlled size successfully. Both factors are crucial for their applications. The solar cell devices based upon GQDs have shown good efficiency and have emerged as the new cost-effective materials for the organic photovoltaic (OPV) devices.

Acknowledgments This work was supported by Indo-UK project "Advancing the effectiveness and production potential of excitonic solar cells (APEX)." The authors would like to thank the director, NPL, for his support. Tanvi Upreti would like to thank the Department of Science and Technology for Senior Research Fellowship. Thanks are due to R. Srivastava, N. Chaudhary, G.D. Sharma, and R. Bhardwaj for their support in this work.

References

1. Novoselov, K.S., et al.: Electric field effect in atomically thin carbon films. Science **306**, 666–669 (2004)
2. Novoselov, K.S., et al.: Unconventional quantum hall effect and Berry's phase of 2π in bilayer graphene. Nat. Phys. **2**, 177 (2006)
3. Jiang, Z., et al.: Quantum hall effect in graphene. Solid State Commun. **143**, 14–19 (2007)
4. Jiang, Z., et al.: Quantum hall states near the charge-neutral Dirac point in graphene. Phys. Rev. Lett. **99**(106802), 1–4 (2007)
5. Zhang, Y.B., et al.: Experimental observation of the quantum hall effect and Berry's phase in graphene. Nature **438**, 201–204 (2005)
6. Novoselov, K.S., et al.: Room-temperature quantum hall effect in graphene. Science **315**, 1379 (2007)
7. Ozyilmaz, B., et al.: Electronic transport and quantum hall effect in bipolar graphene p-n-p Junctions. Phys. Rev. Lett. **99**(186804), 1–4 (2007)
8. Bolotin, K.I., et al.: Ultrahigh electron mobility in suspended graphene. Solid State Commun. **146**, 351–355 (2008)
9. Novoselov, K.S., et al.: Two-dimensional gas of massless Dirac fermions in graphene. Nature **438**, 197–200 (2005)
10. Morozov, S.V., et al.: Giant intrinsic carrier mobilities in graphene and its bilayer. Phys. Rev. Lett. **100**(016602), 1–4 (2008)
11. Han, M., et al.: Electronic transport measurements in graphene nanoribbons. Phys. Status Solidi B Basic Solid State Phys. **244**, 4134–4137 (2007)
12. Nair, R.R., et al.: Fine structure constant defines visual transparency of graphene. Science **320**, 1308 (2008)
13. Yan, X., et al.: Large, solution-processable graphene quantum dots as light absorbers for photovoltaics. Nano Lett. **10**(5), 1869–1873 (2010)
14. Nakada, K., et al.: Edge state in graphene ribbons: nanometer size effect and edge shape dependence. Phys. Rev. B **54**, 17954–17961 (1996)
15. Son, Y.W., et al.: Energy gaps in graphene nanoribbons. Phys. Rev. Lett. **97**, 216803 (2006)
16. Li, X.L., et al.: Large-area synthesis of high-quality and uniform graphene films on copper foils. Science **324**, 1312–1314 (2009)
17. Bai, J., et al.: Graphene nanomash. Nat. Nanotechnol. **5**, 190–194 (2010)
18. Shen, J., et al.: Graphene quantum dots: emergent nanolights for bioimaging, sensors, catalysis and photovoltaic devices. Chem. Commun. **48**(31), 3686–3699 (2012)
19. Pan, D., et al.: Hydrothermal route for cutting graphene sheets into blue-luminescent graphene quantum dots. Adv. Mater. **22**, 734–738 (2010)
20. Yan, X., et al.: Synthesis of large, stable colloidal graphene quantum dots with tunable size. J. Am. Chem. Soc. **132**, 5944–5945 (2010)
21. Gupta, V., et al.: Luminscent graphene quantum dots for organic photovoltaic devices. J. Am. Chem. Soc. **133**, 9960–9963 (2011)
22. Li, Y., et al.: An electrochemical avenue to green-luminescent graphene quantum dots as potential electron-acceptors for photovoltaics. Adv. Mater. **23**, 776–780 (2011)

23. Kim, S., et al.: Anomalous behaviors of visible luminescence from graphene quantum dots: interplay between size and shape. ACS Nano **6**(9), 8203–8208 (2012)
24. Shen, J., et al.: One-pot hydrothermal synthesis of graphene quantum dots surface-passivated by polyethylene glycol and their photoelectric conversion under near-infrared light. New J. Chem. **36**, 97–101 (2012)
25. Dutta, M., et al.: ZnO/Graphene quantum dot solid state solar cell. J. Phys. Chem. C **116**(38), 20127–20131 (2012)
26. Hu, C., et al.: One-step preparation of nitrogen-doped graphene quantum dots from oxidized debris of graphene oxide. J. Mater. Chem. B **1**, 39–42 (2013)
27. Iwan, A., Chuchmała, A.: Perspectives of applied graphene: polymer solar cells. Prog. Polym. Sci. **37**, 1805–1828 (2012)
28. Lee, J., et al.: Uniform graphene quantum dots patterned from self-assembled silica nanodots. Nano Lett. **12**(12), 6078–6083 (2012)
29. Fan, L.: Fluorescence resonance energy transfer quenching at the surface of graphene quantum dots for ultrasensitive detection of TNT. Talanta **101**, 192–197 (2012)
30. Giangregorio, M.M., et al.: Synthesis and characterization of plasmon resonant gold nanoparticles and graphene for photovoltaics. Mater. Sci. Eng. B (2012, in press). doi:10.1016/j.mseb.2012.10.034
31. Zhang, X., et al.: Highly sensitive humidity sensing properties of carbon quantum dots films. Mater. Res. Bull. **48**(2), 790–794 (2013)
32. Cao, L., et al.: Photoluminescence properties of graphene versus other carbon nanomaterials. Acc. Chem. Res. **46**(1), 171–180 (2013)
33. Tetsuka, H., et al.: Optically tunable amino-functionalized graphene quantum dots. Adv. Mater. **24**(39), 5333–5338 (2012)
34. Shinde, D.B., Pillai, V.K.: Electrochemical preparation of luminescent graphene quantum dots from multiwalled carbon nanotubes. Chem. Eur. J. **18**(39), 12522–12528 (2012)
35. Long, R., et al.: Photo-induced charge separation across the graphene–TiO_2 interface is faster than energy losses: a time-domain ab initio analysis. J. Am. Chem. Soc. **134**(34), 14238–14248 (2012)
36. Xu, M., et al.: Unique synthesis of graphene-based materials for clean energy and biological sensing applications. Chin. Sci. Bull. **57**(23), 3000–3009 (2012)
37. Wei, W., Qu, X.: Extraordinary physical properties of functionalized graphene. Small **8**(14), 2138–2151 (2012)
38. Zhu, S., et al.: Surface chemistry routes to modulate the photoluminescence of graphene quantum dots: from fluorescence mechanism to up-conversion bioimaging applications. Adv. Funct. Mater. **22**(22), 4732–4740 (2012)
39. Pan, S., Liu, X.: ZnS–graphene nanocomposite: synthesis, characterization and optical properties. J. Solid State Chem. **191**, 51–56 (2012)
40. Tang, L., et al.: Deep ultraviolet photoluminescence of water-soluble self-passivated graphene quantum dots. ACS Nano **6**(6), 5102–5110 (2012)
41. Pan, Z., et al.: Graphene-based functional materials for organic solar cells. Opt. Mater. Express **2**(6), 814–824 (2012)
42. Chernozatonskii, L.A., et al.: Formation of graphene quantum dots by "Planting" hydrogen atoms at a graphene nanoribbon. JETP Lett. **95**(5), 266–270 (2012)
43. Mirtchev, P., et al.: Solution phase synthesis of carbon quantum dots as sensitizers for nanocrystalline TiO_2 solar cells. J. Mater. Chem. **22**(4), 1265–1269 (2012)
44. Zhang, Z., Wu, P.: Hydrothermal aggregation induced crystallization: a facial route towards polycrystalline graphite quantum dots with blue photoluminescence. Cryst. Eng. Commun. **14**(21), 7149–7152 (2012)
45. Chen, S., et al.: Unusual emission transformation of graphene quantum dots induced by self-assembled aggregation. Chem. Commun. **48**(61), 7637–7639 (2012)
46. Kwon, W., et al.: Formation of highly luminescent nearly monodisperse carbon quantum dots via emulsion-templated carbonization of carbohydrates. RSC Adv. **2**(30), 11223–11226 (2012)

47. Kwon, W., Rhee, S.: Facile synthesis of graphitic carbon quantum dots with size tunability and uniformity using reverse micelles. Chem. Commun. **48**(43), 5256–5258 (2012)
48. Dong, Y., et al.: One-step and high yield simultaneous preparation of single- and multi-layer graphene quantum dots from CX-72 carbon black. J. Mater. Chem. **22**(18), 8764–8766 (2012)
49. Zhu, S., et al.: Graphene quantum dots with controllable surface oxidation, tunable fluorescence and up-conversion emission. RSC Adv. **2**(7), 2717–2720 (2012)
50. Song, T., et al.: Prospects and challenges of organic/group IV nanomaterial solar cells. J. Mater. Chem. **22**(10), 4216–4232 (2012)
51. Xin, G., et al.: Investigation of charge-transfer complexes formation between photoluminescent graphene oxide and organic molecules. Nanoscale **4**(2), 405–407 (2012)
52. Zhang, Z., et al.: Graphene quantum dots: an emerging material for energy-related applications and beyond. Energy Environ. Sci. **5**(10), 8869–8890 (2012)
53. Li, M., et al.: Synthesis and upconversion luminescence of N-doped graphene quantum dots. Appl. Phys. Lett. **101**(10), 103107 (2012)
54. Pan, D., et al.: Cutting sp^2 clusters in graphene sheets into colloidal grapheme quantum dots with strong green fluorescence. J. Mater. Chem. **22**(8), 3314–3318 (2012)
55. Zhu, S., et al.: Control the size and surface chemistry of graphene for the rising fluorescent materials. Chem. Commun. **48**(38), 4527–4539 (2012)
56. Li, H., et al.: Carbon nanodots: synthesis, properties and applications. J. Mater. Chem. **22**(46), 24230–24253 (2012)
57. Luo, P., et al.: Synthesis of gold@carbon dots composite nanoparticles for surface enhanced Raman scattering. Phys. Chem. Chem. Phys. **14**(20), 7360–7366 (2012)
58. Wan, X., et al.: Graphene – a promising material for organic photovoltaic cells. Adv. Mater. **23**(45), 5342–5358 (2011)
59. Pan, S., et al.: Preparation of Ag_2S–Graphene nanocomposite from a single source precursor and its surface-enhanced Raman scattering and photoluminescent activity. Mater. Char. **62**(11), 1094–1101 (2011)
60. Liu, R., et al.: Bottom-up fabrication of photoluminescent graphene quantum dots with uniform morphology. J. Am. Chem. Soc. **133**(39), 15221–15223 (2011)

Chapter 11
Graphene and Quantum Dot Nanocomposites for Photovoltaic Devices

Xukai Xin

Abstract As one of the most attracting nanoscale materials in the past decade, graphene and composites based on graphene have boosted the efficiency of catalytic reaction in energy conversion applications. Charge collection and transport are facilitated with the presence of graphene and reduced graphene oxide in quantum dot solar cells. The 2-D graphene and its derivatives serve as a scaffold for the immobilization of catalytic and photovoltaic nanoparticles as well as promoting selectivity and efficiency of the reaction process. Small size graphene quantum dots, which have quantum confinement effect, are promising for the applications in photovoltaic devices, due to their outstanding properties and advantages, including high optical absorption, tunable bandgap, and earth abundant chemical composition. In this chapter, the synthesis and application of both graphene sheets and graphene quantum dots are highlighted, followed by a perspective of future research in this area.

11.1 Introduction

Since the latter part of the twentieth century, carbon-based nanostructures have attracted most attentions in nanoscience and nanotechnology field. Two Nobel Prizes have been awarded to the researchers in this area during the past decades: the 1996 Nobel Prize in Chemistry awarded jointly to Robert F. Curl, Jr., Sir Harold W. Kroto, and Richard E. Smalley "for their discovery of fullerenes" and the 2010 Nobel Prize in Physics awarded jointly to Andre Geim and Konstantin Novoselov "for groundbreaking experiments regarding the two dimensional material

X. Xin (✉)
School of Materials Science and Engineering, Georgia Institute of Technology,
3131 MoSE Building, 901 Atlantic Drive, Atlanta, GA 30332, USA
e-mail: xxin@gatech.edu

J. Wu and Z.M. Wang (eds.), *Quantum Dot Solar Cells*, Lecture Notes in Nanoscale Science and Technology 15, DOI 10.1007/978-1-4614-8148-5_11,

graphene," which signified the scientific and technological importance of carbon-based nanostructures. Though retaining similar carbon atom networks, fullerenes, graphene, carbon nanotubes, nanofibers, and quantum dots represent significant different properties and applications. As a 2-D carbon-based material, graphene consists extremely of high conductivity and catalytic activity, which facilitates the application of graphene in electronic devices [1].

The first attempt to cleave graphite into exfoliated sheets was reported by Hummers in 1958 [2]. However the importance and potential of graphene was not recognized until 2004, after Novoselov developed a mechanical exfoliation method to synthesize graphene [3, 4]. Thereafter a boom of research on graphene bursts. Up-to-date common synthesis routes of graphene include mechanical cleavage [3, 4], epitaxial growth [5, 6], thermal exfoliation [7], and chemical exfoliation [8–10]. Chemical exfoliation method via solution-based processes is convenient. However carboxylate, hydroxyl, and epoxide functional groups may attach onto graphene during the exfoliation. The functionalized graphene is commonly referred to graphene oxide (GO) [11]. The presence of these functional groups enables suspension of individual GO sheets in solvents. However, damage of the conjugated sp^2 network results in drastic decrease of the conductivity. Partial restoration of the sp^2 network can be achieved by reduction reaction with hydrazine [12] and $NaBH_4$ [13]. Other restoring methods include laser irradiation [14], microwave irradiation [15, 16], sonolysis [10], photocatalysis [17], and thermal [7], electrochemical [18], and photocatalytic [19] reductions. The products of the reduction are commonly referred to as reduced graphene oxide (RGO).

Graphene with 2-D carbon network and single-atom-thick layer preserves high specific surface area for a large interface, high mobility up to 10,000 $cm^2\ V^{-1}\ s^{-1}$, and tunable band gap. As a platform to anchor metal and semiconductor nanoparticles [20], graphene provides a simple way to disperse catalyst particles which played a crucial role in increasing the electrocatalytic area in quantum dot solar cells [21, 22]. In addition to that, graphene improves the charge separation and facilitates collection and transportation of electrons [19].

Larger graphene sheets have zero band gap and infinite exciton Bohr radius [23] without obvious prominent photoactivity. However with a finite size, quantum confinement in graphene exhibits molecular-level activities [24, 25] and singlet and triplet properties [26] similar to a semiconductor quantum dots. Small dimensional graphene with quantum confinement is usually referred to as graphene quantum dot (GQD). As other QDs, GQDs have size-dependent bandgap and large optical absorption [24, 27], which make them promising materials for solar energy conversion. One of the key issues of preparation of colloidal GQDs is to increase solubility in common solvents for fabrication GQD-based solar cells via inexpensive solution-based processing.

Both large size graphene and small size graphene play important roles in photovoltaic applications as nanoparticle scaffold, electron acceptor, charge transport pathway, and photoactive materials, due to their high conductivity, large surface area, and tunable band gap.

In this chapter, the synthesis and application of both large size graphene and small size graphene in quantum dot solar cells will be introduced. The interaction between graphene and quantum dots manifests the advantages of incorporation of graphene into photovoltaic devices. Further improvement of quantum dot solar cell is promising as the research on graphene base composites continues in the near future.

11.2 Interaction Between Graphene and Quantum Dots

The quenching of excited quantum dots follows two routes: resonant energy transfer from and charge transfer from excited quantum dots to single and few-layer graphene. These two transfer routes compete with recombination of the excited electron–hole pairs inside the quantum dots. The energy and charge transfer rate is significant when compared to the recombination rate. Thus the recombination in excited quantum dots is reduced when attached with graphene.

11.2.1 Energy Transfer from QDs to Graphene

Resonant (Förster) energy transfer from quantum dots to single- and few-layer graphene is expected to occur, since graphene exhibits broad absorption across the visible spectral range. For example, single-layer graphene has a linear band dispersion around the corners of its Brillouin zone (K and K′ points) [28] and a nearly constant optical absorption. Thus, electronically excited quantum dots near graphene can be quenched by resonant energy transfer, exciting electron–hole pairs in the graphene [29].

To examine the interaction of graphene with QDs, graphene layers were deposited on quartz substrates by mechanical exfoliation [3] of graphite. Isolated CdSe/ZnS nanocrystals were then spun cast onto graphene layers. Fluorescence from individual QDs could be observed for QDs located both on the bare quartz substrate and on the graphene layers (Fig. 11.1). Comparing to the bare substrate, strong fluorescence quenching can be observed for particles deposited on graphene sheets. The average quenching factor ρ can be defined as $\rho = I^{Q}/I^{G}$, where I^{Q} and I^{G} are the fluorescence intensities on quartz and on graphene, respectively. For single-layer graphene, the measured average intensities after correction give a quenching factor $\rho = {\sim}70$.

According to the calculation by Swathi and Sebastian [30, 31] which considered the relevant Coulomb matrix element between the excited molecule and the electron system of single-layer graphene, the quenching factor can be calculated by:

$$\rho_1 = \frac{\pi}{16}\frac{\alpha}{\varepsilon^{5/2}}\left(\frac{c}{v_{\mathrm{F}}}\right)^4 \int_0^1 \exp\left(-\frac{2\Delta Ezt}{\hbar v_F}\right)\frac{t^3}{\sqrt{1-t^2}}dt + 1 \qquad (11.1)$$

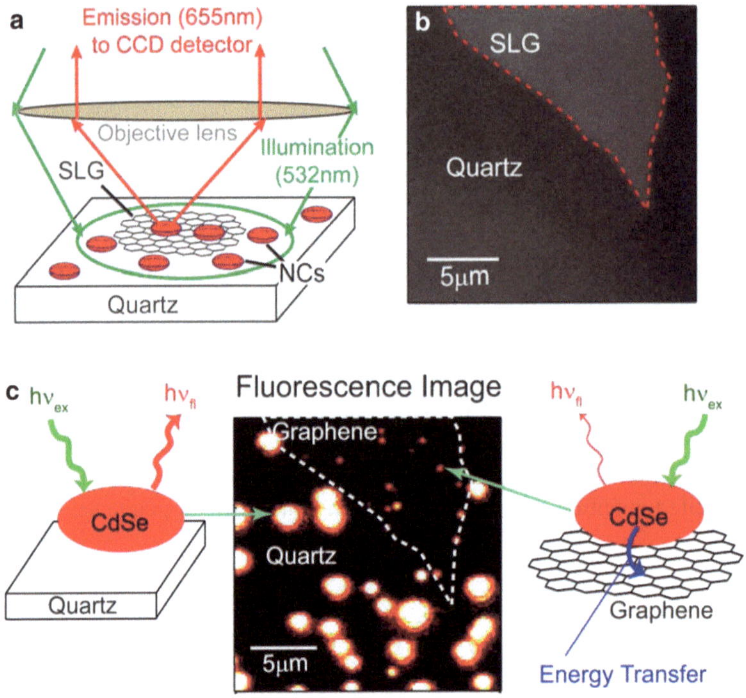

Fig. 11.1 Fluorescence images of CdSe/ZnS nanocrystals on single-layer graphene and on the quartz substrate. (**a**) Schematic diagram of the experimental setup; (**b**) optical reflectivity image in the emission range of our nanocrystals; (**c**) wide-field fluorescence image of individual CdSe/ZnS nanocrystals on quartz (*left*) and on single-layer graphene (*right*). Reprinted with permission from [32]; Copyright © 2010 The American Chemical Society

in which α is the fine structure constant, ε is the dielectric constant of the environment, c is the light speed in vacuum, v_F is the Fermi velocity in single-layer graphene determined experimentally, ΔE is the energy of the emitted photons, z is the distance from the QD center to the graphene plane. Equation (11.1) can be extended to few layers of graphene, since the interactions between the layers of graphene are relatively weak [28]. The few-layer graphene system can be simplified as a stack of decoupled single-layer graphene sheets, and each layer is treated as an independent energy transfer channel with a spacing of $d = 0.34$ nm between each other. Assuming a constant dielectric screening from upper layers of graphene from that of the QDs ligands, the quenching factor of n-layer graphene is then given by:

$$\rho_n = \frac{\pi}{16}\frac{\alpha}{\varepsilon^{5/2}}\left(\frac{c}{v_F}\right)^4 \sum_{i=1}^{n}\left[\int_0^1 \exp\left[-\frac{2\Delta E t(z_1+(i-1)d)}{\hbar v_F}\right]\frac{t^3}{\sqrt{1-t^2}}dt\right]+1 \tag{11.2}$$

in which $z_1 + (i-1)d$ is the distance from the QDs center to the ith graphene plane.

Calculated quenching factor from Eqs. (11.1) and (11.2) is in good agreement with experimental results for single-layer (~70) and bilayer graphene (~115). However for bulk graphite, calculated quenching factor (~200) is much less than experiment values (~600). Alternative calculation of the quenching factor with Persson energy transfer theory provided a closer result (~607) to experiment values [29].

In the particular system of hydrocarbon ligand-coated CdSe/ZnS core/shell QDs adsorbed on graphene layers, resonant energy transfer is much faster than photoexcited electron transfer [32]. One method to increase the relative rates of electron transfer to energy transfer for solar cells is to strengthen the electronic coupling between QDs and graphene. Strong electronic coupling between QDs and graphene can be achieved by removal of insulating ZnS shell or photochemical covalent functionalization of graphene [33].

11.2.2 Charge Transfer from QDs to Graphene

The unique linear electronic dispersion near the K-point of the Brillouin zone of graphene enables the electron to travel along the sheet at a very fast speed. The high electron drift speed in graphene enables a fast dissociation of electrons from the location of injection. However, as mentioned in the previous section, strong electronic coupling is required between QDs and graphene for efficient extraction of the charge carriers [34]. Direct growth of QDs over graphene during the synthesis process is one method to achieve good electronic contact [35]. With strong electronic coupling, QDs–graphene assemblies have shown potential application in photovoltaic measurements [35, 36].

To investigate the electron extraction time from photoexcited QD to graphene, ultrafast transient absorption spectroscopy and time-resolved emission spectroscopy were carried out on a water-soluble CdTe decorated graphene system [34]. In this specific case, 2.2 nm CdTe QDs were attached to RGO. According to the photoluminescence spectroscopy, 90 % of emission CdTe QD is quenched on the graphene surface as compared to that of bare CdTe, which is due to transfer of electrons from photoexcited CdTe to graphene [34]. Time-resolved emission decay of CdTe can be fitted exponentially with time constants of $\tau_1 = 0.27$ ns (10.5 %), $\tau_2 = 2.3$ ns (28.2 %), and $\tau_3 = 14.6$ ns (61.3 %), while on the graphene surface the emission decay can be fitted with time constants $\tau_1 = 0.19$ ns (57 %), $\tau_2 = 0.75$ ns (25 %), and $\tau_3 = 3.4$ ns (18 %). The much faster decay of CdTe on graphene surface clearly indicates electron transfer from photoexcited CdTe to graphene layers. Transient absorption spectroscopy was used to understand the electron transfer dynamics in CdTe/graphene system. Figure 11.2a shows the bleach recovery for both CdTe QD and CdTe/graphene composite probing at 460 nm while pumping at 400 nm. Fitted time constants of $\tau_1 = 0.2$ ps (49 %), $\tau_2 = 3$ ps (23 %), and $\tau_3 = 17$ ps (28 %) for CdTe QD and $\tau_1 = 6$ ps (25.4 %), $\tau_2 = 50$ ps (20.2 %), and $\tau_3 = >200$ ps (54.4 %) for CdTe/graphene composite clearly indicate that

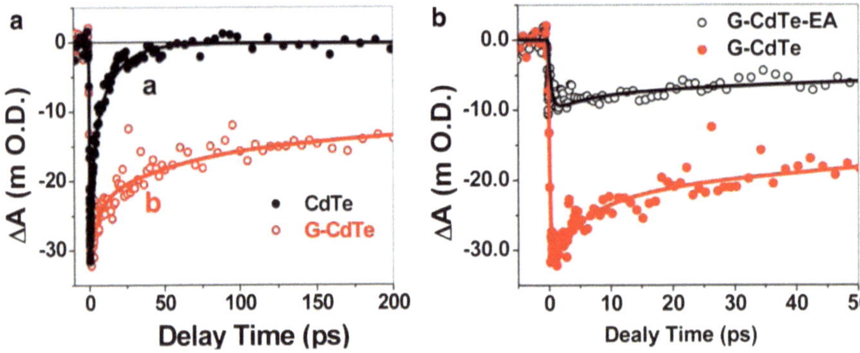

Fig. 11.2 (**a**) Bleach recovery kinetics at 460 nm for (*a*) CdTe and (*b*) CdTe/graphene pumping at 400 nm and (**b**) kinetics decay traces at 460 nm for CdTe/graphene (G-CdTe) and CdTe/graphene with electron quencher (G-CdTe-EA) pumping at 400 nm. Reprinted with permission from [34]; Copyright © 2012 The American Chemical Society

recombination is much faster in CdTe QD as compared to that in the CdTe/graphene composite [34].

By adding electron quencher (benzoquinone, BQ in this case) into the system, further insight of the charge transfer process from QDs to graphene can be achieved. Electron quencher can induce a fast quenching of electrons by accepting the electrons from the conduction band of QDs and then donating the electrons back to the valence band of QDs within a few picoseconds. Bleach recovery of CdTe/graphene composite with and without the presence of BQ at 460 nm is shown in Fig. 11.2b. The bleaching signal reduced by 60 % with the presence of BQ; however, the time constant $\tau_1 = 6$ ps (22.4 %), $\tau_2 = 50$ ps (24.2 %), and $\tau_3 = >200$ ps (51.6 %) is very similar to that without BQ. The reduction of the bleach signal in the presence of BQ can be attributed to the competition of quenching by BQ and injection into graphene of a photoexcited electron in the QD [34].

Transient absorption spectroscopy of bleach recovery clearly shows that exciton recombination happens within 50 ps for pure CdTe QD while longer than 200 ps for QDs on graphene surface, which indicates that charge separation of CdTe/graphene composite drastically improves when comparing to pure CdTe QD.

11.3 Synthesis of Graphene and Graphene/Quantum Dot Composite

The first successful attempt to obtain graphene is reported in 2004 via mechanical exfoliation [4]. Since then, people developed epitaxial grown graphene which came from substrate-related preparation procedures [5, 37]. High quality graphene with few layers can be synthesized by both mechanical exfoliation and epitaxial growth methods. An alternative chemical exfoliation method starts from the formation of graphene oxide under highly oxidizing conditions via modified Hummer's method [2]. Due to its disrupted sp^2-networks, graphene oxide lacks the electronic properties of graphene. Therefore chemical reduction to restore the sp^2-network is of the essence. However, RGO exhibits similar but not identical properties comparing to that of pristine graphene prepared by mechanical exfoliation and epitaxial growth. Functional groups can be introduced onto graphene during the oxidization and reduction process, which makes this method the most common strategy to produce chemically modified graphene [38].

11.3.1 Synthesis Method of Graphene

11.3.1.1 Mechanical Exfoliation

Mechanical exfoliation is the earliest reported method to synthesize graphene [4]. One-millimeter-thick platelets of highly oriented pyrolytic graphite (HOPG) were etching in oxygen plasma to form 5 mm-deep mesas on top of the platelets. One-millimeter-thick photoresist layer was spun coated on glass substrate and then pressed against the platelets. The mesas were attached to the photoresist layer after baking. The rest of the HOPG was cleaved off and scotch tape was used to repeatedly peeling graphite flakes off the mesas. After releasing the flakes left in the photoresist layer in acetone, the flakes can be captured on the Si surface when dipping Si wafer into the solution [4].

11.3.1.2 Epitaxial Growth

Epitaxial growth is another method to obtain single-layer graphene. Few-layer graphene films were grown via chemical vapor deposition (CVD) on a $Ni/SiO_2/Si$ substrate. After sequential etching of SiO_2 and Ni, graphene can be transferred to arbitrary substrates without intense mechanical and chemical treatments. Thus the high crystalline quality of the graphene was preserved and enhanced electrical and mechanical properties were observed [5].

11.3.1.3 Chemical Exfoliation

The first step of chemical exfoliation is oxidization of graphite to obtain graphene oxide (GO) via a modified Hummer's method [2]. Graphite powder, $NaNO_3$, and concentrated H_2SO_4 were stirred in an ice bath while slowly adding $KMnO_4$ over a period of ~30 min. After stirred for 1 h at 35 °C, de-ionized water was added very slowly with care, followed by adding 3 % H_2O_2. Finally, the mixture was filtered to obtain GO [39].

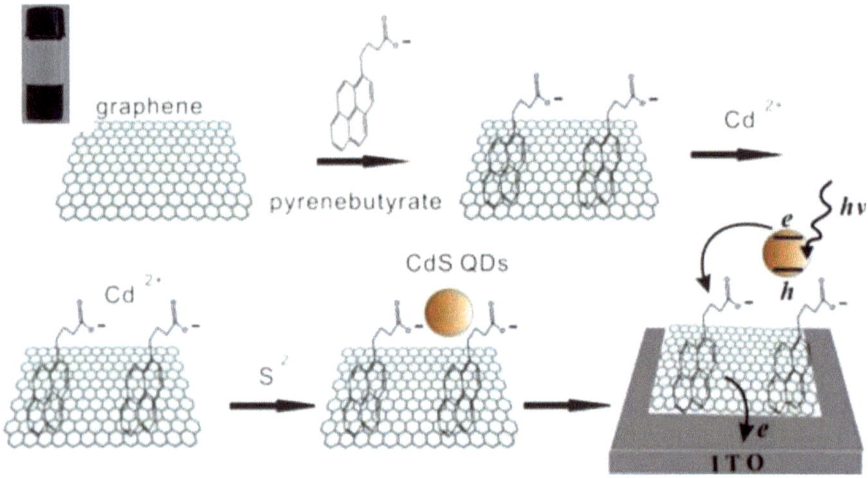

Fig. 11.3 Scheme of in situ growth of CdS QDs on graphene. *Inset* shows 0.5 mg mL^{-1} PB-graphene solution in methanol. Reprinted with permission from [40]; Copyright © 2010 Elsevier

Hydrazine and $NaBH_4$ are two reductive chemicals typically used for reduction of GO. Before adding hydrazine, the pH of GO solution was tuned to 10.0–10.5 with ammonia solution. The reduction finished after 1 h stirring [36]. While adding $NaBH_4$, the pH of GO solution was kept at 14 and the temperature was kept at 0 °C. The solution then was maintained at 0 °C for 2 h and 50 °C for another 2 h [34].

11.3.2 Synthesis Method of Graphene and Quantum Dot Composite

11.3.2.1 In Situ Growth of Quantum Dots on Graphene

In situ growth is one of the most commonly used methods to decorate quantum dots on graphene surface. The quantum dots were produced with the presence of graphene which was added to the quantum dot precursor before the reaction. In one case [40], RGO was synthesized separately by hydrazine reduction of exfoliated graphite oxide obtained by Hummers method [2] and then functionalized with pyrenebutyrate (PB). To grow CdS QDs in situ on graphene, 250 μL 20 mM Cd^{2+} was added to 1 mL 0.5 mg mL^{-1} PB-graphene solution, followed by slowly adding 100 μL 50 mM S^{2-}. CdS QDs will grow in situ on graphene and then centrifuged to separate abundant CdS QDs (Fig. 11.3). The QD loading on graphene was approximately 0.65 wt.%. The QD/graphene composite was then dispersed in methanol for casting on ITO glass to prepare solar cells [40].

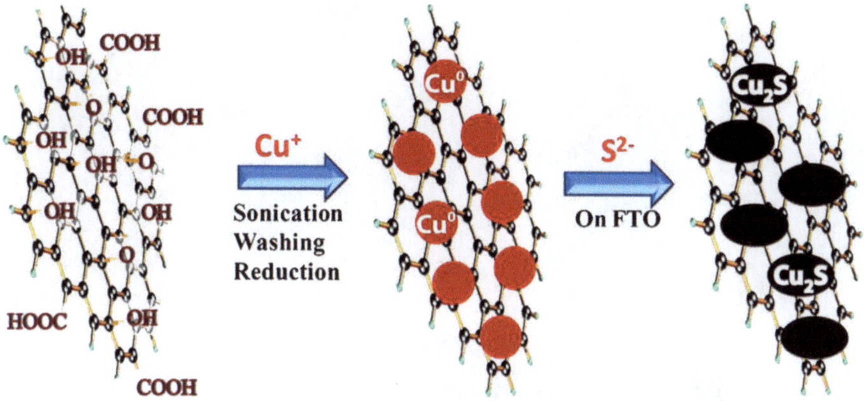

Fig. 11.4 Synthetic steps of Cu_2S/RGO composite material. Reprinted with permission from [9]; Copyright © 2011 The American Chemical Society

In another approach, 10 mmol of Cd precursor and 24 mmol mercaptopropionic acid (MPA) was added to the reduced graphene solution (1 g L^{-1}). The tellurium precursor NaHTe was prepared by the reaction between tellurium powder and $NaBH_4$ at 0 °C for 8 h. Then 10.5 mmol of NaHTe was added to the Cd precursor solution and refluxed for 4 h. Excess precursors were removed by centrifuging [34].

An uncommon in situ growth of CuS QDs on graphene contains two steps: synthesis of Cu^0 nanoparticles on graphene and then react with sulfur ions. Graphene oxide (GO) and RGO have the ability to stabilize metal [41]. This property can be used to synthesize Cu^0 nanoparticles on graphene via disproportionation reaction of Cu^+. In the synthesis process, 30–50 mg of graphene oxide was prepared via Hummer's method [2] and mixed with 30–450 mg of copper actate (CuAc) in ethanol and sonicating for 60 min in an ice bath. Electrostatic interactions combined positively charged copper cations to highly electronegative oxygen moieties on graphene oxide surface. Cu^+ ion undergoes disproportionation in ethanol, with graphene oxide as nucleation sites for the Cu^0. This reaction happened after 30 min of sonication, which not only exfoliates graphene oxide sheets and combines Cu^+ to graphene oxide but also has a possibility to facilitate the reduction of Cu^+ via sonolytic reduction [10]. UV irradiation was used to reduce Cu^0/GO composite after spin coated on FOT glass. Cu^0/Cu^+ bound to RGO were converted to Cu_2S when immersed in polysulfide solution. The synthetic steps are illustrated in Fig. 11.4 [9].

11.3.2.2 Electrophoretic Deposition

Drop cast method of QD and graphene composites always results in film inhomogeneity and aggregation of QDs and graphene. To get homogeneous composite film, electrophoretic deposition (EPD) has been demonstrated to be a good choice

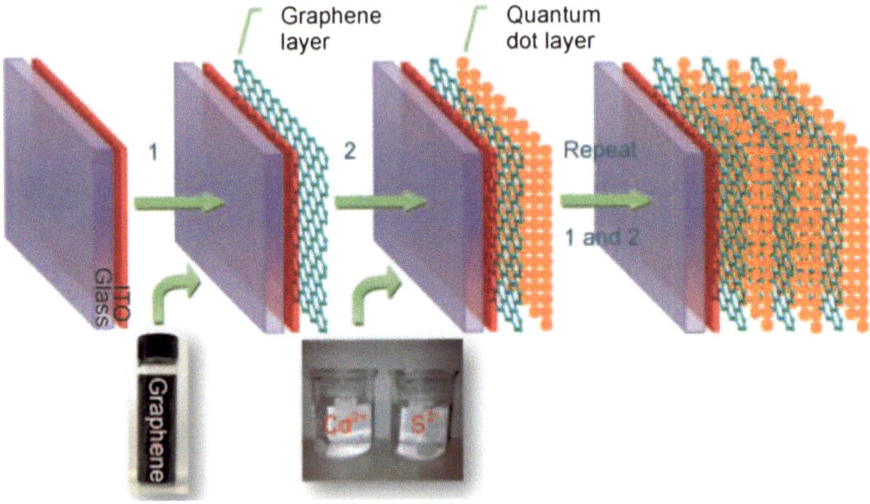

Fig. 11.5 Fabrication of the layered graphene/QDs on ITO glass. (1) Electrophoretic deposition of graphene; (2) sequential SILAR process to synthesize a layer of CdS QDs. The layered graphene/QDs device was obtained by repeating steps 1 and 2. Reprinted with permission from [36]; Copyright © John Wiley and Sons

in preparation of QDs and graphene [39] or C_{60} [42]. Before EPD process, QDs solution and RGO were synthesized separately and mixed together. QDs with hydrophobic ligand can be synthesized with the method proposed by Peng et al. [43]. And reduced graphene oxide can be prepared by modified Hummer's method [2] followed by reduction [14, 44]. Mesoporous anatase TiO_2 electrode were prepared by coating a layer of TiO_2 nanoparticle paste onto fluorine-doped tin oxide (FTO) glass and annealed at 500 °C. Then TiO_2 electrode was immersed in a mixture of QDs and graphene solution (e.g., 50 μL of 9.0×10^{-6} M CdSe QDs in toluene and 30 μL of 0.5 mg mL^{-1} graphene in ethanol, combined in a 0.8:1.0 mixture of toluene/acetonitrile) and a 50 V bias voltage was applied for 1 min followed by remixing.

11.3.2.3 Layer-by-Layer Deposition

Figure 11.5 shows the fabrication process of the layered graphene/QDs device. EPD was applied to keep good contact of graphene with ITO glass or quantum dot layers [36]. A glass container with ITO glass electrode and platinum gauze electrode facing each other with 3 mm distance was used. By applying a constant current density of 0.5 mA cm^{-2} in 0.2 mg mL^{-1} 9.5 Å thick graphene nanosheets aqueous solution for 75 s, 30 nm thick graphene was coated on ITO glass. The current density is critical in the EPD process. Low current density usually leads to an insufficient deposition while large current density results agglomeration nonuniformity of graphene.

Direct synthesis of CdS QDs from aqueous solution on as-prepared graphene layer was achieved by successive ionic layer adsorption and reaction (SILAR) method. Graphene deposited ITO glass was successively immersed in 0.05 M $CdCl_2$ and 0.05 M Na_2S aqueous solution for 30 s each cycle. Excess Cd or S ion was rinsed by de-ionized water after each immersion. A QDs layer with yellow color was obtained after ten cycles of immersion. Layered graphene/QDs films were prepared by repeating deposition of graphene and QDs. With SILAR method, CdS QDs with ~5 nm diameters and good crystal structures were uniformly distributed on the graphene sheet. The thickness of the film linearly increases with the number of graphene/QD bilayers. Cross-sectional view indicates that the thicknesses for each layer are ~30 and 120 nm for graphene and QDs, respectively. The results were verified by TEM, SEM, AFM, and XPS [36].

11.4 Application of Graphene and Quantum Dot Composite

11.4.1 Counter Electrode

Carbon nanocomposites, such as C_{60} and carbon nanotubes (CNTs), can be used as electron acceptors and electron transportation pathways in QD solar cells [8, 42, 45]. As a rising star of carbon nanocomposites family, graphene has been extensively investigated for the applications in solar cell. According to its high conductivity, large surface area, and tunable band gap [36], graphene has usually been considered as counter electrodes or conducting scaffold incorporated with QDs in solar cells [9, 39, 46, 47].

One attractive challenge currently is to utilize 2-D graphene with QDSSC. QDSSC is one type of quantum dot solar cells in which the QDs are used to sensitize wide bandgap semiconductor metal oxide (e.g., TiO_2, SnO_2, and ZnO). These metal oxides usually contain lower conduction band and high conductivity comparing to QDs, thus have the ability to accept excited electrons from QDs and facilitate charge separation and transport. Incorporation of graphene into metal oxide layers, or even substitution of these metal oxides, may further increase the conductivity and acceptance of the electrons.

11.4.1.1 Cu_2S/Graphene Counter Electrode

In principle, QDSSC adopts the design of dye-sensitized solar cell (DSSC) [48]. Photoexcited electrons are injected into wide bandgap semiconductor metal oxide (e.g., TiO_2, SnO_2, and ZnO) from excited QDs. Comparing to dyes, QDs have the advantages of tunable band gaps [49], hot electron injection [50], high molar extinction coefficients [49, 51], and large intrinsic dipole moments [52, 53], which

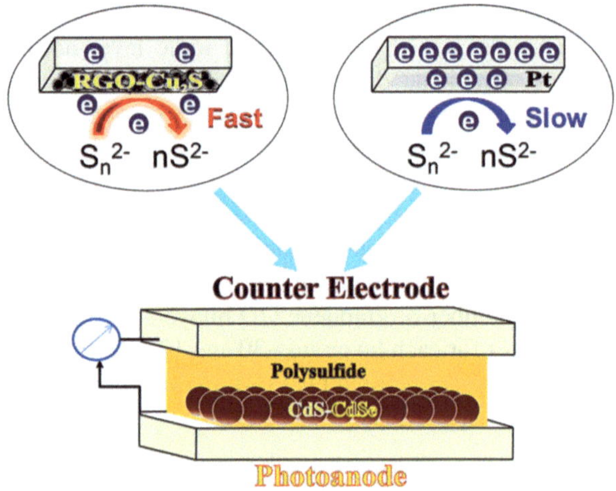

Fig. 11.6 Scheme of a QDSSC with the comparison of the kinetics of redox reaction at two different counter electrodes. Reprinted with permission from [9]; Copyright © 2011 The American Chemical Society

are desirable for power conversion in solar cells. Sulfide/polysulfide couple is used as the electrolyte in QDSSC due to the instability of QDs in iodine/iodide electrolyte. However, commonly used Pt counter electrode shows poor redox activity toward sulfide/polysulfide electrolyte [54]. To improve the redox activity of the counter electrode in the sulfide/polysulfide electrolyte and maximize power conversion efficiency (PCE) of the solar cells, metal-chalcogenides, such as CoS, PbS, and Cu_2S, were used due to their high electrocatalytic activity for polysulfide reduction [55]. Metal-chalcogenides can be prepared by directly exposing the metal foils of Cu, Co, or Pb to sulfide solution. However such method suffers problems from continual corrosion and ultimately mechanical instability [9]. To overcome these problems, a composite material consisting of RGO and Cu_2S was developed. Graphene contains high surface area and high conductivity, which promotes high numbers of the Cu_2S reactive sites and rapid electron transport to the reactive sites where the electrons can reduce the oxidized polysulfide. The response of RGO/Cu_2S and platinum electrodes toward polysulfide reduction during the operation of QDSSC is shown in Fig. 11.6.

In situ growth of Cu_2S/RGO on FTO glass can be made by immersing Cu^0/RGO-coated FTO glass into sulfide solution. Poly(vinylidene) fluoride (PVDF) binder is added into Cu^0/RGO composite before casting on the FTO glass to overcome the stability issue. To fabricate QDSSCs, TiO_2/CdS/CdSe photoanode and polysulfide electrolyte (1 M Na_2S and 1 M S in aqueous solution) are used as working electrode and electrolyte, respectively, with Cu_2S/RGO on FTO glass as counter electrodes.

In the QDSSC with Cu_2S/RGO counter electrode, a short circuit current density (J_{sc}) of 18.2 mA cm^{-2}, open circuit voltage (V_{oc}) of 0.52 V, and fill factor (FF) of 0.46 was obtained. Comparing to that with Pt counter electrode ($J_{sc} = 11.3$ mA cm^{-2}; $V_{oc} = 0.46$; FF $= 0.31$), the values of QDSSC with Cu_2S/RGO counter electrode significantly increased. Therefore the PCE of QDSSC with Cu_2S/RGO counter electrode increases to 4.4 % from 1.6 % of that with Pt counter electrode. The maximum incident photon to current conversion efficiency (IPCE) at 400 nm is

Table 11.1 Performance of QDSSCs with different counter electrode

Counter electrode	I_{sc} (mA cm^{-2})	V_{oc} (V)	FF	PCE (%)	Reference
Cu_2S/RGO	18.4	0.52	0.46	4.40	[9]
Au/RGO	7.07	0.51	0.381	1.38	[56]
Cu_2S/carbon	10.7	0.5	0.58	3.08	[57]
Pt	11.3	0.46	0.31	1.60	[9]
Pt	10.5	0.66	0.4	2.80	[58]
Brass	11.1	0.52	0.6	3.47	[59]
Brass	13.7	0.58	0.63	4.92	[60]

nearly 90 % for QDSSC with Cu_2S/RGO counter electrode. This high IPCE value indicates a nearly ideal conversion of incident photons to current at low incident light intensities. The charge transfer resistance R_{ct} was extracted from electrochemical impedance spectroscopy (EIS) data of the counter electrode. The value for Pt counter electrode was 998 Ω cm^2, whereas that of the Cu_2S/RGO electrode was only 1.61 Ω cm^2. This tremendous decrease of R_{ct} is the sign of superior catalytic activity for S^{2-}/S_n^{2-} redox couple. When used as counter electrode, Cu_2S/RGO composite can rapidly shuttle electrons to the oxidized redox polysulfide species. RGO here is essential for the dispersion and stability of Cu_2S reactive sites and electron transport to the S^{2-}/S_n^{2-} redox couple.

11.4.1.2 Au/Graphene Counter Electrode

Au nanoparticle with RGO is another material for counter electrode in QDSSCs [56]. GO solution (1.5 mg mL^{-1}) was placed in a 35 mL microwave tube followed by adding 0.5 mL 1 wt.% $HAuCl_4$. Then the pH of the solution was adjusted to 9 and heated up to 150 °C with a microwave pure RG was also synthesized by direct microwave. The Au/RGO composite films were deposited onto graphite substrates with EPD technique [56]. QDSSCs with Au/RGO counter electrode provide higher photovoltaic performance than those with Au or Pt electrodes. The open circuit potential (V_{oc}), short circuit current density (J_{sc}), fill factor (FF), and PCE of QDSSCs with Au/RGO counter electrode are 7.07 mA cm^{-2}, 0.51 V, and 1.36 %, respectively. This is attributed to the high electrical conductivity of the RGO and high electrocatalytic activity of Au NPs [56]. Table 11.1 listed photoelectrochemical performance of QDSSCs with different counter electrode.

11.4.2 Working Electrode

In addition to the counter electrode, graphene can be used as the electron acceptor and charge transfer pathway in working electrode of QD solar cells. Working electrode in QD solar cells is where photoexcited electrons or holes generate and

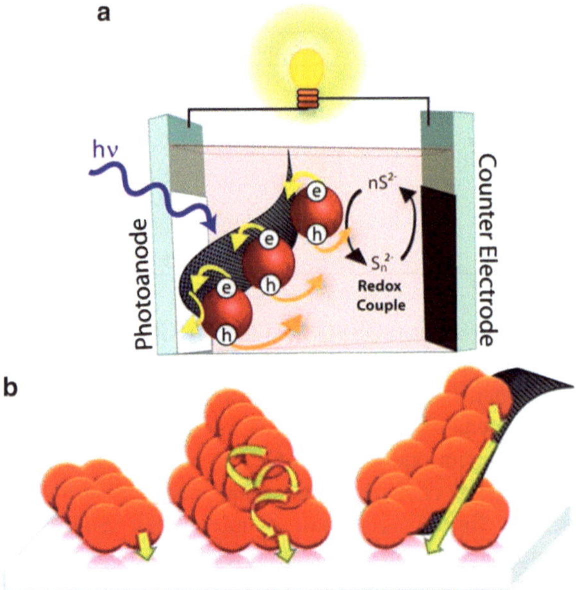

Fig. 11.7 (**a**) Scheme of the solar cells in which photocurrent generation occurs when CdSe QDs absorb light. (**b**) Three electron transfer routes from excited CdSe QDs to collecting electrode. Reprinted with permission from [39]; Copyright © 2012 The American Chemical Society

separate. With the existence of graphene, QDs upon photoexcitation undergo electron transfer to graphene. If a high degree of graphene dispersion within a QD film is achieved, electron transport will be facilitated along the highly conductive graphene pathway toward the electrode. Incorporation of graphene in QD films thus provides a significant step to overcome conductivity problems inherent to QD films and increase the performance of QD solar cells.

11.4.2.1 Electrophoretic Deposition of CdSe/Graphene

To get uniformly disperse graphene in the CdSe QD film, EPD was used to preserve the homogeneity of the film [39]. QDs and RGO were synthesized separately and mixed together before EPD. TiO_2 electrode was immersed into the mixture and a 50 V bias voltage was applied for 1 min followed by remixing. The deposited electrode was then used as working electrode in solar cells with Cu_2S/graphene films as counter electrodes and $S^{2-}/S_n{}^{2-}$ as the redox electrolyte (Fig. 11.7a). Comparing to CdSe QD films, CdSe/graphene composite shows significant enhancement in photocurrent response. The maximum photocurrent was achieved at a loading of ~1.4×10^{15} QDs cm^{-2} and ~20 μg cm^{-2} graphene. IPCE at 530 nm (i.e., the absorption peak of CdSe QDs) was 13.8 % for CdSe/graphene film, which is 3.6 times of that for CdSe QD films (3.8 %) [39]. This enhancement in IPCE can be attributed to the presence of graphene as electron acceptors and pathway for electron shuttling to the electrode surface. This enhancement observed in CdSe/graphene composites has also been reported in CdS and TiO_2 composites with C_{60} and carbon nanotubes [42, 61]. When illuminated with visible light, the excited electron and hole pairs separated throughout CdSe QDs. Without the presence of graphene, photogenerated electrons in CdSe QDs close to the remote end of the film are likely to experience recombination while traveling in the QDs network before reaching the electrode surface (Fig. 11.7b). Thus the charge recombination at the grain boundaries of QDs limits the IPCE of the solar cells. The presence of graphene dispersed inside QDs layers provides an electron collection pathway and minimizes charge recombination loss. The important role of graphene in interacting with the excited QDs and facilitating electron transport to the electrode is endorsed by the high IPCE values obtained from CdSe/graphene solar cells [39].

11.4.2.2 Layered CdSe/Graphene

In addition to the mixture of QDs and graphene, layered graphene and QD films can be fabricated on indium tin oxide (ITO) glass from aqueous solutions. The structure and favorable work function of graphene enable it for the collection and transfer of photogenerated charges from QD to the electrode, resulting in a high-performance QD solar cells [47].

Layered graphene and QD films can be fabricated by sequential EPD of graphene and SILAR deposition QDs on ITO glass. A QD solar cell consisting of layered graphene/CdSe QDs was built up with Pt and 0.1 M Na_2S aqueous solution as counter electrode and electrolyte, respectively. Quite constant open circuit voltage (V_{oc}) of 0.68 V was observed in devices with more than one bilayer of graphene/QD, while the short circuit photocurrent density (I_{sc}) increases with the number of bilayers until a maximum of 1.08 mA cm^2 in devices with eight bilayers of graphene/QDs. Comparing with devices without graphene, i.e., CdSe QDs directly deposited on ITO glass, the I_{sc} was enhanced by three times (0.26 mA cm^{-2} for devices without graphene). The stability was also improved. After illumination of 36 h, layered graphene/QD solar cell remained 92 % of its original I_{sc}, while only 32 % of initial I_{sc} was left for QD solar cell without graphene. Further comparison of layered graphene/QD solar cell and solar cells with single wall carbon nanotube (SWNT) and QD composites revealed 0.1 V increase of V_{oc} and 2.5 times enhancement of I_{sc}. IPCE of layered graphene/QD reached a maximum of 16 %, which is almost twice of the maximum IPCE of 9 % with SWNT/QD [36]. Thus layered graphene/QD composite can significantly improve photoresponses, especially photocurrent, herein confirm that graphene is a good candidate for the collection and transport of photogenerated charges, and the layered structure can improve charge collection and transport in QD solar cells, thus provide a new and promising direction toward developing high-performance light-harvesting devices for the next-generation solar cells.

11.4.2.3 Mixture of TiO_2/Graphene

The concept of incorporation of graphene into metal oxides has been demonstrated in DSSCs, which have almost same components with QD sensitized solar cells, except the sensitizer. For example, Kim et al. [62] deposited an intermediate layer between FTO glass and TiO_2 film with a mixture of graphene and TiO_2, which improved the conversion efficiency from 4.89 to 5.26 % in DSSCs. Yang et al. [63] reported an increase of conversion efficiency by 39 % when introduced graphene into the TiO_2 for DSSCs with an increase of conversion efficiency by 39 %. Same improvement is expected for QD sensitized solar cells. By directly mixing graphene and TiO_2 nanoparticles, the PCE of the QD sensitized solar cell increased by 56 %.

Exfoliated graphite oxide was reduced by hydrazine reduction to form graphene [64]. Graphene/TiO_2 aqueous solution was made by directly mixing a certain amount of graphene into TiO_2 nanoparticles (P25, Degussa) solution. The graphene/TiO_2 solution was then screen printed on FTO glass and sintered at 500 °C for 30 min. CdS QD was deposited via SILAR.

The short circuit current (I_{sc}) and PCE increased remarkably from 3.8 mA cm^{-2} and 0.92 % for the QDSSC without graphene to 7.1 mA cm^{-2} and 1.44 % for the cell with 0.8 wt.% graphene. A similar trend can be observed in the IPCE curves. The maximum of IPCE increased from 36 to 42 % when adding 0.8 wt.% graphene. However additional graphene more than 0.8 wt.% decreased the performance of the solar cell. Due to the lower conduction band of graphene comparing to TiO_2, graphene in TiO_2 film can capture photogenerated electrons and transport them to FTO glass, which decrease the recombination and back-transport, therefore gave rise to increased photocurrent [17, 63, 65, 66]. However, excess graphene may reduce the QDs adsorption and recombination center for charge recombination, leading to the low efficiency of QDSSCs [67].

11.4.2.4 TiO_2/Graphene/Au Composites

Attaching noble metal nanoparticles, such as Au or Ag, on a TiO_2 substrate is another strategy commonly adopted for increasing charge separation [68]. At the semiconductor–metal interface, a space charge layer is formed according to Fermi level equilibration. Thus the excited electron is rapidly transferred to the metal nanoparticle which behaves as an electron sink and storage center [69].

Mixture solution of TiO_2 anatase nanoparticles and functionalized graphene oxide (FGO) was coated on FTO glass and annealed at 150 °C for 40 min to dry the film and attachTiO_2/graphene to the FTO glass. Before sensitized with QD, the TiO_2/graphene film was immersed in a 0.1 M MPA linker aqueous solution for 10–12 h followed by immersing in 4 mM Au NPs and 200 mM TOP/TOPO capped CdSe QDs mixture for another 10–12 h for anchoring CdSe and Au NPs.

The conduction band level of CdSe QDs lies at −3.9 eV, which is more positive to the conduction band of TiO_2 (−4.2 eV). Thus the photogenerated electrons in

Table 11.2 Performance of graphene-based QDSSCs with different working electrode

Working electrode	V_{oc} (V)	I_{sc} (mA cm^{-2})	FF	PCE	IPCE (%)	Reference
EPD CdSe/graphene					14	[39]
Layered CdSe/graphene	0.68	1.08			16	[36]
TiO_2/graphene/CdSe	0.58	7.1	35 %	1.44 %	42	[67]
TiO_2/CdSe/graphene/Au	0.7				27	[68]
Graphene/ZnO	0.598	4.65	61.80 %	1.72 %	27	[70]

CdSe QDs can transfer to the conduction band of TiO_2. Due to the offset of 0.22 eV between the conduction band of TiO_2 and the work function of FGO (−4.42 eV), the electron will further transfer to FGO. The work function of neat Au lies at about −5.1 eV; however, the apparent Fermi level is expected to increase when contact with the TiO_2/CdSe/FGO. Therefore Au NPs can be a conduction bridge between FGO and FTO glass, thus allowing the transport of electrons to the external circuit. Platinum sheet and 0.1 M Na_2S aqueous solution are used as the counter electrode and electrolyte, respectively [68].

Upon illumination, the composite film TiO_2/CdSe/FGO/Au-based solar cell showed a fivefold enhancement in photovoltage comparing to those without FGO and Au. The maximum V_{oc} of TiO_2/CdSe/FGO/Au-based solar cell is about 0.7 V, while only 0.15 V for TiO_2/CdSe solar cell. The IPCE of TiO_2/CdSe/FGO/Au-based solar cell is 27 %, which is almost 4.5 times more than that of the solar cells without graphene [68].

11.4.2.5 Graphene with ZnO

Despite TiO_2, ZnO is also common electron transport material in QD solar cells. The use of graphene/ZnO composite architecture has been demonstrated to enhance the electron transport in QD solar cells [70].

In particular, graphene oxide was prepared by chemical exfoliation based on modified Mummer's method [2], followed by spin-coating onto FTO glass and reduction by hydrazine. ZnO seed layer was then deposited on the graphene sheet by ultrasonic spray pyrolysis. ZnO nanorods grew from the seed by immersing into a 0.01 M zinc nitrate and 0.01 M hexamethylenetetramine solution at 95 °C for 10 h, then sensitized with one-pot grown CdSe [71] by electrophoretic deposition. The best performance of QD solar cell was achieved with 1 nm thick graphene sheet with V_{oc}, J_{sc}, FF, and PCE of 0.598 V, 4.65 mA cm^{-2}, 61.8 %, and 1.72 % under 100 mW cm^{-2} irradiation, respectively, while the performance of QD solar cell without graphene exhibited a V_{oc}, J_{sc}, FF, and PCE of 0.566 V, 3.37 mA cm^{-2}, 40.9 %, and 0.78 %, respectively. The maximum IPCE at 400 nm are 18 and 27 % for solar cells with graphene layer and without graphene, respectively, which is in agreement with the difference of J_{sc} [70]. Some important performance of the solar cells discussed above is summarized in Table 11.2.

11.5 Graphene Quantum Dot

GQDs contain size-dependent bandgap and large optical absorption, making them particularly promising for the applications in solar cells and electronic devices [24, 25, 72]. However, preparation of colloidal GQDs with excellent solubility in common solvents remains undeveloped and becoming a key factor to fabricate GQD-based solar cells. Without sufficient surface protection, strong attraction between small graphene sheets usually leads to the formation of graphite particles [73].

11.5.1 Synthesis of Colloidal Graphene Quantum Dot

So far, two approaches have been exploited to prepare colloidal GQDs, i.e., top-down cutting of graphene sheets and bottom-up oxidative condensation reactions. Top-down approach refers to break large pieces of graphene sheets into small GQDs, while bottom-up method starts from the synthesis of graphene species containing a certain number of conjugated carbon atoms [74].

11.5.1.1 Top-Down Approach

Pan et al. [75] reported a hydrothermal method to cutting graphene sheets into GQDs. Briefly, graphene sheets were oxidized in a mixture of sulfuric acid and nitric acid under careful control with mild ultrasonication. The oxidized graphene sheets were then collected and cut into GQDs via hydrothermal method in a Teflon autoclave at high temperature. The average diameter of obtained GQDs is 9.6 nm, which consists of 1–3 layer graphene. The quantum yield of emitting blue light is 6.9 % using quinine sulfate as a reference. The cutting mechanism involved the complete breakup of mixed epoxy chains composed of fewer epoxy groups and more carbonyl groups under hydrothermal conditions.

Later, Li et al. developed an electrochemical cutting method which produces green-luminescent GQDs and utilized the prepared GQDs as electron acceptors in GQD/P3HT hybrid thin film solar cells [72]. In the cutting process, GQDs were formed by electrochemical oxidation of a graphene electrode in phosphate buffer solution. The size of the obtained GQDs is 3–5 nm. The oxygen containing groups on the surface of GQDs provided an aqueous solubility and facilitated further surface functionalization. Similar to GQDs obtained via hydrothermal method [75], GQDs synthesized by electrochemical oxidation also consist of 1–3 graphene layers, which is attributed to strong attraction between graphene sheets.

11.5.1.2 Bottom-Up Approach

On the other hand, the bottom-up method successfully synthesized GQDs via solution chemistry by oxidative condensation of aryl groups (Scholl reaction)

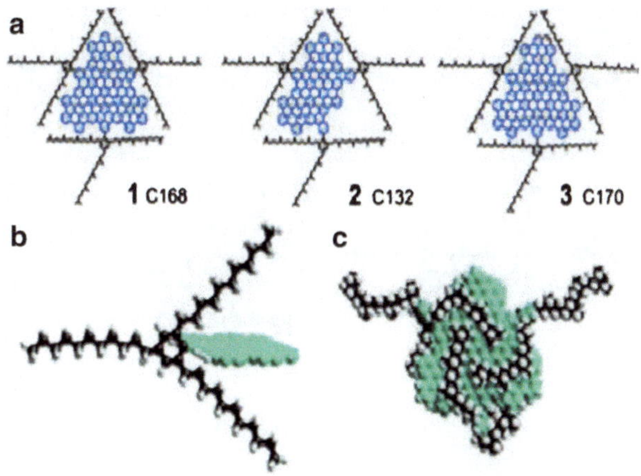

Fig. 11.8 (a) Schemes of GQDs with tunable size. (b, c) Strategies to make GQDs soluble. Reprinted with permission from [24, 80]; Copyright © 2010 The American Chemical Society

[25, 73, 76, 77]. Intramolecular variation has been demonstrated to be useful for the synthesis of large polycyclic aromatic hydrocarbons (PAHs) from dendritic arene precursors [73, 78, 79]. GQDs with up to 222 conjugated carbon atoms have been successfully prepared, though the solubility of such GQDs was unfortunately low and they tend to aggregate due to strong attraction between each other. Figure 11.8a shows GQDs containing 168, 132, and 170 conjugated carbon atoms, respectively.

Recently, a 2′, 4′, 6′-triakyl phenyl groups was used as stabilizing agent to form 3-D protection at the edge of graphene pieces by Li et al., resulting in GQDs with tunable size and high stability [24, 80]. The formation of 3-D protection was attributed to the strong covalent binding and twisted phenyl groups due to crowdedness at the edge of graphene pieces. The mechanism for 3-D protection by twisted ligands that covalently attach to the edge of GQDs is depicted in Fig. 11.8b. Covalently attached trialkyl phenyl groups to the edge of graphene moieties rendered them with excellent solubility in common organic solvents, such as toluene, THF, and chloroform. With one of the most effective methods of identifying GQDs, isotope-resolved mass spectroscopy with mild ionization methods (e.g., MALDI-TOF) suggested that GQDs had high uniformity in size. Dynamic light scattering (DLS) on GQDs showed an average size more than twice as high as that obtained from molecular modeling (i.e., 13.5 nm versus 5 nm for GQDs with 168 conjugated carbon atoms), indicating oligomerization of GQDs in solution due to residual attraction between each other, which also explained the reason for failure of conventional liquid NMR spectroscopy in characterizing GQDs [24, 80].

Yan et al. [24] developed a solution-based synthetic route to GQDs. Polyphenylene dendritic precursors (1–3 in Fig. 11.9) were prepared via a stepwise solution chemistry shown in the scheme below to synthesize GQDs based on oxidation. Size and shape-dependent absorption spectrum in these GQDs was revealed by UV–vis absorption spectroscopy (Fig. 11.10) [24]. The absorption edges of GQDs with larger size (i.e., 168 and 170 conjugated carbon atoms) exhibited a significantly red shift comparing to smaller GQDs (i.e., 132 conjugated carbon atoms). Such size-dependent absorption confirmed the quantum confinement in GQDs, which is consistent with the size-dependent effects in other QDs

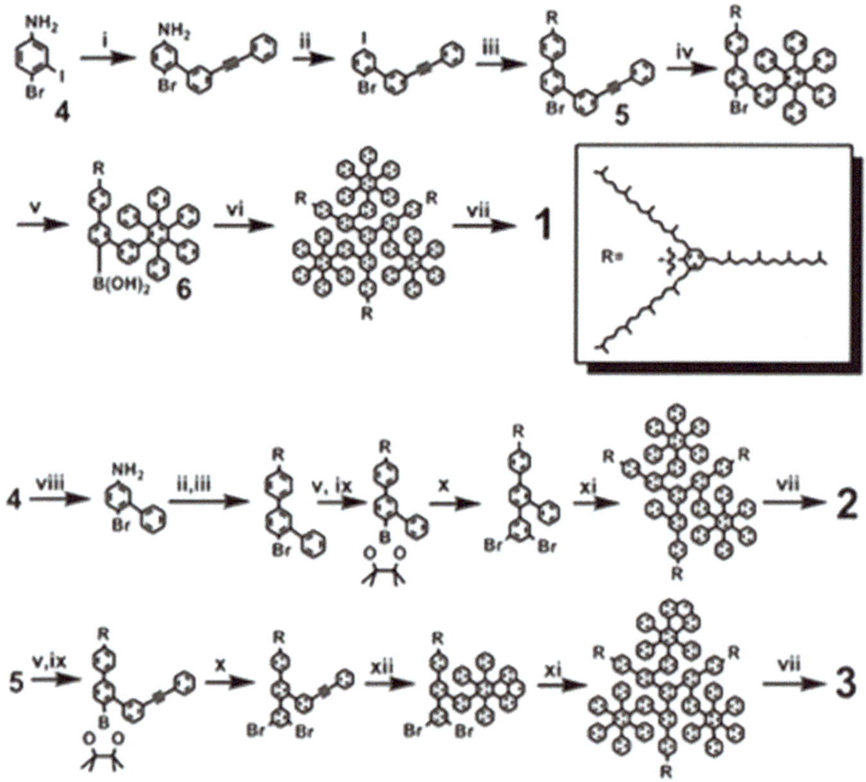

Fig. 11.9 Synthesis of GQDs via a wet chemistry approach. Reprinted with permission from [24]; Copyright © 2010 The American Chemical Society

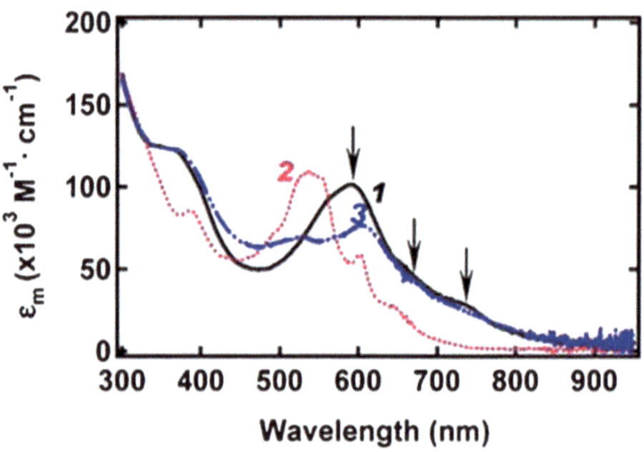

Fig. 11.10 UV–vis absorption of colloidal GQDs. α, p, and β bands of GQDs with 168 conjugated carbon atoms are marked by *arrows* (i.e., 1 in Figs. 11.8a and 11.9), from *right* to *left*, respectively. Reprinted with permission from [24]; Copyright © 2010 The American Chemical Society

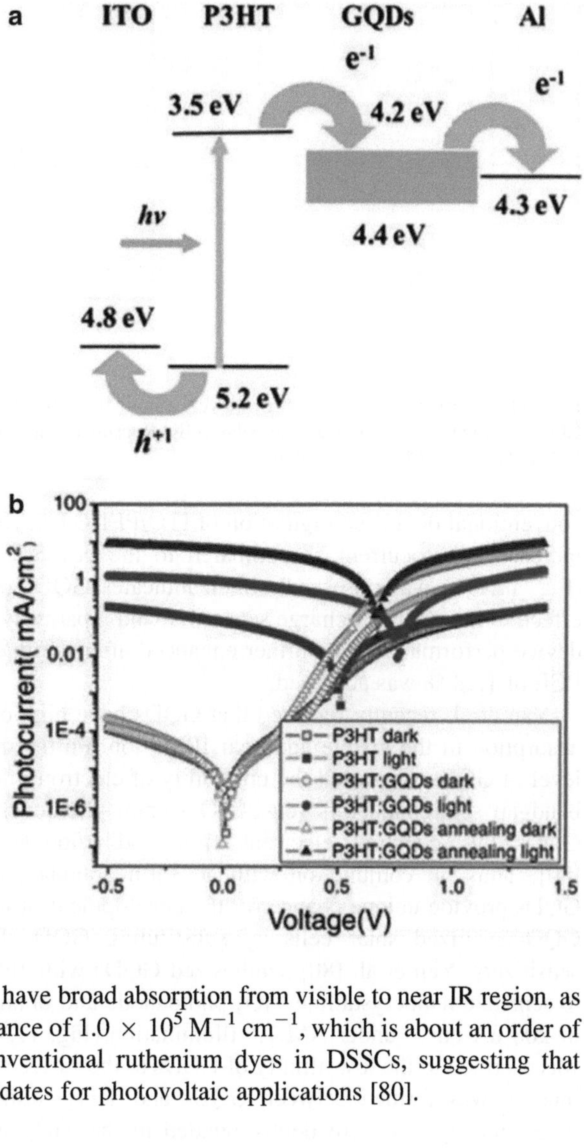

Fig. 11.11 (**a**) Energy level alignment diagram of the ITO/PEDOT:PSS/P3HT: GQDs/Al solar cell device. (**b**) *J–V* curves of the ITO/PEDOT:PSS/P3HT/ Al, ITO/PEDOT:PSS/ P3HT:GQDs/Al and ITO/PEDOT:PSS/P3HT: GQDs/Al devices after annealed at 140 °C for 10 min. Reprinted with permission from [72]; Copyright © Wiley

[81]. Notably, these GQDs have broad absorption from visible to near IR region, as well as a maximum absorbance of $1.0 \times 10^5\ M^{-1}\ cm^{-1}$, which is about an order of magnitude larger than conventional ruthenium dyes in DSSCs, suggesting that GQDs are promising candidates for photovoltaic applications [80].

11.5.2 Photovoltaics Based on Colloidal GQDs

Due to their attracting optoelectronic properties and solution processability, colloidal GQDs are ideal candidates for the applications in low-cost, high-performance photovoltaic devices.

Li et al. [72] prepared colloidal GQDs with green emission as an electron collector in conjugated polymer, poly(3-hexylthiophene) (P3HT)-based thin film solar cells instead of PCBM. Figure 11.11a shows the energy level alignment with a

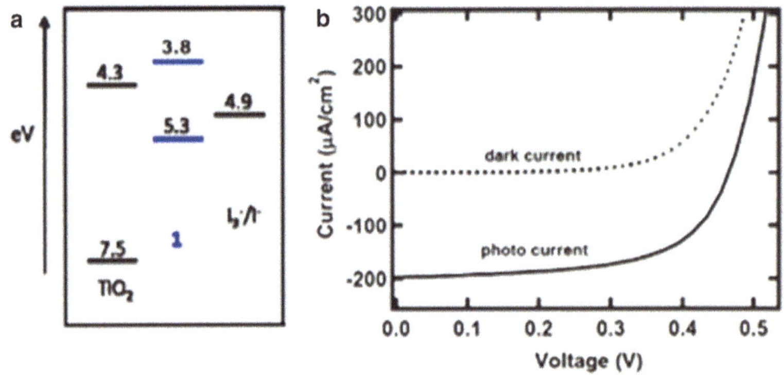

Fig. 11.12 (**a**) Calculated energy level alignment. (**b**) *J–V* characteristics of colloidal GQD-sensitized TiO_2 nanoparticle solar cells. Reprinted with permission from [80]; Copyright © The American Chemical Society

conventional device configuration of ITO/PEDOT:PSS/P3HT:GQDs/Al. A greatly increased photocurrent as compared to the device fabricated with P3HT only (Fig. 11.11b) was observed, which indicates GQDs ($c = 10$ wt.%) provided an effective interface for charge separation and a pathway for electron transport. The device performance was further enhanced after thermal annealing, and an overall PCE of 1.28 % was achieved.

Yan et al. recently reported that GQDs have high optical absorption and ideal absorption in the visible and near IR region. Furthermore, the calculated energy level in GQDs suggested the possibility of electron injection from GQDs to large bandgap semiconductors (e.g., TiO_2) upon photoexcitation and regeneration of GQDs by accepting electrons from iodine/iodide electrolyte (Fig. 11.12a) [80]. Thus, in conjunction with an earth abundant composition characteristic, GQDs provide unique chance for the development of inexpensive, high efficiency GQD-sensitized solar cells by substituting GQDs for metal-organic dyes as sensitizers. Yan et al. [80] synthesized GQDs with 168 conjugated carbon atoms to sensitize nanocrystalline TiO_2 photoanode and obtained a photocurrent density of 200 $\mu A\ cm^{-2}$ under AM 1.5 illumination (Fig. 11.12b). The low performance was ascribed to the low affinity of GQDs to the surface of TiO_2. Since no chemical binding was formed between GQDs and TiO_2, while in conventional DSSCs, ruthenium dyes are covalently bonded to the TiO_2 surface by carboxyl groups which enable much higher optical absorption and faster exciton dissociation.

11.6 Conclusion and Outlook

Graphene provides us promising opportunities to design hybrid assemblies composed of different quantum dots for energy conversion. Further understanding of its high conductivity and large surface interactions will be the key to the development of next-generation high-performance photovoltaic devices. As a scaffold to anchor QDs and nanoparticles, graphene boosts the efficiency of quantum dot solar cells and other devices. The major barrier for commercial development is the synthesis of large graphene sheets with minimum surface defects, and small GQDs with good dispersion in solvent. Due to the absorption properties of graphene, the concentration of graphene must be fine-tuned when applied in a photoactive assembly.

Further researches on graphene and QDs composites will focus on obtaining better controlled synthesis, surface functionalization and novel device fabrication process. These researches will shed light on increasing the efficiency of the solar devices towards practical applications.

Acknowledgments The author wishes to thank the financial support from DOE Ames Laboratory Seed Fund.

References

1. Prezhdo, O.V., Kamat, P.V., Schatz, G.C.: Virtual issue: graphene and functionalized graphene. J. Phys. Chem. C **115**, 3195–3197 (2011)
2. Hummers, W.S., Offeman, R.E.: Preparation of graphitic oxide. J. Am. Chem. Soc. **80**, 1339–1339 (1958)
3. Novoselov, K.S., Jiang, D., Schedin, F., Booth, T.J., Khotkevich, V.V., Morozov, S.V., Geim, A.K.: Two-dimensional atomic crystals. Proc. Natl. Acad. Sci. USA **102**, 10451–10453 (2005)
4. Novoselov, K.S., Geim, A.K., Morozov, S.V., Jiang, D., Zhang, Y., Dubonos, S.V., Grigorieva, I.V., Firsov, A.A.: Electric field effect in atomically thin carbon films. Science **306**, 666–669 (2004)
5. Kim, K.S., Zhao, Y., Jang, H., Lee, S.Y., Kim, J.M., Kim, K.S., Ahn, J.-H., Kim, P., Choi, J.-Y., Hong, B.H.: Large-scale pattern growth of graphene films for stretchable transparent electrodes. Nature **457**, 706–710 (2009)
6. Wei, D., Liu, Y., Wang, Y., Zhang, H., Huang, L., Yu, G.: Synthesis of N-doped graphene by chemical vapor deposition and its electrical properties. Nano Lett. **9**, 1752–1758 (2009)
7. Schniepp, H.C., Li, J.-L., McAllister, M.J., Sai, H., Herrera-Alonso, M., Adamson, D.H., Prud'homme, R.K., Car, R., Saville, D.A., Aksay, I.A.: Functionalized single graphene sheets derived from splitting graphite oxide. J. Phys. Chem. B **110**, 8535–8539 (2006)
8. Kongkanand, A., Martínez Domínguez, R., Kamat, P.V.: Single wall carbon nanotube scaffolds for photoelectrochemical solar cells. capture and transport of photogenerated electrons. Nano Lett. **7**, 676–680 (2007)
9. Radich, J.G., Dwyer, R., Kamat, P.V.: Cu_2S reduced graphene oxide composite for high-efficiency quantum dot solar cells. Overcoming the redox limitations of S-2(−)/S-n(2-) at the counter electrode. J. Phys. Chem. Lett. **2**, 2453–2460 (2011)
10. Vinodgopal, K., Neppolian, B., Lightcap, I.V., Grieser, F., Ashokkumar, M., Kamat, P.V.: Sonolytic design of graphene−Au nanocomposites. Simultaneous and sequential reduction of graphene oxide and Au(III). J. Phys. Chem. Lett. **1**, 1987–1993 (2010)
11. Stankovich, S., Dikin, D.A., Piner, R.D., Kohlhaas, K.A., Kleinhammes, A., Jia, Y., Wu, Y., Nguyen, S.T., Ruoff, R.S.: Synthesis of graphene-based nanosheets via chemical reduction of exfoliated graphite oxide. Carbon **45**, 1558–1565 (2007)
12. Bourlinos, A.B.., Gournis, D., Petridis, D., Szabó, T., Szeri, A., Dékány, I.: Graphite oxide: chemical reduction to graphite and surface modification with primary aliphatic amines and amino acids. Langmuir **19**, 6050–6055 (2003)

13. Si, Y., Samulski, E.T.: Synthesis of water soluble graphene. Nano Lett. **8**, 1679–1682 (2008)
14. Abdelsayed, V., Moussa, S., Hassan, H.M., Aluri, H.S., Collinson, M.M., El-Shall, M.S.: Photothermal deoxygenation of graphite oxide with laser excitation in solution and graphene-aided increase in water temperature. J. Phys. Chem. Lett. **1**, 2804–2809 (2010)
15. Sokolov, D.A., Shepperd, K.R., Orlando, T.M.: Formation of graphene features from direct laser-induced reduction of graphite oxide. J. Phys. Chem. Lett. **1**, 2633–2636 (2010)
16. Hassan, H.M.A., Abdelsayed, V., Khder, A.E.R.S., AbouZeid, K.M., Terner, J., El-Shall, M.S., Al-Resayes, S.I., El-Azhary, A.A.: Microwave synthesis of graphene sheets supporting metal nanocrystals in aqueous and organic media. J. Mater. Chem. **19**, 3832–3837 (2009)
17. Williams, G., Seger, B., Kamat, P.V.: TiO_2-graphene nanocomposites. UV-assisted photocatalytic reduction of graphene oxide. ACS Nano **2**, 1487–1491 (2008)
18. Ramesha, G.K., Sampath, S.: Electrochemical reduction of oriented graphene oxide films: an in situ Raman spectroelectrochemical study. J. Phys. Chem. C **113**, 7985–7989 (2009)
19. Ng, Y.H., Lightcap, I.V., Goodwin, K., Matsumura, M., Kamat, P.V.: To what extent do graphene scaffolds improve the photovoltaic and photocatalytic response of TiO_2 nanostructured films? J. Phys. Chem. Lett. **1**, 2222–2227 (2010)
20. Kamat, P.V.: Graphene-based nanoarchitectures. Anchoring semiconductor and metal nanoparticles on a two-dimensional carbon support. J. Phys. Chem. Lett. **1**, 520–527 (2009)
21. Kamat, P.V.: Graphene-based nanoassemblies for energy conversion. J. Phys. Chem. Lett. **2**, 242–251 (2011)
22. Yu, D., Nagelli, E., Du, F., Dai, L.: Metal-free carbon nanomaterials become more active than metal catalysts and last longer. J. Phys. Chem. Lett. **1**, 2165–2173 (2010)
23. Geim, A.K., Novoselov, K.S.: The rise of graphene. Nat. Mater. **6**, 183–191 (2007)
24. Yan, X., Cui, X., Li, L.: Synthesis of large, stable colloidal graphene quantum dots with tunable size. J. Am. Chem. Soc. **132**, 5944–5945 (2010)
25. Li, L., Yan, X.: Colloidal graphene quantum dots. J. Phys. Chem. Lett. **1**, 2572–2576 (2010)
26. Mueller, M.L., Yan, X., McGuire, J.A., Li, L.: Triplet states and electronic relaxation in photoexcited graphene quantum dots. Nano Lett. **10**, 2679–2682 (2010)
27. Nair, R.R., Blake, P., Grigorenko, A.N., Novoselov, K.S., Booth, T.J., Stauber, T., Peres, N.M. R., Geim, A.K.: Fine structure constant defines visual transparency of graphene. Science **320**, 1308 (2008)
28. Castro Neto, A.H., Guinea, F., Peres, N.M.R., Novoselov, K.S., Geim, A.K.: The electronic properties of graphene. Rev. Mod. Phys. **81**, 109–162 (2009)
29. Persson, B.N.J., Lang, N.D.: Electron–hole-pair quenching of excited states near a metal. Phys. Rev. B **26**, 5409–5415 (1982)
30. Swathi, R.S., Sebastian, K.L.: Long range resonance energy transfer from a dye molecule to graphene has (distance)[sup −4] dependence. J. Chem. Phys. **130**, 086101 (2009)
31. Swathi, R.S., Sebastian, K.L.: Resonance energy transfer from a dye molecule to graphene. J. Chem. Phys. **129**, 054703 (2008)
32. Chen, Z.Y., Berciaud, S., Nuckolls, C., Heinz, T.F., Brus, L.E.: Energy transfer from individual semiconductor nanocrystals to graphene. ACS Nano **4**, 2964–2968 (2010)
33. Liu, H., Ryu, S., Chen, Z., Steigerwald, M.L., Nuckolls, C., Brus, L.E.: Photochemical reactivity of graphene. J. Am. Chem. Soc. **131**, 17099–17101 (2009)
34. Kaniyankandy, S., Rawalekar, S., Ghosh, H.N.: Ultrafast charge transfer dynamics in photo-excited CdTe quantum dot decorated on graphene. J. Phys. Chem. C **116**, 16271–16275 (2012)
35. Lin, Y., Zhang, K., Chen, W., Liu, Y., Geng, Z., Zeng, J., Pan, N., Yan, L., Wang, X., Hou, J.G.: Dramatically enhanced photoresponse of reduced graphene oxide with linker-free anchored CdSe nanoparticles. ACS Nano **4**, 3033–3038 (2010)
36. Guo, C.X., Yang, H.B., Sheng, Z.M., Lu, Z.S., Song, Q.L., Li, C.M.: Layered graphene/quantum dots for photovoltaic devices. Angew. Chem. Int. Ed. **49**, 3014–3017 (2010)
37. Berger, C., Song, Z., Li, X., Wu, X., Brown, N., Naud, C., Mayou, D., Li, T., Hass, J., Marchenkov, A.N., Conrad, E.H., First, P.N., de Heer, W.A.: Electronic confinement and coherence in patterned epitaxial graphene. Science **312**, 1191–1196 (2006)

38. Dreyer, D.R., Park, S., Bielawski, C.W., Ruoff, R.S.: The chemistry of graphene oxide. Chem. Soc. Rev. **39**, 228–240 (2010)
39. Lightcap, I.V., Kamat, P.V.: Fortification of CdSe quantum dots with graphene oxide. Excited state interactions and light energy conversion. J. Am. Chem. Soc. **134**, 7109–7116 (2012)
40. Chang, H.X., Lv, X.J., Zhang, H., Li, J.H.: Quantum dots sensitized graphene: in situ growth and application in photoelectrochemical cells. Electrochem. Commun. **12**, 483–487 (2010)
41. Muszynski, R., Seger, B., Kamat, P.V.: Decorating graphene sheets with gold nanoparticles. J. Phys. Chem. C **112**, 5263–5266 (2008)
42. Brown, P., Kamat, P.V.: Quantum dot solar cells. Electrophoretic deposition of CdSe−C60 composite films and capture of photogenerated electrons with nC60 cluster shell. J. Am. Chem. Soc. **130**, 8890–8891 (2008)
43. Peng, Z.A., Peng, X.: Formation of high-quality CdTe, CdSe, and CdS nanocrystals using CdO as precursor. J. Am. Chem. Soc. **123**, 183–184 (2000)
44. Wang, H., Robinson, J.T., Li, X., Dai, H.: Solvothermal reduction of chemically exfoliated graphene sheets. J. Am. Chem. Soc. **131**, 9910–9911 (2009)
45. Cai, D., Blair, D., Dufort, F.J., Gumina, M.R., Huang, Z., Hong, G., Wagner, D., Canahan, D., Kempa, K., Ren, Z.F., Chiles, T.C.: Interaction between carbon nanotubes and mammalian cells: characterization by flow cytometry and application. Nanotechnology **19**, 1–10 (2008)
46. Yan, J.F., Ye, Q., Wang, X.L., Yu, B., Zhou, F.: CdS/CdSe quantum dot co-sensitized graphene nanocomposites via polymer brush templated synthesis for potential photovoltaic applications. Nanoscale **4**, 2109–2116 (2012)
47. Dai, L.M.: Layered graphene/quantum dots: nanoassemblies for highly efficient solar cells. ChemSusChem **3**, 797–799 (2010)
48. Gratzel, M.: Photoelectrochemical cells. Nature **414**, 338–344 (2001)
49. Yu, W.W., Qu, L., Guo, W., Peng, X.: Experimental determination of the extinction coefficient of CdTe, CdSe, and CdS nanocrystals. Chem. Mater. **15**, 2854–2860 (2003)
50. Pandey, A., Guyot-Sionnest, P.: Hot electron extraction from colloidal quantum dots. J. Phys. Chem. Lett. **1**, 45–47 (2009)
51. Wang, P., Zakeeruddin, S.M., Moser, J.E., Humphry-Baker, R., Comte, P., Aranyos, V., Hagfeldt, A., Nazeeruddin, M.K., Grätzel, M.: Stable new sensitizer with improved light harvesting for nanocrystalline dye-sensitized solar cells. Adv. Mater. **16**, 1806–1811 (2004)
52. Vogel, R., Pohl, K., Weller, H.: Sensitization of highly porous, polycrystalline TiO_2 electrodes by quantum sized CdS. Chem. Phys. Lett. **174**, 241–246 (1990)
53. Vogel, R., Hoyer, P., Weller, H.: Quantum-sized PbS, CdS, Ag_2S, Sb_2S_3, and Bi_2S_3 particles as sensitizers for various nanoporous wide-bandgap semiconductors. J. Phys. Chem. **98**, 3183–3188 (1994)
54. Chakrapani, V., Baker, D., Kamat, P.V.: Understanding the role of the sulfide redox couple (S2–/Sn2–) in quantum dot-sensitized solar cells. J. Am. Chem. Soc. **133**, 9607–9615 (2011)
55. Hodes, G., Manassen, J., Cahen, D.: Electrocatalytic electrodes for the polysulfide redox system. J. Electrochem. Soc. **127**, 544–549 (1980)
56. Zhu, G., Pan, L.K., Sun, H.C., Liu, X.J., Lv, T., Lu, T., Yang, J., Sun, Z.: Electrophoretic deposition of a reduced graphene-Au nanoparticle composite film as counter electrode for CdS quantum dot-sensitized solar cells. Chemphyschem **13**, 769–773 (2012)
57. Deng, M., Zhang, Q., Huang, S., Li, D., Luo, Y., Shen, Q., Toyoda, T., Meng, Q.: Low-cost flexible nano-sulfide/carbon composite counter electrode for quantum-dot-sensitized solar cell. Nanoscale Res. Lett. **5**, 986–990 (2010)
58. Niitsoo, O., Sarkar, S.K., Pejoux, C., Rühle, S., Cahen, D., Hodes, G.: Chemical bath deposited CdS/CdSe-sensitized porous TiO_2 solar cells. J. Photochem. Photobiol. A **181**, 306–313 (2006)
59. Huang, X., Huang, S., Zhang, Q., Guo, X., Li, D., Luo, Y., Shen, Q., Toyoda, T., Meng, Q.: A flexible photoelectrode for CdS/CdSe quantum dot-sensitized solar cells (QDSSCs). Chem. Commun. **47**, 2664–2666 (2011)
60. Zhang, Q., Guo, X., Huang, X., Huang, S., Li, D., Luo, Y., Shen, Q., Toyoda, T., Meng, Q.: Highly efficient CdS/CdSe-sensitized solar cells controlled by the structural properties of compact porous TiO_2 photoelectrodes. Phys. Chem. Chem. Phys. **13**, 4659–4667 (2011)

61. Robel, I., Bunker, B.A., Kamat, P.V.: Single-walled carbon nanotube–CdS nanocomposites as light-harvesting assemblies: photoinduced charge-transfer interactions. Adv. Mater. **17**, 2458–2463 (2005)
62. Kim, S.R., Parvez, M.K., Chhowalla, M.: UV-reduction of graphene oxide and its application as an interfacial layer to reduce the back-transport reactions in dye-sensitized solar cells. Chem. Phys. Lett. **483**, 124–127 (2009)
63. Yang, N., Zhai, J., Wang, D., Chen, Y., Jiang, L.: Two-dimensional graphene bridges enhanced photoinduced charge transport in dye-sensitized solar cells. ACS Nano **4**, 887–894 (2010)
64. Li, H., Lu, T., Pan, L., Zhang, Y., Sun, Z.: Electrosorption behavior of graphene in NaCl solutions. J. Mater. Chem. **19**, 6773–6779 (2009)
65. Sun, S., Gao, L., Liu, Y.: Enhanced dye-sensitized solar cell using graphene-TiO[sub 2] photoanode prepared by heterogeneous coagulation. Appl. Phys. Lett. **96**, 083113 (2010)
66. Manga, K.K., Zhou, Y., Yan, Y., Loh, K.P.: Multilayer hybrid films consisting of alternating graphene and titania nanosheets with ultrafast electron transfer and photoconversion properties. Adv. Funct. Mater. **19**, 3638–3643 (2009)
67. Zhu, G., Xu, T., Lv, T.A., Pan, L.K., Zhao, Q.F., Sun, Z.: Graphene-incorporated nanocrystalline TiO_2 films for CdS quantum dot-sensitized solar cells. J. Electroanal. Chem. **650**, 248–251 (2011)
68. Narayanan, R., Deepa, M., Srivastava, A.K.: Nanoscale connectivity in a TiO_2/CdSe quantum dots/functionalized graphene oxide nanosheets/Au nanoparticles composite for enhanced photoelectrochemical solar cell performance. Phys. Chem. Chem. Phys. **14**, 767–778 (2012)
69. Jakob, M., Levanon, H., Kamat, P.V.: Charge distribution between UV-irradiated TiO_2 and gold nanoparticles: determination of shift in the Fermi level. Nano Lett. **3**, 353–358 (2003)
70. Chen, J., Li, C., Eda, G., Zhang, Y., Lei, W., Chhowalla, M., Milne, W.I., Deng, W.Q.: Incorporation of graphene in quantum dot sensitized solar cells based on ZnO nanorods. Chem. Commun. **47**, 6084–6086 (2011)
71. Chen, J., Song, J.L., Sun, X.W., Deng, W.Q., Jiang, C.Y., Lei, W., Huang, J.H., Liu, R.S.: An oleic acid-capped CdSe quantum-dot sensitized solar cell. Appl. Phys. Lett. **94**, 153115 (2009)
72. Li, Y., Hu, Y., Zhao, Y., Shi, G.Q., Deng, L.E., Hou, Y.B., Qu, L.T.: An electrochemical avenue to green-luminescent graphene quantum dots as potential electron-acceptors for photovoltaics. Adv. Mater. **23**, 776-+ (2011)
73. Simpson, C.D., Brand, J.D., Berresheim, A.J., Przybilla, L., Räder, H.J., Müllen, K.: Synthesis of a Giant 222 carbon graphite sheet. Chem. Eur. J. **8**, 1424–1429 (2002)
74. Wang, J., Xin, X., Lin, Z.: Cu_2ZnSnS_4 nanocrystals and graphene quantum dots for photovoltaics. Nanoscale **3**, 3040–3048 (2011)
75. Pan, D., Zhang, J., Li, Z., Wu, M.: Hydrothermal route for cutting graphene sheets into blue-luminescent graphene quantum dots. Adv. Mater. **22**, 734–738 (2010)
76. Rempala, P., Kroulík, J., King, B.T.: A slippery slope: mechanistic analysis of the intramolecular scholl reaction of hexaphenylbenzene. J. Am. Chem. Soc. **126**, 15002–15003 (2004)
77. Hamilton, I.P., Li, B., Yan, X., Li, L.: Alignment of colloidal graphene quantum dots on polar surfaces. Nano Lett. **11**, 1524–1529 (2011)
78. Wu, J., Pisula, W., Müllen, K.: Graphenes as potential material for electronics. Chem. Rev. **107**, 718–747 (2007)
79. Wu, J., Tomović, Ž., Enkelmann, V., Müllen, K.: From branched hydrocarbon propellers to C3-symmetric graphite disks. J. Org. Chem. **69**, 5179–5186 (2004)
80. Yan, X., Cui, X., Li, B., Li, L.: Large, solution-processable graphene quantum dots as light absorbers for photovoltaics. Nano Lett. **10**, 1869–1873 (2010)
81. Medintz, I.L., Uyeda, H.T., Goldman, E.R., Mattoussi, H.: Quantum dot bioconjugates for imaging, labelling and sensing. Nat. Mater. **4**, 435–446 (2005)

Chapter 12
The Dynamics of Multiple Exciton Generation in Semiconductor Quantum Dots

Qing Shen, Kenji Katayama, and Taro Toyoda

Abstract In this paper, the dynamics of multiple exciton generation (MEG) in colloidal PbS quantum dots (QDs) characterized using a transient grating (TG) technique are reported. Only one peak, soon after the absorption of light, can be observed in the TG kinetics when the photon energy of the pump light, $h\nu$, is smaller than $2.7E_g$ (E_g: the band gap between the LUMO and HOMO in the QDs). When $h\nu$ is greater than $2.7E_g$, however, after the initial peak, the TG signal first decreases and then (after as little as 200 fs) increases, and a new peak appears at about 3 ps. The second peak intensity becomes larger with increasing $h\nu$. The first peak is considered to be related to hot carriers and the second one is believed to result from MEG. We found that MEG begins after just 200 fs and that the carrier density due to MEG increases and finally saturates after about 2 ps. Then, following the second peak, a fast decay, due to Auger recombination of the generated multiexcitons with a lifetime of about 30 ps, together with a slow decay, due to

Q. Shen (✉)
Faculty of Informatics and Engineering, Department of Engineering Science, The University of Electro-Communications, 1-5-1 Chofugaoka, Chofu, Tokyo 182-8585, Japan

PRESTO, Japan Science and Technology Agency, 4-1-8 Honcho, Kawaguchi, Saitama 332-0012, Japan

CREST, Japan Science and Technology Agency, 4-1-8 Honcho, Kawaguchi, Saitama 332-0012, Japan
e-mail: shen@pc.uec.ac.jp

K. Katayama
Faculty of Science and Engineering, Department of Applied Chemistry, Chuo University, 1-13-27 Kasuga, Bunkyo, Tokyo 112-8551, Japan

T. Toyoda
Faculty of Informatics and Engineering, Department of Engineering Science, The University of Electro-Communications, 1-5-1 Chofugaoka, Chofu, Tokyo 182-8585, Japan

CREST, Japan Science and Technology Agency, 4-1-8 Honcho, Kawaguchi, Saitama 332-0012, Japan

J. Wu and Z.M. Wang (eds.), *Quantum Dot Solar Cells*, Lecture Notes in Nanoscale Science and Technology 15, DOI 10.1007/978-1-4614-8148-5_12,

single exciton relaxation with a lifetime longer than nanoseconds, is observed in the TG kinetics. We succeeded in directly observing the occurrence and dynamics of MEG in semiconductor QDs using the TG technique.

12.1 Introduction

Semiconductor quantum dots (QDs) have been attracting increasing attention as potential light-harvesting materials in cheap and highly efficient next-generation solar cells because of the following unique properties [1–4]. Firstly, the energy gap of the QDs can be tuned by controlling their size, and therefore their absorption spectra can be tuned to match the spectral distribution of sunlight. Secondly, semiconductor QDs have large extinction coefficients due to the quantum confinement effect. Thirdly, these QDs have large intrinsic dipole moments, which can lead to rapid charge separation. Finally, semiconductor QDs have the potential to generate multiple electron–hole pairs with the absorption of a single photon (multiple exciton generation: MEG) [1, 4], which would lead to incident-photon-to-current conversion efficiencies (IPCE) of more than 100 %. Therefore, it is predicted theoretically that the maximum thermodynamic efficiency for photovoltaic devices with a single sensitizer could be improved to as much as ~44 % by employing semiconductor QDs [5]. Since QDSCs can be made very cheaply using simple chemical methods, they have the potential to be among the most promising next-generation cost-effective high efficiency solar cells [1–4].

As shown in Fig. 12.1, in a bulk semiconductor, absorption of a photon with energy greater than the semiconductor band gap, E_g, creates free electron–hole pairs that have energies in excess of the band gap. These electron–hole pairs are called hot

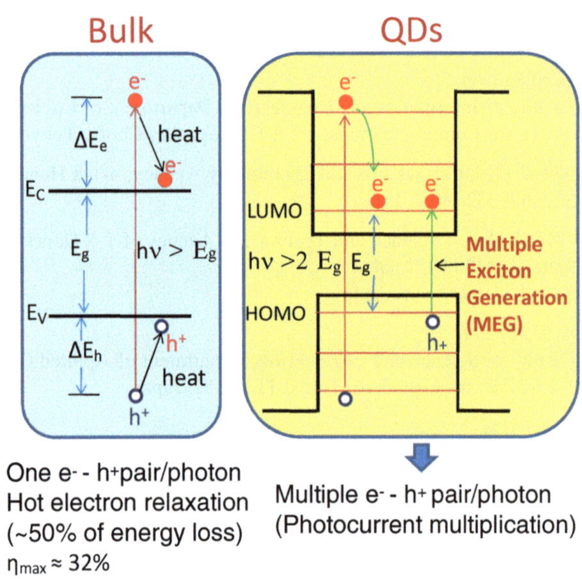

Fig. 12.1 Schematic illustration of multiple exciton generation (MEG) in quantum dots (QDs)

carriers or hot excitons. This excess electron energy is kinetic free energy and is lost quickly in a timescale from subpicoseconds to picoseconds through electron–phonon scattering, converting the excess kinetic energy into heat. So, mostly only one electron–hole pair can be created by the absorption of a single photon. This is the main reason for the limitation to the energy conversion efficiency of present day solar cells. As a result, the maximum thermodynamic efficiency is theoretically predicted to be about 32 % for a single junction solar cell. However, 10 years ago, Nozik proposed that the production of more than one electron–hole pair in semiconductor QDs by the absorption of photons with energies larger than $2E_g$, which is called multiple exciton generation (MEG), would achieve higher efficiencies compared to those in bulk semiconductors [4]. This is because semiconductor QDs have several unique properties due to the spatial confinement of electrons and holes in the QD. For example, the recombination of hot electrons and holes can be slowed because of the formation of discrete electronic states. The need to conserve crystal momentum is relaxed because the crystal momentum is not a good quantum number for 3-dimensionally confined carriers. Auger processes are greatly enhanced because of the increased electron–hole Coulomb interaction.

Since the first experimental evidence of efficient MEG in colloidal PbSe QDs was obtained by Schaller and Klimov [6], enhanced MEG has been observed in several other types of semiconductor QDs, such as PbS, CdSe, PbTe, InAs, and Si, at threshold photon energies of 2–3 times the HOMO–LUMO transition energy (E_g) of the QDs, based on transient absorption (TA) spectroscopy and the time-resolved photoluminescence (TRPL) technique [6–11]. Encouragingly, most recently, more direct signatures of MEG have been observed in PbS QD photodetectors and PbS QD-sensitized solar cells for excitation photon energies exceeding $2.7E_g$ [12, 13].

However, several recent reports have questioned the reported higher MEG quantum yield (QY) in semiconductor QDs (e.g., PbSe QDs) and even the existence of MEG in some cases (e.g., in CdSe QDs) [14, 15]. To clarify this debate and effectively apply MEG to solar cells, further experimental and theoretical studies are necessary. An understanding of the MEG phenomenon in QDs, including the occurrence, the dynamics, and relaxation of MEG, the MEG mechanism and the MEG QY are all needed. In particular, new characterization methods are needed to enable more direct observations of the occurrence of MEG on a fast timescale. Till now, the most common and useful methods that have been used to characterize MEG in QDs are the TA and TRPL techniques [1, 7, 11]. In these methods, the recombination dynamics of excitons in the lowest excited state can be monitored by measuring the transient absorption or transient luminescence kinetics between the LUMO and HOMO in the QDs. One condition for these measurements is that the pump light intensity is sufficiently low for less than one exciton to be excited in each QD. Then, if there is no MEG, the lifetime of a single exciton would be of the order of nanoseconds, and only a slow decay process due to the relaxation of a single exciton with a lifetime of the order of nanoseconds can be observed. However, if MEG occurs after absorption of the light, then more than one exciton will be produced in the QD as hot carriers cool to the lowest excited state. As a result, in addition to the slow decay due to single exciton relaxation in the TA or

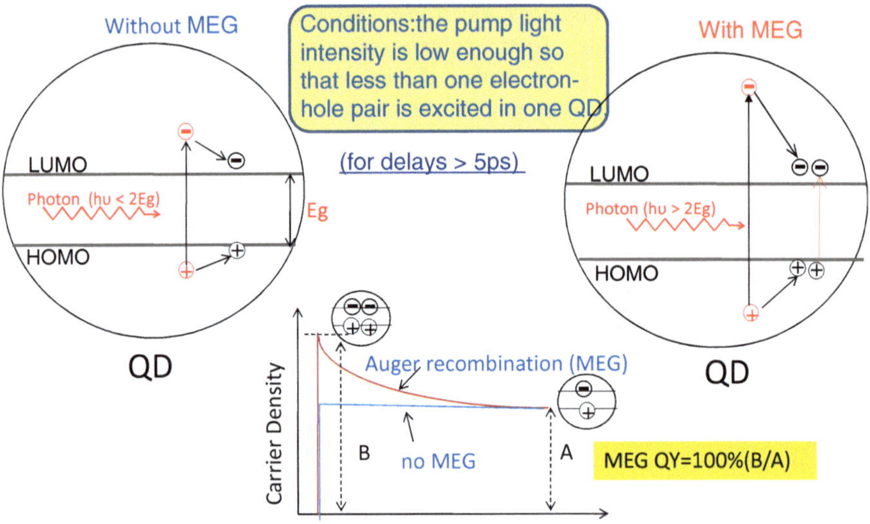

Fig. 12.2 Schematic illustration of MEG characterization using a transient absorption (TA) method

TRPL data, a fast decay appears due to Auger recombination of the generated multiexcitons [1, 9, 11] (Fig. 12.2). Thus the MEG signature contains fast decay dynamics with lifetimes in the range between 10 and 100 ps appearing in the TA or TRPL data. For characterizing MEG in QDs using TA or TRPL, such as confirming the occurrence of MEG and evaluating the quantum yield (QY), only carrier dynamics over times longer than 3–5 ps after absorption of the light have been utilized [1, 9, 11], by which time hot carrier cooling and MEG are complete [1, 9, 11]. However, the dynamics of MEG, which begins within a few picoseconds of the absorption, have not been observed nor clearly understood till now.

In this paper, we report on the first observation of the dynamics of MEG, including the beginning and completion of MEG, together with the hot exciton relaxation dynamics in PbS QDs [16], using an improved transient grating (TG) technique [17–22]. The relaxation dynamics of the generated multiple excitons to a single exciton in the QDs through Auger recombination is also shown.

12.2 Dynamics of MEG

12.2.1 An Improved Transient Grating Technique

The transient grating (TG) method is a well-established laser spectroscopic technique of four-wave mixing [23, 24]. In the TG method, two time-coincident short laser pulses (pump beams) with the same wavelength and intensity intersect at an angle in a sample to generate an optical interference pattern at the intersection.

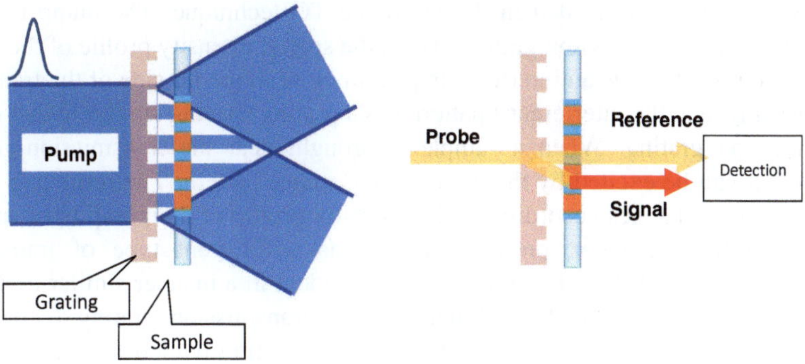

Fig. 12.3 Schematic illustration of the principles of generation and detection of a transient grating in a sample using the transient grating technique

Interaction between the light field and the material results in a spatially periodic modulation of the complex refractive index, which works like a transient diffraction grating for a third laser pulse (probe beam) incident on the photoexcited region. Then, by measuring the time dependence of the diffracted light of the probe beam, the dynamics of the transient grating produced in the sample can be monitored. The TG technique is a powerful tool for detecting population dynamics [25], thermal diffusion [26], diffusion of photoexcited species [27], energy transfer from photoexcited species to liquids [28], structural or orientational relaxation [26], the velocity of sound in liquids [29], and so on. Although this technique provides valuable information, it presents some technical difficulties for general researchers [24]. Firstly, the three beams must overlap at a small spot, typically within a spot diameter of less than 100 μm on the sample, and each beam must be temporally controlled. This is very difficult for pulsed laser beams, especially for those with pulse widths of ~100 fs. Secondly, since the diffraction of the probe beam, i.e., the signal, is quite weak, it is difficult to find the diffracted beam during the measurements. Thirdly, for a solid sample, the surface must be optically smooth. It is almost impossible to measure a sample with a rough surface using the conventional TG technique.

In 2003, Katayama and co-workers [17] proposed an improved TG technique (it was also called a lens-free heterodyne TG (LF-HD-TG) or a near field heterodyne TG (NF-HD-TG) technique in some papers), which overcomes the difficulties that exist in the conventional TG technique. The improved TG technique features (1) simple and compact optical equipment and easy optical alignment and (2) high phase stability due to the short optical path length of the probe and reference beams. This method is thought to be very versatile with applicability to many kinds of samples, such as opaque solids, solids with rough surfaces, transparent solids, and liquids, because both transmission and reflection-type measurements can be made. The principle of the improved TG technique has been explained in detail in a previous paper [17] and is only described briefly here (as shown in Fig. 12.3). Unlike the conventional TG technique, only one pump beam and one probe beam without focusing are needed in the improved TG technique. The pump beam is incident on the transmission grating. Then, the spatial intensity profile of the pump beam is known to have an interference pattern close to the far side of the transmission grating, and the interference pattern has a grating spacing similar to that of the transmission grating. When a sample is brought near the transmission-grating surface, it can be excited by the optical interference pattern. The refractive index of the sample changes according to the intensity profile of the pump light and the induced refractive index profile functions as a different type of transiently generated grating. When the probe beam is incident in a manner similar to that of the pump beam, it is diffracted both by the transmission grating (called the reference light) and by the transiently generated grating (called the signal light). In principle, the two diffractions progress along the same direction; therefore, these two diffractions interfere, and this is detected by a detector positioned at a visible diffraction spot of the reference beam.

Both the TA and TG techniques are powerful tools for studying photoexcited carrier dynamics in various materials such as dye–semiconductor systems and semiconductor quantum dots [30, 31]. However, there are some differences between these two techniques. For the TA technique, the photoexcited electron dynamics in QDs, such as electron relaxation processes from higher excited states to the lowest excited state or those from the lowest excited state to the ground state, can be monitored by detecting the changes in electronic absorption (i.e., the imaginary part of the complex refractive index) in the sample in the different states (i.e., higher excited states, lowest excited state, and ground state), whereas, for the TG technique, such photoexcited electron dynamics can be monitored by detecting the changes in the refractive index (i.e., the real part of the complex refractive index) of the sample in different electronic states. Sometimes, similar information on the photoexcited electron dynamics can be obtained from both the TA and TG techniques [30, 31]. However, for the TA technique, the sensitivity is not so high and usually the probe wavelength has to be selected around some optical absorption peak. On the other hand, the TG technique has higher sensitivity and the range of the probe wavelength is greater [24], since the real part of the complex refractive index can change over a large wavelength range. Up till now, the TG technique has been proved to have the unique ability to measure some dynamics which are difficult or impossible to detect using the conventional TA technique [22, 24, 30–37]. As an example, the improved TG technique has been applied to study the photoexcited electron and hole dynamics of CdSe QDs, including electron injection into TiO_2 [18–22].

In the improved TG technique used for studying the ultrafast carrier dynamics of semiconductor QDs, the laser source was a titanium/sapphire laser (CPA-2010, Clark-MXR Inc.) with a wavelength of 775 nm, a repetition rate of 1 kHz, and a pulse width of 150 fs. The light was separated into two parts. As shown in Fig. 12.4, one part was used as a probe pulse. The other was used to pump an optical parametric amplifier (OPA) (a TOAPS from Quantronix) to generate light pulses with a wavelength tunable from 290 nm to 3 μm. The probe pulse wavelength was 775 nm. It has been confirmed that the changes in the real and imaginary parts of the refractive index in a sample can be selectively detected by controlling the distance

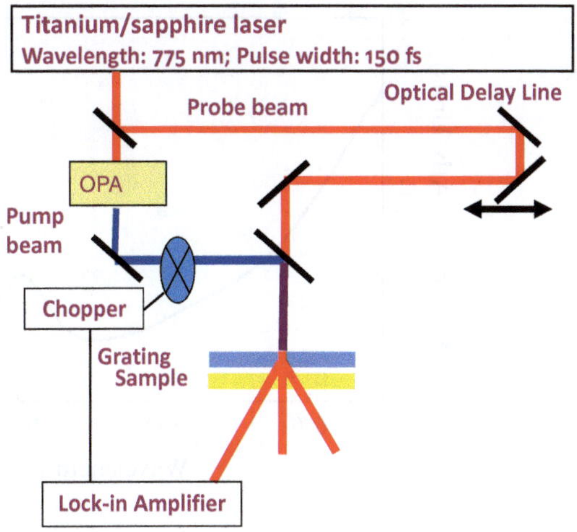

Fig. 12.4 Setup of the improved transient grating technique

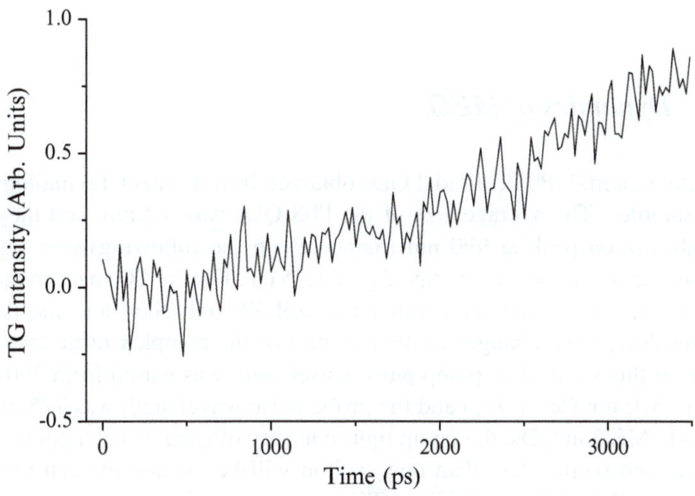

Fig. 12.5 TG signal of nitrobenzene measured at the sample positions that only real part of the complex refractive index change in the sample was selectively measured

between the transmission grating and the sample [38]. In this experiment, nitrobenzene was used as a reference sample to determine the sample positions for selectively detecting changes in the real part of the complex refractive index due to excitation by UV pump light. Nitrobenzene is well known for use as a thermal reference material because all the energy absorbed by it is released in the form of heat within 1 ns [38]. Thus, the sample position for selective detection of the real part of the refractive index change was determined when the maximum TG signal for nitrobenzene was obtained, at a longer time such as 3 ns (Fig. 12.5). Similarly, a

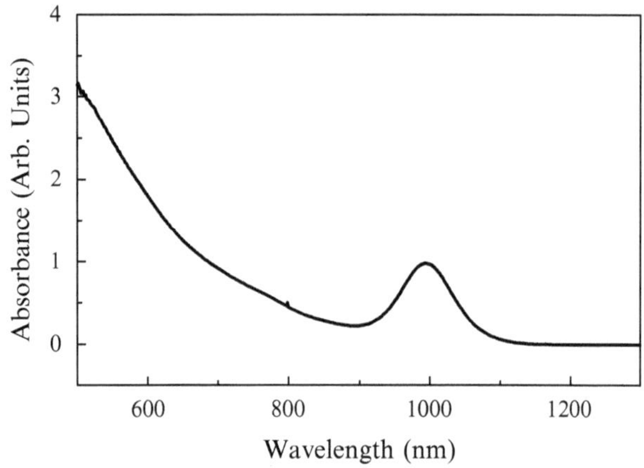

Fig. 12.6 Optical absorption spectrum of PbS colloidal QDs

Sunset Yellow solution was used as a reference sample to determine the position for selective detection of the change in the real part of the complex refractive index for visible light excitation.

12.2.2 Dynamics of MEG

As a model material, PbS colloidal QDs obtained from Evident Technologies were used as samples. The average size of the PbS QDs was 3.2 nm, and they had an optical absorption peak at 990 nm (named P_1 in the following) (see Fig. 12.6), corresponding to a band gap energy, E_g, of 1.25 eV. For the TG measurements, the PbS QD solution was put into a 1-mm quartz cell. We only show and discuss the TG signals resulting from changes in the real part of the complex refractive index in PbS QDs in this study. The pump pulse wavelength was varied from 290 nm ($h\nu$: $3.42E_g$) to 520 nm ($h\nu$: $1.9E_g$) and the probe pulse wavelength was 775 nm.

To study MEG in QDs, the pump light intensity of each wavelength used has to be low enough so that less than one exciton will be excited in each QD. If this condition is satisfied, the normalized TG kinetics on a time scale as long as a few hundred picoseconds is expected to be independent of the pump light intensity. Thus, we first measured the pump light intensity dependence of the TG kinetics over a time scale as long as 400 ps. Figure 12.7a shows the normalized TG kinetics of the PbS QDs for a wavelength of 520 nm ($h\nu$: $1.9E_g$), in which the pump light intensity was decreased from 4.2 to 0.84 μJ/pulse. It is known that MEG should not occur at this wavelength, since the photon energy of $1.9E_g$ is less than two times E_g [1, 6–11]. We found that the dependence of the TG signal intensity on the pump light intensity is linear, and that the waveforms of the TG kinetics overlap each other very well when they are normalized, as shown in Fig. 12.7a. This result means that the relaxation time constants are independent of the pump light intensity and only single excitons with lifetimes as long as nanoseconds were excited, even

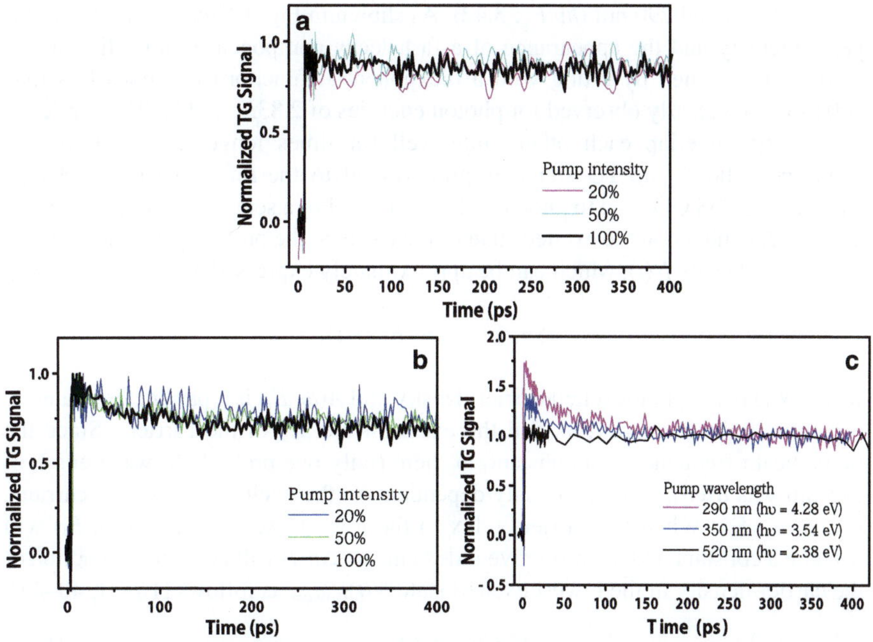

Fig. 12.7 (**a**) Normalized TG kinetics of the PbS QDs for the wavelength of 520 nm ($h\nu$: $1.9E_g$), in which the pump light intensity decreased from 4.2 μJ/pulse (100 %) to 0.84 μJ/pulse (20 %); (**b**) normalized TG kinetics of the PbS QDs for the wavelength of 350 nm ($h\nu$: $2.83E_g$), in which the pump light intensity decreased from 0.9 μJ/pulse (100 %) to 0.18 μJ/pulse (20 %); (**c**) a comparison of the TG kinetics of the PbS QDs normalized at 400 ps for three different wavelengths of 520 nm ($h\nu$: $1.9E_g$), 350 nm ($h\nu$: $2.83E_g$) and 290 nm ($h\nu$: $3.42E_g$)

for the highest pump intensity, and no "many-body" recombination (such as Auger) processes occurred under these experimental conditions. This is due to the smaller initial average number of excitons per QD, $\langle N_0 \rangle$, which was calculated to be less than 0.1 from the equation $\langle N_0 \rangle = j_p \sigma_0$ (where j_p is the pump photon fluence and σ_0 is the QD absorption cross section) for the highest pump intensity [39]. Figure 12.7b shows the normalized TG kinetics of the PbS QDs for a wavelength of 350 nm ($h\nu$: $2.83E_g$), in which the pump light intensity was decreased from 0.9 to 0.18 μJ/pulse. The waveforms of the TG kinetics overlap each other very well when they are normalized, which indicates that the carrier relaxation times were independent of the pump light intensity under the experimental conditions. In this case, $\langle N_0 \rangle$ was calculated to be smaller than 0.15 for the highest pump intensity. Similarly, low pump light intensities were used for all other pump wavelengths from 290 to 520 nm and it was confirmed that the normalized TG kinetics were independent of the pump light intensity for each wavelength. These results mean that the pump light intensities that were used were sufficiently low and thus significantly less than one exciton was excited in each PbS QD on average for each pump light wavelength. Figure 12.7c shows a comparison of the TG kinetics of the PbS QDs normalized at 400 ps for three different wavelengths: 520 nm ($h\nu/E_g$: 1.9), 350 nm

($h\nu/E_g$: 2.83), and 290 nm ($h\nu/E_g$: 3.42). As shown in Fig. 12.7c, an increase in the peak intensity and the appearance of a fast decay component (with a lifetime of ~30 ps determined by fitting with a biexponential function) for times less than 100 ps can be clearly observed for photon energies of $2.83E_g$ and $3.42E_g$, while the TG kinetics overlap each other quite well for times longer than 150 ps. As mentioned above, the TG signal is proportional to the refractive index change $\Delta n(t)$ of the PbS QDs due to photoexcited carriers. For a semiconductor, in general, $\Delta n(t)$ is dependent on the excited state of the carriers, the probe light frequency and the carrier density $N(t)$, which can be approximately expressed as follows [23, 40]:

$$\Delta n(t) = A(\omega, E_i)N(t), \tag{12.1}$$

where $N(t)$ is the photoexcited carrier density and $A(\omega, E_i)$ is a parameter related to the probe light frequency, ω, and the excited state, E_i, of the carriers. Since the probe beam frequency is a constant, ω_0 here (only one probe light wavelength of 775 nm was used), $A(\omega, E_i)$ is only dependent on the excited state of the carriers, i.e., $A(\omega_0, E_i)$. When the carriers relax to the lowest excited state, $A(\omega, E_i)$ will become a constant and the refractive index change $\Delta n(t)$ will be a linear function of the carrier density in the lowest excited state $N(t)_{\mathrm{Lowest}}$ as follows [16–24, 40–42]:

$$\Delta n(t) \propto N(t)_{\mathrm{Lowest}}. \tag{12.2}$$

Similar to TA methods applied to characterizing MEG [1, 6, 9, 11], comparison of the relative TG signal intensity at 5 ps (by which time hot carrier cooling and MEG are complete) shown in Fig. 12.7c (the TG kinetics are normalized at 400 ps) can be used to compare the initial carrier density in the lowest excited state for different excitation photon energies. Since considerably less than one exciton was excited in one QD under our experimental conditions (as mentioned above), the increase in the TG signal intensity at 5 ps with increasing photon energy and the appearance of the fast decay shown in Fig. 12.7c suggest the occurrence of MEG for photon energies of $2.83E_g$ and $3.42E_g$. We believe that the fast decay process with a lifetime of ~30 ps corresponds to Auger recombination of generated biexcitons due to MEG, while the slow process corresponds to single exciton relaxation after the completion of the Auger recombination of biexcitons. This is in good agreement with the experimental results obtained by the TA technique [1, 6, 9, 11]. We also confirmed that there is no difference between the TG kinetics measured both with and without stirring the PbS QD colloidal suspension under our experimental conditions (an example is shown in Fig. 12.8). This result strongly supports the proposition that the increase in the peak intensity for higher photon energies shown in Fig. 12.7c originates from MEG rather than photo charging [1].

To clarify the ultrafast carrier dynamics, including the expected occurrence of MEG, and carrier cooling from the higher excited states to the lowest excited state, we next investigated the TG kinetics of PbS QDs over a short time scale (up to 16 ps) with a higher temporal resolution for different pump light wavelengths.

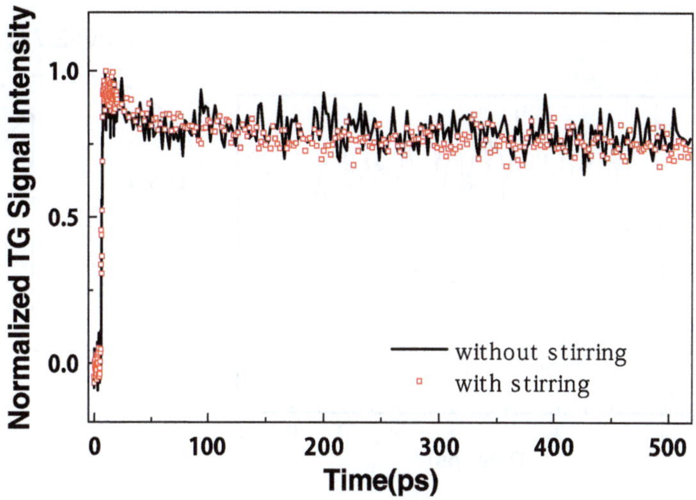

Fig. 12.8 Normalized TG kinetics of the PbS QDs for the wavelength of 360 nm (*hν*: $2.83E_g$) measured with stirring and without stirring the PbS QD solution

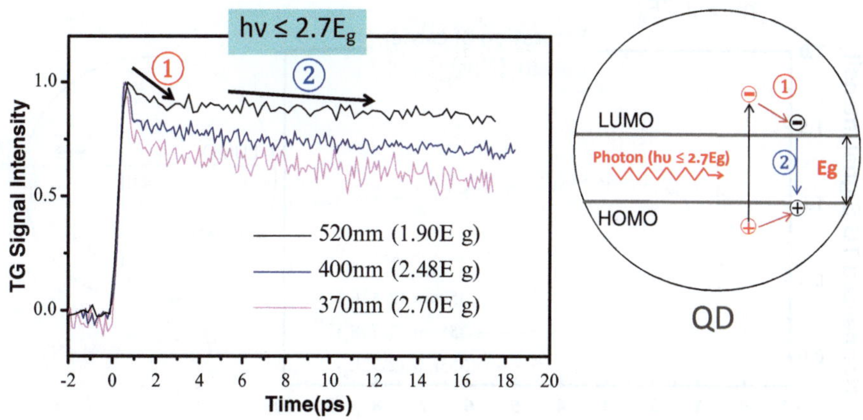

Fig. 12.9 Normalized TG kinetics for pump light wavelengths decreasing from 520 nm (*hν*: $1.9E_g$) to 370 nm (*hν*: $2.70E_g$)

Figure 12.9 shows the TG kinetics for pump light wavelengths decreasing from 520 to 370 nm, i.e., the photon energy increases from $1.9E_g$ to $2.70E_g$. As shown in Fig. 12.9, the TG signal decays after an initial peak soon after absorption of the light. To compare the decay processes of the TG kinetics measured with different photon energies more clearly, the TG kinetics are normalized at the peak position. A fast decay process with a time constant of less than 1 ps (about 400 fs) and a subsequent slow decay process can be clearly observed. Figure 12.10 shows the TG kinetics for pump light wavelengths decreasing from 365 to 290 nm, i.e., the photon energy increases from $2.72E_g$ to $3.42E_g$. It is very interesting that, besides an initial

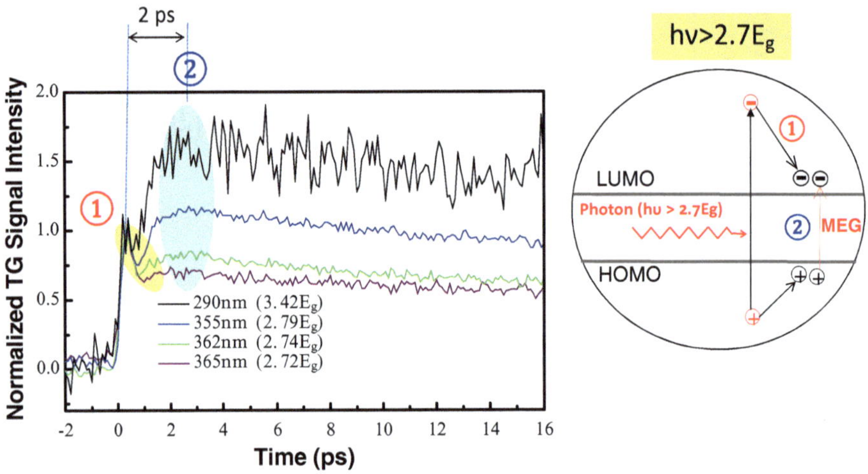

Fig. 12.10 Normalized TG kinetics for pump light wavelengths decreasing from 365 nm ($h\nu$: $2.72E_g$) to 290 nm ($h\nu$: $3.42E_g$), where the TG signals are normalized at the initial peaks

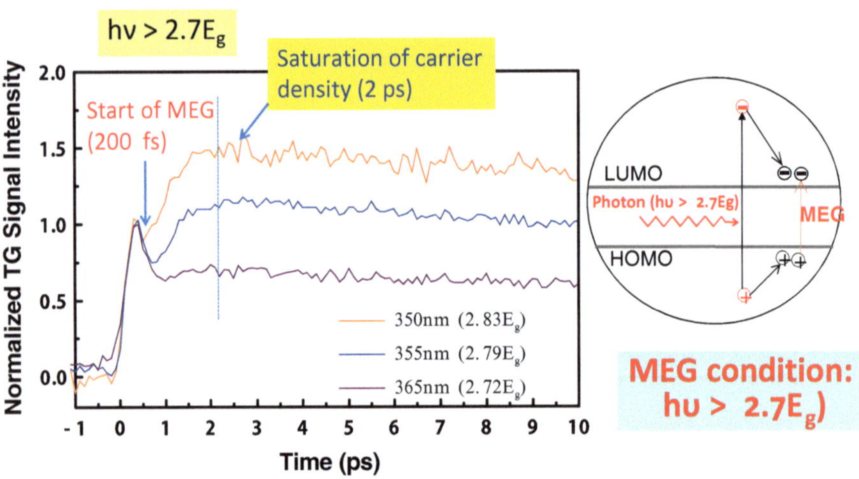

Fig. 12.11 Detailed initial TG kinetics of a part of Fig. 12.10

sharp peak, a new broad peak appears at around 2 ps. Figure 12.11 shows the detailed initial TG kinetics of part of Fig. 12.10. In Fig. 12.10, the TG kinetics are also normalized to the initial peaks to compare the relative intensities of the second peaks. To confirm that the obtained TG kinetics in Fig. 12.10 are not due to some artifact from the measurements, we measured colloidal CdSe QDs with a band gap of 2.18 eV (i.e., an optical absorption peak of 570 nm), obtained from Evident Technologies, under the same experimental conditions. We found that the TG kinetics of the CdSe QDs are different from those of the PbS QDs (Fig. 12.12).

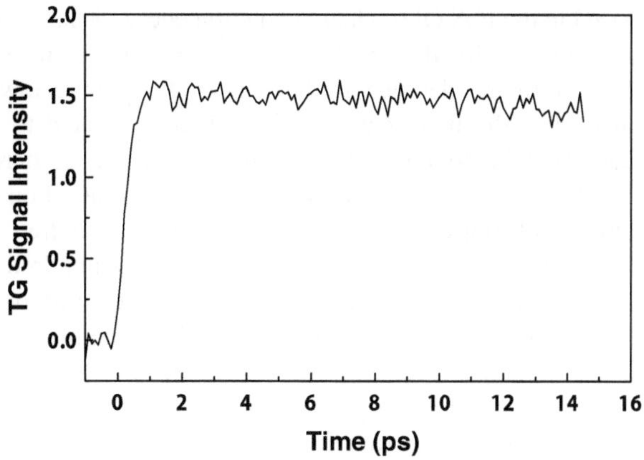

Fig. 12.12 TG kinetics of colloidal CdSe QDs with a bandgap of 2.18 eV (i.e., an optical absorption peak of 570 nm) obtained from Evident Technologies measured under the pump light wavelength of 420 nm

We didn't observe any second peaks in the TG kinetics of the CdSe QDs for pump beam wavelengths varying from 290 to 420 nm, i.e., the pump photon energy changed from $1.96E_g$ to $1.36E_g$. Therefore, we believe that the TG kinetics of the PbS QDs shown in Fig. 12.10 are due to the real carrier dynamics of the PbS QDs, rather than any artifact of the measurements.

For the TG kinetics shown in Fig. 12.9 for photon energies from $1.9E_g$ to $2.7E_g$, we think that the fast decay with a time constant of a few hundred femtoseconds corresponds to hot carrier cooling from higher excited states to the lowest excited state, and the slow one corresponds to the exciton relaxation dynamics in the lowest excited state (i.e., the recombination between LUMO and HOMO). For hot carrier cooling from higher excited states to the lowest excited state, the refractive index change $\Delta n(t)$ depends on both the change in the excited state (i.e., $A(\omega_0, E_i)$) and the carrier density $N(t)$, as given by Eq. 12.1. For photon energies which are too small to result in MEG (e.g., $1.9E_g$), the carrier density $N(t)$ can be considered to be the same after the hot carriers have cooled from higher excited states to the lowest excited state. Then, the decrease in the TG signal intensity after hot carrier cooling shown in Fig. 12.10 suggests that $A(\omega_0, E_i)$ of the lowest excited state becomes smaller compared to that of the higher excited states in this case. The relative intensity of the fast decay component in the TG kinetics increases as the photon energy increases from $1.9E_g$ to $2.7E_g$, which suggested that $A(\omega_0, E_i)$ increases with the increasing excited state energy of the PbS QDs for the probe light frequency used in this study. From these results, we can suppose that the refractive index change due to electrons in higher excited states (hot carriers) is higher than that due to electrons in lower excited states in this case. Thus, the TG signal intensity should also decrease monotonously for photon energies from $2.72E_g$ to $3.42E_g$ if the exciton density in the lowest excited state is the same as that in the higher excited states, just as those shown in Fig. 12.9. However, in Fig. 12.10, we observe the appearance of a second peak, whose intensity increases with increasing photon energy. Therefore, we think that this second peak is due to the increase in exciton density in the lowest excited state. These results clearly confirm the occurrence of

MEG in the PbS QDs when the photon energy is larger than $2.7E_g$, which is in good agreement with the results reported by Nozik and other researchers [1, 7, 12, 13]. From the detailed initial carrier dynamics shown in Fig. 12.11, we found that, after the first peak, considered to be related to hot carriers, the TG signal decreases due to hot carrier cooling, and then soon increases again (after as little as 200 fs), and finally saturates at around 2 ps. Based on these results, we think that, along with hot carrier cooling within a few hundred femtoseconds (as shown in Fig. 12.9), MEG also commences within a few hundred femtoseconds, resulting in a saturated exciton density (i.e., a maximum exciton density) in the lowest excited state at around 2 ps. The carrier saturation time in the lowest excited state of about 2 ps is consistent with that observed by Nozik and co-workers using TA measurements [7]. Very interestingly, this experimental result is consistent with the theoretical prediction that a saturated carrier density due to MEG would appear on a picosecond time scale in PbSe QDs reported very recently by Efros and co-workers [43]. Our study provides a very clear experimental demonstration of the occurrence of MEG together with hot carrier cooling. For the first time, we experimentally demonstrated that MEG occurs within 200 fs for PbS QDs when the photon energy is larger than $2.7E_g$. These experimental results are very important for further understanding of the mechanism and conditions for MEG in QDs, as well as for developing applications utilizing MEG, such as QD-based solar cells. The hot carriers need to be picked up within 1 ps in order to utilize them for improving the open circuit voltage in solar cells. On the other hand, to make great improvements in the photocurrent in solar cells by employing the MEG phenomenon, the generated multiexcitons should be separated and the multiple electrons transferred to an electrode such as a metal oxide semiconductor within a few tens of picoseconds before the Auger recombination is complete. This would be realized by controlling the interface between the QDs and the electrode surface. Characterization of the ultrafast carrier dynamics of semiconductor QDs deposited on substrates, including the occurrence and relaxation of multiple excitons will be the next step in our study for efficiently applying MEG to solar cells.

12.3 Conclusions

In summary, we have demonstrated that the improved TG technique is a useful tool for studying ultrafast carrier dynamics in QDs including hot carrier cooling and relaxation at the lowest excited state. We have clearly confirmed the occurrence of MEG in PbS colloidal QDs when the excitation photon energy becomes larger than 2.7 times the band gap energy, E_g, of the PbS QDs from the TG kinetics. For the first time, we have succeeded in directly observing the occurrence of MEG on a timescale of 200 fs (which is close to the temporal resolution of $\sim$150 fs of the TG technique used in this study) and the completion of MEG (i.e., the saturation of carriers in the lowest excited state) at around 2 ps. We also confirmed the Auger recombination due to the generated multiple excitons with a lifetime of a few tens of picoseconds. To clarify the mechanism of MEG in QDs, quantitative analysis of the occurrence and dynamics of MEG together with the hot carrier process, as well as a study of the QD size dependence of MEG, are now in progress.

Acknowledgments This research was supported by the PRESTO program, Japan Science and Technology Agency (JST) and by a Grant in Aid for Scientific Research (No. 21310073) from the Ministry of Education, Sports, Science and Technology of the Japanese Government.

References

1. Nozik, A.J., Beard, M.C., Luther, J.M., Law, M., Ellingson, R.J., Johnson, J.C.: Chem. Rev. **110**, 6873 (2010)
2. Kamat, P.V.: J. Phys. Chem. C **112**, 18737 (2008)
3. Mora-Sero, I., Bisquert, J.: J. Phys. Chem. Lett. **1**, 3046 (2010)
4. Nozik, A.J.: Physica E **14**, 115 (2002)
5. Hanna, M.C., Nozik, A.J.: J. Appl. Phys. **100**, 074510 (2006)
6. Schaller, R.D., Klimov, V.I.: Phys. Rev. Lett. **92**, 186601 (2004)
7. Ellingson, R.J., Beard, M.C., Johnson, J.C., Yu, P.R., Micic, O.I., Nozik, A.J., Shabaev, A., Efros, A.L.: Nano Lett. **5**, 865 (2005)
8. Schaller, R.D., Petruska, M.A., Klimov, V.I.: Appl. Phys. Lett. **87**, 253102 (2005)
9. Murphy, J.E., Beard, M.C., Norman, A.G., Ahrenkiel, A.P., Johnson, J.C., Yu, P.R., Micic, O. I., Ellingson, R.J., Nozik, A.J.: J. Am. Chem. Soc. **128**, 3241 (2006)
10. Schaller, R.D., Petruska, M.A., Klimov, V.I.: Nano Lett. **7**, 3469 (2007)
11. Beard, M.C., Knutsen, K.P., Yu, P.R., Luther, J.M., Song, Q., Metzger, W.K., Ellingson, R.J., Nozik, A.J.: J. Phys. Chem. B **109**, 2625 (2005)
12. Sukhovatkin, V., Hinds, S., Brzozowski, L., Sargent, E.H.: Science **324**, 1542 (2009)
13. Sambur, J.B., Novet, T., Parkinson, B.A.: Science **330**, 63 (2010)
14. Nair, G., Geyer, S.M., Chang, L.-Y., Bawendi, M.G.: Phys. Rev. B **78**, 125325 (2008)
15. Ben-Lulu, M., Mocatta, D., Bonn, M., Banin, U., Ruhman, S.: Nano Lett. **8**, 1207 (2008)
16. Shen, Q., Katayama, K., Sawada, T., Hachiya, S., Toyoda, T.: Chem. Phys. Lett. **542**, 89 (2012)
17. Katayama, K., Yamaguchi, M., Sawada, T.: Appl. Phys. Lett. **82**, 2775 (2003)
18. Shen, Q., Katayama, K., Yamaguchi, M., Sawada, T., Toyoda, T.: Thin Solid Films **486**, 15 (2005)
19. Shen, Q., Katayama, K., Sawada, T., Yamaguchi, M., Toyoda, T.: Jpn. J. Appl. Phys. **45**, 5569 (2006)
20. Shen, Q., Yanai, M., Katayama, K., Sawada, T., Toyoda, T.: Chem. Phys. Lett. **442**, 89 (2007)
21. Shen, Q., Katayama, K., Sawada, T., Toyoda, T.: Thin Solid Films **516**, 5927 (2008)
22. Shen, Q., Ayuzawa, Y., Katayama, K., Sawada, T., Toyoda, T.: Appl. Phys. Lett. **97**, 263113 (2010)
23. Eichler, H.J., Gunter, P., Pohl, D.W.: Laser-Induced Dynamic Gratings. Springer, Berlin (1986)
24. Harata, A., Shen, Q., Sawada, T.: Annu. Rev. Phys. Chem. **50**, 193 (1999)
25. Rajesh, R.J., Bisht, P.B.: Chem. Phys. Lett. **357**, 420 (2002)
26. Glorieux, C., Nelson, K.A., Hinze, G., Fayer, M.D.: J. Chem. Phys. **116**, 3384 (2002)
27. Terazima, M., Nogami, Y., Tominaga, T.: Chem. Phys. Lett. **332**, 503 (2000)
28. Miyata, R., Kimura, Y., Terazima, M.: Chem. Phys. Lett. **365**, 406 (2002)
29. Ohmor, T., Kimura, Y., Hirota, N., Terazima, M.: Phys. Chem. Chem. Phys. **3**, 3994 (2001)
30. Shen, Q., Ogomi, Y., Park, B., Inoue, T., Pandey, S.S., Miyamoto, A., Fujita, S., Katayama, K., Toyoda, T., Hayase, S.: Phys. Chem. Chem. Phys. **14**, 4605 (2012)

31. Hata, H., Katayama, K., Shen, Q., Toyoda, T.: Jpn. J. Appl. Phys. **51**, 042601 (2012)
32. Katayama, K., Inagaki, Y., Sawada, T.: Phys. Rev. B **61**, 7332 (2000)
33. Shibamoto, K., Katayama, K., Sawada, T.: Chem. Phys. Lett. **433**, 385 (2007)
34. Okuda, M., Tsuruta, T., Katayama, K.: Phys. Chem. Chem. Phys. **11**, 2287 (2009)
35. Nakazato, Y., Taniguchi, K., Ono, S., Eitoku, T., Katayama, K.: Phys. Chem. Chem. Phys. **11**, 10064 (2009)
36. Inoue, H., Katayama, K., Iwai, K., Miura, A., Masuhara, H.: Phys. Chem. Chem. Phys. **14**, 5620 (2012)
37. Eitoku, T., Nakasone, Y., Matsuoka, D., Tokutomi, S., Terazima, M.: J. Am. Chem. Soc. **127**, 13238 (2005)
38. Okuda, M., Katayama, K.: Chem. Phys. Lett. **443**, 158 (2007)
39. Klimov, V.I.: J. Phys. Chem. B **104**, 6112 (2000)
40. Uskov, A.V., O'Reilly, E.P., McPeake, D., Ledentsov, N.N., Bimberg, D., Huyet, G.: Appl. Phys. Lett. **84**, 272 (2004)
41. Kashinski, J.J., Gomez-Jahn, L.A., Faran, K.J., Gracewski, S.M., Miller, R.J.D.: J. Chem. Phys. **90**, 1253 (1989)
42. Kim, J., Meuer, C., Bimberg, D., Eisenstein, G.: Appl. Phys. Lett. **94**, 041112 (2009)
43. Witzel, W.M., Shabaev, A., Hellberg, C.S., Jacobs, V.L., Efros, A.L.: Phys. Rev. Lett. **105**, 137401 (2010)

Chapter 13
Light-Induced Charge Carrier Dynamics at Nanostructured Interfaces Investigated by Ultrafast Electron Diffractive Photovoltammetry

Kiseok Chang, Ryan A. Murdick, Tzong-Ru T. Han, Fei Yuan, and Chong-Yu Ruan

Abstract We present an ultrafast photovoltammetry framework to investigate the surface charge carrier dynamics at the nanometer scale. This diffraction-based method utilizes the feature-gated nanomaterial diffraction pattern to identify the scattering sites and to deduce the associated charge dynamics from the nanocrystallographic refraction-shift observed in the ultrafast electron diffraction patterns. From applying this methodology on SiO_2/Si interface, and surfaces decorated with nanoparticles and water–ice adsorbed layer, we are able to elucidate the localized charge injection, dielectric relaxation, and carrier diffusion, with direct resolution in the charge state and possibly correlated structural dynamics at these interfaces, which are central to nanoelectronics, photovoltaics, and photocatalysis development. These new results highlight the high sensitivity of the interfacial charge transfer to the nanoscale modification, environment, and surface plasmonics enhancement and demonstrate the diffraction-based ultrafast surface voltage probe as a unique and powerful method to resolve the nanometer scale charge carrier dynamics.

13.1 Introduction

Characterizing the photoinduced charge dynamics in nanostructures and interfaces is relevant to understand fundamental processes in the new types of solar cells and photocatalysts incorporating nanomaterials [1, 2] and molecularly engineered interfaces [3]. These emerging nanoscale-engineered devices utilize the interface-induced carrier separation to reduce the recombination of the photogenerated electron/hole pairs and improve sensitization by visible light spectrum through

K. Chang • R.A. Murdick • T.-R.T. Han • F. Yuan • C.-Y. Ruan (✉)
Department of Physics and Astronomy, Michigan State University, East Lansing, MI 48824, USA
e-mail: ruan@pa.msu.edu

J. Wu and Z.M. Wang (eds.), *Quantum Dot Solar Cells*, Lecture Notes in Nanoscale Science and Technology 15, DOI 10.1007/978-1-4614-8148-5_13,

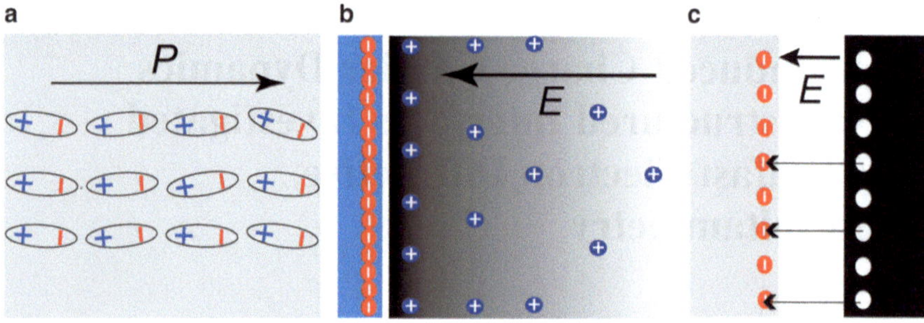

Fig. 13.1 Charge redistribution at nanomaterials' interfaces subject to photoexcitation. (**a**) Dielectric realignment; (**b**) carrier diffusion; (**c**) interfacial charge transfer

controlling the size, dopant, and surface plasmonic effect [4–12]. At nanomaterials' interface, the migration of carriers from the sensitized interface region to the metal contact is driven not only by the local electrochemical potential difference but also by the survival rate of such photoactivated carriers in the transport process. As the device dimension approaches 1 nm [13–16], shorter than the mean free path of electron, ballistic transport becomes the prevalent channel, but carrier trapping and recombination at the interface are central issues to deal with [17–19]. These processes are nonetheless difficult to characterize in real time due to challenges associated with forming the contact to allow ultrafast photocurrent measurement. While several ultrafast optical techniques have been applied to investigate electron dynamics at the interfaces [20–24], they are usually not well suited to measure the photoconductivity at the interface. In this chapter, we describe a contact-free photovoltammetry approach, based on the ultrafast electron diffraction technique [25–27], for investigating the ultrafast charge dynamics at interfaces and in nanostructures. Through the sensitivity of the diffracted electron beams to the local electric fields, three prominent charge redistribution processes induced by photoexcitation can be observed as described in Fig. 13.1: (1) Dielectric realignment: The alignment of the dipolar elements within the dielectrics changes, resulting in displacement field even though there is no actual carrier current in the materials. (2) Carriers diffusion: The photocarriers in the excited region diffuse to the unexcited region, inducing internal photocurrent. (3) Interfacial charge transfer: The photoexcitation changes the balance of chemical potential at interface, resulting in charge transfer to counteract the shift in free energy. Decay of these photovoltages might involve drift, dipolar relaxation, in addition to carrier recombination, diffusion, and radiative decays.

To demonstrate the proof of principle, systematic investigations are conducted at interfaces with progressively grown complexities. First, planar SiO_2/Si surface is investigated with infrared laser pulse to drive the hot carriers through the SiO_2 slab. In the sub-10 nm dielectric SiO_2 slab, ultrafast surface charge dynamics is shown to strongly couple to the sub-surface carrier dynamics, exhibiting drift-diffusion relaxation behavior. Second, gold nanoparticles decoration is incorporated with molecular linkers connecting the nanoparticles to the surface, The large curvature of the nanoparticle surface facilitates the field focussing effect that can enhance the carrier dynamics across the interface. Field-assisted tunneling and high-field dielectric breakdown processes have been observed. Third, surface plasmon enhancement effects are investigated. Strong dielectric-induced spectral shift in the response function highlights the mode-selective optical antenna effects between the nanoparticles and their surroundings, including silicon substrate and water–ice adsorbed layer. These studies elucidate the key features of light-driven charge transfer dynamics around these interfaces, which help understand the fundamental processes relevant to nanoelectronics, photovoltaics, and photocatalysis.

13.2 Ultrafast Diffractive Photovoltammetry Methodology

13.2.1 Framework

The ultrafast photovoltammetry setup is schematically shown in Fig. 13.2. The experiments are driven by a Ti:Sapphire femtosecond laser system, which delivers 2.5 mJ/cm^2 sub-100 fs, 800 nm laser pulses at 1 kHz repetition rate. The output pulses are split into two paths (pump and probe) by a beam splitter. The laser pulses along the pump path drive an optical parametric amplifier (OPA) that generates a broad range of selectable optical wavelength to initiate the carrier excitations in nanomaterials. The laser pulse along the probe path is frequency-tripled into an ultraviolet pulse (266 nm, $h\upsilon \approx 4.7$ eV) that efficiently drives the fs photoemission from an Ag photocathode to form the initial probing electron beam. The photoelectrons are consequently accelerated to 20–40 keV and focussed by a short-focal distance magnetic lens, into a 5–30 μm narrow probe to interrogate the nanointerfaces by way of electron diffraction [28]. The ultrafast electron diffraction takes place in a vacuum chamber with a sample mounted on a goniometer and the incidence geometry of the electron beam is typically at a grazing incidence ($\approx 1°$) to achieve the nanometer scale sensitivity. As timing control is

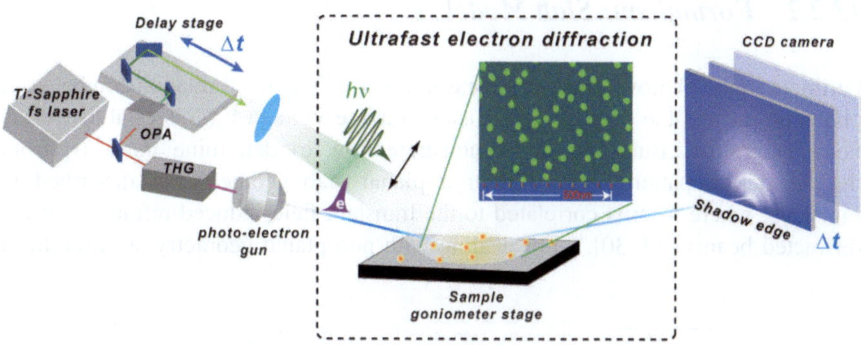

Fig. 13.2 The setup of ultrafast electron diffraction experiment used to measurement the surface voltages of nanoparticles and interfaces

essential, in our experiment, the time (Δt) between the pump (laser) and probe (electron) pulses is controlled by an optical delay stage employed in the pump path with ≈10 fs precision. We use the convention that the zero-of-time ($t = 0$) is defined at the time when the pump pulse arrives at the interface. A sequence of probing electron pulses at different times t are deployed to establish the dynamical evolution at a relative low repetition rate of 1 kHz or less to ensure that there is no accumulative build-up of charges during each pump-probe cycle. Besides studying single-crystal surface (Sect. 13.3.1), the nanoparticle-decorated surface (Sect. 13.3.2) and coverage of molecular adlayers, including water–ice (Sect. 13.3.4), have also been investigated. The diffraction patterns can range from ordered Bragg reflection arrays to powder diffraction from randomly oriented nanoparticles, where, due to the existence of a supporting substrate, the diffraction cone is cut into half by the shadow edge. The electron probe is employed as an ultrafast local probe to detect the surface potential changes following fs photoexcitation. By way of local field-induced refractive effect on the diffraction pattern, we can quantify the ultrafast charge redistribution in the active region. Formalism for integrating over the slab (Sect. 13.2.2) and non-planar geometries (Sect. 13.4.3) to deduce the effective photovoltage V_s has been developed. As a first-order approximation, an effective circuit model is used to estimate the transport timescale based on local capacitance and integrated conductance (Sect. 13.4.1). The *emf* can be determined from the surface voltage through deconvolution using the linear response theory based on the local resistance and capacitance. At a more sophisticated level, considering the nonequilibrium nature of the hot carriers, microscopic description involving differential conductance is further implemented to describe transient photocurrent driven by a time-dependent electromotive force (*emf*) from the source to the acceptor sites or its reversal at different energy states. Electron and hole energy distribution calculated using the two-temperature model (TTM) (Sect. 13.4.2) provides a way to estimate the transient quasi-Fermi levels of the photogenerated carriers as the source of *emf* and is compared with the experimental observation.

13.2.2 Formalism: Slab Model

Pertinent to the photovoltammetry measurement, a functional relationship between photocurrent and the responding transient surface voltage $V_s(t)$ is established via field integration. But, first, to demonstrate how to determine the $V_s(t)$ from the diffraction pattern, we consider a planar slab geometry, as described in Fig. 13.3, where $V_s(t)$ is correlated to the transient-field-induced refraction of the diffracted beams [29, 30].[1] For dealing with non-planar geometry, a generalized

[1] The shift in diffraction due to electron wavlength change in the materials, which is on the level of 3×10^{-5} per 1 V in photovoltage for 30 keV electron beam, is small compared to refraction shift, which is on the 10^{-2} level under the same condition.

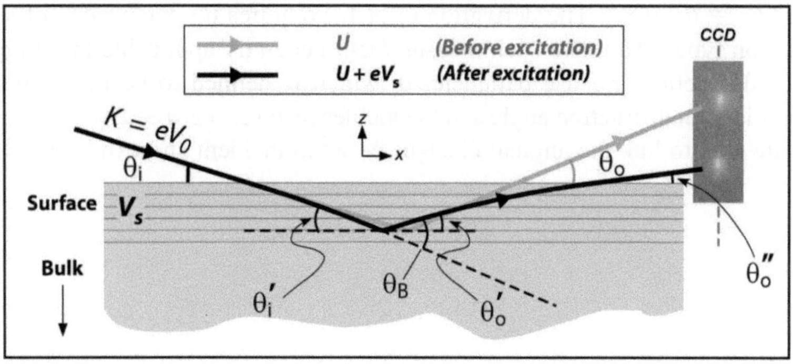

Fig. 13.3 The idealized slab model for considering the transient surface voltage. The top trajectory is the electron scattering from the crystal planes with the presence of a surface field. The electron beam, incident at θ_i, is Bragg scattered at θ_B, exiting the surface at θ_o. Introducing an attractive surface potential V_s will cause the electron beam to be "refracted" deeper into the crystal (θ_i') and the same for the Bragg diffracted beam that would ultimately exit the crystal at θ_o" with a net shift Δ_B relative to θ_o. Note here the shift is highly exaggerated for getting the point through [33]

formalism is derived using the skeleton of the slab formalism with angle-dependent correction factors $\Theta(\theta, V_s)$, see Sect. 13.4.3. In the slab geometry, the electric field is along the surface normal (z axis) and changes the course of the incident and outgoing beams accumulatively over the probed depth from z_0 to z_1. The surface voltage $V_s = \int_{z_0}^{z_1} E_z(z)dz$, which is correlated to the overall change of the diffracted beam Δ_B. The changes in the incident and diffracted beam angles, θ_i and θ_o, respectively, accrue to Δ_B. Considering the effect on the incident beam first, we use the energy relationship along the z direction:

$$p_{z_1}^{i\ 2} - p_{z_0}^{i\ 2} = 2m_e eV_s, \tag{13.1}$$

where $p_{z_1}^i$ and $p_{z_0}^i$ are the momenta of the incident beam projected along z at z_0 and z_1. Expressed in terms of angle θ, Eq. (13.1) can be rewritten as:

$$\tan^2\theta_i' = \tan^2\theta_i + \frac{\chi}{\cos^2\theta_i}, \tag{13.2}$$

where $\chi = V_s/V_o$, by utilizing $\tan\theta_i = p_{z_1}^i/p_x$, $\tan\theta_i' = p_{z_0}^i/p_x$, and $e\,V_0$ is the beam energy prior entering the field region. Similarly for the outgoing beams, we have:

$$\tan^2\theta_o' = \tan^2\theta_o'' + \frac{\chi}{\cos^2\theta_o''}. \tag{13.3}$$

The surface voltage measurement is established when V_s can be deduced as a function of the observable Δ_B, which is defined as the angular shift of the diffracted beam ($\Delta_B \equiv \theta_o'' - \theta_o$). The derivation of $\chi(\Delta_B)$ requires the knowledge of surface diffraction types. To make such measurement generally applicable to all types of surface diffraction, we use parameter a, which is defined to be the ratio of the changes in total diffraction angle and the incidence angle, i.e. $a \equiv \Delta_B/\Delta\theta_i$. Through a we are able to link the angular changes between incident and exit beams:

$$\Delta\theta_o = (a - 1)\Delta\theta_i. \tag{13.4}$$

Therefore,

$$\theta_o' = (a - 1)(\theta_i' - \theta_i) + \theta_o. \tag{13.5}$$

We define D in terms of θ_i and θ_o:

$$D \equiv \tan\theta_o' = \frac{\tan[\theta_o + (1-a)\theta_i] - \tan[(1-a)\theta_i']}{1 + \tan[\theta_o + (1-a)\theta_i]\tan[(1-a)\theta_i']}, \tag{13.6}$$

and, from Eq. (13.2), $\theta_i' = \tan^{-1}\left(\sqrt{\frac{\sin^2\theta_i + \chi}{1 - \sin^2\theta_i}}\right)$. Therefore, we can calculate θ_o'', from Eq. (13.3), at given V_s, θ_i and θ_o:

$$\theta_o'' = \sin^{-1}\sqrt{\frac{D^2 - \chi}{1 + D^2}}. \tag{13.7}$$

To get V_s-induced angular shift in the diffraction pattern:

$$\Delta_B = \sin^{-1}\sqrt{\frac{D^2 - \chi}{1 + D^2}} - \theta_o. \tag{13.8}$$

One salient feature of refraction (with the exception of $a = 2$) is that Δ_B increases as θ_o decreases. This is easily seen in Fig. 13.4, where Δ_B is calculated for $a = 0, 1, 2$ at $V_s = 1$ V, following Eq. (13.8). We note that the corresponding change in Δ_B at $a = 0$ is nearly twice the value at $a = 1$, whereas at $a = 2$, Δ_B remains to be 0 for all θ_o. We note that the nonreciprocal relationship as shown in Fig. 13.4 is important to distinguish the refraction-induced shift and the structurally relevant shift, which, in contrast, follow the reciprocal relationship given by Bragg's law, i.e. the shift in the reflection peak position should linearly increase with the scattering order in the reciprocal space. For interested reader, we provide the methodology for analyzing a under different diffraction scenarios in the following section. Those interested primarily on the applications can skip to Sect. 13.3.

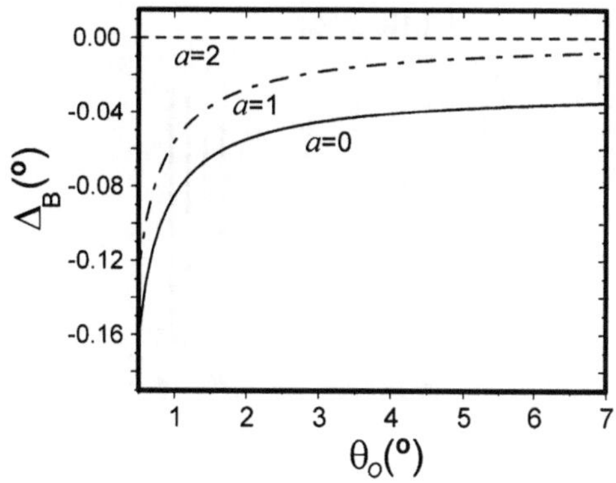

Fig. 13.4 The refraction-induced shift (Δ_B) for diffraction peak located at exit angle θ_o at V_s=1 V calculated from voltammetry formalism for different surface diffraction condition characterized by $a = 0, 1, 2$. The incidence angle θ_i is set at 2.01° [27]

13.2.3 Characterization of Surface Diffraction Types

While the canonical reflection relationship, i.e. $\theta_i = \theta_o$, roughly holds for ultrafast electron diffraction, the differential changes of θ_i and θ_o can deviate significantly from the canonical reflection law in nanomaterial diffraction due to finite size broadening of the reciprocal lattice nodes from different types of surface diffraction. To understand this phenomenon, we resort to the Ewald construction, which predicts the location of constructive diffraction patterns according to the intercept between the reciprocal lattice and the spherical Ewald surface in the reciprocal space. In the case of high-energy electron diffraction, the Ewald sphere is significantly larger than the reciprocal lattice spacings as the Ewald radius $R_{EW} = 2\pi/\lambda_e$, where $\lambda_e = 12.2643/(V_0 + 0.97845 \times 10^{-6})^{1/2}$ (in Å), easily exceeds 100 Å $^{-1}$ for electron energy ($e\ V_0$) larger than 30 keV, which makes the Ewald sphere nearly tangential to the reciprocal lattice along surface normal at small angles. The shape of the reciprocal lattice nodes is sensitive to the dimensions, or the persistence lengths of the materials that coherently diffract electron waves. As clearly shown in Fig. 13.5, the intercept can be extended into regions around the canonical reflection point as the dimensionality of sample reduces. The size of the reciprocal node (δ) in each direction can be estimated based on the persistent length L: $\delta \approx 2\pi / L$, which, in the case of nanomaterial diffraction, is limited by the material size or probe depth.[2] For a smooth single crystal surface, the in-plane persistent length can be very large compared to the penetration depth of the electron, producing very thin reciprocal rods (relrods). In the limit of single layer, the reciprocal lattice becomes two-dimensional (2D) array of relrods, and the diffracted beams, defined by the intercept between the Ewald sphere and the continuous relrod network, render circular diffraction patterns, generally described as the Laue zones in reflective

[2] The persistence length here refers to the length of the crystal in the sample that allow the probing electron to scatter coherently to form diffraction pattern. The persistence length can be limited by the size of the crystal, the coherence length of the probing electron, or the penetration depth of the probing electron, which ever is the smallest.

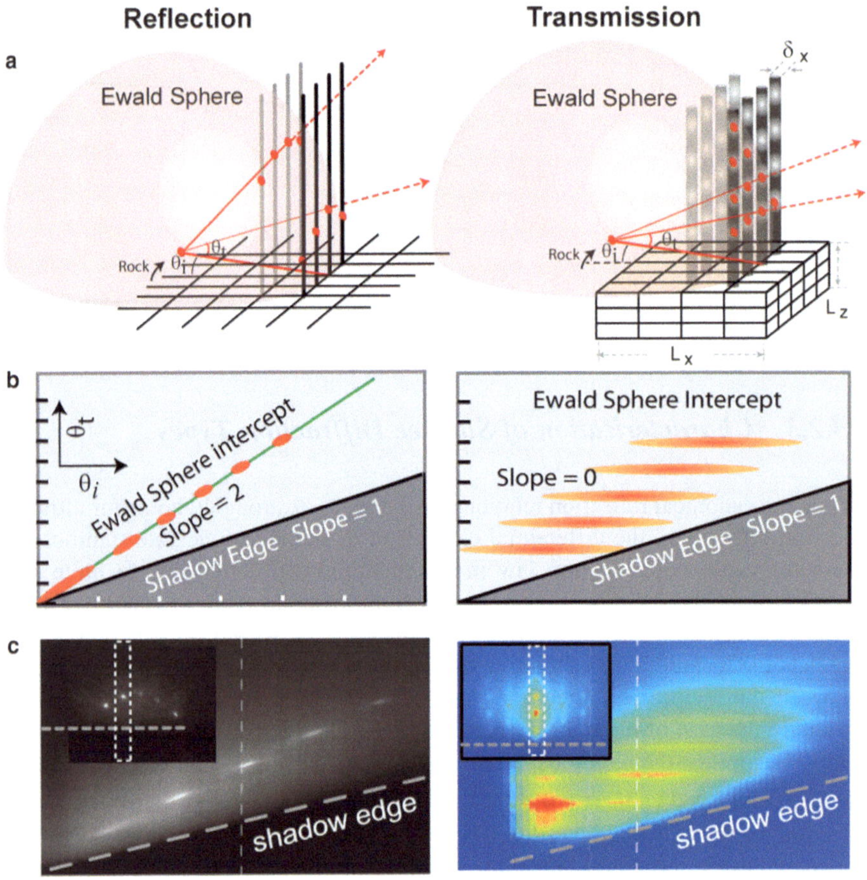

Fig. 13.5 Surface electron diffraction pattern in different conditions. (**a**) Ewald sphere construction in the grazing incidence angle geometry. By tilting (rocking) the angle of incidence between the electron beam and the sample, the Ewald sphere intercepts the reciprocal lattice rods (relrods) at different height. The coherent diffraction condition is satisfied when the intercept is at the reciprocal lattice node. The *left panel* shows diffraction from 2D surface. Because there is almost no structure within each relrod, the diffraction exhibits ring patterns simply at the locations where Ewald sphere intercept with relrods. The *right panel* shows diffraction from 3D finite-size materials with reciprocal nodes broadened by the finite persistence length L of the sample. Sampled by the elongated Ewald sphere intercepts, the diffraction pattern exposes the multiple periodic modulations along the relrods, characteristic of the crystal structure. (**b**) The simulated rocking maps for the two diffraction scenarios described in (**a**). (**c**) Experimental rocking maps taken from a smooth Si/SiO_2 surface and micrograined highly oriented pyrolytic graphite (HOPG) surface [27]

high-energy electron diffraction (RHEED) [31]. For nanostructured surfaces, $L_z \approx L_x$, the relrod broadens significantly in 3D, and the diffraction pattern can deviate significantly from circular Laue patterns, and observing more than one diffraction peaks along a single relrod is possible at a given θ_i.

We can construct the so-called rocking map to characterize the nature of nanoscale diffraction, from which the parameter a used in Eq. (13.4) can be readily determined. By rocking azimuthally the sample against the incident electron beam, different portion of the reciprocal lattice rods along z can be examined. We select a stripe region in the diffraction pattern (dashed region in the insets of Fig. 13.5c) that records the constructive interference along a relrod at a given θ_i. The strong constructive interference typically occurs along the Ewald sphere cut with total scattering angle θ_t near $2\theta_i$. By stitching together such stripes over different θ_is a rocking map is constructed that tracks the trend of constructive interference in the coordinates of (θ_i, θ_t). Figure 13.5b (simulation) and Fig. 13.5c (experiment) show such rocking maps constructed for two types of diffraction. Clear differences can be seen in the map for thin surface diffraction (left column, reflection type) and for nanomaterials (right column, transmission type). Most notable is that the slope of the coherent features in the map follows the trajectory of the Ewald sphere cut with a slope of 2 in the reflection type, while for transmission type it is flat. The simulations using the Ewald construction over the respective reciprocal lattice nodes, defined by the samples' geometry, provide a good guide to predict the pattern. In the case of quasi-2D lattice represented by surface diffraction from Si, the Ewald sphere intercepts with an array of nearly continuous relrods. In the rocking map, the extent of the constructive region along the $a = 2$ slope grows longer as θ_i approaches 0, due to the decrease of penetration depth (L_z). In contrast, for the transmission type, represented by diffraction from micrograined graphite, the multiple diffraction spots can appear with the symmetry of the atomic lattice in the microstructure, and constructed from which the rocking map shows parallel stripes. We would also like to point out that it might be presumptuous to always depend on Ewald construction to predict a, as when multiple scattering is involved as in the example of resonance diffraction occurring within a Kikuchi line [32], a could range anywhere between 0 and 2 [27]. For this reason, it is prudent to always conduct rocking experiment analysis prior to an ultrafast voltammetry measurement.

13.3 Prototypical Photovoltammetry Experiments on Nanostructures and Heterointerfaces

13.3.1 Surface Charging of SiO_2/Si Interface

SiO_2/Si interface is the fundamental building block of CMOS device. The nature of charge transfer, trapping, and detrapping at the SiO_2/Si interface has gained notable interest as the dielectric oxide layer necessarily becomes thinner as the elemental feature size continues to shrink. Moreover, it has also been tapped as the electronic grade interface to grow molecular electronic devices with increasing effort exploring different avenues of fabricating interface using self-assembled monolayers (SAMs) [34] and nanoparticles [16, 34–37]. For validating the diffractive photovoltammetry methodology, the Si/SiO_2 interface provides a simplest test ground. Under ultrafast laser irradiation, a fraction of the valence band (VB) electrons in Si are promoted to the conduction band (CB), followed by charge rearrangement at the interface, depending on the nature of the excitation (laser intensity, energy, etc.). The very large band gap of SiO_2 (8.9 eV) ensures that SiO_2 slab is transparent to the incident visible light. The field arises from charge separation at the interface as carriers are transported from the bulk to the surface. At the ultrathin SiO_2 layer (1–10 nm), the charge transport across the SiO_2 slab is predominantly via tunneling, which makes both surface charging and discharging relatively efficient processes. However, the details of the dynamics remain nontrivial for photocarriers as both direct or field-induced (Fowler–Nordheim) tunneling [38–40] are driven by the increase of the quasi-Fermi level of hot electrons (holes) above the conduction (valence) band offset. The possibility of dielectric enhancement of the interface dipole, as well as thermionic and multiphoton-mediated photoemission can also add to this near-surface field [41–43].

Several techniques have been employed to examine charge transfer mechanisms associated with this interface, most notably, EFISH (electric field-induced second harmonic generation), which has been applied to characterize leakage currents through oxide layers, long-lived trap states, and band offsets [44, 45]. Photoelectron spectroscopic methods have been effective in elucidating the interfacial electronic structures [46, 47], as well as monitoring the surface state populations directly [48–50]. Bulk carrier dynamics in Si have been studied extensively on the pico- to femtosecond time scales with a variety of time-resolved techniques [49, 51–56]. In many of the surface-sensitive studies, the experiments are performed with high repetition rate lasers, on the order of 80 MHz, implying that the system is pumped every $\sim$ 13 ns, which is before trapped charge can relax, such that the residual charge level is continuously pumped. This of course depends on the integrity of the interface, or, more directly, the density of interface states.

Using a standard RCA cleaning protocols [57], a thin (2–5 nm) insulating SiO_2 layer can be reliably grown on single crystalline Si substrate. Because for Si the penetration depth (l) of the infrared pump laser (800 nm) is significantly longer (1 μm) [58] than that of the electron beam ($\sim$ 5 nm, incident angle 6. 8°), characteristic carrier lifetime (τ_i) probed at the surface ($\sim$ 700 ps), which can be estimated based on $\tau_i = l^2/D$, where $D \approx 15\ cm^2s^{-1}$ [59] is the diffusivity of the electrons, is significantly longer than electron–phonon coupling time ($\sim$ 5 ps), thus allowing the stored photon energy to be maintained within the photoexcited region. Meanwhile, due to the large disparity in the electronic and lattice heat capacities, even at a relatively high fluence of 65 mJ/cm^2, the lattice temperature rise is very small ($\approx$ 40 K), as revealed from the TTM calculation shown in Fig. 13.6. This is in sharp contrast to the more than three orders of magnitude increase of the carrier density in the surface excited region from the intrinsic level [59] by photoexcitation, thereby creating a favorable condition for studying hot electron-driven interfacial charge transfer across the SiO_2 layer without worrying too much of lattice effect.

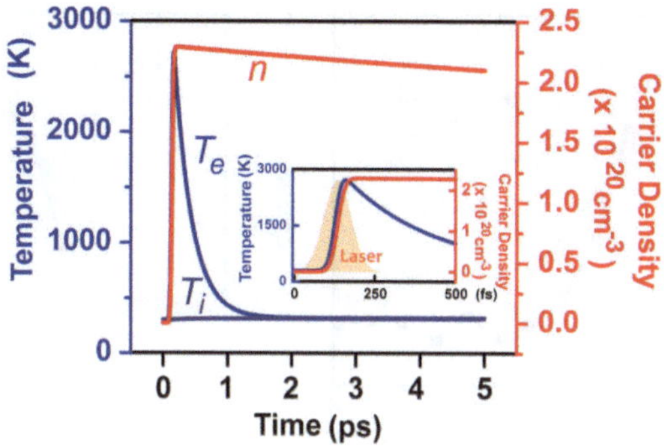

Fig. 13.6 Non-thermal Boltzmann transport coupled two-temperature model calculations of electronic/ionic temperature (T_e, T_i) and carrier density induced by 800 nm femtosecond laser in silicon. The laser fluence and pulse width are 65 mJ / cm^2 and 45 fs, and the silicon sample thickness is 50 μm [33]. *Inset*: Closeup view of the first 500 fs [33]

Using the formalism from Sect. 13.2 the changes in diffraction angle are converted to surface voltage, which is presented in Fig. 13.7. The objective here is to analyze the Bragg peak dynamics from multiple peaks. This was to see if the agreement is robust as incidence angle and diffraction order are varied. In principle, the transient surface voltage V_s should remain the same as different rods are examined. We choose to investigate (0,3,24), (0,1,21), (0,1,24) on the (0,3) and (0,1) relrods with $N = 7$ and 8 (crystallographic notation for diffraction order is multiplied by a factor of 3 due to the ABC layering of the Si(111) surface), corresponding to θ_i of 6.24°, 4.15°, & 4.70°, and θ_o of 3.50°, 4.02°, and 4.63°, respectively. The excitation fluence is fixed at 65 mJ/cm^2. Under photo-illumination, the transient movement of the three diffracted beams, depicted in Fig. 13.7a, indeed exhibits nonreciprocal signatures as described previously, i.e. the higher order Bragg peak shifts less than the lower order one, which is characteristic of the surface voltage-induced effect [25]. Closer examination of the shifts shows that nonreciprocity applies only to θ_o, but not to θ_i, as the maximally shifted beam is (0,3,24), which has a θ_i=6.24° larger than the rest, whereas its corresponding θ_o is 3.50°, which is smaller than the rest. This angular dependence is confirmed by the rocking map analysis for the relrods exhibiting $a = 1$ near the in-phase diffraction, as shown in Fig. 13.7b for (0,3) relrod. By applying $a = 1$ in Eq. (13.8), we deduce V_s for the three diffracted beams and successfully reproduce V_s largely independent of θ_i, indicating the slab model description is a good model for describing the essential photovoltage response observed by diffraction.

In Fig. 13.8, we show a comparison of the surface voltage dynamics determined by diffractive photovoltammetry technique on silicon surface with results obtained using time-resolved photoemission technique [49]. The V_s has been rescaled to match with the amplitude of the data extracted from [49]. A striking similarity in the relaxation rates is clear. If the observed field here were because of charge transfer to deep trap states, the timescale for the TSV to decay would be on the order of ns-μs or possibly longer, and the observed 100 ps would not make sense. There would

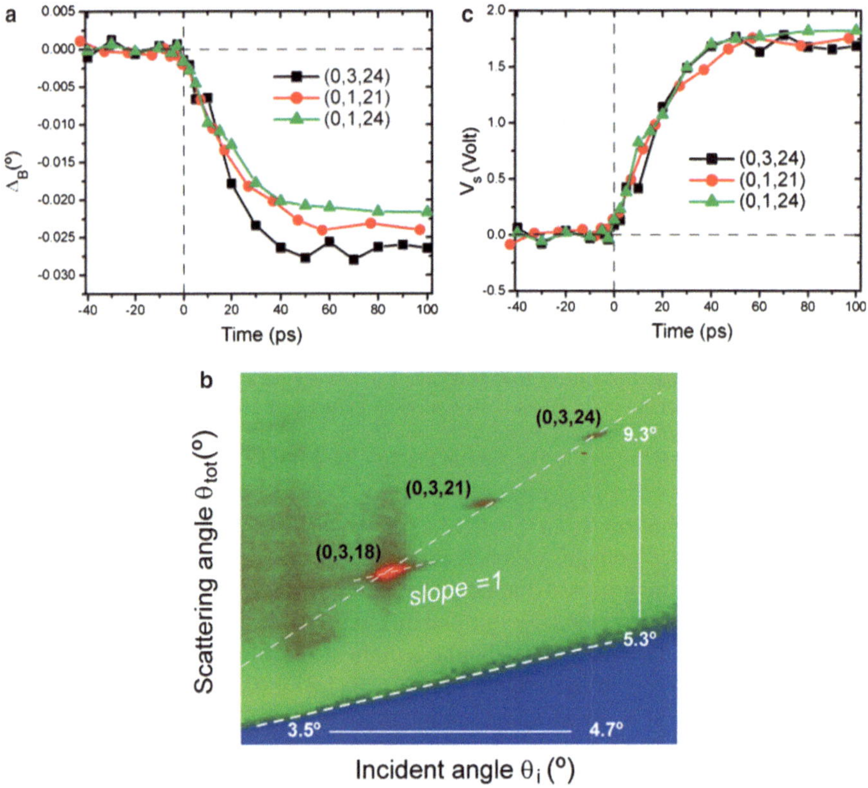

Fig. 13.7 Transient voltammetry from three diffracted beams from Si/SiO_2 interface. (**a**) The angular shift of (0,3,24), (0,1,21), and (0,1,24) beams excited at $F = 65$ mJ/cm^2. (**b**) The rocking map characterization of (0,3)-relrod, showing a surface diffraction condition $a = 1$. (**c**) The photovoltage deduced from (0,3,24), (0,1,21), and (0,1,24) beams based on Eq. (13.8) using $a = 1$ [27]

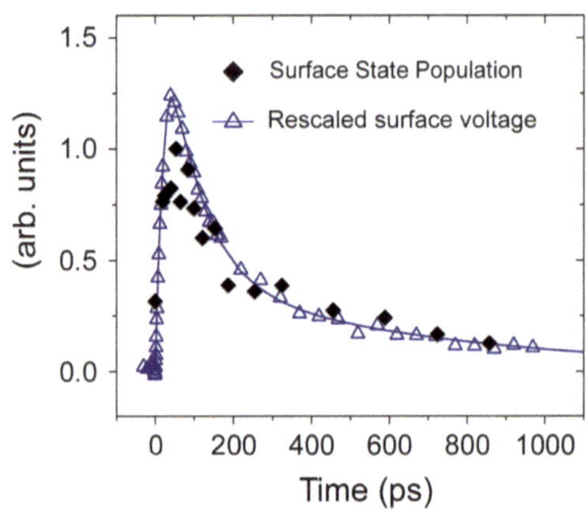

Fig. 13.8 Transient surface state population measured by photoemission [49] (*diamonds*) compared with the measured transient surface voltage, demonstrating very similar decay rates [25]

also be a residual charging effect since deep trap state occupation is largely irreversible. No residual effect was observed. Since it is known that the generated photovoltage involves the surface (bulk Dember fields are far too weak to detect the beam to this extent), it is very reasonable to presume the same strong surface-bulk coupling as in [49], ultimately dictating the decay rate of the field.

Our diffractive voltammetry results contain sufficient information to understand the origin of the photovoltage, which might include fields along the surface dielectric SiO_2 layer, the subsurface bulk space charge region, and even the near-surface vacuum region. In the case of the band photovoltage induced in the bulk space charge region, the limit is the flat-band value of approximately 300 mV [60]. As our V_s exceeds such a limit, we rule out that the band bending induced by the space-charge layer is mainly responsible for the V_s observed here. This is further evidenced by the fact that the space-charge layer in Si is rather thick (≈ 1 μm) as compared to the penetration depth of the electron probe (≤ 10 nm). We turn our attention to the surface charging across SiO_2 dielectric layer, where the surface field is fully probed by the electron beam. The subsurface carrier dynamics is strongly coupled to the net surface charges, which, through the elevated carrier screening, defines a dynamical space charge region. In the previous investigation of surface charging processes by EFISH [44, 61–65], the sub-surface electric field is deduced based on modeling the field-enhancement of optical second harmonic generation signal as well as photoemission [48]. We find that the field strength $E \sim 1$ V/nm obtained in our study is very similar to what was found in EFISH studies under similar excitation conditions [44, 64], but because of a lower laser repetition rate applied here (1 kHz, compared to $\sim$ 80 MHz in EFISH) cyclic residual charge accumulation from deep trap states [66–69] is avoided, allowing ps interfacial charging dynamics to be resolved directly.

Because of the extremely high ionization potential (≈ 10 eV) in SiO_2 layer, the photoemission from SiO_2 surface is rather weak and is plausible only at high fluence via the thermionic emission or multiphoton photoionization [70]. Thermionic emission occurs when the high-energy tail of the Fermi–Dirac distribution rises above the ionization potential ϕ, whereas multiphoton ionization can produce highly energetic photoelectrons with kinetic energy $K = nh\upsilon - \phi$, where n is the number of photons being absorbed by the electrons which can range from 2 to 5 at a high intensity. As the process evolves, a space-charge cloud develops, which serves as a hindrance to further electron emission. While the free carriers density that can be librated into the vacuum is expected to be extremely small ($< 10^{-6}$) compared to the surface charge density, the surface voltage build-up can gain strength as the integration depth is increased as the photoelectrons expand into the space. The ballistic speed of photoelectron expansion has been measured [27, 42] to be on the order of 1 μm/ps, which sets the limit in the rate of the photovoltage build-up associated with photoemission. Indeed, detailed space charge measurement in the free space using shadow projection imaging technique has verified such a contribution in a set of controlled experiments designed to evaluate the relative contribution of photoelectrons to the measured $V_s(t)$ [27]. The experiments, combined with numerical N-particle modeling, have shown that photoelectron contribution is at

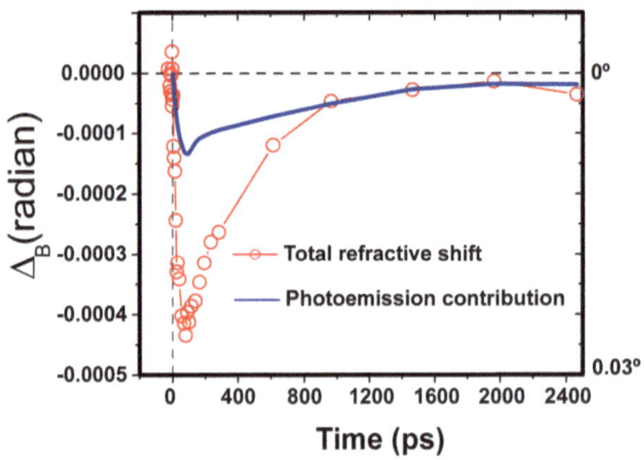

Fig. 13.9 Experiments to characterize surface photovoltage and photoelectron dynamics performed at $F = 65$ mJ/cm^2. Data (*symbols*, colored in *red*) show the refraction shift of a selected diffracted beam. An *N*-particle simulation to estimate the refraction contribution associated with photoemission is shown (*solid line*, colored in *blue*) for comparison [27]

maximally 30 % level even at a high fluence ($\geq$ 60 mJ/cm^2) within the 100 ps timescale, whereas it might play a more important role at the ns timescale for some systems. The relative contributions in the case of SiO_2/Si are depicted in Fig. 13.9 for comparison. Perhaps, the more profound impact of the photoelectrons is on controlling the subsurface charge dynamics after the surface charges largely dissipate. Nonetheless, such long-range effect can be avoided by experiments at a significantly lower fluence where the photoemission contribution will generally be negligible.

13.3.2 Charge Transfer Between Surface-Supported Nanoparticles and the Substrate

The scope of nanoparticle research spans many disciplines with vast possibilities of incorporation into practical use, including nanoelectronics [71, 72], photovoltaics [1, 73–76], and even quantum-dot-based lasers [77, 78]. On the biological/life sciences side, it has recently been shown that gold and silver nanoparticles, when delivered to site-selective cancerous cells, can destroy them upon exposure to a pulsed-laser source, without damaging surrounding tissue [79, 80]. Noble metal nanoparticle also serves as electron trap that enhances charge separation in the near junction region [81]. Of recent interest is the concept of using a nanoparticle as an optical antenna by exploiting the plasmon-based near-field enhancement, which is capable of surpassing the diffraction limit for spatial resolution [82, 83]. It has been observed that the same near-field optical-enhancement effect improves the carrier generation in the nearby semiconductor substrate and boosts the efficiency of the

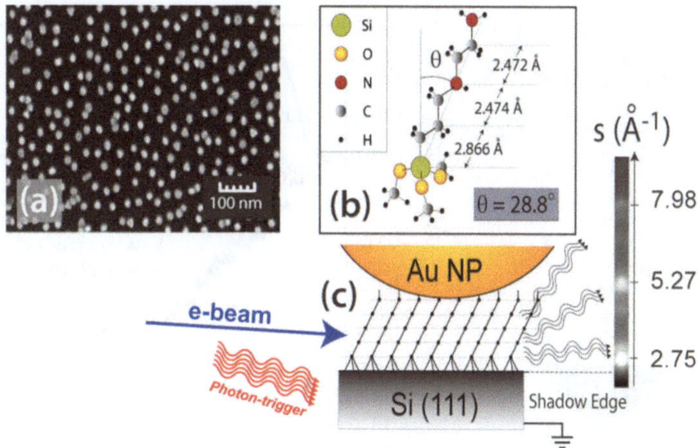

Fig. 13.10 (**a**) A sample of Au nanoparticles (NPs) immobilized on a functionalized Si substrate. (**b**) The chemical form of the AEAPTMS linker molecule. (**c**) Schematic of an electron beam scattering from the ordered self-assembled monolayer chain and the corresponding diffraction pattern [26]

solar cell [84] or photocatalyst [85]. At a fundamental level, the photoinduced carrier dynamics and possible plasmonic enhancement effects at the contact region can be investigated via diffractive voltammetry by monitoring the photovoltage between the nanoparticles and the substrate, which in this case is directly related to the charge state of the nanoparticle: $V_s = CQ^2/2$, where Q is the surface charge of the nanoparticle and C is the capacitance of the nanoparticle/SAM/Si interface. Because of the large curvature of nanoparticle surface, the surface charges on SiO_2/Si surface can strongly interact with the injected charges in the nanoparticle through field focussing at the contact region, leading to enhanced carrier transport and charge separation. Two different mechanisms should be considered, namely: (I) The hot electron-driven process, where the strong light sensitization at metallic nanoparticles generates high-temperature carriers with higher efficiency to tunnel through the SAM energy barrier, creating a photoelectron current from nanoparticle to the substrate; and (II) The photovoltage-driven dynamics initiated by electrons at Si CB edge, created by photoexcitation. While the net effect of enhancing the hot carriers concentration at metal-semiconductor interface is the same, the net photocurrent flow in the two cases is opposite, which can be measured directly with diffractive photovoltammetry. When channel (I) is dominant, it leads to a positive charging of the nanoparticle, thereby causing the diffracted beam to shift upward. If channel (II) is dominant, the nanoparticle will be negatively charged, and consequently the diffraction peaks will move downward in addition to the movement induced by the surface charging of the SiO_2.

Monodisperse metallic nanoparticles (Fig. 13.10) can be deposited on Si substrates by way of SAM of aminosilanes AEAPTMS [26, 86]. The aminosilane chains, which are anchored to the Si substrate, immobilize the nanoparticles

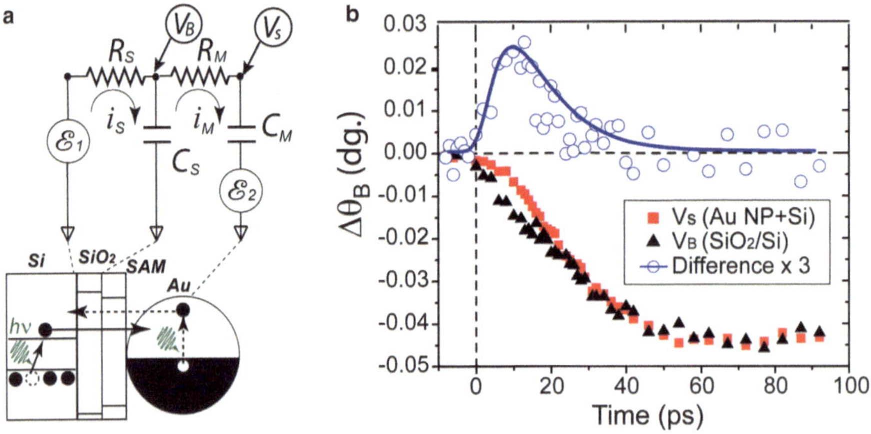

Fig. 13.11 (**a**) An effective circuit model depicting the transient surface voltage $V_s(t)$ measurement via the refraction shift of the diffracted beams through SAM. The R_S, C_S, R_M, C_M are the effective resistance and capacitance of the substrate (S) and the SAM (M). The electromotive potential that can drive the photocurrent through the SiO_2 layer (i_S) and further through SAM (i_M) is mainly from the hot carriers generated from the Si ($\mathcal{E}_1(t)$); whereas the short-lived photoexcited hot carriers generated within the Au nanoparticle (NP) can also drive the charge transfer in the opposite direction ($\mathcal{E}_2(t)$). (**b**) The overall refraction shift determined by SAM diffracted beam (labeled V_s), the background (labeled V_B) obtained from SiO_2/Si interface, and the molecular charge transport contribution, obtained by subtracting V_B from V_s

through the formation of strong van der Waals bonds. When an ensemble of nanoparticles is sampled by the electron beam, a powder diffraction pattern results [26], which is constituted from the many different crystal faces that the beam samples in its footprint on the surface. In addition, the linker molecules that take part in the anchoring of nanoparticles tend to orient themselves in a manner that is sufficiently ordered, such that very clear diffraction spots from the linker molecules are present in the patterns (Fig. 13.10c). Here, the charging dynamics will be examined for an interfacial structure consisting of the interconnected nanoparticle/SAM/semiconductor geometry discussed here, shown in Fig. 13.11a. The fortuitous discovery of SAM diffraction peaks in conjunction with the extension of diffractive voltammetry to arbitrary geometries (Sect. 13.4.3), fuels an ongoing effort toward using the SAM dynamics for uncovering the charge transport (forward or backward) across the interface from semiconductor to nanoparticle, or vice versa. Following photoexcitation, hot electrons are generated in the substrate and nanoparticles, from which a charge transfer process will ensue depending on the relative offset between the transient electrochemical potentials of the two systems. Quantifying this process would provide an additional technique for the investigation of molecular transport, an area of notable interest [35, 87, 88]. The surface potential associated with these nanostructures can be determined from an extended slab model, with an angular-dependent correction factor (Θ from Sect. 13.4.3) to model the relevant angular shifts observed in the diffraction.

In our first experiment studying the molecular transport, 20 nm gold nanoparticles were used, and the interface is illuminated with 800 nm near-infrared laser pulse [26]. This arrangement avoids the surface plasmon excitation of the nanoparticles ($\approx$ 500 nm), and we expect the interfacial charge dynamics between the nanoparticle and the SiO_2/Si is determined by two competing processes as described in Fig. 13.11a. Specifically, the transient voltage induced by charge transfer between the substrate and the nanoparticle can be monitored by analyzing the SAM diffraction peaks corresponding to momentum transfers, $s = 2.75$, 5.27, 7.98 $Å^{-1}$ (orders $N = 1$–3), as depicted in Fig. 13.10c. The total transient shift Δ_B includes two serial voltage drops: one across the Si/SiO_2 surface (V_B), and the other across the SAM (V_M), as depicted by the effective RC-circuit in Fig. 13.11a. To isolate the voltage across the linker molecule, the contribution from V_B must be subtracted out. Typically, the Δ_B associated with a bare Si/SiO_2 surface has an exponential down turn followed by a $\approx$ 200 ps recovery (Fig. 13.8). The presence of an additional upward swing followed by a downward swing to recovery can be discerned with careful analysis, which is shown in Fig. 13.11b. These additional components sampled by diffraction through SAM suggest a new channel of photocurrent emerges between the silicon substrate and the nanoparticles, which creates a more positive potential at the interface to account for the initial upswing of the diffracted beams [26].

These results strongly indicate that both processes (I) and (II) are at play. Process (I) is a key player at 1–40 ps due to the hot electrons generated within the strongly sensitized gold nanoparticle. However, because of the lifetime of the hot electrons in gold nanoparticle is limited by the electron–phonon coupling on the few ps timescale, the *emf* $\mathcal{E}_2(t)$ is short-lived. Meanwhile the photoexcited CB carriers at silicon surface is very long-lived, limited mainly by the drift-diffusion carrier recombination at >200 ps timescale [25]. The $\mathcal{E}_2(t)$ is persistent and ultimately dominant at long times driving the photocurrent across the interface.

It is evident that only at a high fluence that the process (I) will be efficient. Indeed, under a medium high fluence $F = 15$ mJ/cm^2 a more distinct $\mathcal{E}_2(t)$ effect is observed within first 30 ps. This is verified first from Fig. 13.12a that all the long-time behavior of $V_s(t)$ agrees extremely well with the charging potential of SiO_2/Si [$V_B(t)$], signifying that the long time dynamics is driven mainly by $\mathcal{E}_1(t)$. Furthermore, the values of β obtained from comparing V_s and V_B (Fig. 13.12c) shows a nearly linear relationship between β and F. In contrast, the changes seen at the short time are driven completely by $\mathcal{E}_2(t)$. From Fig. 13.12b, we see that V_s first engages a more dramatic upward movement, followed by a sequence of downward swing and a recovery back to V_B. This is in contrast to the dynamics seen at $F \leq 10$ mJ/cm^2) where no such dramatic downward movement is seen. The initial negative $V_M(t)$ across SAM indicates again net positive charging of the gold nanoparticles, driven mainly by $\mathcal{E}_2(t)$, whereas a reversal of nanoparticle charge state begins after $\sim$ 22 ps. At higher fluences as shown in Fig. 13.12d, this trend is heightened up to $F = 32$ mJ/cm^2 when a saturation is reached and the reversal time exhibits a minimum value of 3 ps, limited by our time resolution. The giant reversal is indicative of a field-induced dielectric breakdown of SAM, leading to a near

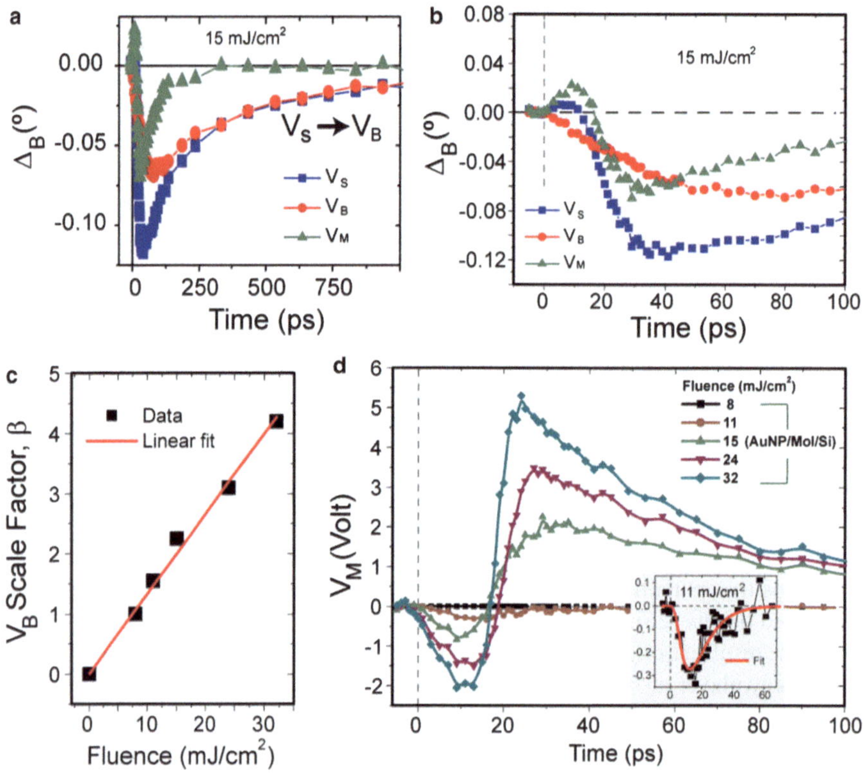

Fig. 13.12 Ultrafast transport at gold nanoparticle/SAM/silicon interface. (**a**) The overall refraction shift determined by SAM diffracted beam (labeled V_s), the background (labeled V_B) scaled to match the long time transient of V_s, and the molecular charge transport contribution, labeled V_M, obtained by subtracting $\Delta_B(V_B)$ from $\Delta_B(V_s)$. The data are obtained at $F = 15$ mJ/cm^2. (**b**) The short time response of (**a**). (**c**) The scale factor β used in scaling V_B to match V_s in the long time for different fluences, showing a linear trend w.r.t fluence. (**d**) V_M determined for different fluences, compared to the transient voltage obtained for SiO_2/Si at $F = 65$ mJ/cm^2. Inset, the V_M obtained at $F = 11$ mJ/cm^2 is fit to an RC model with nearly equal time constants of 8 ± 1 ps for charging/discharging [27]

zero resistance ($R_M \approx 0$) for electrons to flow back, or even overshoot, from the Si substrate to the nanoparticle. From examining this trend a critical value of ≈ 0.8 V/nm can be determined as the threshold for conductance switching. Following the switching, the dissipation of the stored electrons in the nanoparticles is much slower (~ 40 ps), indicating a robust recovery of the dielectric constant to the insulator status in SAM.

Examining the data at $F = 11$ mJ/cm^2 (see inset of Fig. 13.12d) shows a charging and discharging time of 8 ± 1 ps. Based on the effective RC model, we obtain a resistance $R_M = 2.74$ MΩ using $C = 2.92 \times 10^{-18}$ Farad, deduced from finite element modeling of the interface. It is rather interesting to compare this molecular resistance with the steady-state value of 12.5 MΩ obtained by applying a bias voltage across the molecular interface [86] using 10 nm (thus a SAM area ~ 4 times smaller) Au nanoparticle. The R_M obtained using ultrafast voltammetry is four times smaller than the steady-state measurement for 10 nm particle, which shows the molecular resistivity obtained from two different methods are nearly identical.

13.3.3 Charge Transfer Between Gold Nanoparticles and Semiconductor Substrate Enhanced by Surface Plasmon Resonance Excitation

When light impinges a metal, the free electrons can be excited into collective oscillatory motion about the heavy ionic core. This occurs when the conditions are met for surface plasmon resonance (SPR), which depends on the dielectric properties of the metal as well as its size. While such a high-speed modulation of charge density near the nanoparticle surface is too fast to directly draw current to the nanoparticles, the associated intense electromagnetic oscillation intensifies the local optical field, in other words acting like an antenna. Light trapping of resonance photons at SPR increases the absorption of low-intensity photons to boost molecular sensing [89] and solar cell performance [84], or enhance an electric field, which leads to strong light-matter coupling [90], or the formation of energetic charge carriers that can be transferred to the surrounding and to drive the chemical transformations [91].

From previous section, it is shown there are two non-mutually exclusive channels of carrier transfer mechanisms between the nanoparticle–semiconductor interface. In this section, we will discuss how these mechanisms will be affected by the SPR excitation. In the context of field enhancement, the improvement of the light-dielectric coupling at near-field increases the photocurrent at nanojunction, which is the subject of investigation here. The correlation between the SPR intensity and the carrier generation on nearby silicon substrate is used to evaluate the SPR enhancement effect. The carrier generation can be evaluated utilizing the linear relationship between the surface potential of SiO_2/Si and the sub-surface carrier concentration that drives the surface charging process as shown previously (Fig. 13.12c) at off-SPR condition. Near SPR, further enhancement can be evaluated by the deviation from the linear V_s vs. fluence dependence. To evaluate the SPR enhancement effect, the wavelength λ of the incident laser pulses is varied from 800 to 400 nm using an OPA (see Fig. 13.2), while keeping F at a constant value of 4 mJ/cm^2. The spectrum of maximum V_s obtained at different λ exhibits a Gaussian-like function centered at 525 nm, as depicted in Fig. 13.13a. This excitation spectroscopy of V_s has a peak similar in position and bandwidth to that of the light absorption spectrum measured in water [92], and is $\approx$ 30 nm shift from the resonance curve of the same-size spherical nanoparticles estimated using Mie scattering at free space. The strong resonance at $\lambda \sim$ 525 nm is identified as the dipolar resonance mode—a dominant extinction mode for particle smaller than 50 nm. On the other hand, the photovoltage spectrum does not contain the

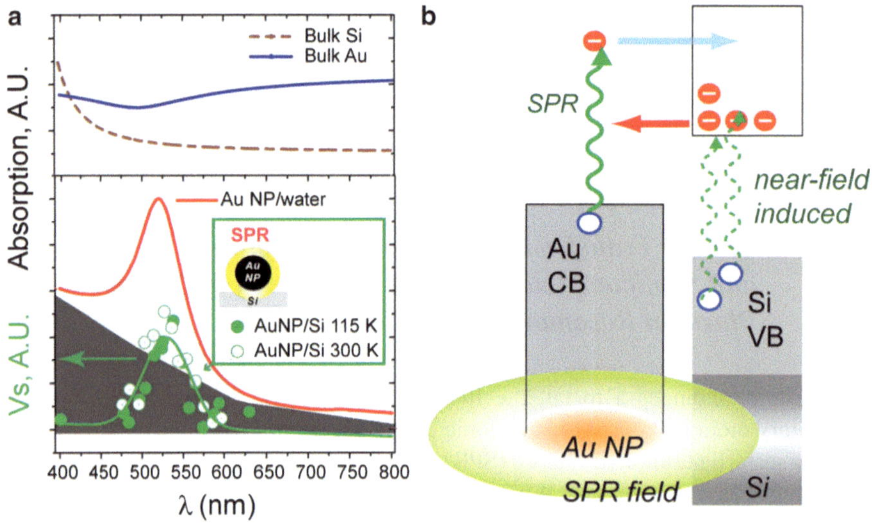

Fig. 13.13 Ultrafast transport at gold nanoparticle (NP)/SAM/silicon interface near surface plasmon resonance (SPR) excitation. (**a**) The maximum transient photovoltage response near SPR excitation, as compared to the absorption of the similar nanoparticles in water, bulk gold, and silicon substrate. It is evident that while both show characteristic SPR peak at ≈ 525 nm the nanoparticle's V_s spectrum at surface lacks the background seen in the optical absorption spectrum of similar nanoparticles in water. (**b**) The enhancement of hot carrier generation at Si surface can be achieved via SPR evanescent field within the Si surface or charge carrier injection from highly excited Au nanoparticle

background seen in the optical measurement. The lack of non-SPR background in photovoltage measurement is a strong indication that surface charge dynamics is driven by dipolar resonance structure highly localized at the interface between Si and the nanoparticle. In contrast, the far-field absorption response obtained in optical spectroscopy has components not necessarily localized at the interface. The photovoltage measurements at 800 and 400 nm, both of which are far off from the dipolar resonance spectral range, are nearly an order of magnitude smaller than that at dipolar resonance (at 525 nm). Therefore, the near-field enhancement factor of photocarrier generation at Si surface by dipolar resonance excitation is at least a factor of 10. To compare, the optical absorption in Si and bulk gold has a rather flat absorption curve, as shown in Fig. 13.13a.

13.3.4 Dielectric Enhancement with the Coverage of Water–Ice

The application of a water–ice layer that completely buries the nanoparticle-decorated interface can alter the incident photon resonance structure with the evanescent SPR field strongly coupled to the dipole moments of the surrounding

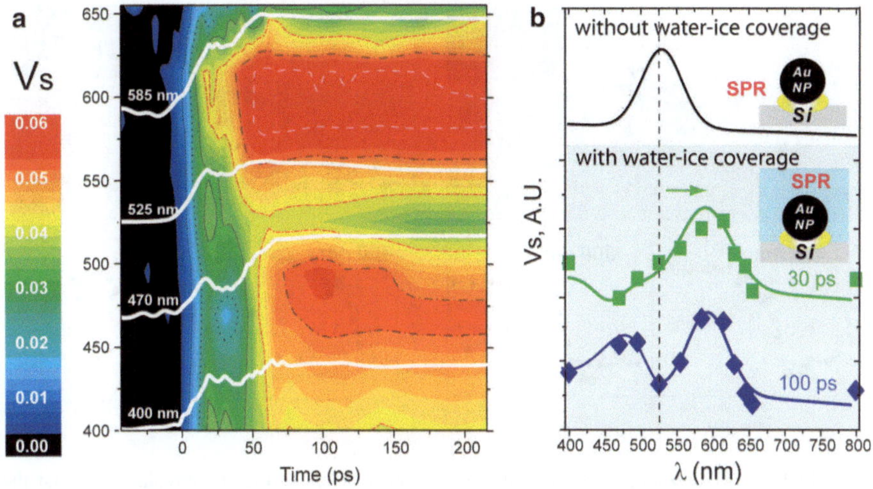

Fig. 13.14 (**a**) Surface photovoltage response map constructed using the diffractive voltammetry conducted on the surface of the water–ice slab dosed on the gold nanoparticle (NP)/SAM/silicon interface at excitation wavelength λ from 400 to 800 nm. Selected surface photovoltage curves $V_s(t)$ at λ=400, 470, 525, and 585 nm show a composition of two dynamics with different timescales. (**b**) Comparisons between the surface voltage response spectra obtained from the interface without the coverage of water–ice (*black line*) and the ones with water–ice covers shows a red shift of the resonance peak (*green line*) at 30 ps, and bifurcation of peaks (*blue line*) at 100 ps [93]

water molecules. The new resonance structure will impact the SPR-induced charge carrier dynamics, which is a core issue to consider in searching the optimal geometry to enhance the performance of composite photocatalysts that contain semiconductor and plasmonic-metal nanostructures. Recent studies found strong positive correlation between SPR and reaction rate, leading to a hypothesis that the metallic SPR enhances rates of photocatalytic reaction at nearby semiconductor surface through transfer energetic charge carriers at the interface [91]. In this section, we examine various routes that energetic charge carriers travel at metal/semiconductor/water–ice interface, for understanding SPR enhancement effects. The experiment builds on the same nanoparticle interface discussed in Sect. 13.3.2 that conveniently serves as a controlled experiment for the studies here. We additionally apply a layer of water–ice coverage on the nanoparticles by vapor dosing at 90 K using the setup described in Fig. 13.2. The steady growth of ice is monitored by the grazing incidence electron beam until the signal of the gold diffraction disappears, giving a rough estimate of the thickness of water–ice layer to be around 50 nm. We then scan the pump laser wavelength from 800 to 400 nm and monitor the surface voltage of the ice layer from the refraction shift of ice diffraction pattern, over a time span from – 50 to 1,000 ps. From these experiments a wavelength *vs.* time photovoltage response map is constructed as shown in Fig. 13.14a.

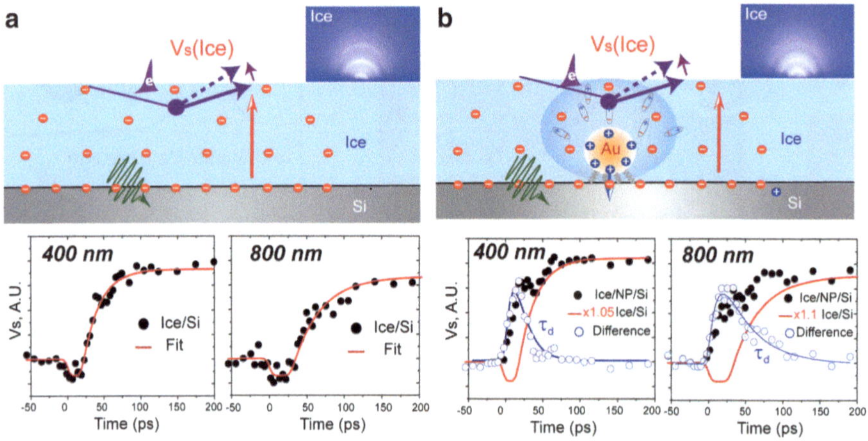

Fig. 13.15 (**a**) Controlled experimental measurements of ice surface photovoltage without the presence of nanoparticles. The rise of the surface voltage is generally delayed by the timescale of the energetic charge carriers migrating to the surface after being launched from the Si substrate. (**b**) Equivalent results obtained the surface with Au nanoparticles decoration. By comparing (**a**) and (**b**), we can deduce the fast components (*blue circles*) unique to nanoparticle (NP)-decorated surface [93]

Inspection of the map shows that the photovoltage spectrum evolves from initially a singular peak at 585 nm to a dual peak structure with an additional peak at 470 nm. From the time-slices across the map at representative $\lambda = 400, 470, 525$, and 585 nm, the spectral evolution can be seen as encompassing two temporally separated processes: one that commences at $t = 0$ and decays largely within 100 ps and the other that rises as the first one decays and doesn't seem to relax at all within the observed timescale. The singular peak response function, from 0 to 50 ps, mainly manifests the fast process; whereas the dual peak structure describes the latter. The shift of the dipolar resonance peak from 525 to 585 nm can be understood by finite difference time domain (FDTD) simulation as due to a strong coupling between the SPR of the nanoparticles and the surrounding water dielectrics [93]. However, the presence of high-energy response at $\approx$ 470 nm cannot be produced from the optical domain calculation alone, as such short-wavelength mode is typically of high-order and is not optically active at particle size less than 50 nm.

Figure 13.15a shows the results of a controlled experiment to elucidate the two different dynamics observed in Fig. 13.14, in which the photovoltage at ice surface is measured without the presence of Au nanoparticles. We observe a very similar slow dynamics seen also with nanoparticles being present. This result thus reveals that the slower process is in fact charge carrier dynamics launched straight from the silicon substrate. Since the dielectric property of ice is unlikely influenced directly by visible light excitation, the apparent delay in the voltage rise on ice surface is indicative of a long-range charge carrier injection from Si into the ice surface. The shortening of such an injection time as the energy of photon increases (from 800 to 400 nm) is particularly interesting. The dependence on photon energy suggests energetic carrier injection from Si as supported by the high charge drift velocity

($\approx$ 1 nm/ps) observed here, which can be rationalized with transport through ice conduction band. Following the energetic electron charging, the dissipation of the charges, after energy relaxation, takes a significantly loner time ($\geq$ ns). We note that a small but nearly instantaneous downward refraction shift is observed in every experiment near time zero. We attribute this phenomenon to the shift in the chemical potential at the ice/silicon interface due to photoexcitation, resulting in a swift change of the dielectric alignment of the ice layer.

The slower surface voltage dynamics can be fully attributed to long-range charging from the substrate, whereas the fast dynamics (blue circles in Fig. 13.15b), which is completely absent in the controlled experiment, must be mediated through the nanoparticles directly. Indeed, such dynamics bear striking resemblance to the photo-induced charging dynamics observed previously in the gold nanoparticles without ice coverage (see inset in Fig. 13.12d at $F = 11$ mJ/cm^2). The surface field induced by charged gold nanoparticle can be felt on ice surface through dielectric realignment of the water molecules surrounding the nanoparticles. We are able to extract this fast local charging dynamics by fully deducing from the overall dynamics the slower charging component using the data from the controlled experiments, shown as the blue curves in Fig. 13.15. Since the spectrum representing such fast local charging dynamics (green curve in Fig. 13.14b) peaks at 585 nm, which is a dipolar resonance, we conclude that charge transfer between the nanoparticles and the Si substrate is predominantly driven by the near-field dipolar resonance fields. Furthermore, this dipolar resonance is also responsible for the elevation in the transfer of energetic electrons to the ice surface, as shown in the enhanced photovoltage also at 585 nm at a longer timescale (blue curve in Fig. 13.14b). This is not surprising as generation of energetic carriers at Si surface can also be enhanced by the evanescent waves of SPR in the region. Finally, it is also important to note that the appearance of the additional 470 nm peak in elevating the charge level on ice surface indicates that the optically inactive, spatially nonhomogeneous higher-moment SPR might also be activated at the nanostructural interface [94–96]. Such a high-energy mode excitation, while much weaker than the dipolar resonance, can play a significant role in the surface charging response spectrum because the high-energy barrier required for the carrier injection into the ice conduction band (3.2 eV above the Si CB edge) significantly suppresses the sole dominance of the dipolar excitation over such a channel.

13.4 Modeling of the Surface Photovoltammetry

13.4.1 Effective Circuit Model

To simulate the photovoltammetry dynamics driven by the *emf* of the photoexcited carriers, we can construct equivalent circuit models. As an example, we examine the interfacial charge transfer from semiconductor photoreceptor (Si) across the

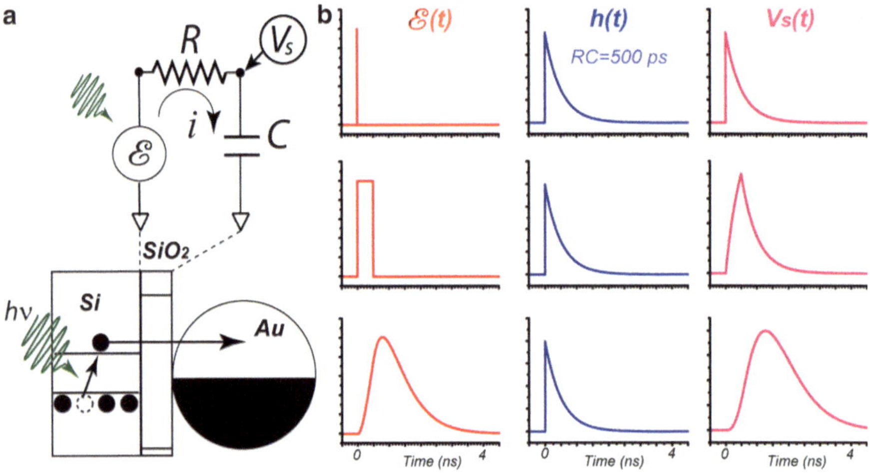

Fig. 13.16 (**a**) An effective circuit model depicting the transient surface voltage $V_s(t)$ measurement via the refraction shift across the SiO_2 slab at SiO_2/Si interface. The electromotive potential $\mathcal{E}(t)$ comes from the hot carriers generated at the Si substrate. (**b**) Examples showing the photovoltage $V_s(t)$ measured at the interface as a function of $\mathcal{E}(t)$. The relationship can be seen as a convolution with kernel function $h(t)$ characteristic of the RC circuit

insulating tunnel junction (SiO_2) layer to the surface with metallic (Au) nanoparticles or with just the surface states. The schematic layout of such a nanocircuit is depicted in Fig. 13.16. The photoexcitation of the semiconductor substrate generates electron/hole pairs that elevate the electrochemical potential at semiconductor surface, which can be a source of *emf* driving carriers across the insulating junction. The injection of hot carriers into the metal surface results in charge separation, which forms a photovoltage V_s between the semiconductor and metal surface. Such a photovoltage generation is fundamentally limited by the charging time, $\tau = R_J C_J$, of the junction, where R_J and C_J are the junction resistance and capacitance. The photovoltage $V_s(t)$ determined here can be used to derive the *emf* if the response time τ is known, based on the linear response of the nanocircuit model:

$$V_s(t) = \int_{-\infty}^{\infty} \mathcal{E}(\tau) \cdot h(t-\tau) d\tau, \tag{13.9}$$

where $h(t) = 1/RCe^{-(t-t_0)/RC}u(t), u(t)$ is the heaviside step function, and t_0 is the zero-of-time. In this formulation, the transient surface voltage $V_s(t)$ is the convoluted response of $\mathcal{E}(t)$ by the kernel function $h(t)$. $h(t)$ is inherent to the circuit setup and can be obtained by the inverse Laplace transformation of the effective reactance ratio γ, which, in the case of the RC circuit describe in Fig. 13.16a, is $Z_C/(Z_R + Z_C)$, where $Z_C = 1/Cs$ is the reactance of capacitor, and $Z_R = R$ is the

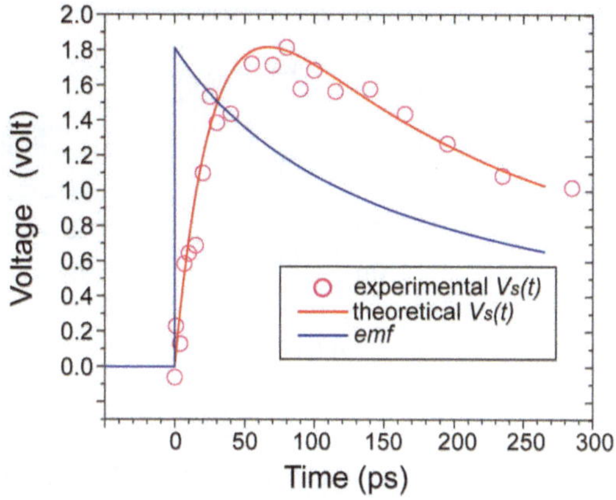

Fig. 13.17 Theoretical modeling of transient surface voltage $V_s(t)$. Using RC time of 30.8 ps, the corresponding electromotive force *emf* is deduced, showing a spontaneous rise and a long decay

reactance of the resistor in the complex frequency (s) domain. Applying Laplace transformation on both sides of Eq. (13.9), we establish the relationship between V_s and $\mathcal{E}$ in the s-domain:

$$\tilde{V}_s(s) = \tilde{\mathcal{E}}(s)\frac{1/RC}{s+1/RC}, \tag{13.10}$$

from which we can derive $\tilde{\mathcal{E}}(s)$, based on $\tilde{V}_s(s)$, and use it to deduce $\mathcal{E}(t)$ through inverse Laplace transform. Figure 13.16b depicts three examples of $V_s(t)$ driven by different $\mathcal{E}(t)$. For an impulse $\mathcal{E}(t)$, the $V_s(t)$ is simply the kernel function $h(t)$, an exponential decay with $\tau = R_J C_J$. For a square pulse, the $V_s(t)$ rises exponentially with $\tau = R_J C_J$. In practice, the carriers electromotive potential will likely have a form of nonexponential rise and decay, as depicted in the bottom panel of Fig. 13.16b. Given a typical nanodevice dimension, such as 10 nm nanoparticle next to a gold electrode with a $\sim$ 2 nm tunneling gap, the effective RC time is on the order of 10 ps. This timescale is fundamentally much shorter than the contemporary integrated device with a typical switching time $\geq$ 50 ps. On the other hand, the RC time is on the upper end of the interfacial charge transfer timescale reported by optical and photoemission techniques, which is at the limit of state-to-state population dynamics, and fundamentally different from the tunneling time [98].

Using Eq. (13.10) we can deduce the effective *emf* dynamics based on photovoltammetry data, which shows rise and decay times of $\tau_c = 30.84$ ps, and $\tau_d = 296.47$ ps, respectively, for surface charging and discharging. Using τ of 30.8 ps, we can deduce the $\mathcal{E}(t)$ as depicted in Fig. 13.17, where a spontaneous rise and long decay is seen which is consistent with the carrier density dynamics (Fig. 13.6) calculated using the Boltzmann transport coupled TTM.

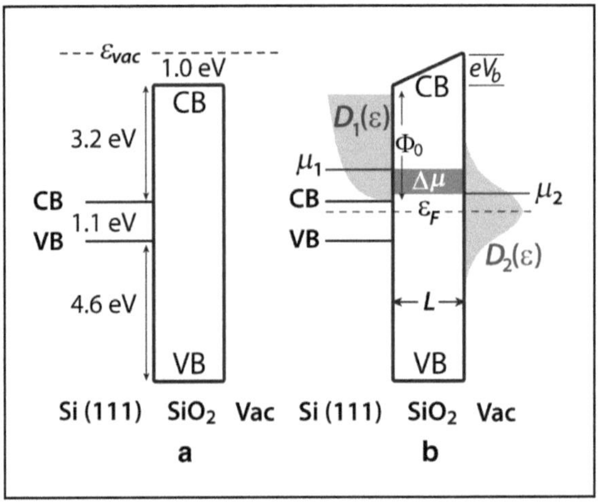

Fig. 13.18 (**a**) A band diagram depicting the Si/SiO2/vacuum interface. (**b**) Schematic of the energy diagram with relevant quantities used for the tunneling calculations. Electrons that tunnel through the oxide to the SiO_2/vac interface cause the quasi-Fermi level μ_2 to raise, making the barrier take on a trapezoidal shape. The electronic density of states is shown in *light-gray* for the silicon (D_1) and the SiO_2/vac interface (D_2). The difference between the two levels, $\Delta\mu$ (*dark gray band*) is the primary factor in determining the tunneling current. For a given energy (ε), when $\Delta\mu$, $D_1(\varepsilon)$, and $D_2(\varepsilon)$ are all $\neq 0$, then there will be a tunneling current through the SiO_2 [33]

13.4.2 The Microscopic Charge Transfer Model

At the microscopic level, the interfacial charge transport can be described by Landauer–Buttiker formula [19, 97–102]:

$$i(t) = \frac{4\pi e}{\hbar} \int D_1(\varepsilon) D_2(\varepsilon) \, |M_{12}(\varepsilon)|^2 \, [f_1(\varepsilon, \mu_1) - f_2(\varepsilon, \mu_2)] \, d\varepsilon, \qquad (13.11)$$

where $D_1(\varepsilon)$ represents the density of states in the bulk Si, $D_2(\varepsilon)$ is the density of states at the Si/SiO_2 interface, f_1 and f_2 are the Fermi functions at the Si/SiO_2 and SiO_2/vacuum interfaces, respectively, and $M_{12}(\varepsilon)$ are the matrix elements of the transmission given by [97, 100, 101]:

$$M_{12} = -\frac{\hbar^2}{2m} \int_0^L (\psi_1^* \nabla \psi_2 - \psi_2 \nabla \psi_1^*) dz, \qquad (13.12)$$

where ψ_1 and ψ_2 are the wave functions on the left ($z \leq 0$) and right sides ($z \geq L$) of the barrier, respectively (see Fig. 13.18). The linear response can be seen in the low temperature limit where the difference in the Fermi functions f_1 and f_2 in Eq. (13.11) becomes $\mu_1 - \mu_2$, which is the *emf* between the two interfaces.

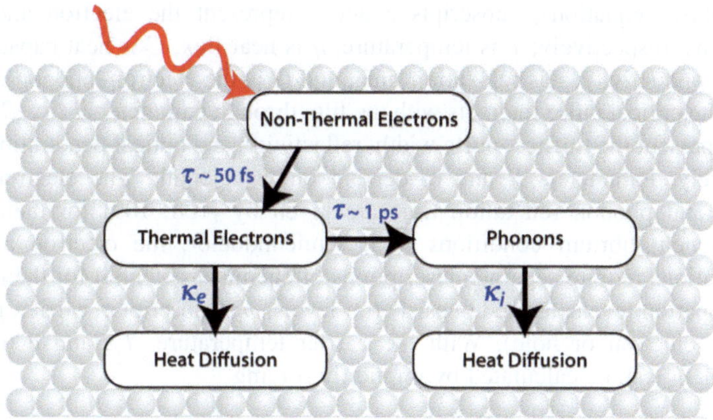

Fig. 13.19 The energy coupling scheme assumed by the TTM. First, the laser pulse is absorbed by the free electrons in the metal, which evolve into a Fermi–Dirac distribution in the first ≈ 50 fs. The thermal electrons dump energy into the lattice, causing them to equilibrate after ≈ 1 ps (depending on the metal). Both electrons and lattice undergo thermal losses from heat diffusion, the rates of which are largely governed by the thermal conductivity, κ (and strictly speaking the heat capacity and sound speed too) [33]

However, for the situation discussed here, the electrochemical potential depends on the carrier temperatures and concentration, which is strongly time dependent. TTM [103–107] have been invoked to calculate transient electronic temperature T_e, and photo-carrier concentration in metal and semiconductor. When an ultrafast optical pulse impinges the surface of a metal or semiconductor, the energy is absorbed by the electrons. If the width of the excitation pulse is shorter than the electron–phonon coupling time, a nonequilibrium is established between the electrons and lattice [108]. The TTM describes such processes, given by

$$C_e(T_e)\frac{\partial T_e}{\partial t} = -\frac{\partial q_e}{\partial z} - G(T_e - T_i) + S(z,t), \quad (13.13)$$

$$C_i\frac{\partial T_i}{\partial t} = -\frac{\partial q_i}{\partial z} + G(T_e - T_i), \quad (13.14)$$

$$q_{e,i} = -\tau_{e,i}\frac{\partial q_{e,i}}{\partial t} - \kappa_{e,i}(T_e, T_i)\frac{\partial T_{e,i}}{\partial z}, \quad (13.15)$$

where the laser heating source term, $S(z, t)$, can be described by a Gaussian in time and an exponential spatial decay into the metal (Fig. 13.19):

$$S(z,t) = \sqrt{\frac{4\ln 2}{\pi}}\frac{(1-R)F}{t_p\delta_s}e^{-z/\delta_s}e^{-4\ln 2\left(\frac{t-t_p}{t_p}\right)^2}. \quad (13.16)$$

In the above equations, subscripts e and i represent the electron and lattice subsystems, respectively; T is temperature, q is heat flux, C is heat capacity, κ is thermal conductivity, τ is relaxation time, and G is the electron–phonon coupling factor. The optical pulse interacts with the film through parameters F, t_p, R, and δ_s, which represent the fluence, pulse width, reflectivity, and optical penetration depth, respectively. Parameters R and δ_s depend on wavelength. The electronic heat capacity and phonon relaxation time are given by [108, 109] for metal films. Under nonequilibrium conditions in a semiconductor, the electron and hole concentrations require separate descriptions of chemical potential for their respective occupation levels, termed quasi-Fermi levels, ϕ_c (subscript c, carrier, means it could be electron or hole). With the carrier temperature, T_c, and density, n_c, evaluated, ϕ_c can be calculated by implicitly solving

$$n_c = N_c(T_c)\frac{2}{\sqrt{\pi}}\int_0^{\infty}\frac{\sqrt{x}}{1+\exp(x-\eta_c)}\,dx, \tag{13.17}$$

where $N_c(T_c)$, the effective density of states, is expressed as

$$N_c(T_c) = 2\left[\frac{m_c{}^{*}k_B T_c}{2\pi\hbar^2}\right]^{3/2}, \tag{13.18}$$

and the scaled reduced Fermi level, η_c, is expressed for electrons (e) and holes (h) (in units of $k_B T_c$) as

$$\eta_e = \frac{\phi_e - E_C}{k_B T_e}, \tag{13.19}$$

and

$$\eta_h = \frac{\phi_h - E_V}{k_B T_e}. \tag{13.20}$$

In principle, it is straightforward to implement the calculation of the quasi-Fermi level based on TTM. In 2005, Chen and colleagues formulated a self-consistent model for transport dynamics in semiconductors subjected to ultrafast laser irradiation, using the relaxation time approximation of the Boltzmann equation [59]. Here, for the brevity of the discussion, the detailed model given by Chen and colleagues [59] will not be discussed. For interested reader, we recommend reading [33, 59] for the framework of the Boltzmann Transport Model (BTM) given by Chen and colleagues and its implementation to calculate quasi-Fermi level in Si. It is suffice to say that BTM is an extension of the TTM, though it involves a third rate equation, in addition to electron and phonon temperature, which is governing equation for carrier generation and recombination. The nonequilibrium regime associated with the internal relaxation of the electron gas to a Fermi function is treated by

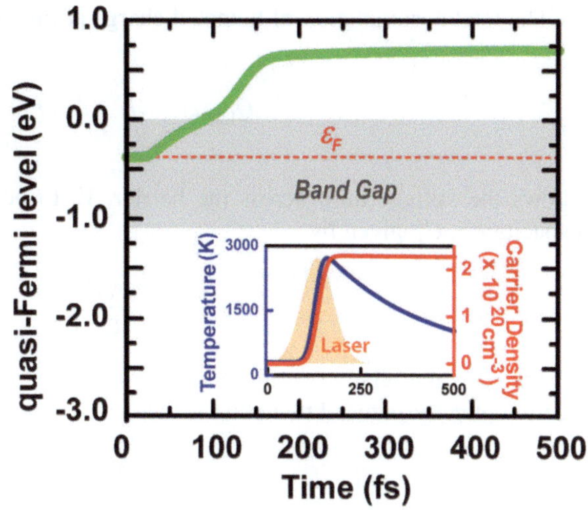

Fig. 13.20 The Boltzmann-Transport Model (BTM) calculation of the quasi-Fermi level of Si following photoexcitation is plotted. The fluence, wavelength, pulse width, and thickness are 35 mJ/cm^2, 800 nm, 45 fs, and 50 μm, respectively. *Inset*: Zoom-in view of the first 500 fs of evolution in the carrier density, and the carrier temperature, together with the laser temporal profile [33]

implementing the formalism of Carpene [110]. An example calculation is given in Fig. 13.6. The carrier density, n, reaches its maximum concurrently with the laser pulse (near $\approx$ 135 fs) and changes very slowly compared to T_e. The carrier density decay should be slower than that of the electron thermal energy. The thermal energy propagates through carrier collisions, which is more efficient than actually displacing carriers, as they will likely collide before traveling very far. A rough estimate of the carrier density speed is given by $v_n \sim \alpha D \approx 450$ m/s. A similar estimate can be made for the electronic thermal propagation speed, $v_T \sim \alpha(\kappa_e / C_{e\ h}) \approx 2,500$ m/s, nearly an order of magnitude higher than that of the carrier density. Simply put, these example calculations show the difference between chemical and heat diffusion speeds. The quasi-Fermi level is plotted in Fig. 13.20, which has a slow decay following the long lifetime of the photoexcited carrier at the conduction band. Optical voltage generation occurs because the quasi-Fermi level of the substrate is raised due to the photoexcitation of (hot) carriers, in conjunction with a finite tunneling probability, which is directly attributable to its nanoscale thickness. While electronic injection to deep trap states at the interface will contribute to the overall transient surface voltage, as revealed from optical pump-probe studies [111, 112], its contribution is negligible on shorter time scales (hundreds of ps) due to the relatively low density of deep trap states compared to the surface charge density accumulated through tunneling. While deep trap states tend to have longer lifetimes ($\geq$ ns), nonetheless they can be discharged within the pump-probe cycle of 1 ms and hence are not observed here. Modeling the hot carrier tunneling dynamics across the SiO_2 barrier (see Fig. 13.18) can be made using the time-dependent Fermi function calculated from BTM in Eq. (13.11). The transmission matrix elements can be found by tunneling across a trapezoidal barrier given by [113].

The total accumulation of tunneled charge at time t,

$$Q_{\text{tun}}(t) = \int_0^t i(t')\, dt', \tag{13.21}$$

allows the voltage drop across the barrier, V_b to be deduced from the junction capacitance, C_J, given by

$$C_J = \frac{\varepsilon_0 \varepsilon_{ox} S}{L}, \tag{13.22}$$

where ε_{ox} is the relative permittivity of SiO_2. The voltage drop across the barrier due to optically-induced tunneling is then

$$V_b(t) = \frac{Q_{\text{tun}}(t)}{C_J}. \tag{13.23}$$

We can fit the data using parameters $L = 1.0$ nm, $\Phi_0 = 3.125$ eV, $\tilde{\sigma} = 2.75 \times 10^{13}$ cm^{-2}, $w = 0.35$ eV, $\varepsilon_F = -0.373$ eV, where L is the SiO_2 thickness, $\tilde{\sigma}$ is the surface charge density, w is the width of the surface charge states.

13.4.3 Correction of Photovoltage Beyond the Slab Model

While the general formalism deduced in Sect. 13.2.2 is limited to a smooth interface in which the transient surface field is modeled to have a slab geometry, the basic concept of diffractive voltammetry is applicable to different types of interfaces beyond the slab geometry. For different geometries, the timescales of the charge redistribution can likely be deduced correctly from $\Delta_B(t)$ with formulation from slab-formalism, whereas $V_s(t)$ deduced this way is merely an effective parameter, which is subjective to corrections from the shape factor and the boundary conditions. To treat diffractive voltammetry beyond an infinite slab model, we extend the slab-formalism described in Eq. (13.8) to consider refraction shift from finite size object with non-planar geometries. Under such circumstances, the location of z_1 are often not the same for incident and diffracted electron beams as shown in Fig. 13.21. The effective V_s in such asymmetric geometry can be modelled by applying the finite-size correction factors, $\Theta_i(\theta_i, \alpha)$ and $\Theta_o(\theta_o, \alpha)$ to separately describe the deviation of electron trajectory from the infinite slab model. This generalized picture, in which we can parameterize the finite interface structure with a nominal lateral length L and vertical height h with an aspect ratio parameter $\alpha \equiv h/L$, is depicted in Fig. 13.21. The corresponding $\Theta_i(\theta_i, \alpha)$ and $\Theta_o(\theta_o, \alpha)$ associated with the interface can be obtained numerically by ray tracing methods.

We can then rewrite Eqs. (13.2)–(13.8) in terms of $\Theta_i(\theta_i, \alpha)$ and $\Theta_o(\theta_o, \alpha)$:

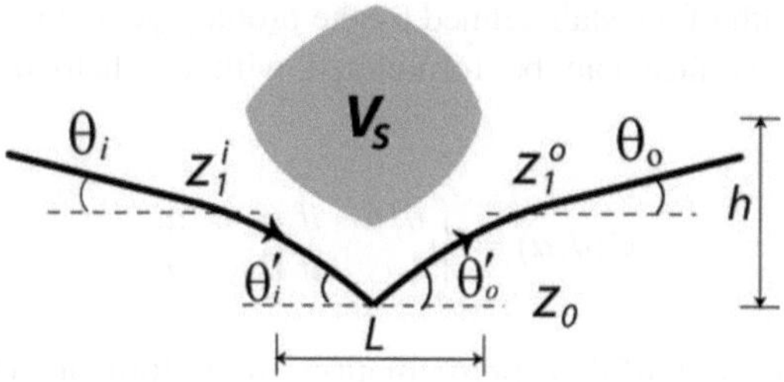

Fig. 13.21 General refraction geometry for grazing incident electron beam. The effective bending of the incident (exit) beam, caused by the local field associated with V_s in the nanostructures, differs depending on the entry (exit) point $z_1{}^i(z_1{}^o)$. Since the relative change of the transverse momentum remains small ($V_s \ll V_0$), the slab model can be extended to treat the general refraction effect considering an angle- and position-dependent correction factor (see text) [27]

$$\tan^2\theta_i' = \tan^2\theta_i + \frac{\Theta_i(\theta_i, \alpha)\chi^2}{\cos\theta_i}, \tag{13.24}$$

$$\tan^2\theta_o' = \tan^2\theta_o'' + \frac{\Theta_o(\theta_o, \alpha)\chi^2}{\cos\theta_o''}, \tag{13.25}$$

and

$$\Delta_B = \sin^{-1}\sqrt{\frac{D^2 - \Theta_o(\theta_o, \alpha)\chi}{1 + D^2}} - \theta_o, \tag{13.26}$$

where

$$D = \frac{\tan(\theta_i + \theta_o) - \sqrt{\tan^2\theta_i + \Theta_i(\theta_i, \alpha)\chi / \cos^2\theta_i}}{1 + \tan(\theta_i + \theta_o)\sqrt{\tan^2\theta_i + \Theta_i(\theta_i, \alpha)\chi / \cos^2\theta_i}}. \tag{13.27}$$

Here, $\chi \equiv V_s / V_0$. Applying small angle approximation allows the voltammetry formalism: $\chi(\Delta_B, \theta_i, \theta_o)$ to be established analytically, which is detailed in the Appendix of [27]. Below we give two examples to show how the correction terms can be implemented for dealing with more complex nanostructured interfaces beyond the infinite slab model.

13.4.4 Correction for a Finite Slab Geometry

The simplest extension of an infinite slab formalism is to consider the edge effects in a finite slab geometry. Using aspect ratio $\alpha = 2h/l$, where h and l are the height and width of the slab defined by the probing geometry, the photovoltage V_s at non-planar surface can be formulated with the help of the generalized correction factor:

$$\Theta(\theta, \alpha) = \begin{cases} \theta/\alpha & \text{if } \theta < \alpha, \\ 1 & \text{if } \theta \geq \alpha. \end{cases} \tag{13.28}$$

To assess the validity of this approximation, we perform an electron ray tracing simulation[3] with a setup shown in Fig. 13.22a to calculate $\Theta(\theta, \alpha)$, which accounts for the fringe fields not included in Eq. (13.28). We calculate a few instances of $\Theta(\theta, \alpha)$ as a function of α and show that the analytical expression described in Eq. (13.28) is a fairly good approximation for $\theta < 3.5°$, as illustrated in Fig. 13.22b. The finite-slab correction effectively suppresses the divergence of Δ_B as the diffracted electron beam approaches the shadow edge (θ_o approaches 0).

13.4.5 Correction for a Spherical Nanoparticle Decorated Interface

A frequently encountered molecular electronic interface involves replacing the top piece of the finite slab with a spherical nanoparticle and has a molecular contact between the nanoparticle and the substrate. In order to model the TSV in this spherical interface, the correction factor Θ is parameterized as a function of angle θ_i (θ_o). Ray tracing, as depicted in Fig. 13.22c, shows that the $\Theta(\theta, \alpha)$ (Fig. 13.22d) determined for this spherical interface is amazingly similar to that of the flat finite slab interface (Fig. 13.22b), if we re-define $\alpha = h/D$, where D is the diameter of the nanoparticle and h is the thickness of the molecular contact layer. For small angle diffraction ($\theta \leq 2°$), the dispersion in Θ due to position-dependent refraction effect can be ignored, as shown in the inset of Fig. 13.22d, calculated for the 20 nm nanoparticle interface.

[3] The electrostatics calculations were performed using the Charged Particle Toolkit software from Field Precision. The geometry was setup according to Fig. 13.22a, c. In both cases, a potential of $V_s = -5$ V was imposed. For nanoparticle simulations, the relative permittivity of the dielectric layer was set to 2.5. The Si(111) was treated as a grounded metal because of the high carrier density under photoexcitation. The potential far away from the nanoparticle was set to 0. Electrons of 30 keV were initialized at different launch angles (1°–5°) and positions along the height of the dielectric layer (1/3, 1/2, and 2/3). The slab model calculations were also carried out with this software, with varying capacitor slab separations.

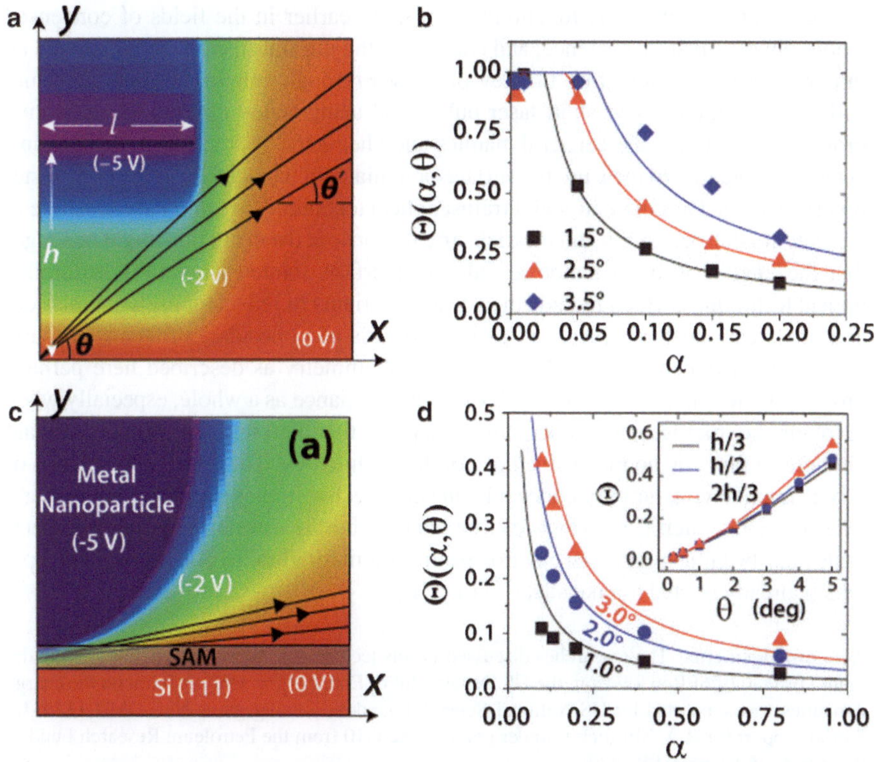

Fig. 13.22 (*Color online*) corrections for the boundary condition associated with finite-size geometries. (**a**) The calculated potential for a slab of length $2l$, fixed at -5 V, a distance h from a grounded base plane. (**b**) The correction factor calculated for various aspect ratios, $\alpha = 2h/l$. *Each curve* represents a different launch angle, θ_i. The *solid lines* are the predicted correction factors given by Eq. (13.28). (**c**) The calculated potential for the case of a 20 nm metallic nanoparticle, charged to -5 V, with a 1 nm thick dielectric layer ($\varepsilon = 2.5$) on the Si substrate. (**d**) The correction factor as a function of aspect ratio (SAM to nanoparticle) is calculated by performing ray-tracing in the field distribution described in (**a**). The *solid symbols* are the simulated data; the *lines* are from Eq. (13.28) using the definition of α from the text. *Inset*: The correction factor as a function of launch angle θ for three different launch positions along the SAM [27]

13.5 Summary and Outlook

In the last decade, we are witnessing a growing interest in developing nanotechnology based on more controlled experiments on nanodevices through integrating chemical synthesis, materials engineering, and the advanced probing techniques. It has become more and more clear as we approach the fundamental length scale of material building block (≈ 1 nm), the quantum nature of the electron is expressed more and more dominantly to influence the function of these devices. On the nanometer scale, the chemistry that is often only a concern at the molecular level is at play, the electron transport has become more ballistic in nature, and the surface states and interfacial contacts start to behave as functioning units. As earlier in the fields of condensed matter physics, material science, and chemistry, the ultrafast technique has played an important role in elucidating the key processes through creating a nonequilibrium collective excitation with short laser pulses and using state- and optically sensitive probes to investigate the carrier dynamics and chemistry. Nevertheless, there hasn't been a reliable tool to measure the surface potential of a nanodevice at the ps or faster timescale [24]. The conventional ultrafast optical techniques are uniquely sensitive to the state-to-state transition and the dielectric response at the depth that might be longer than the lengthscale of the nanomaterials and interface of interest here. While ultrafast optical techniques, which include the versatile variants of reflectance, photoemission, and nonlinear spectroscopies, can give the details of molecular, surface states, and carrier filling dynamics, the ultrafast photovoltammetry as described here perhaps provides better characterization of device's performance as a whole, especially when invoking the microscopic nanocircuit concept for the interpretation. What being laid out here so far is by no means the limit of this technology. The future applications of this diffractive voltammetry concept lie in site-selected studies on nanostructured and heterogeneous interfaces, which can be enabled by the development of nanometer scale high-brightness ultrafast electron beam system for ultrafast electron microscope, which already started to take shape [114–116].

Acknowledgements The researches discussed in this lecture note were largely supported under grant DE-FG02-06ER46309 from the US Department of Energy. The analytic work on the charge dynamics was supported by US National Science Foundation under grant NSF-DMR 1126343. Partial support for R.A. Murdick is under grant 45982-G10 from the Petroleum Research Fund of the American Chemical Society.

References

1. Gratzel, M.: J. Photochem. Photobiol. C Photochem. Rev. **4**, 145 (2003)
2. Haruta, M.: Catal. Today **36**, 153 (1997)
3. Ashkenasy, G., Cahen, D., Cohen, R., Shanzer, A., Vilan, A.: Acc. Chem. Res. **35**, 121 (2002)
4. Shipway, A., Katz, E., Willner, I.: Chem. Phys. Chem. **1**, 18 (2000)
5. Milliron, D.J., Hughes, S.M., Cui, Y., Manna, L., Li, J.B., Wang, L.W., Alivisatos, A.P.: Nature **430**, 190 (2004)
6. Luque, A., Marti, A., Nozik, A.J.: MRS Bull. **32**, 236 (2007)
7. Leschkies, K.S., Divakar, R., Basu, J., Enache-Pommer, E., Boercker, J.E., Carter, C.B., Kortshagen, U.R., Norris, D.J., Aydil, E.S.: Nano Lett. **7**, 1793 (2007)
8. Kongkanand, A., Tvrdy, K., Takechi, K., Kuno, M., Kamat, PV.: J. Am. Chem. Soc. **130**, 4007 (2008)
9. Huang, S.Y., Schlichthorl, G., Nozik, A.J., Gratzel, M., Frank, A.J.: J. Phys. Chem. B **101**, 2576 (1997)
10. Califano, M., Zunger, A., Franceschetti, A.: Appl. Phys. Lett. **84**, 2409 (2004)
11. Bach, U., Lupo, D., Comte, P., Moser, J.E., Weissortel, F., Salbeck, J., Spreitzer, H., Gratzel, M.: Nature **395**, 583 (1998)
12. Aroutiounian, V., Petrosyan, S., Khachatryan, A., Touryan, K.: J. Appl. Phys. **89**, 2268 (2001)

13. Adams, D.M., Brus, L., Chidsey, C.E.D., Creager, S., Creutz, C., Kagan, C.R., Kamat, P.V., Lieberman, M., Lindsay, S., Marcus, R.A., et al.: J. Phys. Chem. B **107**, 6668 (2003)
14. Aviram, A., Ratner, M.A.: Chem. Phys. Lett. **29**, 277 (1974)
15. Avouris, P., Chen, Z., Perebeinos, V.: Nat. Nanotechnol. **2**, 605 (2007)
16. Bezryadin, A., Dekker, C., Schmid, G.: Appl. Phys. Lett. **71**, 1273 (1997)
17. Marcus, R., Sutin, N.: Biochim. Biophys. Acta **811**, 265 (1985)
18. Chen, C.J.: Introduction to Scanning Tunneling Microscopy. Oxford University Press, Oxford (1993)
19. Datta, S.: Electronic Transport in Mescoscopic Systems. Cambridge University Press, Cambridge (1995)
20. Anderson, N., Lian, T.: Annu. Rev. Phys. Chem. **56**, 491 (2005)
21. Miller, R.J.D., McLendon, G.L., Nozik, A.J., Schmickler, W., Willig, F.: Surface Electron Transfer Processes. VCH Publishers, Inc., New York (1995)
22. Kubo, A., Onda, K., Petek, H., et al.: Nano Lett. **5**, 1123 (2005)
23. Peng, P., Milliron, D.J., Hughes, S.M., Johnson, J.C., Alivisatos, A.P., Sakyally, R.J.: Nano Lett. **5**, 1809 (2005)
24. Tanaka, S.-I.: J. Electron Spectros. Relat. Phenom. **185**, 152 (2012)
25. Murdick, R.A., Raman, R.K., Murooka, Y., Ruan, C.-Y.: Phys. Rev. B **77**, 245329 (2008)
26. Ruan, C.-Y., Murooka, Y., Raman, R.K., Murdick, R.A., Worhatch, R.J., Pell, A.: Micros. Microana. **15**, 323 (2009)
27. Chang, K., Murdick, R.A., Tao, Z., Han, T.-R.T., Ruan, C.-Y.: Mod. Phys. Lett. B **25**, 2099 (2011)
28. Ruan, C.-Y., Murooka, Y., Raman, R.K., Murdick, R.A.: Nano Lett. **7**, 1290 (2007)
29. Yamamoto, N., Spence, J.C.H.: Thin Solid Films **104**, 43 (1983)
30. Spence, J.C.H., Poon, H., Saldin, D.: Microsc. Microanal. **10**, 128 (2004)
31. Braun, W.: Applied RHEED. Reflection High Energy Electron Diffraction During Crystal Growth. Springer, Berlin (1999)
32. Wang, Z.L.: Reflection Electron Microscopy and Spectroscopy for Surface Analysis. Cambridge University Press, New York (1996)
33. Ryan A.M.: Ph.D. thesis, Michigan State University, East Lansing (2009)
34. Aswal, D., Lenfant, S., Guerin, D., Yakhmi, J., Vuillaume, D.: Anal. Chim. Acta **568**, 84 (2006)
35. Tian, W.D., Datta, S., Hong, S.H., Reifenberger, R., Henderson, J.I., Kubiak, C.P.: J. Chem. Phys. **109**, 2874 (1998)
36. Chen, J., Reed, M., Rawlett, A., Tour, J.: Science **286**, 1550 (1999)
37. Forrest, S.: Nature **428**, 911 (2004)
38. Ranuarez, J.C., Deen, M.J., Chen, C.-H.: Microelectron. Reliab. **46**, 1939 (2006)
39. Goldman, E.I., Kukharskaya, N.F., Zhdan, A.G.: Solid State Electron. **48**, 831 (2004)
40. Chakraborty, G., Chattopadhyay, S., Sarkar, C.K., Pramanik, C.: J. Appl. Phys. **101**, 024315 (2007)
41. Park, H., Zuo, J.M.: Appl. Phys. Lett. **94**, 251103 (2009)
42. Raman, R.K., Tao, Z., Han, T.-R., Ruan, C.-Y.: Appl. Phys. Lett. **95**, 181108 (2009)
43. Schfera, S., Lianga, W., Zewail, A.H.: Chem. Phys. Lett. **493**, 11 (2010)
44. Dadap, J.I., Xu, Z., Hu, X.F., Downer, M.C., Russell, N.M., Ekerdt, J.G., Aktsipetrov, O.A.: Phys. Rev. B **56**, 13367 (1997)
45. Malevich, V.L., Adomavicius, R., Krotkus, A.: C. R. Phys. **9**, 130 (2008)
46. Alay, J.L., Hirose, M.: J. Appl. Phys. **81**, 1606 (1997)
47. Hattori, T., Takahashi, K., Seman, M.B., Nohira, H., Hirose, K., Kamakura, N., Takata, Y., Shin, S., Kobayashi, K.: Appl. Surf. Sci. **212**, 547 (2003)
48. Marsi, M., Belkhou, R., Grupp, C., Panaccione, G., Taleb-Ibrahimi, A., Nahon, L., Garzella, D., Nutarelli, D., Renault, E., Roux, R., Couprie, M.E., Billardon, M.: Phys. Rev. B **61**, R5070 (2000)
49. Halas, N.J., Bokor, J.: Phys. Rev. Lett. **62**, 1679 (1989)

50. Bokor, J., Halas, N.J.: IEEE J. Quantum Electron. **25**, 2550 (1989)
51. Ruan, C.-Y., Vigliotti, F., Lobastov, V.A., Chen, S., Zewail, A.H.: Proc. Natl. Acad. Sxi. USA **101**, 1123 (2004)
52. Sjodin, T., Petek, H., Dai, H.L.: Phys. Rev. Lett. **81**, 5664 (1998)
53. Sokolowski-Tinten, K., Von Der Linde, D.: Phys. Rev. B **61**, 2643 (2000)
54. Hebeisen, C.T., Sciaini, G., Harb, M., Ernstorfer, R., Kruglik, S.G., Miller, R.J.D.: Phys. Rev. B **78**, 081403(R) (2008)
55. Sundaram, S.K., Mazur, E.: Nat. Mater. **1**, 217 (2002)
56. Harb, M., Ernstorfer, R., Hebeisen, C.T., Sciaini, G., Peng, W., Dartigalongue, T., Eriksson, M.A., Lagally, M.G., Kruglik, S.G., Miller, R.J.D.: Phys. Rev. Lett. **100**, 155504 (2008)
57. Kern, W., Poutinen, D.A.: RCA Rev. **31**, 187 (1970)
58. Jellison, G.E., Modine, F.A.: Appl. Phys. Lett. **41**, 180 (1982)
59. Chen, J.K., Tzou, D.Y., Beraun, J.E.: Int. J. Heat Mass Transf. **48**, 501 (2005)
60. Cohen, R., Zenou, N., Cahen, D., Yitzchaik, S.: Chem. Phys. Lett. **279**, 270 (1997)
61. Lee, C.H., Chang, R.K., Bloembergen, N.: Phys. Rev. Lett. **18**, 167 (1967)
62. Chen, C.K., Heinz, T.F., Ricard, D., Shen, Y.R.: Phys. Rev. B **27**, 1965 (1983)
63. Mihaychuk, J.G., Bloch, J., Liu, Y., van Driel, H.M.: Opt. Lett. **20**, 2063 (1995)
64. Aktsipetrov, O.A., Fedyanin, A.A., Mishina, E.D., Rubtsov, A.N., van Hasselt, C.W., Devillers, M.A.C., Rasing, T.: Phys. Rev. B **54**, 1825 (1996)
65. Baldelli, S., Eppler, A., Anderson, E., Shen, Y., Somorjai, G.: J. Chem. Phys. **113**, 5432 (2000)
66. Mihaychuk, J.G., Shamir, N., van Driel, H.M.: Phys. Rev. B **59**, 2164 (1999)
67. Scheidt, T., Rohwer, E.G., Neethling, P., von Bergmann, H.M., Stafast, H.: J. Appl. Phys. **104**, 083712 (2008)
68. Jun, B., Schrimpf, R., Fleetwood, D., White, Y., Pasternak, R., Rashkeev, S., Brunier, F., Bresson, N., Fouillat, M., Cristoloveanu, S., et al.: IEEE Trans. Nucl. Sci. **51**, 3231 (2004)
69. Tolk, N.H., Alles, M.L., Pasternak, R., Lu, X., Schrimpf, R.D., Fleetwood, D.M., Dolan, R.P., Standley, R.W.: Microelectron. Eng. **84**, 2089 (2007)
70. Paul, S., Rotenberg, N., van Driel, H.M.: Appl. Phys. Lett. **93**, 131102 (2008)
71. Klein, D.L., Roth, R., Lim, A.K.L., Alivisatos, A.P., McEuen, P.L.: Nature **389**, 699 (1997)
72. Xia, Y.N., Yang, P.D., Sun, Y.G., Wu, Y.Y., Mayers, B., Gates, B., Yin, Y.D., Kim, F., Yan, Y.Q.: Adv. Mater. **15**, 353 (2003)
73. Yin, Y., Alivisatos, A.P.: Nature **437**, 664 (2005)
74. Schaller, R.D., Agranovich, V.M., Klimov, V.I.: Nat. Phys. **1**, 189 (2005)
75. Raffaelle, R.P., Castro, S.L., Hepp, A.F., Bailey, S.G.: Prog. Photovolt. **10**, 433 (2002)
76. Nozik, A.J.: Phys. E **14**, 115 (2002)
77. Klimov, V.I., Mikhailovsky, A.A., Xu, S., Malko, A., Hollingsworth, J.A., Leatherdale, C.A., Eisler, H.J., Bawendi, M.G.: Science **290**, 314 (2000)
78. Duan, X.F., Huang, Y., Agarwal, R., Lieber, C.M.: Nature **421**, 241 (2003)
79. Pitsillides, C.M., Joe, E.K., Wei, X.B., Anderson, R.R., Lin, C.P.: Biophys. J. **84**, 4023 (2003)
80. Schwartzberg, A.M., Zhang, J.Z.: J. Phys. Chem. C **112**, 10323 (2008)
81. Li, G.H., Gray, K.A.: Chem. Phys. **339**, 173 (2007)
82. Neacsu, C.C., Dreyer, J., Behr, N., Raschke, M.B.: Phys. Rev. B **73**, 193406 (2006)
83. Novotny, L.: Phys. Today 47 (2011)
84. Catchpole, K.R., Polman, A.: Appl. Phys. Lett. **93**, 191113 (2008)
85. Awazu, K., Fujimoki, M., Rockstuhl, C., Tominaga, J., Murakami, H., Ohki, Y., Yoshida, N., Watanabe, T.: J. Am. Chem. Soc. **130**, 1676 (2008)
86. Sato, T., Ahmed, H., Brown, D., Johnson, B.: J. Appl. Phys. **82**, 696 (1997)
87. Widawsky, J.R., Darancet, P., Neaton, J.B., Venkataraman, L.: Nano Lett. **12**, 354 (2011)
88. Wang, Z., Carter, J.A., Lagutchev, A., Koh, Y.K., Seong, N.-H., Cahill, D.G., Dlott, D.D.: Science **317**, 787 (2007)
89. Nie, S.M., Emery, S.R.: Science **275**, 1102 (1997)
90. Barnes, W.L., Dereux, A., Ebbesen, T.W.: Nature **424**, 824 (2003)

91. Linic, S., Christopher, P., Ingram, D.B.: Nat. Mater. **10**, 911 (2011)
92. Link, S., El-Sayed, M.A.: J. Phys. Chem. B **103**, 4212 (1999)
93. Han, T.-R.T., Yuan, F., Szymanski, T.M., Ruan, C.-Y.: submitted for publication
94. Mahmoud, M.A., Chamanzar, M., Adibi, A., El-Sayed, M.A.: J. Am. Chem. Soc. **134**, 6434 (2012)
95. Schertz, F., Schmelzeisen, M., Mohammadi, R., Kreiter, M., Elmers, H.-J., Schonhense, G.: Nano. Lett. **12**, 1885 (2012)
96. Chen, H., Shao, L., Ming, T., Woo, K.C., Man, Y.C., Wang, J., Lin, H.-Q.: ACS Nano **8**, 6754 (2011)
97. Harrison, W.A.: Phys. Rev. **123**, 85 (1961)
98. Buttiker, M., Landauer, R.: Phys. Rev. Lett. **49**, 1739 (1982)
99. Miranda, E., Sune, J.: Microelectron. Reliab. **44**, 1 (2004)
100. Schenk, A., Heiser, G.: J. Appl. Phys. **81**, 7900 (1997)
101. Cai, J., Sah, C.T.: J. Appl. Phys. **89**, 2272 (2001)
102. Ranuarez, J.C., Deen, M.J., Chen, C.-H.: Microelectron. Reliab. **46**, 1939 (2006)
103. Kaganov, M.I., Lifshitz, I.M., Tanatarov, L.V.: Sov. Phys. JETP USSR **4**, 173 (1957)
104. Anisimov, S.I., Kapeliovich, B., Perelman, T.L.: Teoreticheskoi Fiziki **66**, 776 (1974)
105. Qiu, T.Q., Tien, C.L.: J. Heat Transf. Trans. ASME **115**, 835 (1993)
106. Chen, J.K., Beraun, J.E.: Numer. Heat Transf. Part A Appl. **40**, 1 (2001)
107. Allen, P.B.: Phys. Rev. Lett. **59**, 1460 (1987)
108. Tzou, D.Y.: Macro- to Microscale Heat Transfer: The Lagging Behavior. Taylor & Francis, Washington, DC (1996)
109. Chen, J.K., Beraun, J.E., Tham, C.L.: Numer. Heat Transf. Part A Appl. **44**, 705 (2003)
110. Carpene, E.: Phys. Rev. B **74**, 024301 (2006)
111. Aktsipetrov, O.A., Fedyanin, A.A., Melnikov, A.V., Mishina, E.D., Rubtsov, A.N., Anderson, M.H., Wilson, P.T., Ter Beek, H., Hu, X.F., Dadap, J.I., Downer, M.C.: Phys. Rev. B **60**, 8924 (1999)
112. Mihaychuk, J.G., Shamir, N., van Driel, H.M.: Phys. Rev. B **59**, 2164 (1999)
113. Khairurrijal, Noor, F.A., Sukirno: Solid State Electron. **49**, 923 (2005)
114. Zewail, A.H.: Science **328**, 187 (2010)
115. Kim, J.S., LaGrange, T., Reed, B.W., Taheri, M.L., Armstrong, M.R., King, W.E., Browning, N.D., Campbell, G.H.: Science **321**, 1472 (2008)
116. Tao, Z., Zhang, H., Duxbury, P.M., Berz, M., Ruan, C.-Y.: J. Appl. Phys. **111**, 044316 (2012)

Chapter 14
Photonics and Plasmonics for Enhanced Photovoltaic Performance

Yunlu Xu, Joseph Murray, and Jeremy N. Munday

Abstract Nearly all photovoltaic devices require a compromise between light absorption and carrier collection. For planar structures, a thick film is sought for light absorption, while a thin film is sought for efficient carrier extraction. One way to circumvent these contrary requirements is through the use of photonic and plasmonic structures. These structures allow for strong concentration of light into small volumes—simultaneously attaining large absorption enhancement and efficient carrier collection. The improved optical response is obtained by (1) increasing the optical path length and reducing the reflection through particle scattering, (2) enhancing the local field strength through the excitation of localized resonances, or (3) waveguiding. While these concepts are important to all light collection devices, they are of particular interest to quantum dot solar cells, where the need for thin structures that can absorb nearly all of the incident light is a critical design criterion.

14.1 Introduction

While traditional photovoltaics have relied on relatively thick semiconductor layers to ensure maximum light absorption from the sun, the latest devices rely on thin film structures either out of necessity (due to small carrier diffusion lengths) or for market specific applications (e.g., cost reduction, flexibility, and weight). However,

---

Y. Xu • J. Murray
Department of Electrical and Computer Engineering, University of Maryland, College Park, MD 20742, USA

J.N. Munday (✉)
Department of Electrical and Computer Engineering, University of Maryland, College Park, MD 20742, USA

IREAP, Energy Research Facility, Room 1202L, College Park, MD 20742, USA
e-mail: jnmunday@umd.edu

J. Wu and Z.M. Wang (eds.), *Quantum Dot Solar Cells*, Lecture Notes in Nanoscale Science and Technology 15, DOI 10.1007/978-1-4614-8148-5_14,

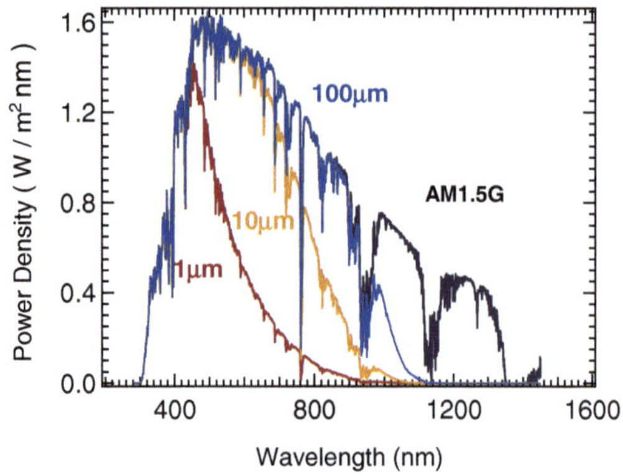

Fig. 14.1 Solar spectrum (AM 1.5G) and representative absorption for thick and thin Si devices. Short wavelength light is absorbed in all structures; however, thin slabs are ineffective at absorbing long wavelength photons

a reduction of the semiconductor's thickness also reduces the amount of incident light that can be absorbed. This leads to a trade-off between absorption, carrier collection, flexibility, etc.

For planar photovoltaic devices, light absorption is exponential. The intensity of the incident illumination decreases from the front surface of the device as:

$$I = I_0 e^{-\alpha L} \tag{14.1}$$

where I_0 is the incident intensity that enters the material, L is optical path length (i.e., the depth traveled within the material), and α is the absorption coefficient which is related to the wavelength λ and imaginary part of the refractive index of the material κ by:

$$\alpha = \frac{4\pi\kappa}{\lambda} \tag{14.2}$$

Similarly, the fraction of the incident power (P_0) that is absorbed (P_{abs}) is given by $A(\alpha, L) = P_{abs}/P_0 = (1 - e^{-\alpha L})$. κ, and hence α, generally tends toward zero at the semiconductor bandgap, where the material becomes transparent. Thus, the short wavelengths are more easily absorbed near the front surface, while longer wavelengths (near the band-edge) are absorbed deeper within the cell. Figure 14.1 shows the power absorbed per unit area for a single optical pass through a silicon slab with thicknesses of 1, 10, and 100 μm. While the 100 μm slab absorbs 85 % of the above bandgap energy photons, the 1 μm slab only absorbs 36 %, making additional light trapping structures necessary to improve the absorption.

For high collection efficiency, the cell generally needs to be significantly thinner than the minority carrier diffusion length, L_D. This condition can be easily met in thin film devices; however, for thick devices, bulk recombination results in reduced carrier collection. Figure 14.2 shows the combined effects of light absorption and carrier collection. For thin devices, nearly all generated carriers are collected; however, a significant fraction of the long wavelength incident solar radiation is not absorbed. Conversely, for thick devices, nearly all of the incident light is

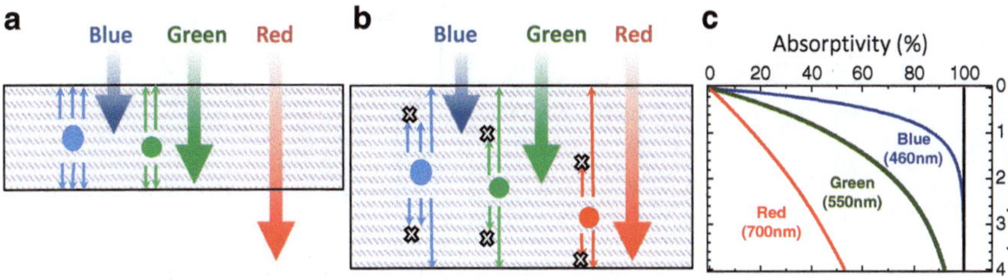

Fig. 14.2 Schematic of absorption depth and carrier collection for different wavelengths of incident light and film thicknesses. (**a**) A thin slab efficiently collects generated carriers; however, long wavelength photons are not absorbed resulting in reduced current. (**b**) A thick device absorbs nearly all of the light; however, the long path for collection increases the likelihood of carrier recombination without collection. (**c**) Absorption profile for different wavelengths of light

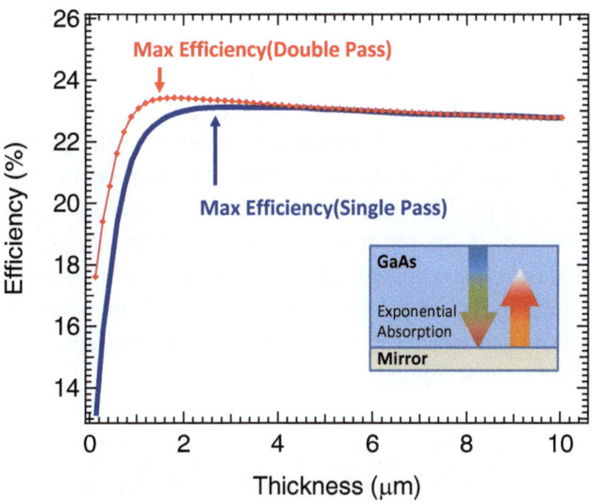

Fig. 14.3 Efficiency as a function of thickness for single and double pass absorption. The maximum efficiency is a trade-off between high absorption efficiency (thick device) and high carrier collection efficiency (thin device). The use of a mirror back reflector allows a device to be half as thick and still absorbs the same amount of light

absorbed; however, carrier collection is impeded by the large distance the minority carriers must travel without recombining with bulk trap states to be collected at the leads. For a given material, this trade-off leads to an optimum thickness for maximum efficiency.

For GaAs, this trade-off results in a device thickness of ~3 μm (see Fig. 14.3). If a reflective back surface (e.g., a mirror) is added, the device can be half as thick and still absorbs the same amount of light. By reducing the thickness, carrier collection is improved without suppressing the absorption, which leads to a thinner device that is more efficient.

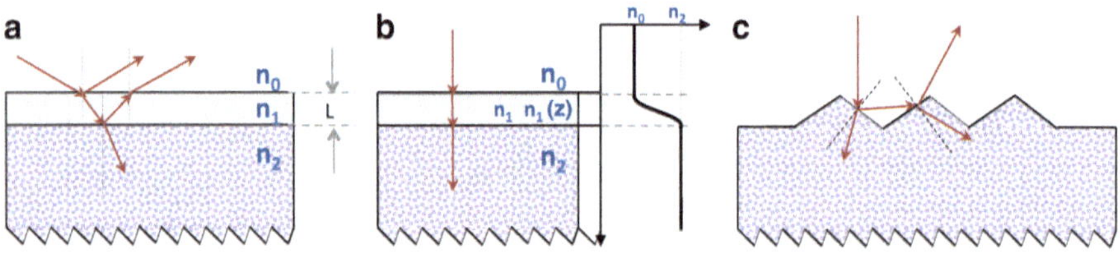

Fig. 14.4 Typical antireflection techniques for macroscopic solar cells: (**a**) destructive interference of reflected plane waves, (**b**) graded-index structures (where $n_2 > n_1 > n_0$), (**c**) pyramidal surface texturing

From a practical point of view, we can first pick a semiconductor thickness that allows for easy carrier collection, and then determine the appropriate optical design to allow for sufficient light absorption. Traditional light management is achieved through antireflection coatings (ARCs) and microscale texturing (Fig. 14.4). For planar structures, ARCs are generally constructed by adding one or two additional (non-absorbing) layers to the top of the device to help couple the incident planewaves into the semiconductor. The thickness and index of refraction of the layers are chosen so that there is destructive interference for the reflected wave. To minimize the reflection, two conditions must be met. First the phase of the reflected light from the surfaces should differ by 180°. For normal incidence this occurs when the thickness of the film is equal to a quarter wavelength of the incident light: $h = \lambda/(4n_1)$, where n_1 is the refractive index of the film. The second condition is that the amplitudes of the reflected waves are equal. These conditions specify that the index of refraction for the ARC should be: $n_1 = \sqrt{n_0\ n_2}$, where n_0 and n_2 are the refractive indices of free space ($n_0 = 1$) and the substrate, respectively. The general expression for the reflectivity using a single-layer ARC is [1]:

$$R = \frac{(n_0 - n_2)^2 + (n_0 n_2/n_1 - n_1)^2 \tan^2\phi}{(n_0 + n_2)^2 + (n_0 n_2/n_1 + n_1)^2 \tan^2\phi} \tag{14.3}$$

where $\phi = 2\pi n_1 h \cos\,\theta_i/\lambda$ is the phase shift of the light incident with an angle θ_i from the normal, which has a free space wavelength of λ. Because the reflectivity is wavelength dependent, a compromise must be reached that allows for the maximum amount of light to enter the cell over a broad range of wavelengths (see Fig. 14.5). Multilayer ARCs can also be used to allow for reduced reflection over a larger bandwidth.

In order to achieve broadband antireflection properties, the index of refraction of the ARC film can be continuously varied so that there is never an abrupt change in the refractive index, as shown by Lord Rayleigh mathematically in the 1880s [2]. These structures, referred to as graded-index coatings, have been demonstrated experimentally by partial filling the ARC layer with air [3, 4], using oblique-angle deposition of thin films [5] or using nano-structured cones or pyramids that mimic the functionality of moth eyes [6].

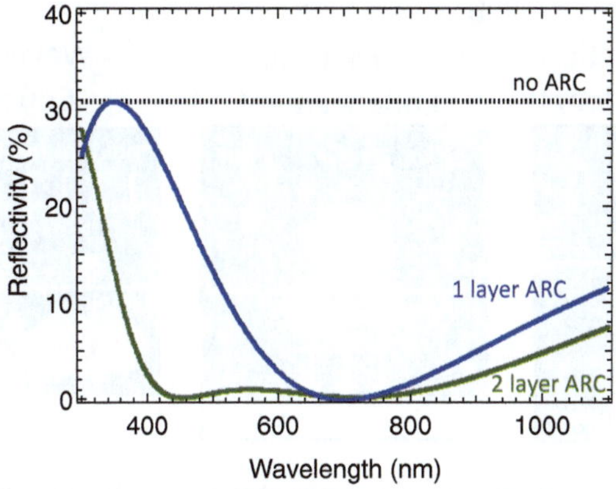

Fig. 14.5 Reflectivity as a function of wavelength from a surface with $n_2 = 3.5$ using no ARC, a single-layer ARC, and a double-layer ARC optimized for $\lambda = 700$ nm

In addition to interference-based thin film ARCs, which are used for nearly all commercial solar cells, pyramidal structuring is also common in silicon (Fig. 14.4). Such structures can be easily created through anisotropic etching of crystalline Si. Texturing typically results in pyramidal structures (~10s μm), which allow incident light multiple entrance attempts at the surface as well as increased path lengths (Fig. 14.4).

As solar cell device thicknesses become comparable to the wavelength of the incident light, new optical structures and scattering processes must be designed to function on the subwavelength scale. The ability of a particle to scatter incident light is strongly dependent upon the index of refraction contrast between the particle and its surrounding, as discussed in Sect. 14.2.1. For this reason metals often make excellent scattering objects, so long as the ohmic loss within the metal is minimized. There are three main mechanisms for increasing light absorption within a semiconductor device using metal nanoparticles (see Fig. 14.6). First, the nanoparticles can be used to increase forward scattering and hence reduce reflection (Fig. 14.6a). Second, metal particles or gratings can be used to concentrate the incident light into a small volume, acting as a subwavelength antenna (Fig. 14.6b). Third, the metal nanostructures can be used to couple the incident free-space light into waveguide modes of the structure, which in turn dramatically increases the optical path length (Fig. 14.6c).

Metallic structures are particularly useful for concentrating and trapping light because of the excitation of surface plasmons [7], i.e. charge density excitations that result from the coupling of the incident electromagnetic radiation and the surface charge density of the metal (Fig. 14.7). For nano-structures, this coupling can result in localized excitations with field enhancements of many orders of magnitude. When light is incident on a planar structure, the excitation can result in a surface plasmon polariton (SPP), which may be capable of traveling several microns along the surface of the metal. Both localized and propagating modes can be tailored to

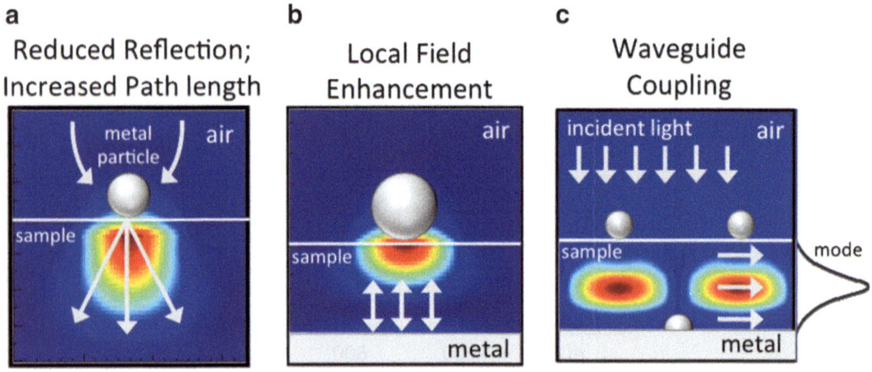

Fig. 14.6 Plasmonic nanoparticles can improve photovoltaics by (**a**) reducing reflection and increasing path length, (**b**) creating high local fields, or (**c**) coupling incident light into waveguide modes of the structure. Background shows the relative E-field intensity for the different situations

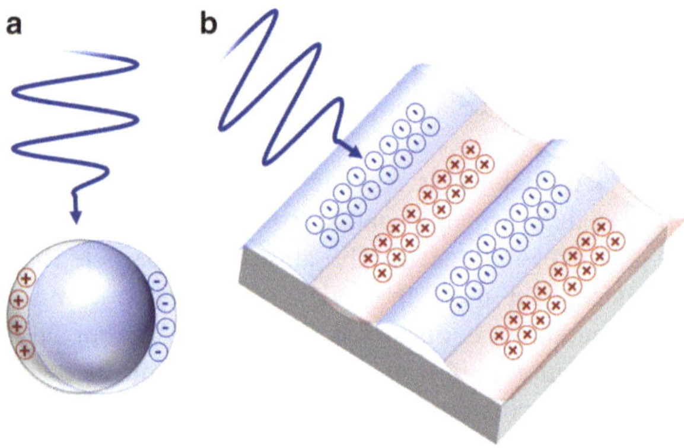

Fig. 14.7 Excitation of surface plasmons. (**a**) The incident light is coupled to the free electrons in a metal nanoparticle resulting in excitation of a localized surface plasmon resonance. (**b**) When light is incident on a planar structure, the coupling results in a propagating surface wave called a surface plasmon polariton. The *circles* with "−" and "+" represent separated charges

meet the specific frequency range required for a particular photovoltaic application through tuning either the geometry of the structure or the materials involved (choice of metal and surrounding dielectrics). This tunability will be explored in further detail in the following sections.

During the past several years, there has been a great deal of research into the use of surface plasmons for increasing the light absorption in thin film photovoltaic devices [8–10]. Early studies were conducted by Stuart and Hall in the mid-1990s and showed nearly a 20× enhancement of the photocurrent for long wavelength incident light in a Si photodetector due to metallic nanoparticles that were formed on top of the device [11]. Since these first studies, improved current generation has been found for a variety of scattering structures placed on top of, within, or on the bottom of photovoltaic devices. In addition to Si solar cells, enhancements have been achieved for a variety of structures and materials including quantum dot/well devices, GaAs solar cells, polymer solar cells, and dye-sensitized solar cells.

Quantum dot solar cells are particularly well suited for plasmonic enhancement due to the thinness of most devices. Excitation of a surface plasmon results in high field intensities near the metal interface, which decay with distance from the metal surface. With appropriate incoupling to a surface plasmon mode, even a monolayer of quantum dots can lead to nearly complete optical absorption. Further, the plasmon resonance can be tuned through geometry and material choice in order to overlap with the absorption spectrum of the quantum dots. In the next section we will discuss the different types of optical modes, the scattering properties, and the ways to tune the resonance.

14.2 Photonics and Plasmonics

In order to increase light absorption within the semiconductor, we must increase the amount of time that the light spends within the active layer. The increased time that the light spends within the semiconductor improves the likelihood of photon absorption and ultimately leads to increased carrier collection and photocurrent. When light is scattered into a waveguide mode of the structure, the optical path length is effectively increased, because the vertical propagation has been converted into horizontal propagation, and the optical path can be many times longer than the thickness of the semiconductor layer. Depending on the boundary conditions at the interfaces, either photonic or plasmonic modes can be excited. Photonic modes are the result of light guiding based on a contrast in the index of refraction as used in fiber optic waveguides and can exist as either transverse electric (TE) or transverse magnetic (TM) modes (as described below). SPP modes are electromagnetic surface waves coupled to oscillations of conduction electrons at the interface of a metal and a dielectric and generally only exist as TM modes in planer structures. In addition to the SPP modes, non-propagating surface plasmons (SP) can be excited in metallic nano-structures, such as nanoparticles or grooves, and lead to various resonance and scattering effects. The next two sections will discuss light scattering by particles and waveguide modes of planar structures.

14.2.1 Scattering

Subwavelength particles can help a photovoltaic device absorb more of the incident spectrum by increasing the amount of light that is incoupled into the device and also by increasing the optical path length of light that enters the cell. For a bulk material, the Beer–Lambert law describes the exponential absorption of photons within the slab. As discussed above, the intensity decays as $I = I_0 e^{-\alpha L}$. Once a material has been chosen for the photovoltaic device, α is fixed. Thus, to increase the absorption within the solar cell, we need to increase the optical path length. As we shall see in this section, nanoparticles can make for excellent scatterers, which can increase the optical path length and lead to enhanced absorption within the solar cell.

Optical scattering is the process by which normally incident light acquires components perpendicular to its original path. When light is transmitted through an optically inhomogeneous material, energy is dispersed throughout the complete solid angle of 4π and leads to the existence of energy propagation in a direction perpendicular to the original propagation direction. As light encounters a small particle, the atomic orbitals of the molecules in the scattering particle tend to oscillate at the frequency of the incident light—inducing a dipole moment. This dipole moment serves as a secondary source and radiates in all directions. This re-radiated light can be either in phase with the incident beam or out of phase depending on the frequency of the incident light and the resonant frequency of the scatterer. When a scattering particle is placed on top of a solar cell, the normally incident light is scattered at an angle θ from the normal. The path length is increased from L to $\frac{L}{\cos\theta}$, causing the intensity to decrease more rapidly with depth due to increased absorption at a given depth. In this way, scattering particles can increase the optical path length and improve absorption within the solar cell.

Light scattering is fully described by Maxwell's equations; however, there are two regimes where approximate solutions can be obtained and are quite useful: Rayleigh scattering (elastic scattering by particles much smaller than the wavelength of the light) and Mie scattering (solutions represented in the form of an infinite series, which can describe scattering when the wavelength of light is comparable to the size of the particle). In order to consider when these approximate solutions can be used, we introduce a size parameter:

$$\alpha_x = \frac{\pi D_p}{\lambda} \tag{14.4}$$

where D_p is the diameter of the scatterer. This parameter describes the relationship between the size of the scatterer and the wavelength of the incident light. Three regimes are possible:

1. If $\alpha_x < 0.1$, the scattering can be described in the domain of Rayleigh scattering, where larger α_x leads to larger scattering intensities.
2. If $0.1 < \alpha_x < 10$, the dependence of scattering intensity on incident wavelength weakens with increasing of α_x; however, maxima and minima occur in the scattering intensity resulting from excitations of resonances within the particle.
3. If $\alpha_x > 10$, scattering intensity becomes less dependent on the incident wavelength, and eventually the realm of geometric optics is valid.

Although we introduced Rayleigh and Mie scattering above, it doesn't necessarily mean that the scattering can be easily labeled as either Rayleigh or Mie. Both of these are approximate solutions that are many times convenient for obtaining analytical solutions; however, in some situations it is necessary to solve Maxwell's equations numerically to determine the appropriate scattering properties. In the following subsections, we solve Maxwell's equations numerically using the method of Finite Difference Time Domain (FDTD). For further details on analytical models for light scattering from particles, see [12].

14.2.1.1 Figures of Merit for Light Scattering

In order to quantify a particle's ability to scatter light, several figures of merit need to be defined. In this section, we describe the various scattering cross sections, scattering efficiencies, and scattering fractions necessary to describe light scattering by particles.

In the realm of geometric optics, we can define a scattering cross section in terms of the geometric size of the particle; however, for subwavelength particle sizes the scattering cross section can often be much larger than the particle's geometric size due to the wave nature of light. The *extinction cross section* of a particle is the sum of two *hypothetical* areas: the effective area over which the particle acts like a scatterer and the effective area over which it acts like an absorber. Because the absorption and scattering properties depend on the wavelength-dependent permittivity, the cross sections are similarly a function of wavelength. The extinction cross section is thus defined as:

$$\sigma_{\text{ext}}(\lambda) = \sigma_{\text{scat}}(\lambda) + \sigma_{\text{abs}}(\lambda) \tag{14.5}$$

where $\sigma_{\text{scat}}(\lambda) = P_{\text{scat}}(\lambda)/I_{\text{source}}(\lambda)$ is the *scattering cross section* and $\sigma_{\text{abs}}(\lambda) = P_{\text{abs}}(\lambda)/I_{\text{source}}(\lambda)$ is the *absorption cross section*. $P_{\text{scat}}(\lambda)$ is the power scattered, $P_{\text{abs}}(\lambda)$ is the power absorbed, and $I_{\text{source}}(\lambda)$ is the intensity of the incident source. Typically, these cross sections are normalized to the geometric cross section of the particle, σ_{geom}, to obtain the *normalized scattering cross sections*:

$$Q_{\text{scat}}(\lambda) = \frac{P_{\text{scat}}(\lambda)/I_{\text{source}}(\lambda)}{\sigma_{\text{geom}}} \tag{14.6}$$

$$Q_{\text{abs}}(\lambda) = \frac{P_{\text{abs}}(\lambda)/I_{\text{source}}(\lambda)}{\sigma_{\text{geom}}} \tag{14.7}$$

If the particle is used to efficiently scatter light into a solar cell, we want a large scattering cross section and a small absorption cross section, because the absorbed energy in the scatterer is converted into heat instead of exciting electron–hole pairs in the surrounding semiconductor and thus leads to the loss of energy. An important figure of merit to describe the fraction of light scattered to the total extinction is the *scattering efficiency*, which is given by the ratio of the scattering cross section to the extinction cross section:

$$\eta = \frac{\sigma_{\text{scat}}}{\sigma_{\text{ext}}} = \frac{\sigma_{\text{scat}}}{\sigma_{\text{scat}} + \sigma_{\text{abs}}} \tag{14.8}$$

Thus, a higher scattering efficiency leads to better performance by more efficiently scattering the light.

In some cases, the cross sections can be described analytically. One example is a plane wave incident on a sphere. For simplicity, we treat the problem in the quasistatic limit, assuming that the sphere is much smaller than the wavelength of incident light so that the field on the particle is uniform. The scattering and absorption cross sections are:

$$\sigma_{\text{scat}} = \frac{1}{6\pi}\left(\frac{2\pi}{\lambda}\right)^4 \left|\alpha_{\text{sp}}\right|^2 \tag{14.9}$$

$$\sigma_{\text{abs}} = \frac{2\pi}{\lambda}\text{Im}\left[\alpha_{\text{sp}}\right] \tag{14.10}$$

where α_{sp} is the polarizability of the sphere:

$$\alpha_{\text{sp}} = 4\pi r^3 \frac{\varepsilon_{\text{m}} - \varepsilon_{\text{s}}}{\varepsilon_{\text{m}} + 2\varepsilon_{\text{s}}} \tag{14.11}$$

and ε_{s} and ε_{m} are the permittivities of the surrounding material and of the material constituting the sphere, respectively.

While having a large scattering cross section ensures that a large fraction of the light is scattered, only light that is scattered in the forward direction will be useful if the particles are placed on top of the solar cell (likewise, backward scattering is important for photovoltaic applications where the particles are placed on the backside of the device). In order to evaluate the portion of the total scattered energy that goes forward into the cell, we introduce another important figure of merit: the *forward scattering fraction* $f_{\text{forward}}(\lambda)$. This parameter is calculated by dividing the forward scattered power $P_{\text{forward\_scatt}}(\lambda)$ by total scattered power $P_{\text{scatt}}(\lambda)$:

$$f_{\text{forward}}(\lambda) = \frac{P_{\text{forward\_scatt}}(\lambda)}{P_{\text{scatt}}(\lambda)} \tag{14.12}$$

For a very small spherical particle, the scattering behavior is similar to that of a dipole emitter. In fact, because the scattering phenomenon is the result of re-emission of atomic dipole moments, the forward scattering fraction is nearly the same for these two cases. Figure 14.8 shows the forward scattering fraction both for a 50 nm particle sitting on the surface of a substrate ($n = 1.4$) and for a dipole

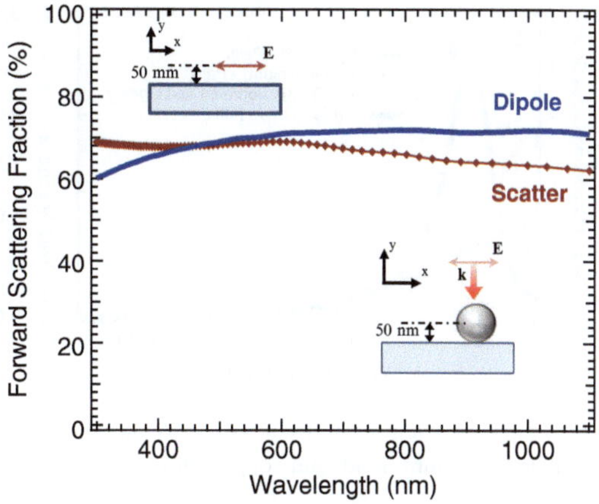

Fig. 14.8 The percentage of forward scattering as a function of wavelength for a particle above a surface (*lower inset*) and for forward emission from a radiating dipole above the surface (*upper inset*). $n = 1.4$ for the substrate

source 50 nm above a substrate. Both structures result in approximately 60–70 % of the incident light scattering in the forward direction, indicating that a dipole source can be used to roughly estimate the forward scattering properties of the particle.

14.2.1.2 Scattering Parameters

A particle's scattering properties depend upon its composition, shape, size, surroundings, and its distance from the surface. In this section, we explore how these variables affect the particle's scattering properties.

Particle Material

Scattering particles can be made from metals, dielectrics, or semiconductors. A large index contrast between the particle and the surrounding environment can increase the optical scattering and suggests that materials whose indices vary significantly from 1 are of interest. In particular, many semiconductors have indices of refraction between 3 and 4 in the visible and thus can be used effectively as Mie scatterers [12, 13]. Metals, on the other hand, have a permittivity that can be negative. For a spherical metal particle with permittivity ε_m embedded in a medium with ε_s, a dipolar surface plasmon resonance can be excited when $\varepsilon_m = -2\varepsilon_s$. This corresponds to a maximum in the polarizability (Eq. 14.11) and hence scattering cross section. Given this unique situation achievable with metals, it is worthwhile to discuss a few typical metals such as silver, gold, copper, and aluminum in more detail.

Using the Drude model, the dielectric permittivity can be expressed as:

$$\varepsilon = 1 - \frac{{\omega_p}^2}{\omega^2 + i\omega\Gamma_p} \quad (14.13)$$

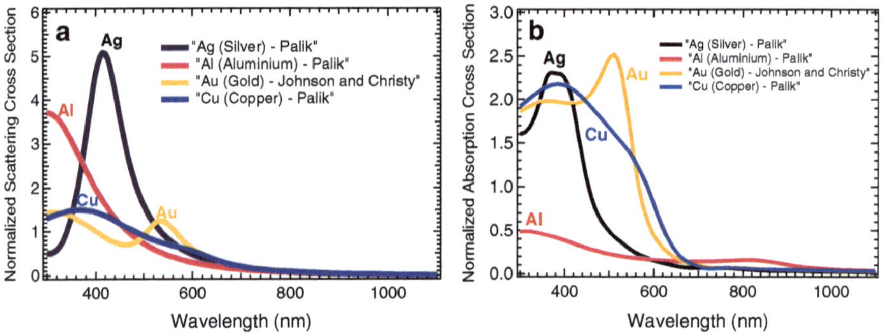

Fig. 14.9 The normalized scattering (**a**) and absorption (**b**) cross section for spherical particles (silver, aluminum, gold, and copper) of radius 50 nm in a vacuum environment ($n = 1$)

where ω_p is the plasmon frequency of the material and Γ_p is the damping rate. Because ω_p is unique to the type of metal, different metals have different optical responses, thus leading to the possibility of having different peaks in the scattering cross section spectrum. When choosing which material to use as the scatterer, it is desirable to choose a material whose scattering peak has sufficient overlap with the solar cell's absorption spectrum. In that case, the particle can efficiently scatter as much power as possible. We consider metallic spheres (Ag, Al, Au, and Cu) with radii of 50 nm in the vacuum and calculate the scattering and absorption cross section (Fig. 14.9).

Plots in Fig. 14.9 clearly demonstrate that in vacuum environment and for a sphere with a radius of 50 nm, the scattering efficiency of silver and aluminum is much higher than that of copper and gold. Therefore, silver and aluminum have a higher potential as scattering material.

Surrounding Material

In the preceding section, the scattering particles were considered to be in free space ($n = 1$). However, these particles are usually embedded within another material of index n_s or are put on top of a substrate, e.g. on top of a solar cell. We can expect that the scattering performance of the particles will be influenced by that layer due to the fact that the effective permittivity of the environment would change. The surroundings in turn can modify the plasmon resonance peak and hence scattering cross sections. To understand how the surroundings will influence the scattering, we consider a spherical silver particle surrounded by dielectric materials with different refractive indices.

Figure 14.10 shows that there is a red shift of the surface plasmon resonance when the refractive index of the surrounding material increases. On the other hand, the absorption cross section decreases with the increasing refractive index of the

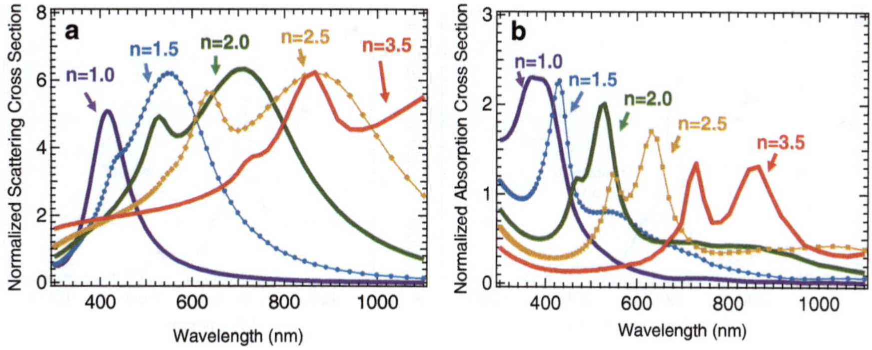

Fig. 14.10 Normalized scattering (**a**) and absorption (**b**) cross sections change with refractive index: $n = 1$ (air), $n = 1.5$ (e.g., SiO_2), $n = 2$ (e.g., Si_3N_4), $n = 2.5$ (e.g., TiO_2), and $n = 3.5$ (e.g., GaAs). The silver scattering particle has a radius of 50 nm

surroundings. Multiple peaks are evident in the cases within a larger index of refraction due to excitation of higher order resonances (e.g., quadruple moments).

Now consider the case of a scattering particle directly on top of the surface rather than in free space or completely embedded. We can make an approximation for the scattering and absorption cross sections if we know the result for a particle embedded in a homogeneous environment. This allows us to quickly find approximate solutions based on cases we already know.

Consider a gold sphere ($r = 25$ nm) half inside an $n = 2$ substrate. Assuming that the substrate is semi-infinite and half of the sphere is in the air, the average permittivity experienced by the sphere is 1.5. Will there be any similarities if we put the sphere inside an $n = 1.5$ material? The answer is yes. As is shown in Fig. 14.11, the differences in the cross sections between the two cases are fairly small.

Particle Shape

The particle shape also plays a crucial role in determining its scattering properties. For simplicity, we consider a few typical examples (sphere, cylinder, and cube). All shapes are made from silver and sit directly on the top of a semi-infinite substrate with $n = 2$. As can be seen in Fig. 14.12, the cube and the cylinder have very similar forward scattering fractions, f_{forward}. However, f_{forward} for a sphere is significantly different. The reason for this effect is that the cube and cylinder have almost the same fraction of their volumes close to the substrate, while the volume of the sphere is centered farther away from the substrate.

Particle Size

The size of the particle is important in determining the scattering cross section (see, for example, that the radius enters into the polarizabilty for spherical particles in Eq. 14.11).

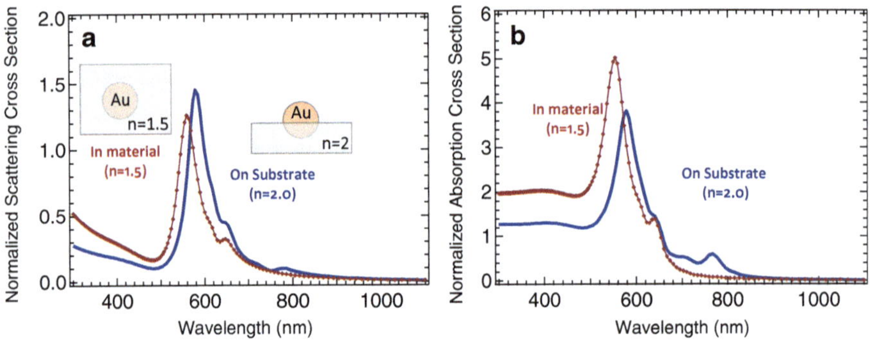

Fig. 14.11 Normalized (**a**) scattering and (**b**) absorption cross sections for two cases: a gold sphere half embedded within a substrate with a refractive index of 2 (*red lines*) and a gold sphere surrounded by a material with a refractive index of 1.5 (*blue lines*). The radii of the spheres are 25 nm in both cases

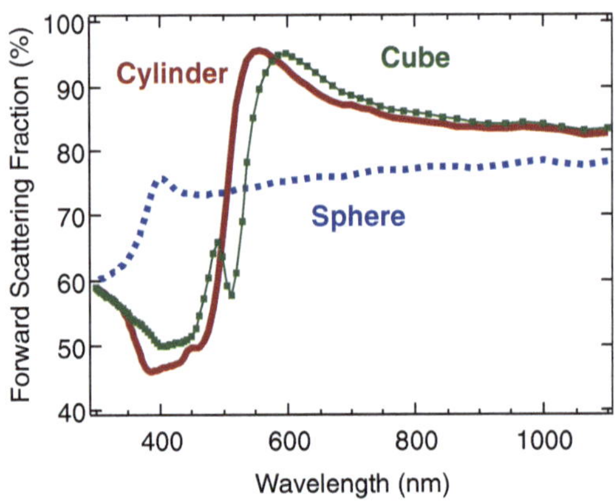

Fig. 14.12 Percentage of forward scattering as a function of wavelength corresponding to different particle shapes. There are three different shapes considered: *sphere* ($d = 100$ nm: *blue line*), *cylinder* ($d = 100$ nm, $h = 100$ nm: *red line*), and *cube* ($d = 100$ nm: *green line*). In all cases, the Ag particles sit on an $n = 2$ dielectric substrate

When the particle is fairly small, σ_{scat} is much smaller than σ_{abs}, which indicates that $\sigma_{\text{scat}} + \sigma_{\text{abs}} \cong \sigma_{\text{abs}}$ and absorption plays the dominant role. However, with the increasing size, σ_{scat} will increase much faster than σ_{abs} and eventually dominates. As the size of the particle approaches the wavelength of light ($\alpha_x \tilde{} 1$), the quasistatic limit will be broken and multipolar modes may become important.

Although the various cross sections increase with radius, the normalized cross section will vary depending on the geometric cross section. Figure 14.13 shows that the normalized scattering cross section increases with radius, while the normalized absorption cross section decreases. Thus, larger scatterers have higher scattering

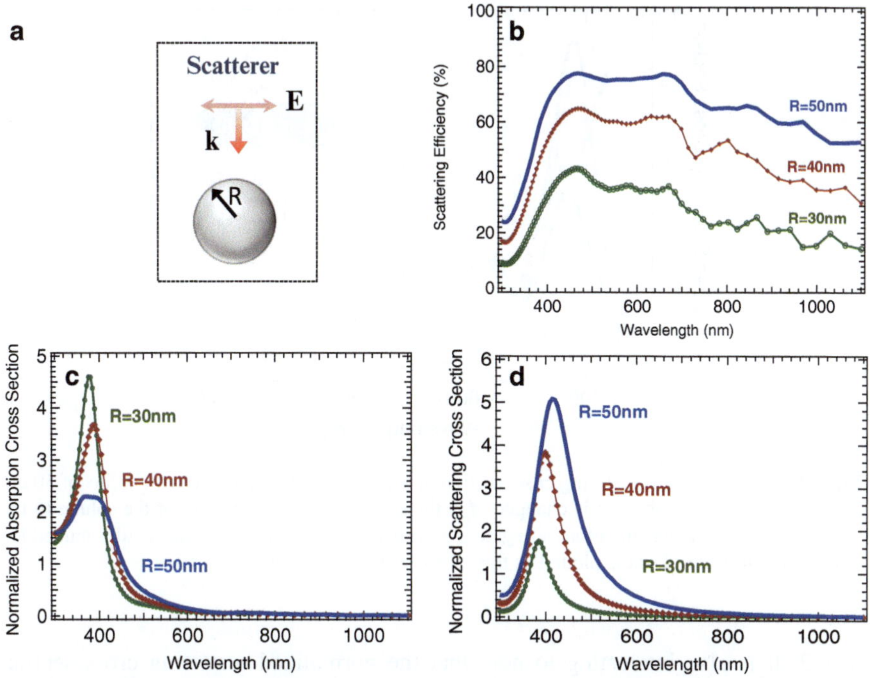

Fig. 14.13 (**a**) Schematic of the simulations for Ag spheres with radii of 30, 40, and 50 nm in vacuum ($n = 1$). (**b**) Scattering efficiency. (**c**) Normalized absorption cross section. (**d**) Normalized scattering cross section

efficiency because scattering plays a dominant role in determining the extinction cross section for larger particles. A sphere with a 50 nm radius has a scattering efficiency of 60–80 %, while a sphere with a radius of 30 nm has a scattering efficiency of only ~30 % (Fig. 14.13b).

Distance from Surface

Previously we considered an isolated particle in free space, completely within a material and half embedded within the material. As a particle approaches a surface, its scattering properties will change as it begins to be influenced by the refractive index of the substrate. Figure 14.14 shows the shift in frequency and magnitude of the normalized scattering cross section of a 50 nm Ag nanoparticle as it approaches the surface of a semi-infinite substrate of index $n = 2$. The nanoparticle begins at a distance $d = 50$ nm above the surface. As the particle approaches the surface, the normalized scattering cross section decreases. When the particle touches the surface ($d = 0$ nm), a slight red shift of the resonance occurs. As the particle enters the dielectric ($d < 0$), a significant red shift occurs until the particle's resonance corresponds to that of the particle surrounded by a uniform material of index

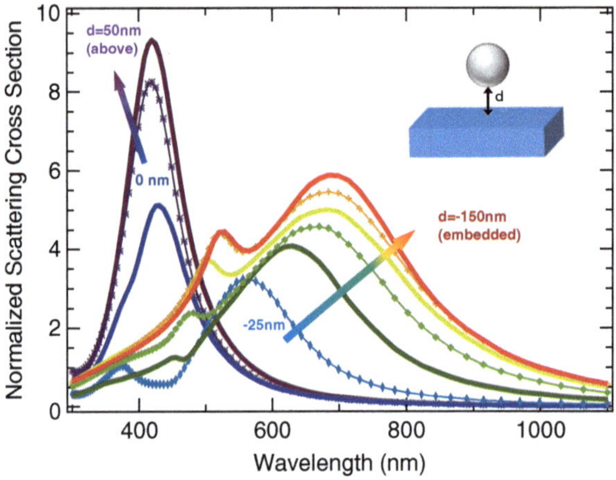

Fig. 14.14 Normalized scattering cross section of a spherical silver particle with radius of 50 nm as distance from the substrate is changed. d is the distance from the bottom of the sphere to the surface of the substrate (note that a negative d corresponds to particle partially/fully embedded within the substrate). Data is plotted as the sphere is moved in steps of 25 nm

$n = 2$. It is also interesting to note that the normalized scattering cross section decreases in magnitude as the sphere approaches the surface from either above or below the interface.

14.2.2 Waveguiding

The simplest waveguide structure consists of a core material surrounded by two cladding dielectric layers. For most guided modes, the refractive index of the core layer is greater than the cladding layers. The solutions to Maxwell's equations in homogeneous media are plane waves of the form:

$$E(x, z, t) \sim E_0 e^{ik_x x} e^{-k_z |z|} \tag{14.14}$$

where x is the direction of propagation, and k_x and k_z are the wave vector components in the x- and z-directions for angular frequency ω. Two independent sets of solutions exist, transverse magnetic (TM) and transverse electric (TE) modes. TE modes have their electric field component completely in the plane, corresponding to the y-direction in Fig. 14.15. Thus, only the H_x, H_z, and E_y field components exist. For TM modes, the magnetic field is completely in the plane, so that only the E_x, E_z, and H_y components are present. For both sets of confined modes, the boundary conditions at the interfaces dictate that the fields decay exponentially outside of the core region.

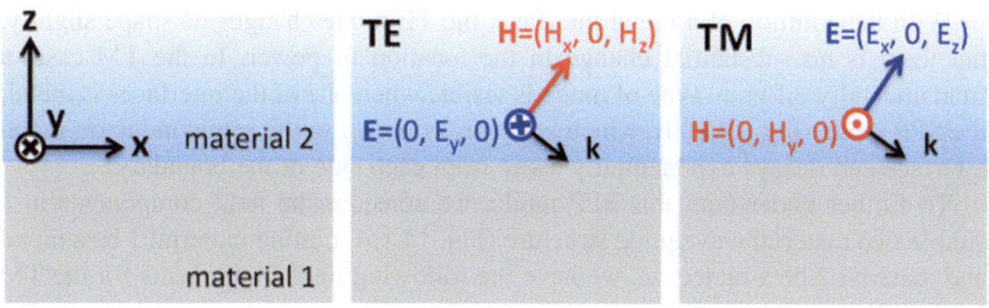

Fig. 14.15 TE and TM modes for a simple two-material waveguide. The interface corresponds to $z = 0$

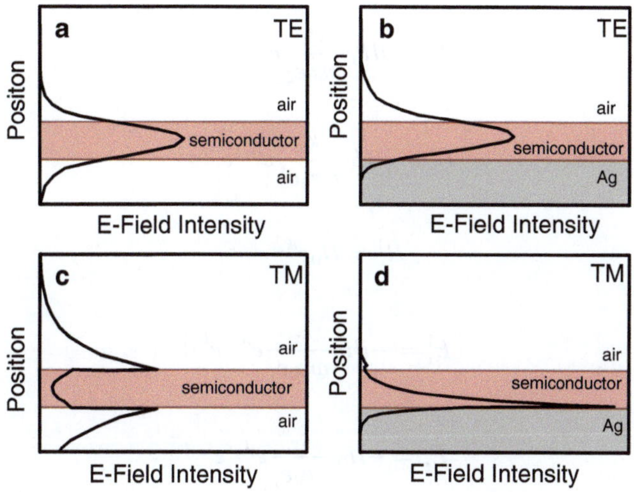

Fig. 14.16 Mode profiles for the lowest order TE [(**a**) and (**b**)] and TM [(**c**) and (**d**)] modes of a simple slab waveguide. When the backside is a metal, it is possible to excite a surface plasmon polariton, which is tightly confined to the metal–semiconductor interface (**d**)

By solving Maxwell's equations subject to the appropriate boundary conditions, we obtain the field intensity profiles for the various waveguide modes of a planer multilayer structure. Figure 14.16 shows the modal profiles (calculated from simulation) at 600 nm with both TE and TM polarizations for a 100 nm thick layer of CdS on the top and either air (a and c) or Ag (b and d) on the backside. These modal profiles are represented by the E-field intensity as a function of position in the waveguide. The modes are largely within the core and decay sharply into the surrounding air or Ag. While we have only depicted a single mode for each case, the number of modes present in the waveguide depends both on the thickness of the slab and on the wavelength of the incident light. These modes can contribute significantly to the absorption within the semiconductor, but the differing overlap ratios within the cladding will change the fraction of power in each mode that contributes to useful absorption and the fraction that is lost to other processes.

With the addition of a metal interface, the TE mode changes its shape slightly, but there is no substantial change in the location of power. In the TM case, a fundamentally different type of mode is visible when one of the interfaces is metal: the SPP mode (Fig. 14.16d), which is tightly confined to the semiconductor–metal interface and decays exponentially away from each side of the boundary.

To further understand this SPP mode, we consider the field components in a simple two material waveguide structure (Fig. 14.15). Letting material 1 be a metal and material 2 be a dielectric, we have the following field components for the TM modes [7]:

$$H_y^2 = H_0 e^{ik_x x} e^{-k_{z_2} z} \tag{14.15}$$

$$E_x^2 = iH_0 \frac{k_{z2}}{\omega \varepsilon_0 \varepsilon_2} e^{ik_x x} e^{-k_{z_2} z} \tag{14.16}$$

$$E_z^2 = -H_0 \frac{k_x}{\omega \varepsilon_0 \varepsilon_2} e^{ik_x x} e^{-k_{z_2} z} \tag{14.17}$$

$$H_y^1 = H_0 e^{ik_x x} e^{k_{z_1} z} \tag{14.18}$$

$$E_x^1 = -iH_0 \frac{k_{z_1}}{\omega \varepsilon_0 \varepsilon_1} e^{ik_x x} e^{k_{z_1} z} \tag{14.19}$$

$$E_z^1 = -H_0 \frac{k_x}{\omega \varepsilon_0 \varepsilon_1} e^{ik_x x} e^{k_{z_1} z} \tag{14.20}$$

where the metal has complex $\varepsilon_1(\omega)$ and corresponding k_{z_1}, and the semiconductor has ε_2 and k_{z_2}. In the above notation, E_x^1 corresponds to the x-component of the field in material 1.

Enforcing continuity of the tangential component of **E** (i.e., $E_x^1 = E_x^2$) and the normal component of **D** (i.e., $\varepsilon_1 E_z^1 = \varepsilon_2 E_z^2$) at the interface yields:

$$\frac{-k_{z_1}}{k_{z_2}} = \frac{\varepsilon_1}{\varepsilon_2} \tag{14.21}$$

which can be satisfied for a metal interface because $\mathrm{Re}[\varepsilon_1] < 0$ and both $\mathrm{Re}[k_{z,1}] > 0$ and $\mathrm{Re}[k_{z,2}] > 0$. Because the H_y component in both materials must satisfy the wave equation, we also have:

$$k_{z_{1,2}}^2 = \varepsilon_{1,2} \left(\frac{\omega}{c}\right)^2 - k_x^2 \tag{14.22}$$

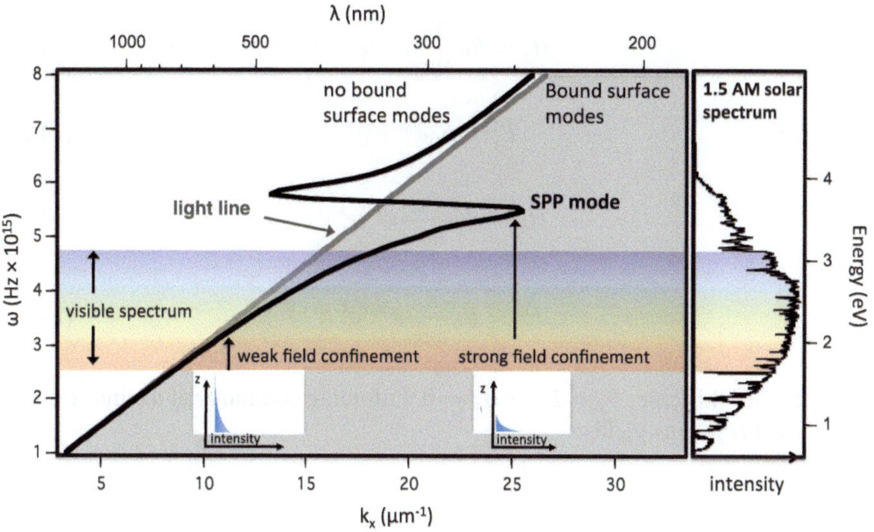

Fig. 14.17 Dispersion relation for the surface plasmon polariton mode at the silver/air interface. A significant fraction of the solar spectrum overlaps with the surface plasmon mode. Figure adapted from [9]

Combining these equations yields the surface plasmon dispersion relation

$$k_x = \frac{\omega}{c}\sqrt{\frac{\varepsilon_1 \varepsilon_2}{\varepsilon_1 + \varepsilon_2}} \tag{14.23}$$

which describes how the wave propagation varies with frequency.

Figure 14.17 shows the dispersion relation for an SPP mode at the Ag/air interface. Modes to the right of the light line, which describes light propagation in material 2, are bound to the interface. At low frequency, the SPP modes are close to, but to the right of, the light line. At resonance, where $\varepsilon_1 = -\varepsilon_2$, the modes are highly confined to large wave vectors, and propagation lengths are very short, ~$1/(2\ \mathrm{Im}(k_x))$. Changes in the optical properties of the surrounding materials can tune the position of resonance throughout the spectrum. Because the modes are confined to the interface, a propagating SPP has most of its power in a small volume of material, and it has been shown that a monolayer of CdSe quantum dots can be made optically thick by SPP absorption [14].

We now show that for the TE case, no bound surface modes exist. From Maxwell's equations for TE polarization,

$$E_y^2 = E_0 e^{ik_x x} e^{-k_{z_2} z} \tag{14.24}$$

$$H_x^2 = -iE_0 \frac{k_{z_2}}{\omega \mu_0} e^{ik_x x} e^{-k_{z_2} z} \tag{14.25}$$

$$H_z^2 = E_0 \frac{k_x}{\omega\mu_0} e^{ik_x x} e^{-k_{z_2} z} \tag{14.26}$$

$$E_y^1 = E_0 e^{ik_x x} e^{k_{z_1} z} \tag{14.27}$$

$$H_x^1 = iE_0 \frac{k_{z_1}}{\omega\mu_0} e^{ik_x x} e^{k_{z_1} z} \tag{14.28}$$

$$H_z^1 = E_0 \frac{k_x}{\omega\mu_0} e^{ik_x x} e^{k_{z_1} z} \tag{14.29}$$

Note that for this case, $E_x = E_z = H_y = 0$. Enforcing continuity at the interface for the E_y and H_x terms yields:

$$E_0(k_{z_1} + k_{z_2}) = 0 \tag{14.30}$$

Because both $\mathrm{Re}[k_{z1}] > 0$ and $\mathrm{Re}[k_{z2}] > 0$, we must have $E_0 = 0$, and thus no surface mode exists under TE polarization.

14.3 Absorption Calculations

In order to determine the effectiveness of any optical structure, it is necessary to calculate the absorption within the active region of the device. For thick planar structures, the absorption is exponential with thickness; however, for more complicated structures, numerical methods [such as FDTD and Finite Difference Frequency Domain (FDFD)] are needed. In this section, we will describe a typical method for calculating absorption in a device and its relation to the short circuit current. As a specific example, we will use the FDTD method to determine the expected photocurrent for a thin film structure with a metallic grating.

The first step is to determine the wavelength-dependent electromagnetic field intensity within the photocurrent generating region. The FDTD method involves dividing the simulation volume into a discrete numerical grid and solving Maxwell's equations in space throughout the grid. The solutions are then evolved in time. A Fourier transform is performed on the time-domain data in order to construct the frequency-domain response of the optical structure to the electromagnetic impulse. This can be done with commercially available software packages, and the result is the steady-state, spatially-dependent electric and magnetic fields over the entire simulation volume.

Once the electric and magnetic fields have been determined, the response must be weighted by the solar spectrum. The absorption per unit volume is obtained from the divergence of the Poynting vector as $-\frac{1}{2}\mathrm{Re}(\vec{\nabla} \cdot \vec{P})$, where $\vec{P}$ represents the

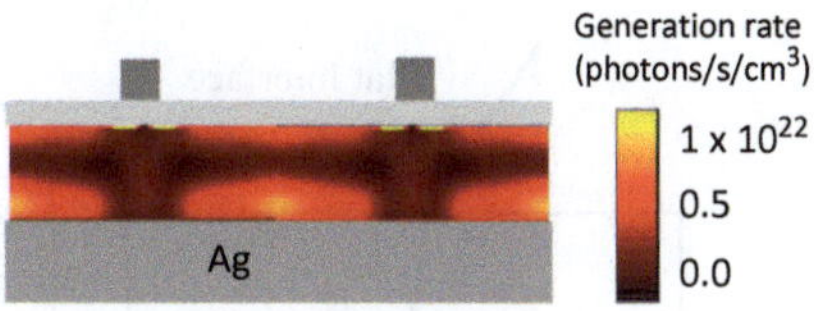

Fig. 14.18 Spatial variation of the generation rate of carriers within a semiconductor. Regions with high field intensities lead to large generation rates

directional energy flux density (i.e., the rate of energy transfer per unit area). If each photon absorbed within the active layer creates one electron–hole pair, the field profiles can be used to calculate the electron generation rate per unit volume:

$$\mathrm{GR} = \int_{\text{solar spectrum}} \left(\frac{\varepsilon'' \left| \vec{E}(\omega) \right|^2}{2\hbar} \right) d\omega \tag{14.31}$$

where $\left|\vec{E}(\omega)\right|^2$ is the magnitude of the electric field squared within the structure resulting from solar illumination and ε'' is the imaginary part of the dielectric function of the absorber. While absorption may occur in other regions (e.g., within the grating or passivation layers), only absorption within the layer that can generate carriers should be used. Figure 14.18 shows the generation rate within a thin film absorber with a silver grating.

In order to calculate the short circuit current density, additional information needs to be known about the collection efficiency of the device. Such information is typically obtained through experimental data. One method is to import the electron generation rate into device simulation software in order to calculate the complete current–voltage characteristic. Alternatively, if the device is expected to easily collect generated carriers, we may assume an internal collection efficiency of unity and simply integrate the generation rate per volume, GR, over the volume of the active layer to obtain the short circuit current density:

$$J_{\mathrm{sc}} = \frac{q}{A} \int \mathrm{GR}\, dV \tag{14.32}$$

Because the efficiency is proportional to the short circuit current, this quantity is a good figure-of-merit. A variety of structures can then be studied to determine which optical structure will lead to the highest power conversion efficiency.

14.4 Light Trapping

As discussed above, One way to increase the photovoltaic conversion efficiency is to absorb nearly all of the incident light within a very small semiconductor layer. For a planar layer, the addition of a back reflecting mirror increases the optical path length

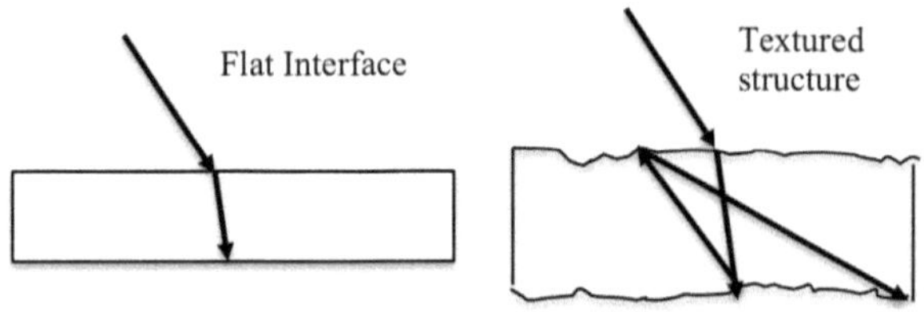

Fig. 14.19 Light path for flat and textured structures

by a factor of 2 yielding a film that absorbs the same amount of light as a film that is twice as thick. The question then becomes, what is the maximum number of path lengths (or similarly what is the maximum absorption enhancement) that is possible based on the laws of thermodynamics? In this section we will address this question.

14.4.1 The Light Trapping Limit for Bulk Materials

One of the first attempts to understand and quantify the absorption limit for solar applications was laid out by Yablonovitch and Cody in 1982 [15]. This result is often referred to as the Yablonovitch limit, the ergodic limit, the ray optics limit, or simply the $4n^2$ limit. The basic approach is to consider a dielectric slab under illumination and to allow the light to become completely randomized (phase and direction) upon entering the material. The randomization is assumed to occur as a result of scattering either from the front or back surface. Without randomization, light entering a nearly non-absorbing material will simply exit the slab after a single or double pass. With appropriate texturing, the light will undergo total internal reflection, resulting in path lengths significantly greater than two (see Fig. 14.19). The principle of detail balance is then employed in direct analogy with the famous Shockley–Queisser derivation to find the internal optical intensity. With this known, the fraction of light absorbed can be calculated.

It is important to note several implicit assumptions that go with this analysis:

1. All modes of the structure are equally occupied as determined by the radiance theorem. It is assumed that this condition is met by having appropriate scattering structures.
2. The material is thick enough that it can be described as having a uniform density of optical states.
3. The material is weakly absorbing.

To approach this calculation, let us first consider a dielectric slab surrounded by black body radiation at temperature T. At equilibrium, the electromagnetic energy density within the slab is given by

$$U(\omega) = \frac{\hbar\omega}{\exp\left(\frac{\hbar\omega}{kT}\right) - 1} \frac{d\Omega k^2}{(2\pi)^3} dk \tag{14.33}$$

and the intensity (power per unit area) is:

$$\begin{aligned} I &= U(\omega)v_g(\omega) \\ &= \frac{\hbar\omega}{\exp\left(\frac{\hbar\omega}{kT}\right)-1} * \rho(\omega) * v_g(\omega) \\ &= \frac{\hbar\omega}{\exp\left(\frac{\hbar\omega}{kT}\right)-1} \frac{2\, d\Omega\, n^2\, \omega^2}{(2\pi)^3\, c^2} d\omega \end{aligned} \tag{14.34}$$

where ω is the angular frequency, $\rho(\omega)$ is the density of optical states, $d\Omega$ is the solid angle, n is the index of refraction of the slab, and $v_g(\omega)$ is the group velocity. For an index of 1 this is simply the Plank blackbody radiation equation. However, because $\rho(\omega)$ scales as n^3 and v_g scales as $1/n$, I is enhanced by a factor of n^2 over the external blackbody radiation. The intensity in the slab can be enhanced further through the addition of a back reflector, which will enhance the internal intensity by an additional factor of 2, yielding a $2n^2$ enhancement of the intensity.

A key assumption, as noted above, is that the modes of the slab are fully occupied so that the intensity is simply proportional to the density of optical states. If we can ensure that all the modes are filled by means of complete internal randomization of the light in the slab, the illumination source need not be a black body. In this case the intensity in the slab is still given by the above equation even if the external radiation is not random. As long as the absorption is low, only a small degree of scattering is needed to provide fully ergodic behavior as many scattering events can take place before the light exits or is appreciatively absorbed. This simple derivation gives insight into why the internal intensity can be enhanced by a factor of $2n^2$ (note: the *absorption enhancement* for this case is actually $4n^2$ as discussed below).

An alternative derivation, also shown by Yablonovitch [16], uses geometric optics and the principle of detail balance (i.e., at steady-state, the power of escaping light must match the power of the incident light) to arrive at this light trapping limit. Figure 14.20 shows the ray optics picture of light incident on a small area dA at the surface of the slab. Assuming the angular distribution of the internal intensity, I_{int}, is uniform, the total escaping intensity is given by:

$$I_{esc} = 2\pi \int_0^{\theta_c} \frac{I_{int}}{2\pi} (T(\theta)\cos(\theta))\sin(\theta) d\theta \tag{14.35}$$

where θ is the angle measured from normal to dA, θ_c is the critical angle given by the equation $n\sin(\theta_c) = 1$, and $T(\theta)$ is the transmission at an angle of θ given by the Fresnel equations. The factor of $\cos(\theta)$ accounts for the reduction in intensity due to the oblique incidence of the light. In general, $T(\theta)$ should be calculated from the Fresnel equation; however, this integral can be simplified if $T(\theta)$ is replaced with a constant T_{esc} which is the average weighted transmission factor. The integral then simply yields:

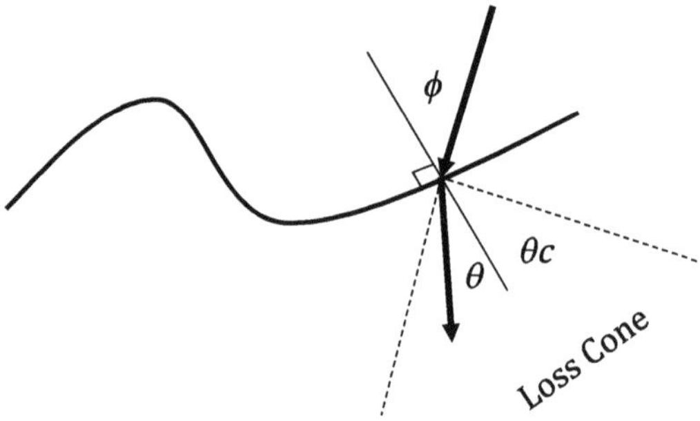

Fig. 14.20 Ray optics depiction of light incident on an arbitrarily oriented section of the surface

$$I_{\mathrm{esc}} = I_{\mathrm{int}} \frac{T_{\mathrm{esc}}}{2n^2} \tag{14.36}$$

By equating this quantity to the incident intensity, I_{inc}, a relationship between the incident and internal intensity is obtained:

$$I_{\mathrm{inc}} T_{\mathrm{inc}}(\phi) = I_{\mathrm{esc}} \tag{14.37}$$

$$I_{\mathrm{int}} = 2n^2 \frac{T_{\mathrm{inc}}(\phi)}{T_{\mathrm{esc}}} I_{\mathrm{inc}} \tag{14.38}$$

where $T_{\mathrm{inc}}(\phi)$ is the incident transmission coefficient for light entering the slab (see Fig. 14.20). Again the factor of $2n^2$ is found with an additional factor given by $\frac{T_{\mathrm{inc}}(\phi)}{T_{\mathrm{esc}}}$. For collimated radiation, it is possible for this factor to be larger than one; however, for isotropic illumination this factor is unity. Further, $T_{\mathrm{inc}}(\phi) = T_{\mathrm{esc}}(\theta)$, as required by time-reversal symmetry. If light were allowed to escape from the back surface, the factor 2 would disappear in Eq. (14.38) because twice the illumination would be need to ensure the same internal intensity.

Finally, given the above, the absorption enhancement due to the randomization of the light may be derived. If the absorption coefficient is sufficiently small, there will be no violation of our assumption of uniform intensity within the slab. This implies that the absorption can be taken as linearly dependent on the light path. The total absorption in a single optical pass of a planar slab with internal intensity I_{p}, area A, and thickness h, illuminated from normal incidence and no scattering can then be taken as:

$$\mathcal{A}_{\mathrm{p}} = I_{\mathrm{p}} \alpha A h \tag{14.39}$$

The absorption in a slab with randomized light is instead given by:

$$\mathcal{A}_{\mathrm{r}} = \int I_{\mathrm{int}} dV d\Omega = I_{\mathrm{int}} \alpha A h \int_0^{\pi} \sin(\theta) d\theta \tag{14.40}$$

$$\mathcal{A}_r = 2I_{int}\alpha Ah \qquad (14.41)$$

Using the expression for the ratio of internal to external intensity:

$$\mathcal{A}_r = 4n^2 I_{inc}\alpha Ah \qquad (14.42)$$

If both slabs have perfect antireflection for the incident light, the scattering slab has $4n^2$ enhancement of absorption. This enhancement can be considerable. For instance, silicon with an index of 3.5 has an enhancement factor of nearly 50. To put it another way, 50 times less material is needed for the same absorption if the light is fully scattered.

This approach is quite useful for calculating the expected absorption enhancement for a weakly absorbing film that is thick compared to the wavelength of light. However, these assumptions no longer hold for a number of interesting cases including thin film solar cells, quantum dot solar cells, plasmonic structures, nano-structured solar cells, etc. The next sections will explore how some of these weaknesses may be overcome.

14.4.2 Revisiting the Thermodynamic Limit for Thin Films

When the absorbing layer becomes comparable to or smaller than the wavelength of the light, it is not clear that the previously mentioned limit is still valid. In fact, as we shall see below, it is even possible to exceed the ergodic limit for thick materials through the use of metals or microtexturing of either the materials themselves or of the surrounding regions.

14.4.2.1 Light Trapping in the Waveguide Limit

One of the first attempts to determine the light trapping limit in the thin film regime was by Stuart and Hall in 1997 [17]. They investigated this limit by looking at all of the modes supported in a thin film including both radiation modes, which can escape, and propagating modes, which are the typical waveguide modes described earlier in the chapter. The absorption in each mode was then calculated and summed to obtain the total absorption for the film. In 2011 Schiff showed how this formalism can be used to consider plasmonic modes [18]. The inclusion of these modes in general should enhance the absorption as the absorption is dependent on the energy density in the absorber. By adding further states that do not radiate, the energy density can be increased. Independently, Munday et al. used a similar approach to analyze structures where the ergodic limit may be exceed by considering the dispersion relations for photonic and plasmonic modes [19]. Each of these will be handled in turn in this section.

The most important simplification used in this approach is that the radiance theorem may be extended to sets of discrete modes and that the energy density derived in the radiance theorem sets an upper limit for discrete modes. Many of the same assumptions are made in this derivation:

1. The absorption is weak so that it may be considered as a small perturbation.
2. Light is coupled perfectly into the absorber.
3. The light is equally coupled into all modes.

First, the energy density of the two sets of modes, radiation and guided (for now, this includes both photonic waveguide modes and SPP modes), are considered separately. The density of radiation modes is the same as for a blackbody times the fraction of the solid angle occupied by radiation modes:

$$\rho_{\mathrm{rad}}(\omega) = \frac{\Omega_{\mathrm{rad}}}{4\pi}\left(\frac{\omega^2 n^3}{\pi^2 c^3}\right) \tag{14.43}$$

where Ω_{rad} the solid angle set by the escape cone. The density of states for the guided modes is given by:

$$\rho_{\mathrm{m}}(\omega) = \frac{R_{\mathrm{m}}}{h} \tag{14.44}$$

where m is the modal index, R_m is the density of states per unit area per unit frequency in the mode, and β_{m} is the modal propagation constant. The thickness dependence arises from the fact that the guided modes are confined to the film and an increase in thickness changes the volume but not the density of guided modes for a given modal number. Assuming all states have equal probability of occupation from scattered light and that the rate of energy loss from a mode is proportional to the energy contained within, the principal of detail balance may be invoked to equate incident power with loss. As above, this gives the ratio of the internal and external intensity. A small rate, γ_{abs}, of absorption can then be added to perturb the balance. Finally, the absorbed fraction, f_{tot}, can be calculated by finding the absorption in each mode and weighting it by the fraction of the total number of states in that mode (see [17] for details):

$$f_{\mathrm{tot}} = \frac{\rho_{\mathrm{rad}}(\omega)}{\rho_{\mathrm{tot}}(\omega)} f_{\mathrm{rad}} + \sum_m \frac{\rho_{\mathrm{m}}(\omega)}{\rho_{\mathrm{tot}}(\omega)} f_{\mathrm{m}} \tag{14.45}$$

where

$$f_{\mathrm{rad}} = \frac{r_{\mathrm{abs}}^{\mathrm{rad}}}{r_{\mathrm{inc}}} = \frac{\alpha}{\alpha + \dfrac{1}{4h\dfrac{\rho_{\mathrm{tot}}(\omega)v_{\mathrm{g}}^{\mathrm{rad}}}{\rho_{\mathrm{inc}}(\omega)v_{\mathrm{g}}^{\mathrm{inc}}}}} \tag{14.46}$$

$$f_{\mathrm{m}} = \frac{r_{\mathrm{abs}}^{\mathrm{m}}}{r_{\mathrm{inc}}} = \frac{\alpha \Gamma_{\mathrm{m}}}{\alpha \Gamma_{\mathrm{m}} + \frac{1}{4h \frac{\rho_{\mathrm{tot}}(\omega) v_{\mathrm{g}}^{\mathrm{m}}}{\rho_{\mathrm{inc}}(\omega) v_{\mathrm{g}}^{\mathrm{inc}}}}} \tag{14.47}$$

Γ_{m} is the modal confinement factor (fraction of energy in the absorber), $v_{\mathrm{g}}^{\mathrm{m}}$ is the appropriate group velocity for the rate $\gamma_{\mathrm{abs}}^{\mathrm{m}} = \alpha v_{\mathrm{g}}^{\mathrm{m}}$ (generally this must be computed as $\gamma_{\mathrm{abs}}^{\mathrm{m}} = \mathrm{Im}(\beta_{\mathrm{m}})$).

This result can be used to recover the ergodic limit when there are no SPP modes and h becomes large. Assuming the incident light comes from air/vacuum, $\rho_{\mathrm{tot}}(\omega)$ approaches $n^3 \rho_{\mathrm{inc}}(\omega)$ (the bulk density of states). $v_{\mathrm{g}}^{\mathrm{m}}$ and $v_{\mathrm{g}}^{\mathrm{rad}}$ also approach the bulk value of $\frac{c}{n}$. Lastly the confinement factor approaches one. This results in:

$$f_{\mathrm{rad}} = f_{\mathrm{m}} = \frac{4\alpha h n^2}{4\alpha h n^2 + 1} \tag{14.48}$$

$$f_{\mathrm{tot}} = f_{\mathrm{rad}} \left(\frac{\rho_{\mathrm{rad}}(\omega)}{\rho_{\mathrm{tot}}(\omega)} + \sum_m \frac{\rho_{\mathrm{m}}(\omega)}{\rho_{\mathrm{tot}}(\omega)} \right) = \frac{4\alpha h n^2}{4\alpha h n^2 + 1} \tag{14.49}$$

As in the calculation above, this simplifies to $4\alpha h n^2$ for weak absorption. This result also shows that the enhancement is typically less than $4n^2$ for thin waveguides surrounded by air.

The inclusion of SPP modes allows for the possibility of exceeding the ergodic limit as noted above. Nothing that Γ_{SPP} goes to unity and that $\frac{v_{\mathrm{g}}^{\mathrm{SPP}}}{v_{\mathrm{g}}^{\mathrm{inc}}}$ goes to n (as we approach the light line), we see that f_{SPP} goes to f_{rad}, β goes to $k_0 n$, and R_{SPP} goes to $\frac{n^2\omega}{2\pi c^2}$. If we then define the classical density of states, $\rho_{\mathrm{cl}}(\omega)$, as the density of states with no SPP, the total absorption can be simplified to (for a full treatment see [18]):

$$f_{\mathrm{tot}} = f_{\mathrm{rad}} \left(\frac{\rho_{\mathrm{cl}}(\omega) + \rho_{\mathrm{SPP}}(\omega)}{\rho_{\mathrm{tot}}(\omega)} \right) = f_{\mathrm{rad}} = \frac{4\alpha h n^2 \left(1 + \frac{\rho_{\mathrm{SPP}}(\omega)}{\rho_{\mathrm{cl}}(\omega)}\right)}{4\alpha h n^2 \left(1 + \frac{\rho_{\mathrm{SPP}}(\omega)}{\rho_{\mathrm{cl}}(\omega)}\right) + 1}$$

$$f_{\mathrm{tot}} = \frac{4\alpha h n^2 (1 + \frac{\pi c}{2hn\omega})}{4\alpha h n^2 (1 + \frac{\pi c}{2hn\omega}) + 1} = \frac{(4\alpha h n^2 + \alpha n \lambda)}{(4\alpha h n^2 + \alpha n \lambda) + 1}$$

In the limit of weak absorption this goes to:

$$f_{\mathrm{rad}} = (4\alpha h n^2 + \alpha n \lambda)$$

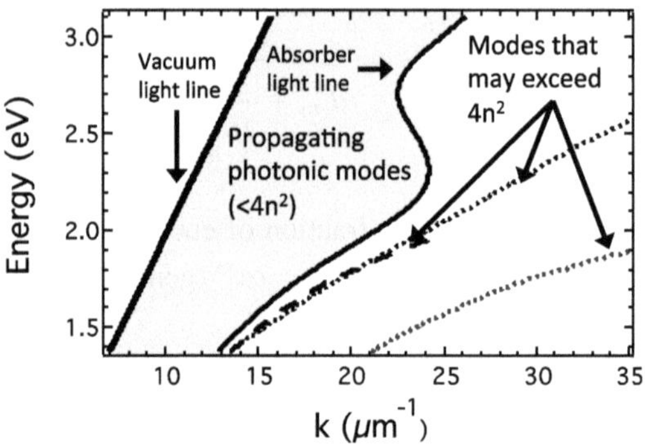

Fig. 14.21 Waveguide dispersion relations. Plasmonic, metal–insulator–metal, and high–low–high index dielectric slot waveguides have dispersion relations that lie to the right of the bulk absorber light line, which have large propagation constants, slow group velocities, and are well confined. Adapted from [19]

Another interesting case is when the thickness is reduced significantly until there is only one mode [19]. Here the $4n^2$ limit may be exceeded. In this case the radiation density of states can be much smaller than the density of states for a single mode and the absorption is given by:

$$f_{\text{tot}} = f_{\text{m}} = \frac{\alpha\Gamma_{\text{m}}}{\alpha\Gamma_{\text{m}} + \frac{1}{4h\frac{\rho_{\text{tot}}(\omega)v_{\text{g}}^{\text{rad}}}{\rho_{\text{inc}}(\omega)v_{\text{g}}^{\text{inc}}}}} = \frac{\alpha\Gamma_{\text{m}}}{\alpha\Gamma_{\text{m}} + \frac{1}{4h\frac{\beta_{\text{m}}\pi c^3}{2\omega^2 h v_{\text{g}}^{\text{m}} n}}} \tag{14.50}$$

The ergodic limit may be exceeded when the following condition is satisfied:

$$\frac{\beta_{\text{m}}}{k_0}\frac{v_{\text{g}}^{\text{rad}}}{v_{\text{g}}^{\text{m}}}\Gamma_{\text{m}} > 4n\frac{h}{\lambda} \tag{14.51}$$

Thus, to exceed the ergodic limit, strong confinement, low modal group velocity, and large propagation constants are needed. For a thin dielectric slab in air these conditions are never met. They can be met, however, by structures such as the metallic waveguides mentioned above or high–low–high index slot waveguides. As a guiding principle, the dispersion relation can be used to pre-select possible waveguide structures that may be capable of absorption enhancements in excess of $4n^2$. Large propagation constants are found for modes that lie to the right of the bulk light line of the material making up the slab. Similarly, slow group velocity modes have a slope that is less steep than those of the bulk material light line. Figure 14.21 shows typical modes from plasmonic waveguides and slot waveguides that are capable of exceeding the ergodic limit. We also note that, while this analysis describes a single mode waveguide, the principles outlined can be extended to multimode structures where the dominant modes satisfy the above criteria [19].

14.4.2.2 Statistical Temporal Coupled-Mode Theory for Light Trapping

The previous methods have described the light trapping limit in terms of either discrete waveguide modes or a continuum found in the bulk. Alternatively, the statistical temporal coupled-mode theory analysis by Yu et al. can handle more complicated cases [20]. Temporal coupled-mode theory breaks the system down into waveguides and resonators and considers channels that couple to these resonances. As an aid for discussion, let us consider a thin weakly absorbing slab with a grating of period L_P. An incoming planewave of some frequency ω will resonantly couple to this thin film based on the periodicity. If the loss mechanisms are small, the resonant behavior will be observed even when L_P is very large. Based on the formalism of temporal coupled-mode theory, the absorption spectrum of the resonance is given as:

$$\mathcal{A}(\omega) = \frac{\gamma_i \gamma_e}{(\omega - \omega_0)^2 + \frac{1}{4}(\gamma_i + N\gamma_e)^2} \tag{14.52}$$

Here γ_i is the instrinsic material absorption of the channel, γ_e is the coupling rate between the incoming plane wave and the resonance, N is the number of channels (i.e., plane waves that can couple to this resonance), and ω_0 is the resonant frequency. However, for solar applications, the spectral width, $\Delta\omega$, of the incident light is much wider than the bandwidth of the resonance. The total absorption, $\mathcal{A}_{tot}^{m}$, for a resonance m can be found by integrating over the absorption spectrum of the resonance and dividing by $\Delta\omega$:

$$\mathcal{A}_{tot}^{m} = \frac{2\pi\gamma_i}{\Delta\omega\left(N + \frac{\gamma_i}{\gamma_e}\right)} \tag{14.53}$$

The total absorption is then the sum of all $\mathcal{A}_{tot}^{m}$:

$$\mathcal{A}_{tot} = \sum_{m} \mathcal{A}_{tot}^{m} \tag{14.54}$$

For illustrative purposes, the ergodic limit can be derived from this formalism when examining the case of a thick slab. To start with, the calculation assumes perfect coupling and low absorption so that $\gamma_i \ll \gamma_e$. The number of resonances per unit frequency, M, for a material with period L_P is given by the density of states (as above) times the volume of the repeated section:

$$M = \frac{L_P{}^2 h n^3\ \omega^2}{\pi^2\ c^3} \tag{14.55}$$

The number of channels is given by the two-dimensional version of M as this is the sum of all the states meeting the periodicity requirement:

$$N = \frac{L_P{}^2\ \omega^2}{2\pi c^2} \tag{14.56}$$

so that the total absorption is:

$$\mathcal{A}_{\text{tot}} = \sum_{m=0}^{M\Delta\omega} \frac{2\pi\gamma_{\text{i}}}{\Delta\omega(N)} \tag{14.57}$$

$$\mathcal{A}_{\text{tot}} = \frac{2\pi\gamma_{\text{i}} M}{N} = 4n^2\gamma_{\text{i}}\left(\frac{n}{c}\right)h = 4n^2 h\alpha \tag{14.58}$$

which is again, the $4n^2$ limit. Here we have used the fact that the absorption coefficient is given by the rate of absorption divided by the group velocity.

The theory can also be applied to the thin film limit where the ergodic limit is no longer applicable. Once the wavelength of light is significantly longer than the thickness of the film, the density of states is described by the two-dimensional density of states times two for the two polarizations. Thus M is given by:

$$M = \frac{L^2 n_{\text{g}}{}^2\ \omega}{\pi c^2} \tag{14.59}$$

where n_{g} is the group index. N is unchanged from the bulk case. This leads to an absorption of:

$$\mathcal{A}_{\text{tot}} = \frac{2\pi\gamma_{\text{i}} M}{N} = 2n_{\text{g}}\lambda\alpha_{\text{g}} \tag{14.60}$$

The lack of dependence on h means that the ergodic limit may be surpassed with an appropriately designed waveguide.

14.4.2.3 Light Trapping in the Local Density of Optical States Approach

A key assumption in the derivation of the ergodic limit is that of uniform density of optical states $\rho(\omega)$. However in many structures the local density of optical states (LDOS), $\rho(\omega, r)$, varies greatly from that of a uniform structure. These need not be exotic materials or geometries. The humble dielectric slab discussed above will have a non-uniform LDOS near the interfaces with the vacuum. For a thick slab, this non-uniformity has minimal impact on the average density of states, but as the slab is made thinner, a greater portion of the slab is near an interface. The analytic

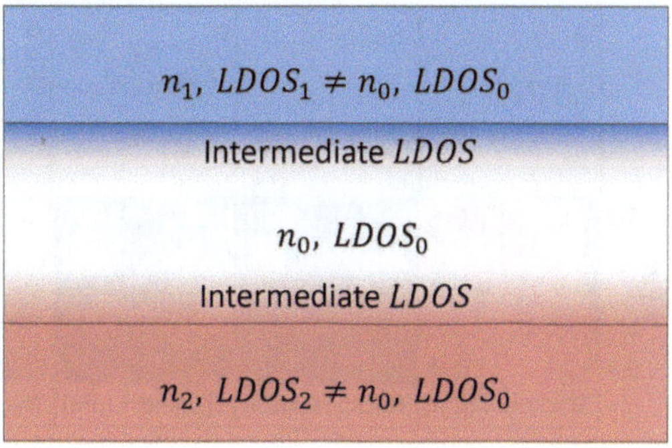

Fig. 14.22 Schematic of a planar structure showing variation of the local density of optical states. Near the mating of the two dielectrics of differing index (and thus differing LDOS) both materials take on an intermediate value LDOS. This can either increase or decrease the LDOS for the slab

expression for the LDOS in this case is complicated but may be intuitively understood by recalling that the density of optical states is dependent on the index of the material and that the density of optical states may be described equivalently as the joint density of states for the molecular optical transition. Thus any material within ~λ will effect the dipole emission. Near the interfaces, the index to be used in the calculation of the density of states must be a weighted average of the bulk index and the index just beyond the interface. Thus, near the interface with a high index material, the LDOS must be increased from that of the bulk and vice versa for a low index material surrounding. This is shown schematically in Fig. 14.22.

While the above case can be determined in a straightforward manner, this may not be possible for more complicated structures. The LDOS can be found by considering the radiative rate of a dipole embedded at the location of interest. In general this can be done using a numerical simulation of the dipole's response to its environment. The main point is that the LDOS will generally be enhanced, for example, by proximity to high index materials, metal layers, or by carefully designing nano-structures where additional surface states may influence the LDOS. For these cases the LDOS is enhanced above its bulk value and can lead to absorption enhancements in excess of the ergodic limit.

The general form for the internal energy density remains the same in this case as it did for the case Yablonovitch considered, but now $\rho(\omega)$ is replaced with the more universal term, $\rho(\omega, r)$:

$$U(\omega, r) = \frac{\hbar\omega}{\exp\left(\frac{\hbar\omega}{kT}\right) - 1} * \rho(\omega, r) \tag{14.61}$$

This can be compared with the bulk value by averaging over the volume, V, and dividing by the energy density of the bulk under the same conditions.

$$\langle U(\omega, r) \rangle = \frac{\hbar\omega}{\exp\left(\frac{\hbar\omega}{kT}\right) - 1} \frac{\int \rho(\omega, r) dV}{V} \tag{14.62}$$

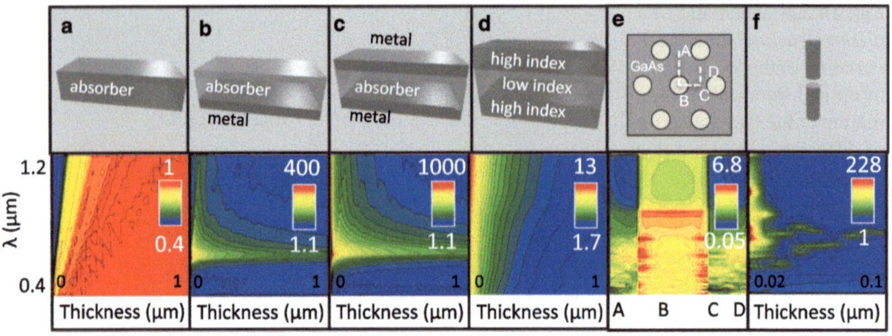

Fig. 14.23 Potential solar cell architectures. (**a**) Planar slab in air in which the LDOS is always below the bulk value. Structures with the potential to beat the ergodic limit. (**b**) Planar slab with an Ag back-reflector, (**c**) metal–absorber–metal planar structure, (**d**) high/low/high index structure with refractive, (**e**) a photonic crystal, and (**f**) a split dipole antenna with an absorber in the gap. The second row shows the LDOS enhancement (on a log scale) over the bulk material for each structure for various cell thicknesses and wavelengths. Values of LDOS enhancement >1 correspond to beating the traditional absorption limit. Adapted from [21]

$$\frac{\langle U(\omega,r)\rangle}{U(\omega)_{\text{bulk}}} = \frac{\langle \rho(\omega,r)\rangle}{\frac{n^3\omega^2}{\pi^2c^3}} \tag{14.63}$$

The condition for exceeding the ray optics limits is thus given by:

$$\frac{\langle \rho(\omega,r)\rangle}{\frac{n^3\omega^2}{\pi^2c^3}} > 1 \tag{14.64}$$

This result shows that we simply want structures where the average density of optical states exceeds that of the bulk for the absorber. A number of geometries have been proposed to meet this criterion [21]. Examples of these are neatly summarized in Fig. 14.23. This plot compares the integrated LDOS for various structures to that of the bulk. As can be seen, enormous enhancements are achievable under the proper conditions.

One final theoretical consideration must be made when attempting to modify the density of states. The density of states sum rules demand that an increase in LDOS over a given wavelength range must be compensated by a similar decrease at another wavelength range. Specifically:

$$\int \frac{\rho_{\text{struct}}(\omega) - \rho_{\text{vac}}(\omega)}{\omega^2} d\omega = 0 \tag{14.65}$$

However this is generally not a limitation as the wavelength range of interest for solar cells is clearly limited and an appropriate material geometry can circumvent this. The idea is depicted schematically in Fig. 14.24.

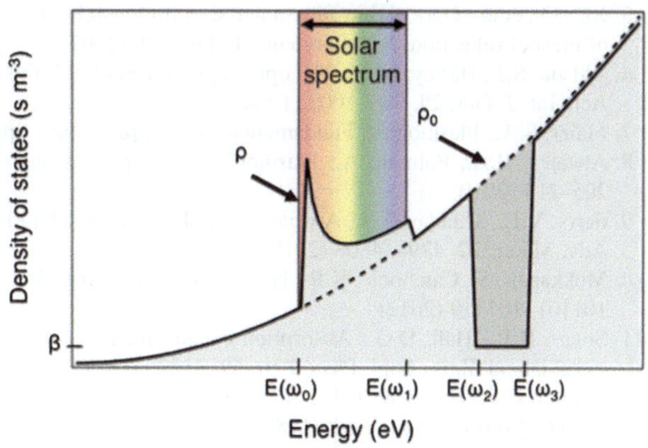

Fig. 14.24 Schematic of the spectral reweighting needed to achieve density of states enhancements over a particular bandwidth while simultaneously decreasing the density of states in another region where the density of states is unimportant for optical absorption. Adapted from [21]

14.5 Conclusions

For thin film photovoltaic devices, such as quantum dot solar cells, light trapping, and optical confinement, are critical to device performance. Small metallic particles can be used as high efficiency scatterers that can reduce reflection, yield high local absorption, and increase the optical path length of the incident light. Further, by coupling into waveguide modes, the incident light can be transformed into confined propagating modes that will increase the absorption probability. All of these effects rely on designing photonic and plasmonic structures that efficiently convert the free space sunlight into localized and propagating modes within the absorber. By careful selection of materials and particle shapes, these resonances can be tailored to most effectively match the absorption properties of the solar cell. Finally, new techniques for light trapping show that the concentration of light can lead to path length enhancements in excess of $4n^2$, which may allow for new solar cell architectures using extremely thin absorbing layers.

Acknowledgments The authors would like to thank H. A. Atwater, M. S. Leite, and D. M. Callahan for insightful discussions and the University of Maryland and the Minta Martin Fund for support.

References

1. Macleod, H.A.: Thin Film Optical Filters. Institute of Physics, Philadelphia (2001)
2. Rayleigh, J.S.: On reflection of vibrations at the confines of two media between which the transition is gradual. Proc. Lond. Math. Soc. **11**, 51–56 (1880)
3. Southwell, W.H.: Gradient-index antireflection coatings. Opt. Lett. **8**, 584–586 (1983)
4. Poitras, D., Dobrowolski, J.A.: Toward perfect antireflection coatings. 2. Theory. Appl. Opt. **43**, 1286–1295 (2004)

5. Xi, J.Q., et al.: Optical thin-film materials with low refractive index for broadband elimination of Fresnel reflection. Nat. Photonics **1**, 176–179 (2007)
6. Wilson, S.J., Hutley, M.C.: The optical properties of "Moth Eye" antireflection surfaces. Opt. Acta Int. J. Opt. **29**, 993–1009 (1982)
7. Maier, S.A.: Plasmonics: Fundamentals and Applications. Springer, New York (2007)
8. Atwater, H.A., Polman, A.: Plasmonics for improved photovoltaic devices. Nat. Mater. **9**, 205–213 (2010)
9. Ferry, V.E., Munday, J.N., Atwater, H.A.: Design considerations for plasmonic photovoltaics. Adv. Mater. **22**, 4794–4808 (2010)
10. Mokkapati, S., Catchpole, K.R.: Nanophotonic light trapping in solar cells. J. Appl. Phys. **112**, 101101–101119 (2012)
11. Stuart, H.R., Hall, D.G.: Absorption enhancement in silicon-on-insulator waveguides using metal island films. Appl. Phys. Lett. **69**, 2327–2329 (1996)
12. Bohren, C.F., Huffman, D.R.: Absorption and Scattering of Light by Small Particles. Wiley Science Editions, New York (1983)
13. Spinelli, P., Verschuuren, M.A., Polman, A.: Broadband omnidirectional antireflection coating based on subwavelength surface Mie resonators. Nat. Commun. **3**, 692 (2012)
14. Pacifici, D., Lezec, H.J., Atwater, H.A.: All-optical modulation by plasmonic excitation of CdSe quantum dots. Nat. Photonics **1**, 402–406 (2007)
15. Yablonovitch, E., Cody, G.D.: Intensity enhancement in textured optical sheets for solar cells. IEEE Trans. Electron Dev. **29**, 300–305 (1982)
16. Yablonovitch, E.: Statistical ray optics. J. Opt. Soc. Am. **72**, 899–907 (1982)
17. Stuart, H.R., Hall, D.G.: Thermodynamic limit to light trapping in thin planar structures. J. Opt. Soc. Am. A **14**, 3001–3008 (1997)
18. Schiff, E.A.: Thermodynamic limit to photonic-plasmonic light-trapping in thin films on metals. J. Appl. Phys. **110**, 104501–104509 (2011)
19. Munday, J.N., Callahan, D.M., Atwater, H.A.: Light trapping beyond the $4n^2$ limit in thin waveguides. Appl. Phys. Lett. **100**, 121121 (2012)
20. Yu, Z., Raman, A., Fan, S.: Fundamental limit of nanophotonic light trapping in solar cells. Proc. Natl. Acad. Sci. **107**, 17491–17496 (2010)
21. Callahan, D.M., Munday, J.N., Atwater, H.A.: Solar cell light trapping beyond the ray optic limit. Nano Lett. **12**, 214–218 (2012)

Index

J. Wu and Z.M. Wang (eds.), *Quantum Dot Solar Cells*, Lecture Notes in Nanoscale Science and Technology 15, DOI 10.1007/978-1-4614-8148-5,

I

L

M

N

O

P

MIX
Papier aus verantwortungsvollen Quellen
Paper from responsible sources
FSC® C105338

If you have any concerns about our products,
you can contact us on
ProductSafety@springernature.com

In case Publisher is established outside the EU,
the EU authorized representative is:
Springer Nature Customer Service Center GmbH
Europaplatz 3, 69115 Heidelberg, Germany

Printed by Libri Plureos GmbH
in Hamburg, Germany